Concepts and Applications of Earth Science

Concepts and Applications of Earth Science

Editor: Russell Sands

R CALLISTO REFERENCE
www.callistoreference.com

Callisto Reference,
118-35 Queens Blvd., Suite 400,
Forest Hills, NY 11375, USA

Visit us on the World Wide Web at:
www.callistoreference.com

ISBN: 978-1-63239-989-2 (Hardback)

Trademark Notice: Registered trademark of products or corporate names are used only for explanation and identification without intent to infringe.

Cataloging-in-Publication Data

Concepts and applications of earth science / edited by Russell Sands.
    p. cm.
Includes bibliographical references and index.
ISBN 978-1-63239-989-2
1. Earth sciences. 2. Geology. I. Sands, Russell.
QE26.3 .C66 2018
550--dc23

Table of Contents

Preface...VII

Chapter 1   **Effects of Landfall Location and Approach Angle of an Idealized Tropical
Cyclone over a Long Mountain Range**..1
Liping Liu, Yuh-Lang Lin and Shu-Hua Chen

Chapter 2   **Inferring Firn Permeability from Pneumatic Testing: A Case Study on the
Greenland Ice Sheet**..15
Aleah N. Sommers, Harihar Rajaram, Eliezer P. Weber, Michael J. MacFerrin,
William T. Colgan and C. Max Stevens

Chapter 3   **Accumulation Rates during 1311–2011 CE in North-Central Greenland Derived
from Air-Borne Radar Data**..27
Nanna B. Karlsson, Olaf Eisen, Dorthe Dahl-Jensen, Johannes Freitag,
Sepp Kipfstuhl, Cameron Lewis, Lisbeth T. Nielsen, John D. Paden,
Anna Winter and Frank Wilhelms

Chapter 4   **The Changing Impact of Snow Conditions and Refreezing on the Mass Balance
of an Idealized Svalbard Glacier**...45
Ward J. J. van Pelt, Veijo A. Pohjola and Carleen H. Reijmer

Chapter 5   **Parameterizing Deep Water Percolation Improves Subsurface Temperature
Simulations by a Multilayer Firn Model**..60
Sergey Marchenko, Ward J. J. van Pelt, Björn Claremar, Veijo Pohjola,
Rickard Pettersson, Horst Machguth and Carleen Reijmer

Chapter 6   **Rock Magnetism of the Offshore Sediments of Lake Qinghai in the Western
China**..80
Peng Zhang, Shan Lin, Hong Ao, Lijuan Wang, Xiaoyan Sun and Zhisheng An

Chapter 7   **Dynamic Changes at Yahtse Glacier, the most Rapidly Advancing Tidewater
Glacier in Alaska**...90
William J. Durkin, Timothy C. Bartholomaus, Michael J. Willis
and Matthew E. Pritchard

Chapter 8   **Circumpolar Mapping of Ground-Fast Lake Ice**..103
Annett Bartsch, Georg Pointner, Marina O. Leibman, Yuri A. Dvornikov,
Artem V. Khomutov and Anna M. Trofaier

Chapter 9   **Inference of Soil Hydrologic Parameters from Electronic Soil Moisture Records**.......119
David G. Chandler, Mark S. Seyfried, James P. McNamara and Kyotaek Hwang

Chapter 10   **Disentangling Natural and Anthropogenic Signals in Lacustrine Records: An
Example from the Ilan Plain, NE Taiwan**..136
Jyh-Jaan Huang, Chih-An Huh, Kuo-Yen Wei, Ludvig Löwemark, Shu-Fen Lin,
Wen-Hsuan Liao, Tien-Nan Yang, Sheng-Rong Song, Meng-Yang Lee,
Chih-Chieh Su and Teh-Quei Lee

Chapter 11   **Millennial-Scale Interaction between Ice Sheets and Ocean Circulation during Marine Isotope Stage 100**...148
Masao Ohno, Tatsuya Hayashi, Masahiko Sato, Yoshihiro Kuwahara, Asami Mizuta, Itsuro Kita, Tokiyuki Sato and Akihiro Kano

Chapter 12   **Extended-Range Ensemble Predictions of Convection in the North Australian Monsoon Region**...157
Wasyl Drosdowsky and Matthew C. Wheeler

Chapter 13   **Challenges of Quantifying Meltwater Retention in Snow and Firn: An Expert Elicitation**...165
Dirk van As, Jason E. Box and Robert S. Fausto

Chapter 14   **Carbon Leaching from Tropical Peat Soils and Consequences for Carbon Balances**.............................170
Tim Rixen, Antje Baum, Francisca Wit and Joko Samiaji

Chapter 15   **Influence of Stress Field Changes on Eruption Initiation and Dynamics**....................................179
Roberto Sulpizio and Silvia Massaro

Chapter 16   **Stress Field Control during Large Caldera-Forming Eruptions**...190
Antonio Costa and Joan Martí

Chapter 17   **Role of Sediment Size and Biostratinomy on the Development of Biofilms in Recent Avian Vertebrate Remains**...203
Joseph E. Peterson, Melissa E. Lenczewski, Steven R. Clawson and Jonathan P. Warnock

Permissions

List of Contributors

Index

Preface

Over the recent decade, advancements and applications have progressed exponentially. This has led to the increased interest in this field and projects are being conducted to enhance knowledge. The main objective of this book is to present some of the critical challenges and provide insights into possible solutions. This book will answer the varied questions that arise in the field and also provide an increased scope for furthering studies.

Earth science as a part of planetary science is a field that includes the study of the geology, ecology, geography, atmosphere, biosphere, hydrosphere, and lithosphere of Earth. Some of the branches of Earth science are physical geography, hydrology, glaciology, geophysics, etc. This book unfolds the innovative aspects of Earth science, which will be crucial for the progress of this field in the future. It includes some of the vital pieces of work being conducted across the world, on various topics related to the subject. It is meant for students who are looking for an elaborate reference text on Earth science.

I hope that this book, with its visionary approach, will be a valuable addition and will promote interest among readers. Each of the authors has provided their extraordinary competence in their specific fields by providing different perspectives as they come from diverse nations and regions. I thank them for their contributions.

Editor

Effects of Landfall Location and Approach Angle of an Idealized Tropical Cyclone over a Long Mountain Range

Liping Liu[1], Yuh-Lang Lin[2,3]* and Shu-Hua Chen[4]

[1] Department of Mathematics, North Carolina A&T State University, Greensboro, NC, USA, [2] Department of Physics, North Carolina A&T State University, Greensboro, NC, USA, [3] Department of Energy and Environmental Systems, North Carolina A&T State University, Greensboro, NC, USA, [4] Department of Land, Air and Water Resources, University of California, Davis, CA, USA

Edited by:
Daniel J. Kirshbaum,
McGill University, Canada

Reviewed by:
Ana María Durán-Quesada,
University of Costa Rica, Costa Rica
Eric Hendricks,
National Park Service, USA

*Correspondence:
Yuh-Lang Lin
ylin@ncat.edu

Specialty section:
This article was submitted to
Atmospheric Science,
a section of the journal
Frontiers in Earth Science

Effects of landfall location and approach angle on track deflection associated with a tropical cyclone (TC) passing over an idealized and Central Appalachian Mountain is investigated by a series of idealized numerical experiments. When the TC landfalls on the central portion of the mountain range, it is deflected to the south upstream, passes over the mountain anticyclonically, and then moves westward downstream. The TC motion is steered by the positive vorticity tendency (VT) which is dominated by horizontal vorticity advection upstream and downstream, but with additional influence from the stretching and residual terms, which are mainly associated with diabatic heating and frictional effects. The track deflection mechanism upstream and downstream is similar to the dry flow in previous study, but is very different in the vicinity of the mountain. When the TC landfalls near the northern (southern) tip, it experiences less (more) southward deflection due to stronger (weaker) vorticity advection around the tip. When the TC approaches the mountain range from the southeast and landfalls on the northern tip, center, or southern tip, the track deflections are similar to those embedded in an easterly flow but with weaker orographic blocking. These results are similar to the cases simulated in the dry flow in previous study, except that there is no track discontinuity due to the weaker orographic blocking associated with strong TC convection. When a TC moves along the north-south mountain range from the south, it tends to deflect toward the mountain and then crosses over to the other side at later time. In these cases, the positive VT is influenced by all horizontal vorticity advection, vorticity stretching (diabatic heating), and residual (friction) terms due to longer and stronger interaction with the mountain range. The vorticity stretching is mainly caused by diabatic heating in the moist flow, instead of by lee slope vorticity stretching in the previous study for dry flow.

Keywords: tropical cyclones, orographic effects, TC track deflection, idealized simulations, approach angle, landfalling location

INTRODUCTION

When a tropical cyclone (TC) passes over a mountain range, its steering flow and cyclonic circulation are often strongly modified by the orography. This tends to enhance the precipitation associated with the storm's rainbands (e.g., Lin, 2007 for a brief review). During the summer and fall seasons, floods over the Appalachians are often a product of enhanced heavy rainfall from hurricanes. This is analogous to flooding associated with the passage of typhoons over Taiwan's Central Mountain Range (CMR) (Lin, 2007). One example is Hurricane Camille (1969), which dumped up to 686 mm of rainfall at one location along the slopes of the Appalachian Mountains in central Virginia (Schwarz, 1970). Another example is Hurricane Ivan (2004) (Stewart, 2004). On September 16, 2004 in Macon County, North Carolina, Ivan produced a debris flow 3.6 km downslope along Peeks Creek to the Cullasaja River, causing five deaths and two injuries along with destruction of 16 buildings. The heavy rainfall areas are closely related to storm tracks, which tend to be strongly influenced by orography. Therefore, it is important to make accurate predictions of the orographic effects on track deflection and understand the underlying mechanisms.

TC track deflections due to orography have occurred over several mesoscale mountain ranges. In particular, deflections have been well documented for typhoons passing over Taiwan's CMR and the Cordillera Central of northern Luzon in the Philippines (e.g., Brand and Blelloch, 1974; Wang, 1980), TC's passing over Madagascar and La Reunion mountains (Rakotomavo et al., 2011), and hurricanes passing over the Cordillera Central of Hispañola (Bender et al., 1987) and the Sierra Madre Mountains of Mexico (Zehnder, 1993; Zehnder and Reeder, 1997). Similar track deflections have also been observed with mid-latitude cyclones passing over the Appalachians (e.g., O'Handley and Bosart, 1996). Effects of orography on TC tracks have been studied extensively for typhoons passing over Taiwan's CMR and the Cordillera Central of northern Luzon in the Philippines.

Due to complicated interactions between orographic and thermal forcing, both idealized numerical simulations (e.g., Chang, 1982; Bender et al., 1987; Yeh and Elsberry, 1993a,b; Lin et al., 1999, 2005, denoted as L05 hereafter, Lin and Savage, 2011; Tang and Chan, 2013, 2014; Wu et al., 2015), real-case simulations (e.g., Wu, 2001; Lin et al., 2006; Jian and Wu, 2008; Huang et al., 2011; Hsu et al., 2013; Wang et al., 2013; Wu et al., 2015), and observations (Brand and Blelloch, 1974; Wang, 1980; Hsu et al., 2013) have shown that when a typhoon approaches Taiwan's CMR, its track may deflect either to the north or south upstream of the mountain. It has been proposed that the track deflection may be due to mean cyclonic circulation (e.g., Chang, 1982; Bender et al., 1987), channeling mechanism (e.g., Lin et al., 1999; Jian and Wu, 2008; Huang et al., 2011), blocking effect (e.g., Yeh and Elsberry, 1993a,b, L05, Lin and Savage, 2011), latent heating (e.g., Hsu et al., 2013; Tang and Chan, 2013, 2014; Wang et al., 2013), terrain-induced gyes (Tang and Chan, 2013, 2014), approaching angle and landing location (Lin and Savage, 2011; Tang and Chan, 2014), and midtropospheric northerly asymmetric flow (Wu et al., 2015). Thus, the basic dynamics

deserve further investigation. In this study, we will focus on the effects of landfall location and impinging angle on TC track deflection in an idealized environment.

In addition to the above mechanisms, the landfall location and approach angle also play important roles in the track deflection of a TC passing over Taiwan's CMR (Wang, 1980). Based on the 6-h Atlantic hurricane best track (HUDAT) of 28 landfalling TCs that impinged on the Appalachians during 1979 and 2006, Harville (2009) (denoted as H09 hereafter) classified the tracks into four types (**Figure 1**): (1) approximately perpendicular to the mountains from east to west (Type A), (2) parallel on the eastern side of the mountain (Type B), (3) parallel on the western side of the mountains (Type C), and (4) similar to Type A, but from west to east. From idealized simulations of a drifting vortex over an idealized version of Taiwan's CMR (Lin and Savage, 2011; denoted as LS11 hereafter), found that the local vorticity generation is dominated more by vorticity advection upstream of the mountain range, vorticity stretching over the lee side and its immediate downstream area, and vorticity advection again far downstream of the mountain as it steers the vortex back to its original direction of movement. The vorticity advection upstream of the mountain range is explained by the flow splitting associated with orographic blocking. Based on a vorticity budget analysis, it was found that jumps in the vortex path are largely governed by stretching on the lee side of the mountain, with the maximum stretching and associated track jump located on the faster side of the vortex. The present study is an extension of Lin and Savage (2011) by adding latent heating and the planetary boundary layer to the numerical simulations.

In idealized simulations, Tang and Chan (2013) found that a pair of terrain-induced gyres is formed when an idealized TC moves closer to Taiwan's CMR. This pair of gyres rotates cyclonically around the TC and the gyre-associated flow near the TC center causes a northward deflection of the TC track prior to landfall. An examination of the potential vorticity tendencies (PVTs) suggests that, in addition to the PVT produced by this flow, the PVT from diabatic heating cannot be ignored in explaining the TC track during landfall. Topography also altered the asymmetric diabatic pattern of the TC by three major mechanisms: an incursion of dry air from the mountain, low-level convergence induced by the terrain-altered wind field in the southwestern part of the TC and the development of convergence cyclonically inward from the eastern side of the mountain to the TC core. In the second part of this series of studies, Tang and Chan (2014) examined the effects of Taiwan's CMR on the tracks of several TCs approaching Taiwan from southeast at different latitudes. They found that TCs approaching the south of Taiwan slow down and are first deflected southwestward and then northward. Moreover, a sharp northward deflection occurs for a TC passing further south of Taiwan, but the deflection is small for a TC passing further north of Taiwan. Although, Tang and Chan (2013, 2014) have advanced our understanding of orographic effects on TC tracks, the TCs inserted in their numerical simulations were steered by the pair of beta gyres far upstream, instead of advected by the southeasterly basic flow. In this study, we will be investigating the orographic effect

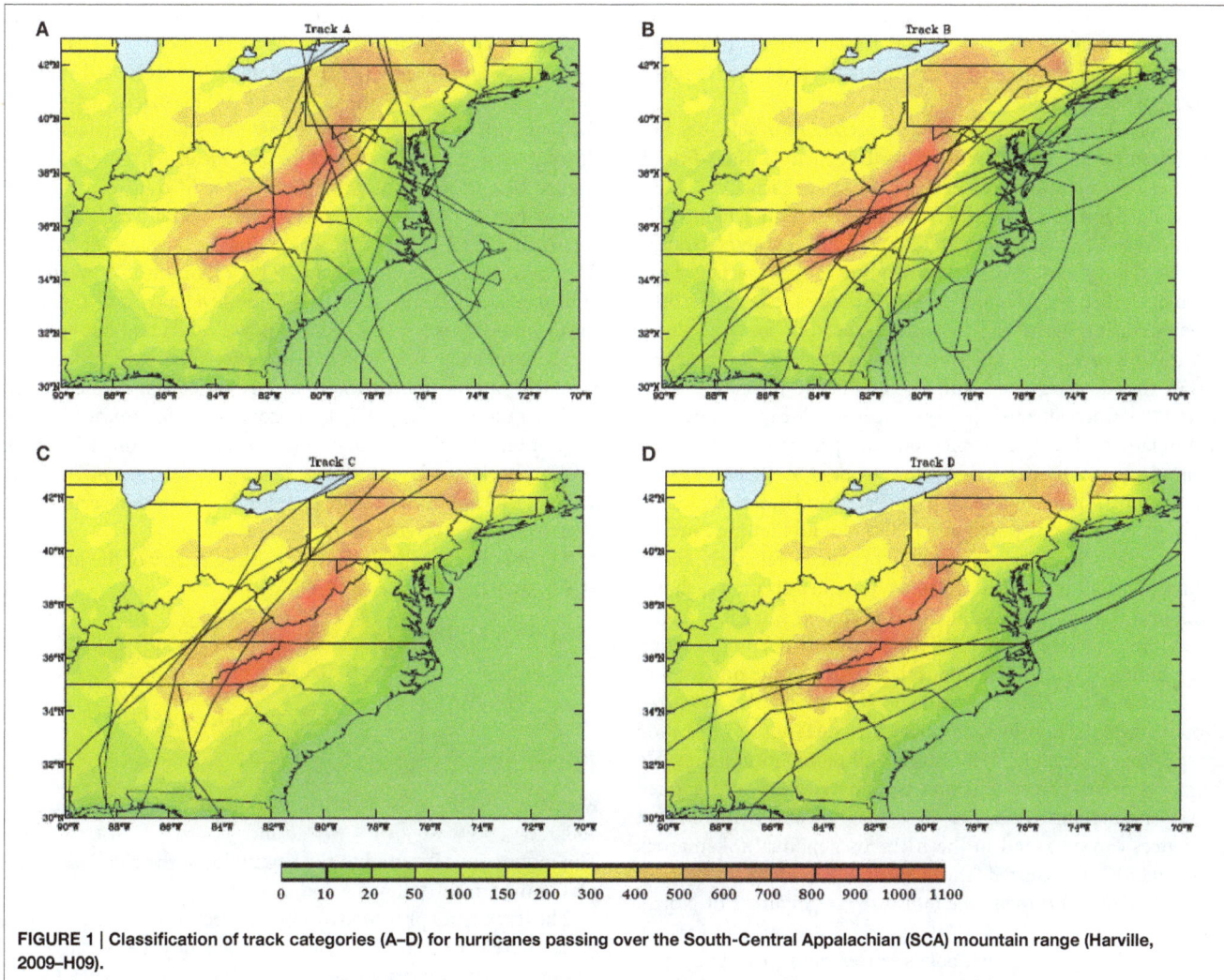

FIGURE 1 | Classification of track categories (A–D) for hurricanes passing over the South-Central Appalachian (SCA) mountain range (Harville, 2009–H09).

on the basic flow and its impact on the TC track. In term of idealized simulations, our current study differs from Huang et al. (2011) and Wu et al. (2015) in focusing on effects of landing location and impinging angle and also including vorticity budget analysis.

In this paper we will extend the study of LS11 to simulate a TC over an idealized Southern and Central Appalachians (SCA) mountain range by using a more realistic, state-of-art numerical weather prediction model. In addition, unlike a dry cyclone vortex used in LS11 and L05, a more realistic bogus TC is initiated in a conditionally unstable atmosphere with land surface, moist and planetary boundary layer (PBL) processes included. A vorticity budget analysis is adopted to help understand the effects of combined forcing of orography and convection, instead of the potential vorticity tendency (PVT) diagnostic approach taken by some previous studies (e.g., Wu and Wang, 2000; Chan et al., 2002; Hsu et al., 2013; Tang and Chan, 2013, 2014; Wang et al., 2013). The direct diabatic heating effects will be investigated along with the rainfall distributions.

The numerical model and its experimental design will be described in Section The Numerical Model and Experimental Design. In Section The Mechanisms for Track Deflection, a vorticity budget analysis will be performed to help understand the fundamental dynamics of the track deflection associated with a numerically simulated TC traversing over an elongated mountain range. Sensitivity of the track deflection to landfall locations and approach angles will be presented in Section Effects of landfall Locations and approach Angles. The concluding remarks can be found in Section Concluding Remarks.

THE NUMERICAL MODEL AND EXPERIMENTAL DESIGN

The model used for this study is the Advanced Research Weather Research and Forecasting (ARW) model version 3.4.1 (Skamarock et al., 2008). The ARW model is a three-dimensional, fully compressible, non-hydrostatic model using terrain-following vertical coordinates. The governing equations for ARW are written in flux-form with conserved mass and dry entropy. In this study, the Runge-Kutta third-order time difference scheme is employed, and the fifth- and third-order advection schemes are used for the horizontal and vertical directions respectively.

The physics parameterization or representation schemes that are chosen for all simulations in this study include: Kain-Fritsch cumulus parameterization scheme (Kain and Fritsch, 1993; Kain, 2004), Purdue-Lin microphysics parameterization scheme (Lin et al., 1983; Chen and Sun, 2002), YSU PBL parameterization scheme (Hong and Pan, 1996), Monin-Obukov surface layer scheme, Unified NOAH land-surface processes scheme, Second-order diffusion term on coordinate surfaces for turbulence and mixing processes, and Horizontal Smagorinsky first-order closure for eddy coefficient option. The Purdue-Lin scheme (Lin et al., 1983; Chen and Sun, 2002) was chosen due to its capability in simulating mixed-phase microphysical processes for practical applications, recommended by Skamarock et al. (2008), as they can account for the interaction of water and ice particles. In idealized simulations, Fovell et al. (2009) found that the Purdue-Lin scheme (Chen and Sun, 2002) produced relatively better results. The Kain-Fritsch cumulus parameterization scheme (Kain and Fritsch, 1993; Kain, 2004) is chosen since it is CAPE based, thus is more suitable in triggering the deep convection in a conditionally unstable atmosphere, such as the one used in the current study. The YSU scheme (Hong and Pan, 1996) is chosen due to the non-local closure, which does not rely on or too sensitive to localized heat, moisture, and momentum fluxes. No longwave or shortwave radiation parameterizations are applied to any simulations. Details of these schemes and their relevant references can be found in the ARW user manual (Skamarock et al., 2008). An idealized "moist tropical sounding" of Dunion (2011) is given to provide the moisture (see Figure 2 of Nolan, 2011).

A preexisting, mid-level, bogus vortex with peak tangential wind speed of $20.8 \, \text{ms}^{-1}$ at a radius of maximum winds of 90 km is set at (5385 km, 3240 km). The vertical structure of the initial vortex is in the radiative-convective-equilibrium state (Nolan, 2011; Nolan et al., 2013). Rather than using the fairly narrow tangential wind profile generated by a Gaussian vorticity distribution, our simulations use a modified Rankine vortex with decay parameter as 0.4, which is more realistic for the development stage (Mallen et al., 2005). The SST is fixed to 28°C. The domain-wide wind profile is purely zonal, with $U(\text{p}) = -5 \, \text{ms}^{-1}$ from the surface to 850 hPa, increasing with the shape of a cosine as a function of log-pressure height to $5 \, \text{ms}^{-1}$ at 200 hPa, then remaining at this value for all greater heights. The zonal wind is assumed to be in thermal wind balance. The model derives the geopotential height using iterative method and then calculates other fields.

All simulations use one single domain (6480 km, 6480 km) with a 15-km horizontal resolution. The number of the horizontal grid intervals is 433 in both the x and y directions. In the vertical, grids are stretched from the surface to the model top (20 km) with a total of 41 levels. The time interval is 45 s. A 4-km deep sponge layer is set from 16 to 20 km to reduce the wave reflection from the model top. The periodic lateral boundary condition is applied to the y-direction and the open boundary conditions are applied to both east and west side boundaries. The model is integrated for 12 days for all cases simulated in this study.

In the control case (CNTL or A2), the westward-moving bogus TC approaches the center of an idealized bell-shaped SCA mountain island from the east. This is the characteristic of track type A2 (see **Figure 2A**). The idealized mountain (see **Figure 2B**), which is oriented in the north-south direction, is approximately 1400 km long and 200 km wide, and the maximum mountain height (h) is 1 km, which mimics the SCA Mountains. The maximum tangential wind (V_{max}) of the TC vortex is approximately 60 m/s after it is fully developed and the basic mean wind (U) is 5 m/s. The basic flow Froude number (U/Nh) is approximately 0.5, which represents moderate orographic blocking on the basic flow. On the other hand, the vortex Froude number (V_{max}/Nh) is approximately 6, indicating weak orographic blocking on the outer circulation of the TC vortex (L05). The initial bogus vortex has $V_{max} = 20.8 \, m/s$ and is located 1100 km from the eastern boundary in the x direction and centered in the y direction.

The following is the formula for the geometry of the idealized SCA mountain range,

$$h(x, y) = \begin{cases} \dfrac{h_0}{\left(1 + \frac{(x-x_0)^2}{a^2}\right)^{3/2}}, & |y - y_0| \leq b \\ \dfrac{h_0}{\left(1 + \frac{(x-x_0)^2}{a^2} + \frac{(y-y_c)^2}{a^2}\right)^{3/2}}, & |y - y_0| > b \end{cases} \quad (1)$$

where $h_0 = 1\text{km}$, $x_0 = 2600\text{km}$, $y_0 = 3240\text{km}$, $a = 50\text{km}$, $b = 600\text{km}$, $y_c = y_0 + b$ for $y \geq y_0 + b$ and $y_c = y_0 - b$ for $y \leq y_0 - b$. The mountain is inserted in the beginning of the simulations, as part of the terrain-following coordinates.

The track types proposed by H09 are reclassified to track types A1-3, B, AB1-3, C, and D1-2, and used to study the sensitivity to landfall locations and approach angles (**Figure 2A**). Types A1-A3 are similar to Track A of H09 except with landfall locations on the northern, central, and southern part of the SCA differentiated. Types AB1-AB3 are similar to A1-A3 respectively, but from south-southeast. Types D1 and D2 are similar to Track D of H09 except the former crosses over the SCA, while the latter skirts around the southern tip of the mountains. Types B and C are the same as those of H09. This study focuses on track types A1-A3, AB1-AB3, B, and C. Their associated flow and orographic parameters are summarized in **Table 1**.

THE MECHANISMS FOR TRACK DEFLECTION

Figure 3A shows the 258 h accumulated rainfall for case CNTL (A2). Since the accumulated rainfall coincides with the TC track (**Figure 3B**), it can also be used as a proxy of the TC path. Note that the TC is deflected to the south upstream, passing over the mountain anticyclonically (clockwise) and westward downstream from the mountain at a latitude south of its original latitude. The continuous track and the southward deflection is similar to the behavior of a drifting TC vortex embedded in a dry, stably stratified fluid flow as found in Lin et al.'s previous

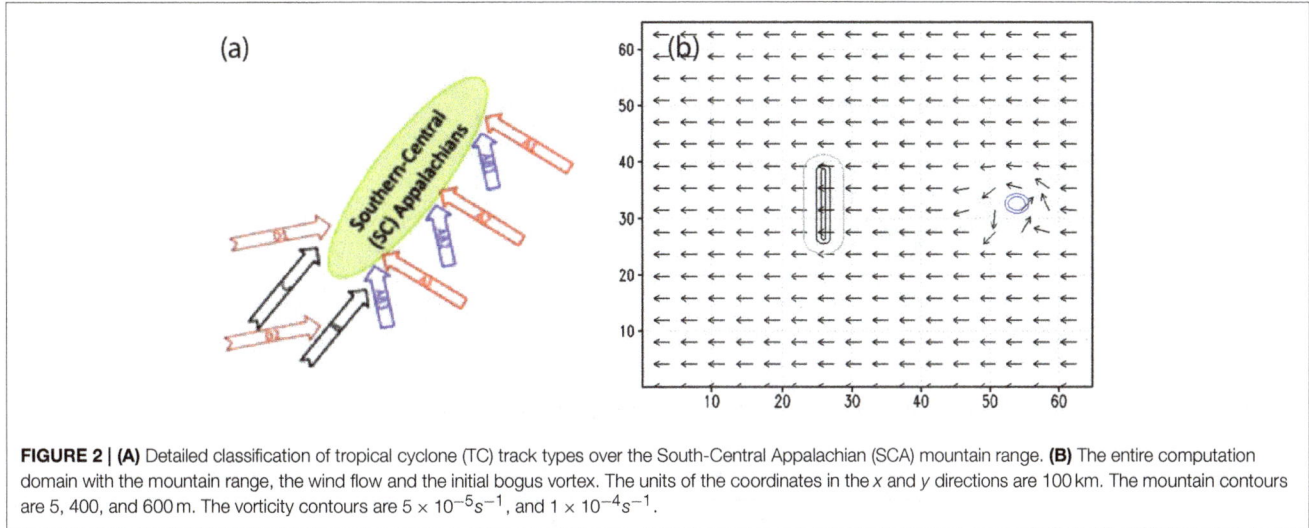

FIGURE 2 | (A) Detailed classification of tropical cyclone (TC) track types over the South-Central Appalachian (SCA) mountain range. **(B)** The entire computation domain with the mountain range, the wind flow and the initial bogus vortex. The units of the coordinates in the x and y directions are 100 km. The mountain contours are 5, 400, and 600 m. The vorticity contours are $5 \times 10^{-5} s^{-1}$, and $1 \times 10^{-4} s^{-1}$.

studies (1999, 2005) and the typhoons observed and simulated by Jian and Wu (2008) and Huang et al. (2011). In a dry flow, they can be explained by the conservation of potential vorticity due to orographic blocking on a drifting vortex embedded in a dry, stably stratified fluid flow (Lin, 2007, Figure 5.36). The mechanisms which control the track deflection in the CNTL case are much more complicated due to the inclusion of moist and PBL processes. In the following, they will be explored by performing a vorticity budget analysis.

It is well-known that a TC tends to move toward an area with a positive vorticity tendency (VT) (i.e., $\partial \zeta / \partial t > 0$; e.g., Holland, 1983, see Lin, 2007, for a brief review). Thus, a vorticity budget analysis of the individual terms of the vorticity equation provides an appropriate approach to reveal the mechanisms which control the track change. In order to do so, we analyze the individual terms of the following vorticity equation on an f-plane in the pressure vertical coordinates, which may be written as below.

$$\frac{\partial \zeta}{\partial t} = -\mathbf{V}_H \cdot \nabla \zeta + \omega \frac{\partial \zeta}{\partial p} + \zeta \frac{\partial \omega}{\partial p} + f_0 \frac{\partial \omega}{\partial p} + \left(\frac{\partial \omega}{\partial y} \frac{\partial u}{\partial p} - \frac{\partial \omega}{\partial x} \frac{\partial v}{\partial p} \right) + R \quad (2)$$
$$1 2 3 4 5 6 7$$

Here $\mathbf{V}_H$ is the horizontal wind vector (u, v), ω the omega vertical motion, f_0 the constant Coriolis parameter ($f_0 = 5.0 \times 10^{-5}\ s^{-1}$). The right-hand side of Equation (2) represents the horizontal vorticity advection (term 2), vertical vorticity advection (term 3), vorticity stretching associated with pre-existing relative vorticity (term 4), vorticity stretching associated with pre-existing planetary vorticity (term 5), tilting term (term 6), and residual term (term 7, mainly contributed by friction). The residual term is obtained by subtracting the VT (term 1) by terms 2–6. The 950–500 hPa layer-averaged VT, horizontal vorticity advection, vorticity stretching with pre-existing relative vorticity, and residual term (i.e., terms 1, 2, 4, and 7, respectively) and the flow and TC fields are shown when the TC is located upstream at 132 h (**Figure 4**), right before crossing the mountain at 168 h (**Figure 5**), just over the mountain at 183 h (**Figure 6**), and downstream from the mountain at 216 h (**Figure 7**). Other

terms are not shown because they are insignificant compared to the above terms.

At 132 h, the TC is located at about 700 km upstream of the mountain ridge, it has already started moving southwestward (**Figure 3**) toward the area with maximum positive VT (**Figure 4A**). This area with positive VT is accompanied by an area with negative VT located to the northeast (**Figure 4A**). Compared to other significant terms (**Figures 4B–D**), the positive VT (**Figure 4A**) is mainly contributed by the horizontal vorticity advection (**Figure 4B**) with additional influence from vorticity stretching (**Figure 4C**) and residual terms (**Figure 4D**). The southwestward advection of the vorticity can be explained by the anticyclonic turning of a continuously stratified flow over a mesoscale mountain due to the generation of high pressure over the mountain (Smith, 1979), which forces the flow to turn southward upstream for an easterly basic flow (see Figure 5.36 of Lin, 2007). This southwestward basic flow then advects the vertical vorticity associated with the TC vortex to give a negative-positive VT dipole toward southwest as shown in **Figure 4B**. Note that in **Figure 4B** could be interpreted as a wavenumber-2 pattern if one takes into account of the small area of positive vorticity advection embedded in the much larger area of negative vorticity advection, however, the overall distribution is dominated by the negative-positive dipole.

The area with maximum vorticity stretching is dominated by the TC convection, especially the diabatic heating associated with the eyewall convection, thus it forms a ring collocated with the eyewall, as approximately represented by the rainfall distribution. The residual term (**Figure 4D**) has multiple (three more specifically) maxima of the same order of magnitude, located at different azimuthal angles, thus it is less certain about its contribution to the VT. Part of the reason is that the residual term may involve some complicated physical processes which are related to viscosity, planetary boundary layer parameterization, turbulent mixing, etc. and numerical smoothing or diffusion. These processes may influence the vorticity tendency differently. For example, viscosity tends to spin down the vertical vorticity, but on the other hand, turbulence mixing may cause turbulence

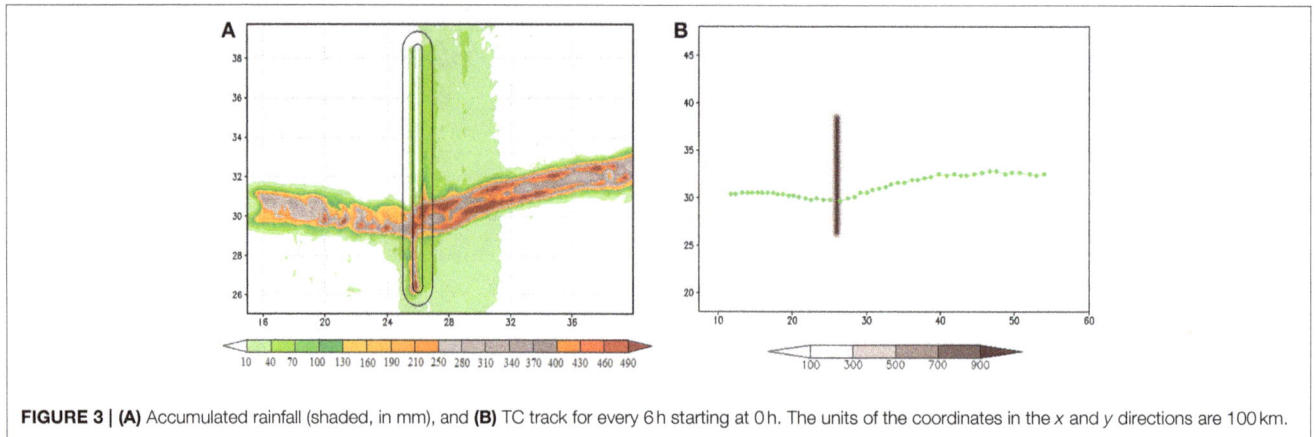

FIGURE 3 | (A) Accumulated rainfall (shaded, in mm), and **(B)** TC track for every 6 h starting at 0 h. The units of the coordinates in the x and y directions are 100 km.

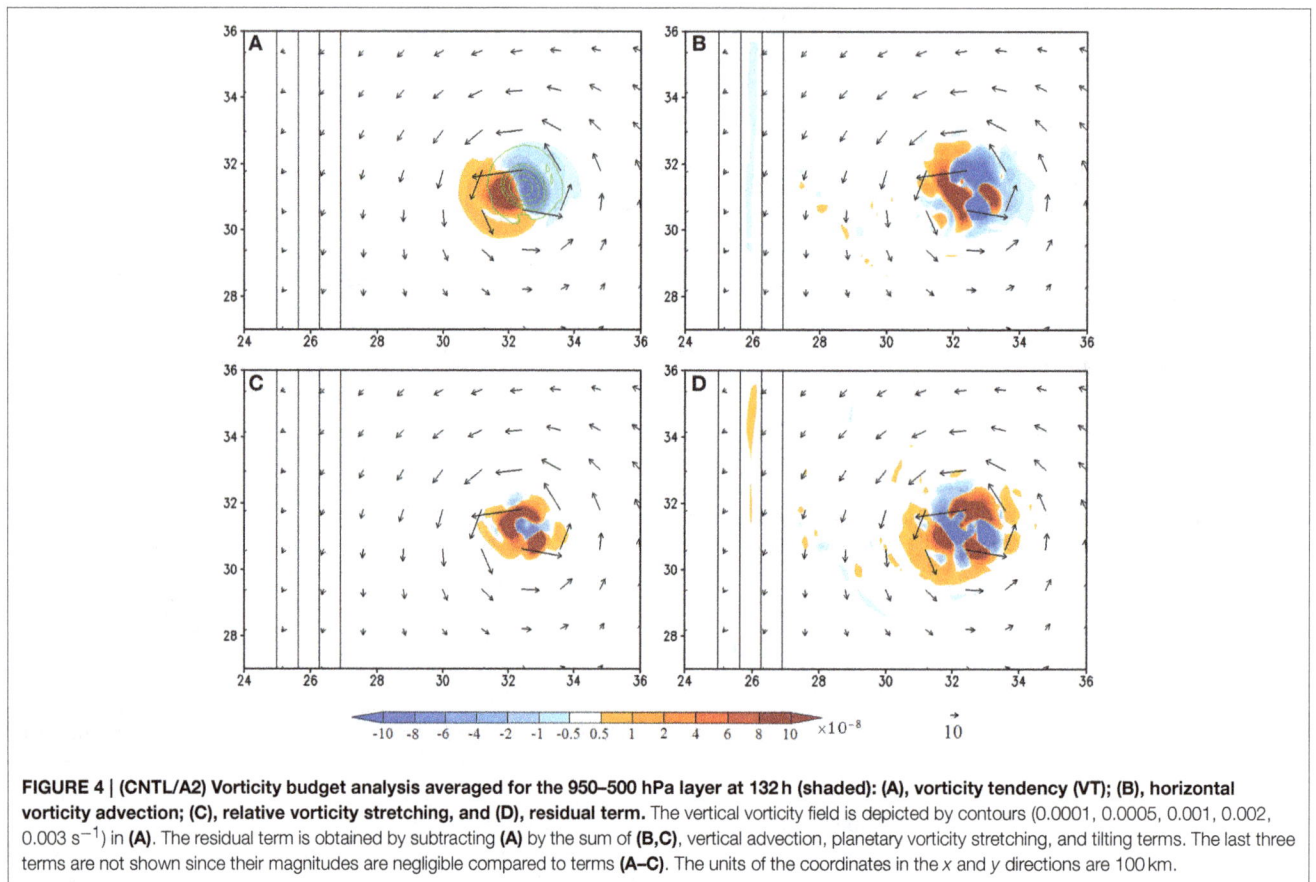

FIGURE 4 | (CNTL/A2) Vorticity budget analysis averaged for the 950–500 hPa layer at 132 h (shaded): (A), vorticity tendency (VT); (B), horizontal vorticity advection; (C), relative vorticity stretching, and (D), residual term. The vertical vorticity field is depicted by contours (0.0001, 0.0005, 0.001, 0.002, 0.003 s^{-1}) in **(A)**. The residual term is obtained by subtracting **(A)** by the sum of **(B,C)**, vertical advection, planetary vorticity stretching, and tilting terms. The last three terms are not shown since their magnitudes are negligible compared to terms **(A–C)**. The units of the coordinates in the x and y directions are 100 km.

dissipation leading to the generation of potential vorticity (e.g., Smith, 1989; Schär and Durran, 1997) and relative vorticity. This problem deserves further investigation, but is out of scope of the present study. In summary, we found that the upstream track deflection is dominated by the vorticity advection with additional contributions from vorticity stretching and viscosity. This is consistent with the studies with a dry flow except the effects of diabatic heating and viscosity (LS11). For real cases, this implies that a TC's motion is mainly steered by the orographically influenced basic flow but slightly modified by convective heating

(mainly represented in the vorticity stretching term) and friction (mainly represented by the residual term).

At 168 h, the TC has moved to the foothill (**Figure 5**) and it turns slightly toward west-southwestward before passing over the mountain (**Figure 3**). Similar to that at 132 h, the VT field shows a positive VT area located to the west-southwest of the TC vortex center, accompanied by a negative VT area to the east-northeast (**Figure 5A**). The area of this VT couplet is expanded slightly larger and exhibits a banded structure to the south of the TC center. The positive VT is mainly contributed

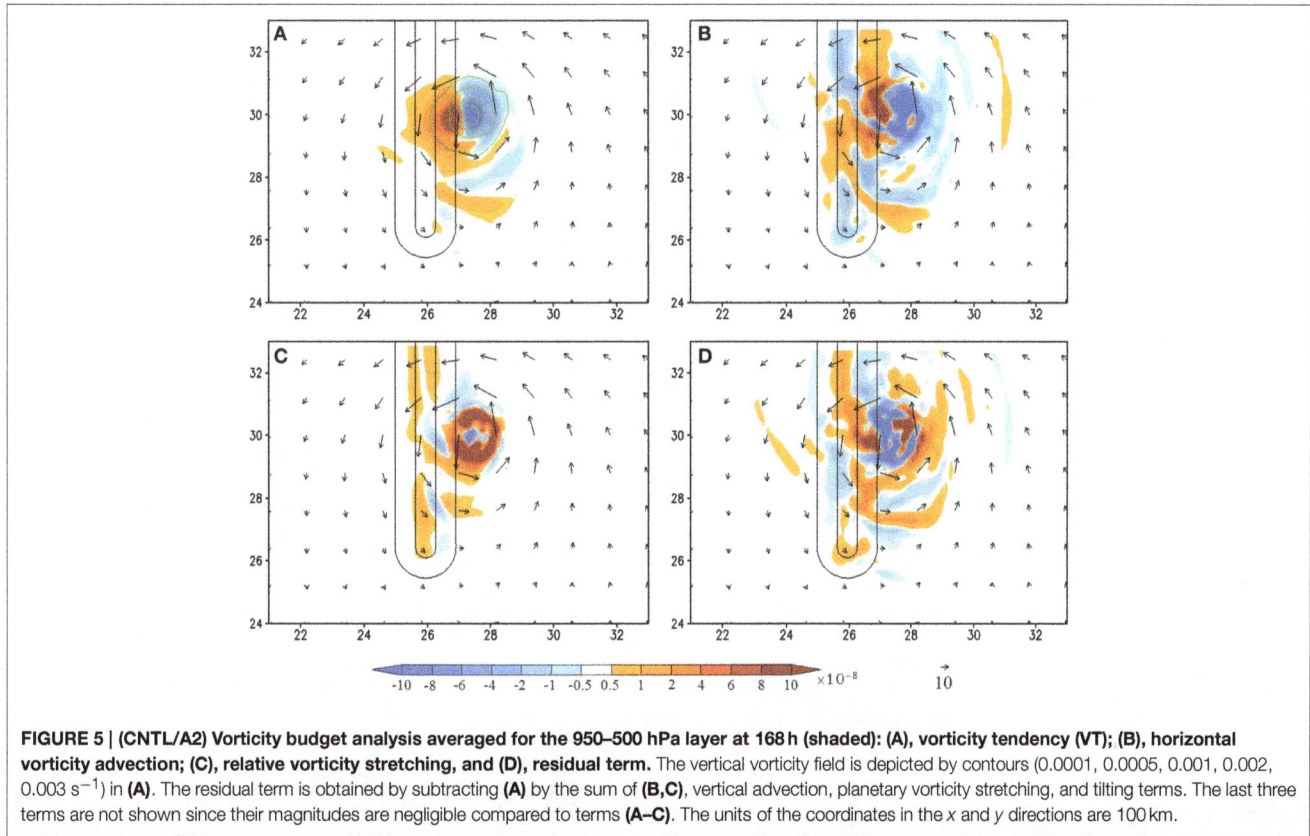

FIGURE 5 | (CNTL/A2) Vorticity budget analysis averaged for the 950–500 hPa layer at 168 h (shaded): (A), vorticity tendency (VT); (B), horizontal vorticity advection; (C), relative vorticity stretching, and (D), residual term. The vertical vorticity field is depicted by contours (0.0001, 0.0005, 0.001, 0.002, 0.003 s^{-1}) in (A). The residual term is obtained by subtracting (A) by the sum of (B,C), vertical advection, planetary vorticity stretching, and tilting terms. The last three terms are not shown since their magnitudes are negligible compared to terms (A–C). The units of the coordinates in the x and y directions are 100 km.

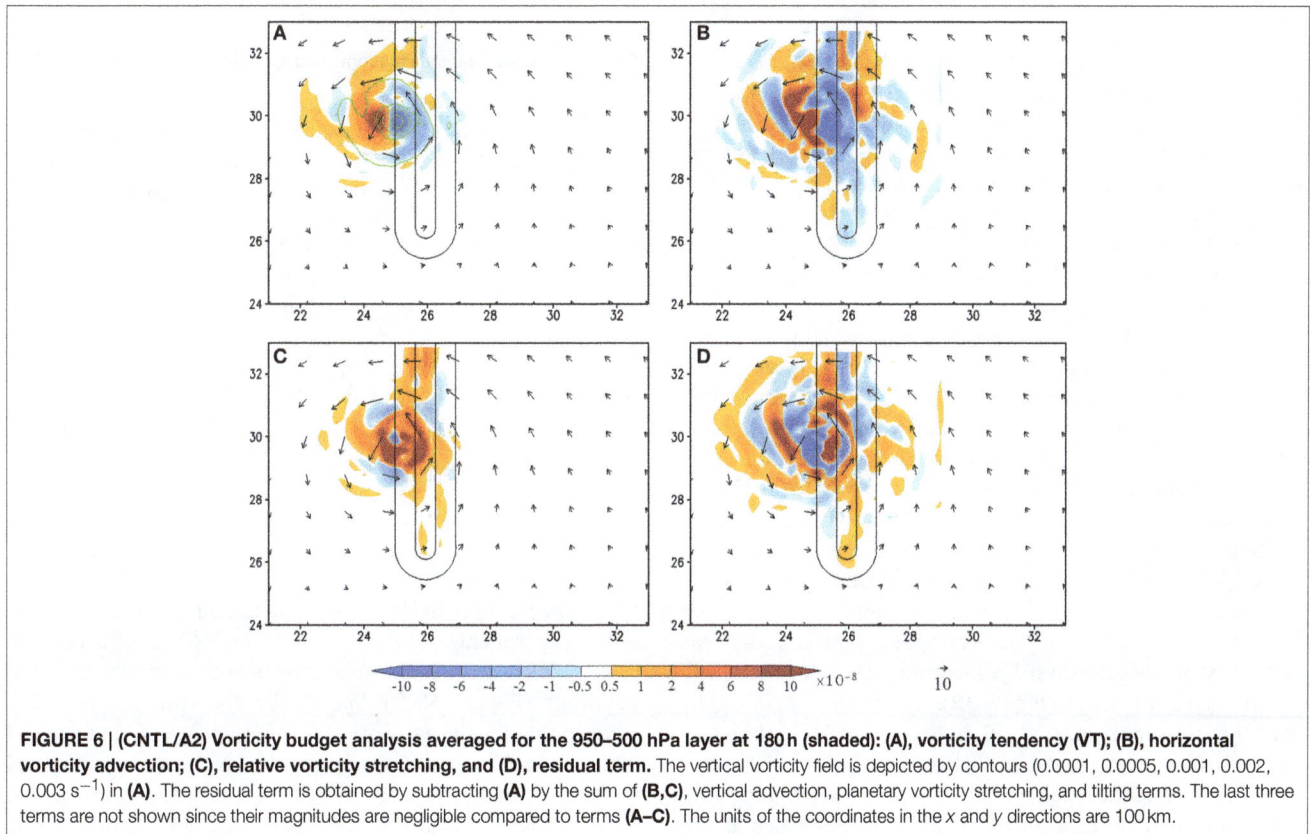

FIGURE 6 | (CNTL/A2) Vorticity budget analysis averaged for the 950–500 hPa layer at 180 h (shaded): (A), vorticity tendency (VT); (B), horizontal vorticity advection; (C), relative vorticity stretching, and (D), residual term. The vertical vorticity field is depicted by contours (0.0001, 0.0005, 0.001, 0.002, 0.003 s^{-1}) in (A). The residual term is obtained by subtracting (A) by the sum of (B,C), vertical advection, planetary vorticity stretching, and tilting terms. The last three terms are not shown since their magnitudes are negligible compared to terms (A–C). The units of the coordinates in the x and y directions are 100 km.

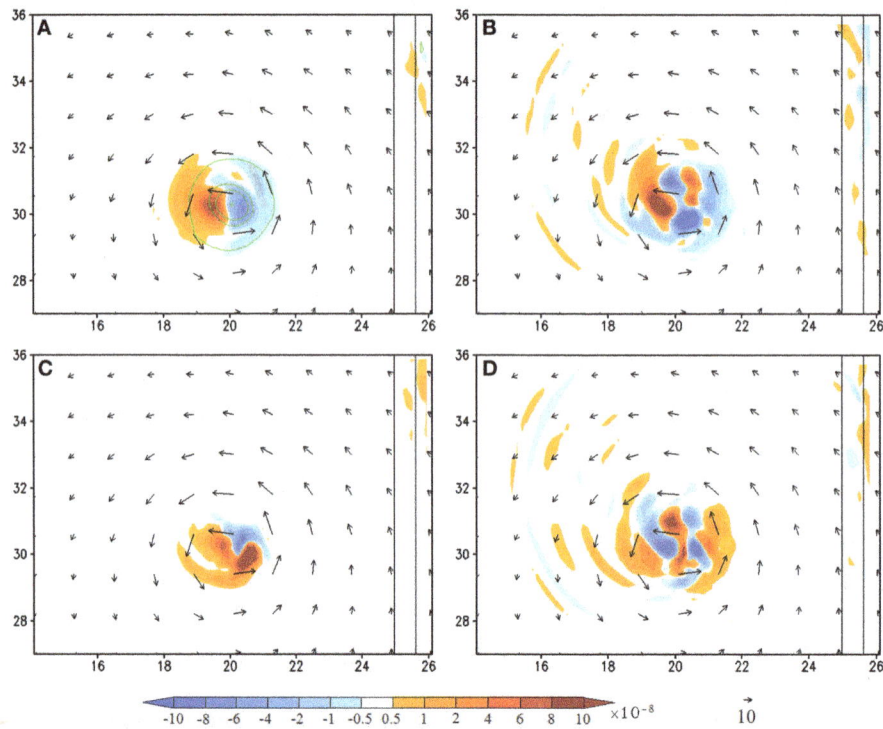

FIGURE 7 | (CNTL/A2) Vorticity budget analysis averaged for the 950–500 hPa layer at 216 h (shaded): (A), vorticity tendency (VT); (B), horizontal vorticity advection; (C), relative vorticity stretching, and (D), residual term. The vertical vorticity field is depicted by contours (0.0001, 0.0005, 0.001, 0.002, 0.003 s^{-1}) in **(A)**. The residual term is obtained by subtracting **(A)** by the sum of **(B,C)**, vertical advection, planetary vorticity stretching, and tilting terms. The last three terms are not shown since their magnitudes are negligible compared to terms **(A–C)**. The units of the coordinates in the x and y directions are 100 km.

by the horizontal vorticity advection, vorticity stretching, and viscosity (residual term). The distribution of positive horizontal vorticity advection is also expanded to a larger area and exhibits a much more complicated pattern (**Figure 5B**) compared to that at 132 h (**Figure 4B**), which, apparently, is caused by orographic forcing. The positive vorticity stretching (**Figure 5C**) and residual term located near the maximum VT, i.e., near $(x, y) = (2700, 3000$ km) (**Figure 5D**), have made contributions to the positive VT (**Figure 5A**). Note that the residual term may play more essential contribution than the horizontal vorticity advection to local positive VT. Similar to those at 132 h (**Figure 4**), the vorticity stretching is associated with the eyewall convection (**Figures 5C, 8B**) and the positive residual term may be associated with viscosity and turbulence mixing. Thus, we may conclude that the maximum positive VT when the TC moves to the foothill is dominated by the horizontal vorticity advection but with additional contributions from vorticity stretching and viscosity.

At 180 h when the TC passes over the mountain peak to lee side of the mountain (**Figure 6**), the area with maximum positive VT is shifted to the west. This indicates that the TC will be moving westward, which indeed is consistent with the simulated TC track (**Figure 3B**). This differs significantly from the dry simulations of L05 in which the TC resumes its westward movement at the same latitude of the initial bogus vortex. It appears that the anticyclonic (clockwise) turning over the mountain requires a longer period of time, which is not

TABLE 1 | Names, landfall locations, and approach angles of simulated cases.

Case	Approach direction (from) and location		Approach angle (between approaching and mountain directions)
A2 (CNTL)	Easterly	On Center	90°
A1		On North	
A3		On South	
AB1	Southeasterly	On South	45°
AB2		On Center	
AB3		On North	
B	Southerly along	East Side	0°
C		West Side	

applicable for a relatively narrower mountain as adopted by the CTNL experiment. Unlike that at 132 and 168 h, the maximum positive VT is mainly contributed by the vorticity advection and stretching (**Figures 6B,C**). Specifically, the vorticity stretching contributing to the positive VT (or say, the TC movement) is the diabatic heating associated with the western portion of the eyewall convection, instead of the convection over the lee slope (upslope with respect to the TC circulation) induced by

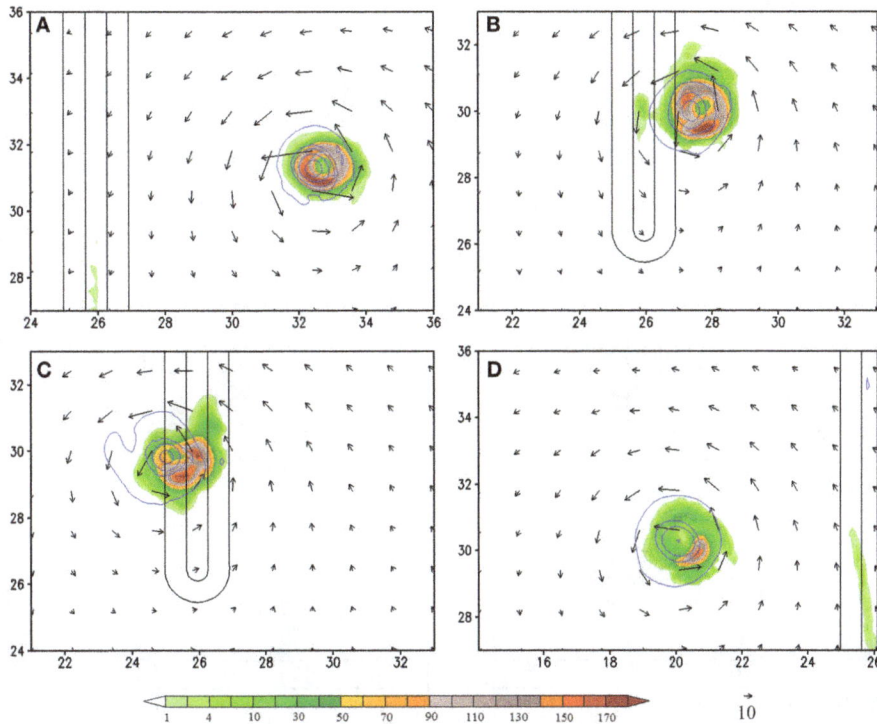

FIGURE 8 | (CNTL/A2) Accumulated 3 h rainfall (shaded, in mm/3 h) at (A) 132 h, (B) 168 h, (C) 180 h, and (D) 216 h. The units of the coordinates in the x and y directions are 100 km. The relative vorticity fields are depicted by contours (0.0001, 0.0005, 0.001, 0.002, 0.003 s^{-1}).

the orography (**Figures 6C, 8C**). The residual term makes a negative contribution to the VT (**Figures 6A,D**), indicating that the frictional effect is spinning down the vorticity over the lee slope. Note that, due to a relatively narrower mountain range, the TC track on the lee side is to the south of the original latitude. This differs significantly from the dry simulations of L05. The anticyclonic (clockwise) turning over the mountain area requires a substantial contribution from the TC convection, which produces strong vorticity stretching over a significant period of time.

At 216 h, the TC vortex moves far away from the mountain to about $(x, y) = (2000, 3000 \text{ km})$ (**Figure 7**). At this time, the maximum positive VT is dominated by vorticity advection, enhanced slightly by vorticity stretching (**Figures 7B,C**), but with almost no contribution from the frictional effects (**Figure 7D**). Note that the diabatic heating does make significant contribution to the vorticity stretching to the southeast of the eyewall (**Figures 7C, 8D**), but not the area with maximum positive VT.

EFFECTS OF LANDFALL LOCATIONS AND APPROACH ANGLES

In this section, the effects of landfall location on track deflection are investigated by performing sensitivity experiments with vorticity budget analyses of the TC making landfall at the northern tip (A1) and the southern tip (track A3) of the mountain. These results are then compared with those of track

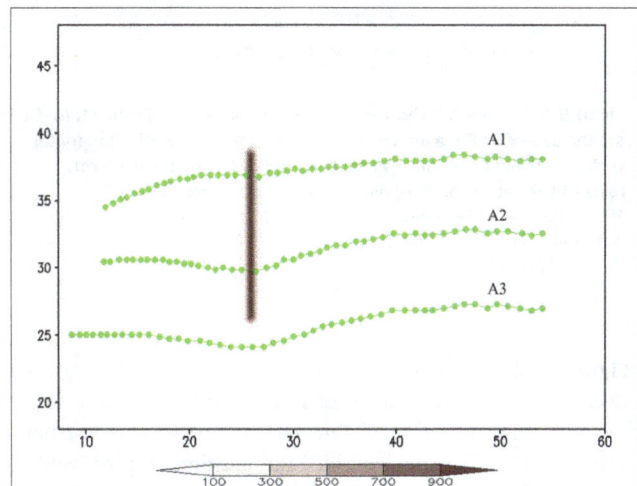

FIGURE 9 | The TC tracks for Tracks A1, A2, and A3. The units of the coordinates in the x and y directions are 100 km.

A2 (case CNTL). **Figure 9** shows the simulated tracks A1, A2, and A3 (track A2 has been discussed and analyzed in the previous section). Note that compared to track A2 (**Figure 3**), track A1 (**Figure 9**) has less southward deflection when it passes over the mountain. Near the foothill, the maximum positive VT is dominated by the vorticity advection without much contribution from vorticity stretching and frictional effects

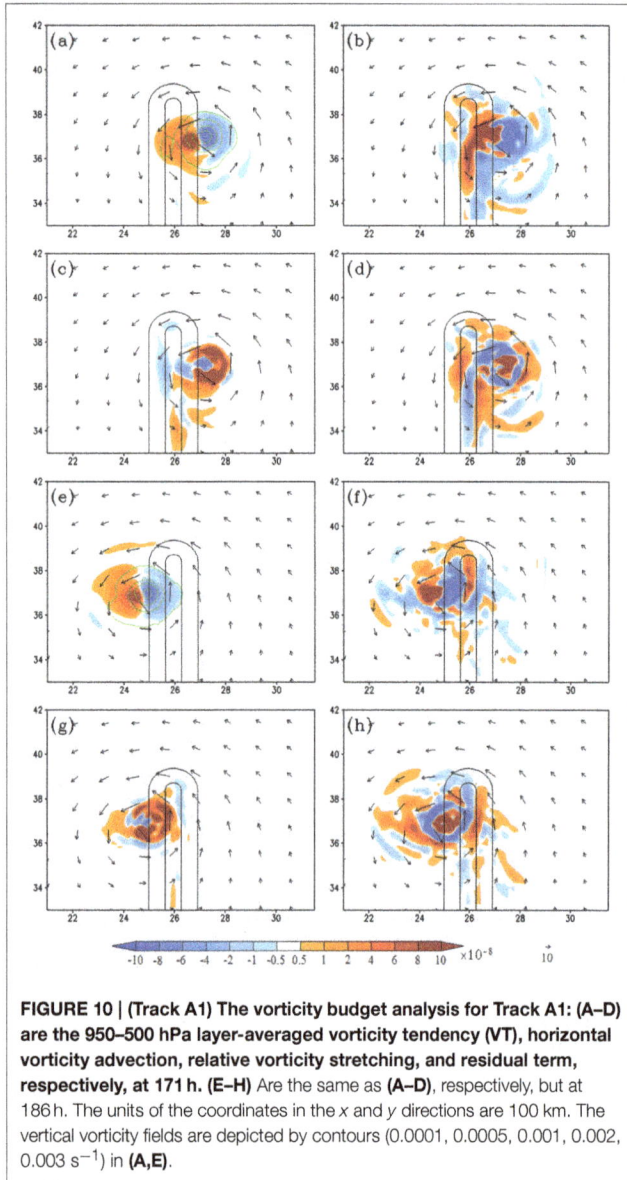

FIGURE 10 | (Track A1) The vorticity budget analysis for Track A1: (A–D) are the 950–500 hPa layer-averaged vorticity tendency (VT), horizontal vorticity advection, relative vorticity stretching, and residual term, respectively, at 171 h. (E–H) Are the same as (A–D), respectively, but at 186 h. The units of the coordinates in the x and y directions are 100 km. The vertical vorticity fields are depicted by contours (0.0001, 0.0005, 0.001, 0.002, 0.003 s⁻¹) in (A,E).

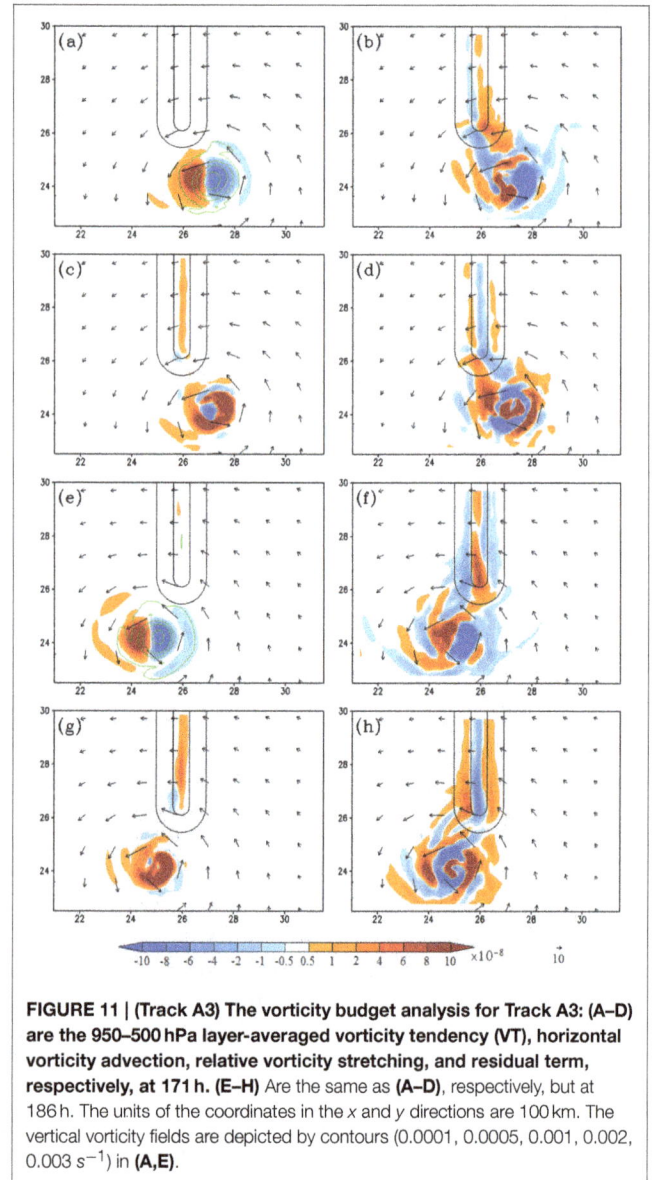

FIGURE 11 | (Track A3) The vorticity budget analysis for Track A3: (A–D) are the 950–500 hPa layer-averaged vorticity tendency (VT), horizontal vorticity advection, relative vorticity stretching, and residual term, respectively, at 171 h. (E–H) Are the same as (A–D), respectively, but at 186 h. The units of the coordinates in the x and y directions are 100 km. The vertical vorticity fields are depicted by contours (0.0001, 0.0005, 0.001, 0.002, 0.003 s⁻¹) in (A,E).

(Figures 10A–D). This is mainly due to stronger vorticity advection toward the northwest induced by the basic easterly flow and the cyclonic TC circulation around the northern mountain tip. On the lee side, the maximum positive VT (Figure 10E) is dominated by the vorticity advection, enhanced slightly by vorticity stretching (Figures 10F,G), but has almost no contribution from the frictional effect (Figure 10H). The dominance of vorticity advection of the maximum positive VT and TC movement is similar to the dry flow (case N of LS11).

Figure 11 shows the vorticity budget analysis when the TC is passing over the southern tip (case A3). When the TC moves to the southeast tip of the mountain, the maximum positive VT (Figure 11A) is mainly contributed by the frictional effects (Figure 11D) and slightly contributed by the vorticity stretching (Figure 11C), but, unlike the above cases (cases A2 and A1), no contributions from vorticity advection. The frictional effects may

be caused by stronger interaction of the northern part of the TC circulation and the southern mountain tip. After passing over the mountain, the track curves northwestward back to its original direction, which is dominated by the vorticity advection, like earlier cases.

In order to investigate the effects of approach angle and landfall locations on track deflection, three sensitivity experiments mimicking track types AB1, AB2, and AB3 have been performed and the results are shown in Figure 12. All of the simulated TCs approach the south-north oriented mountain range from the southeast or 135°. The TC makes landfall near the northern tip, center, and southern tip in cases AB1, AB2, and AB3, respectively. The mechanisms of the track deflection for tracks AB1, AB2, and AB3 are similar to those for tracks A1, A2, and A3 as discussed above. Tracks AB2 and AB3 are less curvy compared to those of A2 an A3 due to less orographic

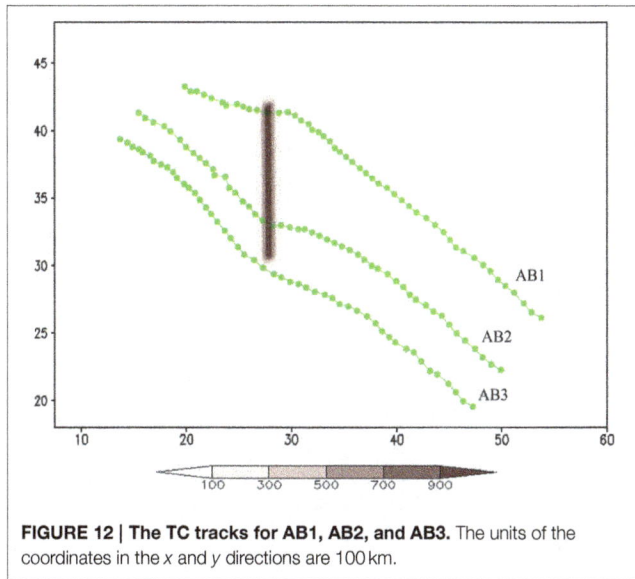

FIGURE 12 | The TC tracks for AB1, AB2, and AB3. The units of the coordinates in the x and y directions are 100 km.

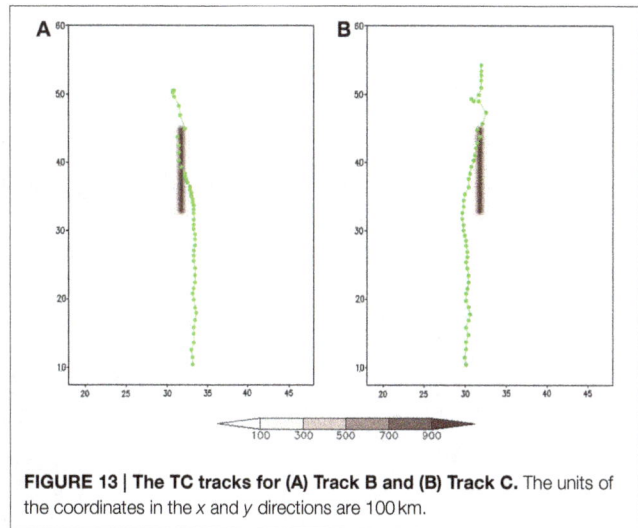

FIGURE 13 | The TC tracks for (A) Track B and (B) Track C. The units of the coordinates in the x and y directions are 100 km.

blocking caused by approach angle. It is interesting to observe that track AB1 is curved opposite to tracks AB2 and AB3 (i.e., turning cyclonically when the TC is passing over the northern tip). This is consistent with that found in dry experiments of LS11 where the vorticity advection upstream of the mountain range is caused by the flow splitting of the basic flow. Track AB1 may be used to explain most of the cyclonic track over northern Taiwan experienced by typhoons (e.g., Wang, 1980; Chang, 1982; see Figure 1A of L05) since the Central Mountain Range of Taiwan has a relatively shorter scale in the north-south direction and most of the typhoons approached the CMR from the southeast, instead of east. Thus, a TC tends to go around the corner due to basic flow splitting and the dominant control of positive vorticity tendency (VT) by vorticity advection.

As classified by H09 and illustrated in Figure 2, hurricanes making landfall on the Gulf coast may transverse along the SCA to the east (type B) or west (type C). This means the approach angle is turned to 180° from the south (southwest) for a south-to-north (southwest-to-northwest) oriented mountain range. Figure 13 shows the TC tracks for sensitivity experiments B and C. These track deflections are quite interesting. For both tracks, the TC turns toward the mountain in the southern portion of the mountain range, crosses over the mountain peak to the foothill on the lee side near the center (track B) or the northern end (track C) of the mountain range, wobbles near the northern tip of the mountain range, and then resumes its original track parallel but to the west of the mountain.

The mechanisms of the track deflection in case B can also be explained by the vorticity budget analysis. Figure 13A shows the TC continuously moving northward along the mountain. When the TC moves along the eastern flank in the southern portion of the mountain range, the vorticity budget analysis at 150 h indicates that the positive VT is influenced by all major terms, including vorticity advection, vorticity stretching, and residual terms (Figures 14A–D). The vorticity stretching is mainly associated with the latent heat released by the eyewall convection,

as evidenced by the 3-h rainfall field (Figure 14I). When the TC moves northward to $y = 3300$ km at 204 h (Figures 14E–H,J), it also moves westward simultaneously toward the eastern slope around $x = 3000$ km. During this period, the maximum positive VT is also influenced by the combined effects of vorticity advection, stretching and viscosity. Since the positive VT is located toward the northwest of the TC center, the TC is deflected toward northwest while it moves northward along the mountain ridge. This continues to be the case as the TC moves farther northward at later time (not shown).

For track C, the track behavior, vorticity budget analysis, and mechanisms for track deflection are similar to those for track B. The simulated track C, which is deflected toward the mountain and crosses over the mountain peak, is similar to the track of Hurricane Ivan (2004) although Ivan's track might have also been influenced by its synoptic environment.

CONCLUDING REMARKS

In this study, a series of idealized numerical experiments are performed using the ARW model to investigate the effects of landfall location and approach angle on track deflection of a tropical cyclone (TC) passing over an idealized South-Central Appalachian (SCA) Mountain. A bogus TC is initialized with a climatological tropical sounding with land surface, moist and PBL processes. A series of sensitivity experiments and vorticity budget analysis are performed to help understand the effects of landfall locations and approach angles on the track deflection.

When the TC landfalls on the central portion of the mountain range (Track A2), it is deflected to the south upstream, passes over the mountain anticyclonically, and then moves westward downstream. The TC motion is steered by the positive vorticity tendency (VT). When the TC is away from the mountain, the positive VT is dominated by horizontal vorticity advection upstream and downstream, but with additional influence from the stretching and residual terms. Physically, the stretching and residual terms are mainly associated with diabatic heating

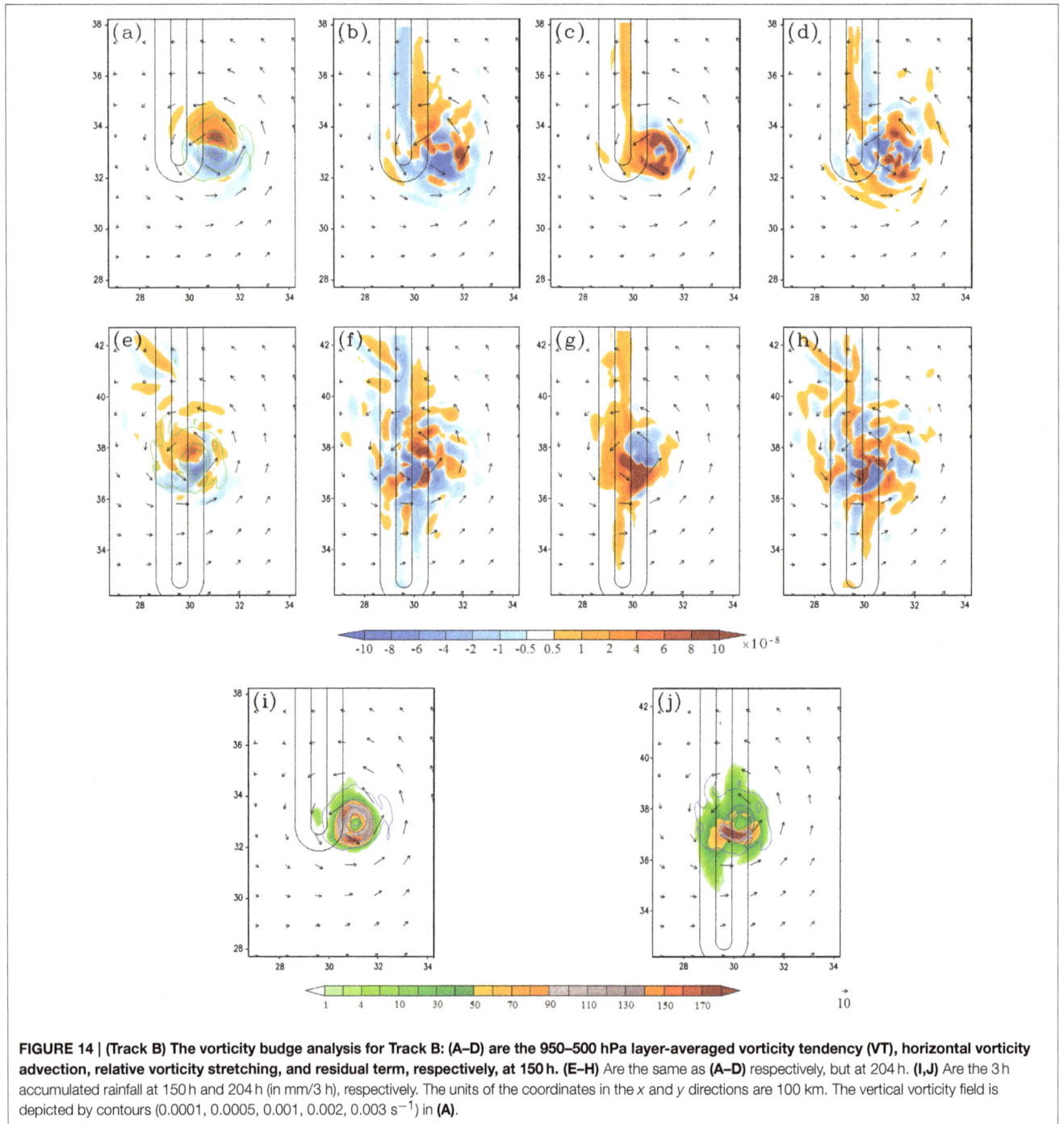

FIGURE 14 | (Track B) The vorticity budge analysis for Track B: (A–D) are the 950–500 hPa layer-averaged vorticity tendency (VT), horizontal vorticity advection, relative vorticity stretching, and residual term, respectively, at 150 h. **(E–H)** Are the same as **(A–D)** respectively, but at 204 h. **(I,J)** Are the 3 h accumulated rainfall at 150 h and 204 h (in mm/3 h), respectively. The units of the coordinates in the x and y directions are 100 km. The vertical vorticity field is depicted by contours (0.0001, 0.0005, 0.001, 0.002, 0.003 s^{-1}) in **(A)**.

and friction effects, respectively. Near the foothill, the TC turns slightly toward west-southwestward before passing over the mountain, contributed by horizontal vorticity advection, vorticity stretching, and residual terms. When the TC passes over the mountain peak to the lee (western) side, it resumes the westward movement. At this time, the area with maximum positive VT is dominated by the horizontal vorticity advection and stretching. When the TC moves farther away from the mountain, the maximum positive VT is dominated by vorticity

advection, enhanced slightly by vorticity stretching. Note that the track deflection upstream and downstream of the mountain, which are dominated by vorticity advection, is similar to the dry flow. The present results are similar to that found in LS 11. However, it is very different in the vicinity of the mountain. In LS11, the vorticity stretching occurs mainly over the lee side and its immediate downstream area, while in this study it makes contribution to the VT over both the upslope and downslope of the mountain due to the diabatic heating associated with moist convection. In addition, the vorticity advection upstream of the mountain is mainly caused by the flow splitting in LS11, which is less significant in the moist flow as simulated in this study. This can be explained by the weaker orographic blocking associated with strong TC convection in the moist flow.

Effects of landfall location on track deflection are investigated by performing a series of sensitivity experiments of the TC making landfall at the northern tip (A1) and the southern tip (A3) of the mountain. When the TC landfalls near the northern (southern) tip (Track A1 (A3)), it experiences less (more) southward deflection due to stronger (weaker) vorticity advection around the tip. The track deflections in the present cases are similar to those in the dry flow (LS11), except that there exists no track discontinuity due to the weaker orographic blocking associated with strong TC convection. The combined effects of approach angle and landfall location on track deflection are investigated by performing three sensitivity experiments, AB1, AB2, and AB3, in which the TC approaches the south-north oriented mountain range from the southeast (or 135°) and makes landfall on the northern tip, center, and southern tip, respectively. The mechanisms of the track deflection for AB1-AB3 are similar to those for A1-A3, but are less curvy due to weaker orographic blocking. The track deflection of case AB2 is similar to that simulated in a dry flow (case SE) in LS11.

For TCs going along the mountain range along the eastern flank (track B) and the western flank (track C), both of them are deflected toward the mountain range, cross over the mountain peak to the other side of the mountain, and then resume their northward movement. In these cases, the positive VT is influenced by all horizontal vorticity advection, vorticity stretching, and residual terms, which are associated with diabatic heating and frictional effects, due to longer and stronger interaction with the mountain range. The major difference of this study from the dry flow is that the vorticity stretching is mainly caused by diabatic heating in the current moist flow, instead of by lee slope vorticity stretching in LS11.

Overall, the TC is steered by the maximum positive VT, which is dominated by the vorticity advection far away from the mountain. The vorticity stretching and residual terms, which are dominated by the diabatic heating and frictional effects, respectively, make additional influences on track deflection when the TC is near or over the mountain area. Note that in the present study, we found that the vorticity tendency (VT) approach is appropriate since the TC track is consistent with the VT, similar to that applied to TC motion, such as northwestward beta drift (e.g., Holland, 1983) and shear effects (e.g., Wu and Emanuel, 1993). The PV Tendency (PVT) diagnostic approach, as originally proposed by Wu and Wang (2000) and Chan et al. (2002) and applied to real-case simulations (e.g., Hsu et al., 2013; Tang and Chan, 2013, 2014; Wang et al., 2013), appears to be an attractive method and will be considered in our future research.

AUTHOR CONTRIBUTIONS

LL: Design and simulate cases, analyze the simulated results, and write the paper. YL: Design the experiments, analyze the simulated results and write the paper. SC: Install the idealized TC model, help design the idealized experiments, run cases, and proofreading the paper.

ACKNOWLEDGMENTS

The help from Dr. David S. Nolan's on idealized WRF modeling with a bogus vortex and the proofreading of the manuscript by Guy Oldaker IV are highly appreciated. The anonymous reviewers' comments have improved the quality of the paper significantly. The first two authors are supported by the NSF Award AGS-1265783, OCI-1126543, and CNS-1429464, while the third author is supported by NSF AGS-1015910.

REFERENCES

Bender, M. A., Tuleya, R. E., and Kurihara, Y. (1987). A numerical study of the effect of island terrain on tropical cyclones. *Mon. Wea. Rev.* 115, 130–155. doi: 10.1175/1520-0493(1987)115<0130:ANSOTE>2.0.CO;2

Brand, S., and Blelloch, J. W. (1974). Changes in the characteristics of typhoons crossing the island of Taiwan. *Mon. Wea. Rev.* 102, 708–713.

Chan, J. C. L., Ko, F. M. F., and Lei, Y. M. (2002). Relationship between potential vorticity tendency and cyclone motion. *J. Atmos. Sci.* 59, 1317–1336. doi: 10.1175/1520-0469(2002)059<1317:RBPVTA>2.0.CO;2

Chang, S. W.-J. (1982). The orographic effects induced by an island mountain range on propagating tropical cyclones. *Mon. Wea. Rev.* 110, 1255–1270.

Chen, S.-H., and Sun, W.-Y. (2002). A one-dimensional time-dependent cloud model. *J. Meteor. Soc. Japan* 80, 99–118. doi: 10.2151/jmsj.80.99

Dunion, J. P. (2011). Re-writing the climatology of the Tropical North Atlantic and Caribbean Sea atmosphere. *J. Clim.* 24, 893–908. doi: 10.1175/2010JCLI3496.1

Fovell, R. G., Corbosiero, K. L., and Kuo, H.-C. (2009). Cloud microphysics impact on hurricane track as revealed in idealized experiments. *J. Atmos. Sci.* 66, 1764–1778. doi: 10.1175/2008JAS2874.1

Harville, S. L. (2009). *(H09): Effects of Appalachian Topography on Precipitation from Landfalling Hurricanes.* MS thesis, North Carolina State University. Available online at: http://repository.lib.ncsu.edu/ir/handle/1840.16/2849

Holland, G. J. (1983). Tropical cyclone motion: environmental interaction plus a beta effect. *J. Atmos. Sci.* 40, 328–342.

Hong, S.-Y., and Pan, H.-L. (1996). Non-local boundary layer vertical diffusion in Medium-Range Forecast model. *Mon. Wea. Rev.* 124, 1215–1238.

Hsu, L.-H., Kuo, H.-C., and Fovell, R. G. (2013). On the geographic asymmetry of typhoon translation speed across the mountainous island of Taiwan. *J. Atmos. Sci.* 70, 1006–1022. doi: 10.1175/JAS-D-12-0173.1

Huang, Y.-H., Wu, C.-C., and Wang, Y. (2011). The influence of island topography on typhoon track deflection. *Mon. Wea. Rev.* 139, 1708–1727. doi: 10.1175/2011MWR3560.1

Jian, G.-J., and Wu, C.-C. (2008). A numerical study of the track deflection of Super-Typhoon *Haitang* (2005) prior to its landfall in Taiwan. *Mon. Wea. Rev.* 136, 598–615. doi: 10.1175/2007MWR2134.1

Kain, J. S. (2004). The Kain-Fritsch convective parameterization: an update. *J. Appl. Meteor.* 43, 170–181. doi: 10.1175/1520-0450(2004)043<0170:TKCPAU>2.0.CO;2

Kain, J. S., and Fritsch, J. M. (1993). Convective parameterization for mesoscale models: the Kain-Fritsch scheme. *Represent. Cumulus Conv. Numerical Models Meteor. Monogr. Amer. Meteor. Soc.* 24, 165–170. doi: 10.1007/978-1-935704-13-3_16

Lin, Y.-L. (2007). *Mesoscale Dynamics.* Cambridge, UK: Cambridge University Press. doi: 10.1017/cbo9780511619649

Lin, Y.-L., Chen, S.-Y., Hill, C. M., and Huang, C.-Y. (2005). (L05): Control parameters for tropical cyclones passing over mesoscale mountains. *J. Atmos. Sci.* 62, 1849–1866. doi: 10.1175/JAS3439.1

Lin, Y.-L., Farley, R. D., and Orville, H. D. (1983). Bulk parameterization of the snow field in a cloud model. *J. Clim. Appl. Meteor.* 22, 1065–1092.

Lin, Y.-L., Han, J., Hamilton, D. W., and Huang, C.-Y. (1999). Orographic influence on a drifting cyclone. *J. Atmos. Sci.* 56, 534–562. doi: 10.1175/1520-0469(1999)056<0534:OIOADC>2.0.CO;2

Lin, Y.-L., and Savage, L. C. (2011). (LS11): Effects of landfall location and the approach angle of a cyclone vortex encountering a mesoscale mountain range. *J. Atmos. Sci.* 68, 2095–2106. doi: 10.1175/2011JAS3720.1

Lin, Y.-L., Witcraft, N. C., and Kuo, Y.-H. (2006). Dynamics of track deflection associated with the passage of tropical cyclones over a mesoscale mountain. *Mon. Wea. Rev.* 134, 3509–3538. doi: 10.1175/MWR3263.1

Mallen, K. J., Montgomery, M. T., and Wang, B. (2005). Reexamining the near-core radial structure of the tropical cyclone primary circulation: implications for vortex resiliency. *J. Atmos. Sci.* 62, 408–425. doi: 10.1175/JAS3377.1

Nolan, D. S. (2011). Evaluating environmental favorableness for tropical cyclone development with the method of point-downscaling. *J. Adv. Model. Earth Syst.* 3, M08001–M08028. doi: 10.1029/2011ms000063

Nolan, D. S., Atlas, R., Bhatia, K. T., and Bucci, L. R. (2013). Development and validation of a hurricane nature run using the Joint OSSE Nature Run and the WRF model. *J. Adv. Earth. Model. Syst.* 5, 1–24. doi: 10.1002/jame.20031

O'Handley, C., and Bosart, L. F. (1996). The impact of the Appalachian Mountains on cyclonic weather systems. Part I: A climatology. *Mon. Wea. Rev.* 124, 1353–1373. doi: 10.1175/1520-0493(1996)124<1353:TIOTAM>2.0.CO;2

Rakotomavo, Z. A. P. H., Raholijao, N., and Lin, Y.-L. (2011). Effects of Madagascar Mountain Range on tropical cyclone tracks. Part I: Classification of cylone tracks reaching the East Coast. 41. Available online at: http://mesolab.ncat.edu/publications%20%28web%29/2011_Rakotomavo%20et%20al._Effects%20of%20Madagascar%20mountains%20on%20TC%20tracks.pdf or http://mesolab.ncat.edu

Schär, C., and Durran, D. R. (1997). Vortex formation and vortex shedding in continuously stratified flows past isolated topography. *J. Atmos. Sci.* 54, 534–554.

Schwarz, F. K. (1970). The unprecedented rains in Virginia associated with the remnants of Hurricane Camille. *Mon. Wea. Rev.* 98, 851–859. doi: 10.1175/1520-0493(1970)098<0851:TURIVA>2.3.CO;2

Skamarock, W. C., Klemp, J. B., Dudhia, J., Gill, D. O., Barker, D. M., Duda, M. G., et al. (2008). *A Description of the Advanced Research WRF version 3.* NCAR technical note. Available online at: http://www2.mmm.ucar.edu/wrf/users/docs/arw_v3.pdf

Smith, R. B. (1979). The influence of mountains on the atmosphere. *Adv. Geophys.* 21, 87–230. doi: 10.1016/S0065-2687(08)60262-9

Smith, R. B. (1989). "Hydrostatic airflow over mountains," in *Advances in Geophysics, Vol. 31.*, ed B. Saltzman (New York, NY: Academic Press), 1–41.

Stewart, S. R. (2004). *Hurricane Ivan, 2-24 September 2004.* Tropical Cyclone Report, National Hurricane Center, NOAA, 1–44. Available online at: http://www.nhc.noaa.gov/data/tcr/AL092004_Ivan.pdf

Tang, C. K., and Chan, J. C. (2013). Idealized simulations of the effect of Taiwan and Philippines topographies on tropical cyclone tracks. *Q. J. Roy. Meteor. Soc.* 140, 1578–1589. doi: 10.1002/qj.2240

Tang, C. K., and Chan, J. C. (2014). Idealized simulations of the effect of local and remote topographies on tropical cyclone tracks. *Q. J. Roy. Meteor. Soc.* 141, 2045–2056. doi: 10.1002/qj.2498

Wang, C.-C., Chen, Y.-H., Kuo, H.-C., and Huang, S.-Y. (2013). Sensitivity of typhoon track to asymmetric latent heating/rainfall induced by Taiwan topography: a numerical study of Typhoon *Fanapi* (2010). *J. Geophys. Res. Atmos.* 118, 3292–3308. doi: 10.1002/jgrd.50351

Wang, S.-T. (1980). *Prediction of the Movement and Strength of Typhoons in Taiwan and its Vicinity.* Research Report, 108. Taipei: National Science Council.

Wu, C.-C. (2001). Numerical simulation of Typhoon Gladys (1994) and its interaction with Taiwan terrain using the GFDL hurricane model. *Mon. Wea. Rev.* 129, 1533–1549. doi: 10.1175/1520-0493(2001)129<1533:NSOTGA>2.0.CO;2

Wu, C.-C., and Emanuel, K. A. (1993). Interaction of a baroclinic vortex with background shear: application to hurricane movement. *J. Atmos. Sci.* 50, 62–76.

Wu, C.-C., Li, T.-H., and Huang, Y.-H. (2015). Influence of mesoscale topography on tropical cyclone tracks: further examination of the channeling effect. *J. Atmos. Sci.* 72, 3032–3050. doi: 10.1175/JAS-D-14-0168.1

Wu, L., and Wang, B. (2000). A potential vorticity tendency diagnostic approach for tropical cyclone motion. *Mon. Wea. Rev.* 128, 1899–1911. doi: 10.1175/1520-0493(2000)128<1899:APVTDA>2.0.CO;2

Yeh, T.-C., and Elsberry, R. L. (1993a). Interaction of typhoons with the Taiwan topography. Part I: upstream track deflection. *Mon. Wea. Rev. 121*, 3193–3212.

Yeh, T.-C., and Elsberry, R. L. (1993b). Interaction of typhoons with the Taiwan topography. Part II: continuous and discontinuous tracks across the island. *Mon. Wea. Rev.* 121, 3213–3233.

Zehnder, J. A. (1993). The influence of large-scale topography on barotropic vortex motion. *J. Atmos. Sci.* 50, 2519–2532. doi: 10.1175/1520-0469(1993)050<2519:TIOLST>2.0.CO;2

Zehnder, J. A., and Reeder, M. J. (1997). A numerical study of barotropic vortex motion near a large-scale mountain range with application to the motion of tropical cyclones approaching the Sierra Madre. *Meteor. Atmos. Phys.* 64, 1–19. doi: 10.1007/BF01044127

Conflict of Interest Statement: The authors declare that the research was conducted in the absence of any commercial or financial relationships that could be construed as a potential conflict of interest.

2

Inferring Firn Permeability from Pneumatic Testing: A Case Study on the Greenland Ice Sheet

Aleah N. Sommers[1]*, Harihar Rajaram[1], Eliezer P. Weber[2], Michael J. MacFerrin[3], William T. Colgan[4] and C. Max Stevens[5]

[1] Department of Civil, Environmental, and Architectural Engineering, University of Colorado at Boulder, Boulder, CO, USA, [2] Apex Companies, LLC, Boulder, CO, USA, [3] Cooperative Institute for Research in Environmental Science, University of Colorado at Boulder, Boulder, CO, USA, [4] Department of Earth and Space Science and Engineering, York University, Toronto, ON, Canada, [5] Department of Earth and Space Sciences, University of Washington, Seattle, WA, USA

Edited by:
Alun Hubbard,
Aberystwyth University, UK

Reviewed by:
Nander Wever,
École Polytechnique Fédérale de
Lausanne (EPFL), Switzerland
Stefan Ligtenberg,
Utrecht University, Netherlands

*Correspondence:
Aleah N. Sommers
aleah.sommers@colorado.edu

Specialty section:
This article was submitted to
Cryospheric Sciences,
a section of the journal
Frontiers in Earth Science

Across the accumulation zone of the Greenland ice sheet, summer temperatures can be sufficiently warm to cause widespread melting, as was the case in July 2012 when the entire ice sheet experienced a brief episode of enhanced surface ablation. The resulting meltwater percolates into the firn and refreezes, to create ice lenses, and layers within the firn column. This is an important process to consider when estimating the surface mass balance of the ice sheet. The rate of meltwater percolation depends on the permeability of the firn, a property that is not well constrained in the presence of refrozen ice layers and lenses. We present a novel, inexpensive method for measuring *in-situ* firn permeability using pneumatic testing, a well-established technique used in environmental engineering and hydrology. To illustrate the capabilities of this method, we estimate both horizontal and vertical permeability from pilot tests at six sites on the Greenland ice sheet: KAN-U, DYE-2, EKT, NASA-SE, Saddle, and EastGRIP. These sites cover a range of conditions from mostly dry firn (EastGRIP), to firn with several ice layers and lenses from refrozen meltwater (Saddle, NASA-SE, EKT), to firn with extensive ice layers (DYE-2 and KAN-U). The estimated permeability in firn without refrozen ice layers at EastGRIP agrees well with the range previously reported using an air permeameter to measure permeability through firn core samples at Summit, Greenland. At sites with ice lenses or layers, we find high degrees of anisotropy, with vertical permeability much lower than horizontal permeability. Pneumatic testing is a promising and low-cost technique for measuring firn permeability, particularly as meltwater production increases in the accumulation zone and ice layers and lenses from refrozen melt layers become more prevalent. In these initial proof-of-concept tests, the estimated permeabilities represent effective permeability at the meter scale. With appropriately higher vacuum pressures and more detailed monitoring, effective permeabilities over a larger scale may be quantified reliably, and multiple measurements during a season and across multiple years could improve understanding of the evolving firn structure and permeability. The technique is also suitable for broad application in Antarctica and other glaciers and ice caps.

Keywords: permeability, firn, snow, Greenland, pneumatic testing, ice lenses, anisotropy

INTRODUCTION

The Greenland Ice Sheet has been losing mass at an accelerated rate in recent years and is a significant contributor to sea level rise (e.g., Krabill et al., 2004; Velicogna and Wahr, 2006; Rignot et al., 2011; Fettweis et al., 2013). Accurate modeling of the surface mass balance of the ice sheet is essential for predicting its contribution to sea level rise over the next century. Percolation and refreezing of meltwater in firn occurs across much of the ice sheet accumulation area and influences runoff, which is a significant component of the surface mass balance. Accurately modeling percolation and refreezing requires an estimate of firn permeability (e.g., Colbeck, 1975; Ambach et al., 1981; Marsh and Woo, 1985).

Permeability (units of length^2) is an intrinsic property of a porous medium that characterizes the connectivity of pore spaces and describes the ability of a fluid (such as air or water) to flow through the material. The hydraulic conductivity (units of length time^{-1}) of a porous medium for flow of water is, in fact, related to the intrinsic permeability as follows (Bear, 1972):

$$K = k\frac{\rho g}{\mu} \qquad (1)$$

Where K is the hydraulic conductivity, k is the intrinsic permeability of the material, ρ is bulk density of water, g is acceleration due to gravity, and μ is the dynamic viscosity of water. More generally, Equation (1) also provides a relationship for the conductivity to any fluid, including air. In firn, meltwater percolation can change the material structure through melting and refreezing, along with grain metamorphic processes. Firn permeability, then, is a material property that evolves with time, and a measurement of air permeability in firn from a short-duration pneumatic test reflects a snapshot in time of intrinsic permeability of the current material structure (which is useful and necessary for modeling meltwater percolation).

In heterogeneous media, the large-scale effective permeability typically exhibits anisotropy, i.e., direction-dependent permeability that requires a tensorial representation (e.g., Bear, 1972; Colbeck, 1975; Albert et al., 1996; Calonne et al., 2012). Firn is an anisotropic porous medium, particularly due to the presence of solid refrozen ice layers, ice lenses, or distinct wind crust layers (ice layers defined as being continuous on a snow pit scale, ice lenses as discontinuous; Fierz et al., 2009).

Although hydraulic conductivity (K) is most directly relevant to meltwater infiltration and transport models, it can be inferred from intrinsic permeability (k) measurements according to Equation (1). It is common to measure k using air as a fluid. Previous field studies have employed various techniques to measure the air permeability of seasonal snow, as well as firn samples collected in polar regions. Early permeability measurements using progressive compression were conducted by Bader (1939) and Bender (1957). Shimizu (1970) developed a dual-cylinder air permeameter to characterize permeability of snow in Hokkaido, Japan. Sommerfeld and Rocchio (1993) used a permeameter with a conditioning column to measure permeability of seasonal snow in Wyoming, USA. A similar

design to the dual-cylinder permeameter of Shimizu (1970) was adopted to measure permeability of snow and firn at different depths at Summit, Greenland (Albert et al., 1996; Albert and Shultz, 2002; Luciano and Albert, 2002; Adolph and Albert, 2014) and in Antarctica (Courville et al., 2007, 2010; Hörhold et al., 2009). Recent studies have included other approaches as well: Courville et al. (2010) compared measured permeability with permeability estimated from microtomography and Lattice-Boltzmann modeling, and Calonne et al. (2012) used three-dimensional images of snow microstructure to estimate the full permeability tensor at the core scale for numerous samples of seasonal snow and Antarctic firn. These studies all measured or estimated permeability of extracted core samples, which would not necessarily accurately represent the field-scale effective permeability in the presence of ice layers and lenses of varying horizontal extent.

Given the implications of increased meltwater percolation in Greenland's firn and to account for ice lenses of varying horizontal extent, it is important to consider *in-situ* field permeability measurements rather than measurements from isolated core samples, and whether these measurements can resolve anisotropy directly. Most of the aforementioned studies involving Greenland firn considered only vertical permeability, with the exceptions of Albert et al. (1996), in which striking differences between horizontal and vertical permeability were observed in samples from the top 1 m of snow at Summit, and Luciano and Albert (2002), who measured both horizontal and vertical permeability in selected sections of a 13 m core at Summit, with distinct directional differences throughout the depth. Since Summit rarely experiences melt (i.e., refrozen ice lenses do not occur), these studies provide excellent reference datasets for permeability measurements of dry Greenland firn without the influence of refrozen ice lenses. In Vermont, USA, Albert and Perron (2000) found that vertical permeability across ice layers and surface crust was significantly lower than in the rest of the snowpack, and we expect that refrozen ice lenses in Greenland firn would exhibit similar properties.

As meltwater production on the Greenland Ice Sheet increases with warming air temperatures, new ice lenses will likely form at higher elevations, affecting the large-scale effective permeability of Greenland firn. This directly affects its percolation and retention capacity, and accordingly its sea level rise buffering capacity (Pfeffer et al., 1991; Harper et al., 2012). An open question associated with climate change is: *How will firn permeability, meltwater percolation, and refreezing rates change with increased meltwater production and more, or thicker, ice lenses, and layers?* Extending that line of thought, *how much meltwater can be retained in the firn as opposed to draining from the ice sheet?* Recent observations have shown that at locations experiencing increased melt, ice lenses can build up to form sufficiently thick layers that become impermeable to further meltwater percolation, yielding surface runoff in areas that could previously absorb the summer meltwater by means of vertical percolation (Charalampidis et al., 2016; Machguth et al., 2016; Mikkelsen et al., 2016). To better infer the consequences of these changes, cost-effective approaches to measuring *in-situ* field permeability are needed. Multiple measurements within a season and annually will also help to improve understanding of evolving permeability, and will help to constrain refreezing and the formation of ice lenses and layers.

Here we present a relatively simple and inexpensive method for measuring *in-situ* firn permeability using pneumatic testing. This technique is frequently used to characterize permeability of anisotropic soil, gravel, and rock formations in the design of soil vapor extraction systems for environmental remediation (Weinig, 1992). In April–May 2016, before the onset of the melt season, we conducted a field campaign to test the method at six sites on the Greenland Ice Sheet. The method is described below, followed by an explanation of how the tests were analyzed and permeability values inferred. The results presented here are primarily intended as proof of concept to demonstrate the utility of our method, while also highlighting the potential for more expansive application to explore the spatial and temporal distribution of poorly constrained firn permeability values used in surface mass balance models for Greenland, Antarctica, and other ice caps and alpine glaciers.

METHODS
Pneumatic Testing Method and Field Setup

Pneumatic testing is a well-established technique for characterizing the permeability of soil, gravel, and rock, in the context of designing soil vapor extraction systems for environmental remediation (e.g., Johnson et al., 1990; Weinig, 1992). The method involves drilling boreholes, applying suction with a vacuum at known depth below an airtight seal in one borehole, and monitoring the air pressure response below a seal at known depth in another borehole some distance from the applied vacuum. A similar procedure can be customized for application to snow and firn. A schematic of this setup is shown in **Figure 1**, and a photograph of the field setup is shown in **Figure 2**.

We conducted tests at six locations on the Greenland Ice Sheet, with varying prevalence of ice lenses and layers in the stratigraphy, at various vacuum and monitoring point depths down to approximately 4 m depth at each site. The monitoring borehole was drilled at a radial distance of 1 m from the vacuum hole. We drilled boreholes by hand using a Kovacs Mark II coring drill for the vacuum hole and a Kovacs ice auger for the monitoring hole. Inflatable rubber plugs (Cherne Muni-Ball models 262,010 and 262,064) were used to seal the holes at depth. The plugs were inflated through tubes connected to a bicycle pump at the surface. Airflow through the snow and firn was induced using a portable wet/dry vacuum (Ridgid model WD4522) powered by a generator (Honda model EU2000T1A1). The flow rate of air was measured using a thermal anemometer (Dwyer model 471B). Pressure responses were recorded using analog differential pressure gauges (Dwyer Series 2000 Magnehelic) as well as a digital pressure transducer (Dwyer Series 607-3) connected to a Campbell Scientific CR1000 datalogger.

Alternative Setup—Drilling into a Firn Pit Wall

A modified experimental orientation was also employed at DYE-2 in south Greenland. In this case, boreholes were drilled horizontally into a large firn pit wall. A schematic of the vacuum and monitoring borehole orientation is shown in **Figure 3**, and a photograph is shown in **Figure 4**. The vacuum was applied 1 m deep into the wall, in a layer of firn bounded above by a 5 cm-thick ice layer and below by a 20 cm-thick ice layer (horizontal extent of the ice layers into the wall is unknown). For these tests, pressure responses were measured in sealed monitoring boreholes drilled 1 m deep into the wall at three locations: 1 m away in the same layer of firn as the vacuum (Monitoring Borehole 1), 20 cm above the upper ice layer (Monitoring Borehole 2), and 20 cm below the thicker, lower ice layer (Monitoring Borehole 3).

Analytical Solution for Air Flow through Firn

Permeability can be inferred from pneumatic field tests by considering analytical solutions for steady-state air flow through a porous, anisotropic medium (e.g., Shan et al., 1992). We treat air flow through snow and firn as Darcian flow of an ideal gas with constant viscosity and composition, and we assume the system to be isothermal. For the pressure solution, it is convenient to define a transformed variable u related to pressure (Collins, 1961):

$$u = P_a^2 - P^2 \tag{2}$$

where P_a is the ambient atmospheric pressure, and P is pressure at a given point in the snow or firn. It can be shown that under steady flow conditions, u satisfies the Laplace equation (Bear, 1972):

$$\nabla \cdot (\boldsymbol{k}\nabla u) = 0 \tag{3}$$

where $\boldsymbol{k}$ is the air permeability tensor. The analytical solution for u during our pneumatic tests is obtained from the solution of the three-dimensional Laplace equation with the vacuum treated as a point sink in spherical coordinates. To account for the constant-pressure surface boundary open to the atmosphere, we use the method of images, a well-established technique for the solution of the Laplace equation for flow toward a source/sink in semi-infinite or finite domains (e.g., Jeans, 1908; Milne-Thomson, 1938; Bear, 1972). With the vacuum (a pressure "sink") applied at depth z_0, adding an "image source" at height z_0 above the surface forms a line of symmetry to ensure that $u = 0$ at the surface, where atmospheric pressure acts. Using a rescaled vertical coordinate to transform an anisotropic medium to an equivalent isotropic medium (Bear, 1972; Shan et al., 1992) yields:

$$u(r,z) = \frac{Q}{4\pi\sqrt{k_r k_z}} \left[\frac{1}{\sqrt{r^2 + \frac{k_r}{k_z}(z-z_0)^2}} - \frac{1}{\sqrt{r^2 + \frac{k_r}{k_z}(z+z_0)^2}} \right] \tag{4}$$

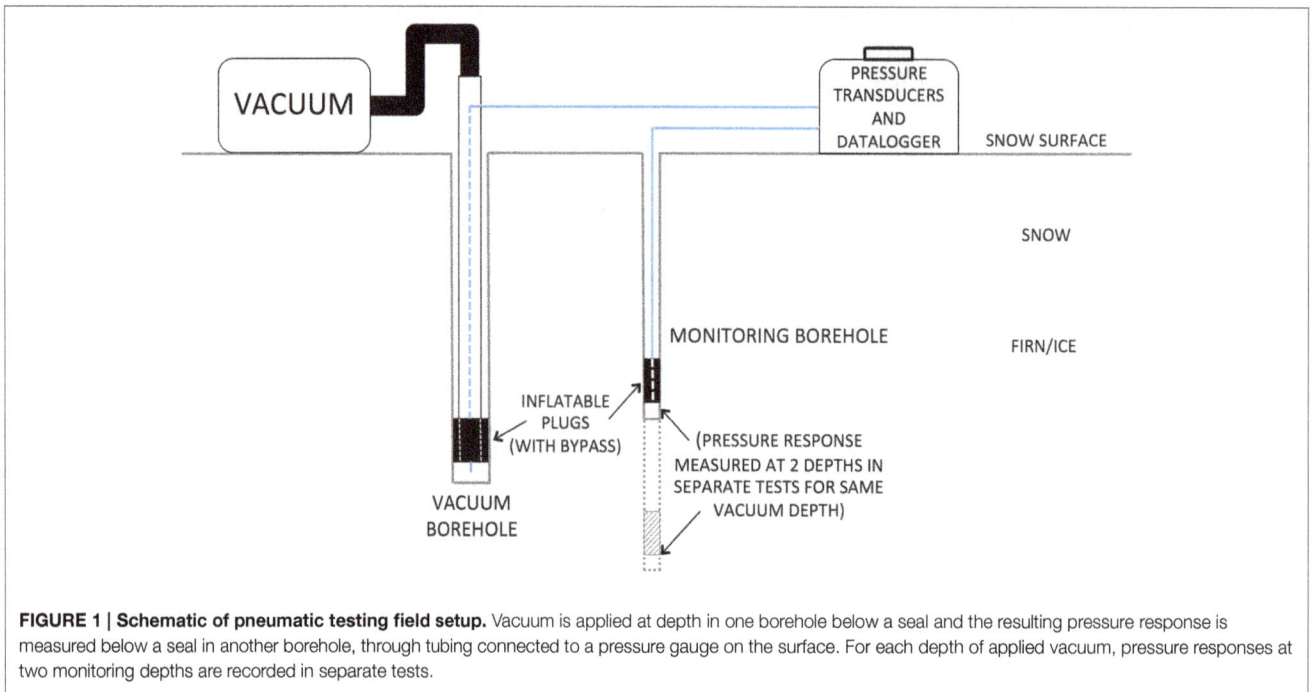

FIGURE 1 | Schematic of pneumatic testing field setup. Vacuum is applied at depth in one borehole below a seal and the resulting pressure response is measured below a seal in another borehole, through tubing connected to a pressure gauge on the surface. For each depth of applied vacuum, pressure responses at two monitoring depths are recorded in separate tests.

where r is radial distance from the vacuum hole, z is depth below the surface, Q is volumetric air flow rate, k_r is radial (horizontal) permeability, k_z is vertical permeability, and z_0 is the depth at which vacuum is applied. Note that this solution assumes a tensorial permeability that is constant (not spatially variable) within the domain, corresponding to an assumed constant heterogeneous firn pack for each test. In reality, firn structure may vary substantially across a depth range of a few meters; tests at multiple depths at each site capture the influence of these variations. **Figure 5** presents an example pressure response field and air flow streamlines based on this solution for one of our pneumatic tests. Using two monitoring points for each vacuum depth, horizontal, and vertical permeability values are inferred by fitting the analytical $u(r,z)$ field (Equation 4) to observed differential pressure responses by employing a least-squares optimization scheme (MATLAB nonlinear least-squares solver *lsqnonlin*). By dividing the equations for these two monitoring points, $u_1(r_1,z_1)/u_2(r_2,z_2)$, it is also possible to directly solve for the anisotropy ratio k_r/k_z. Comparing the value of k_r/k_z obtained directly with that inferred from optimization serves as a useful check of the reliability of optimization results.

The solution for u can be modified to incorporate an impermeable layer at some depth rather than treating the domain as semi-infinite below the surface; with multiple boundaries, the method of images becomes more complicated. In this case, we need to apply an image source to account for the constant-pressure surface boundary as before, and also apply an image sink below the impermeable boundary to effectively cancel out the flux at that boundary. But the image source above the surface then creates an unbalanced flux across the lower boundary, and the image sink below the impermeable boundary creates a pressure drop at the surface, violating the boundary conditions, requiring

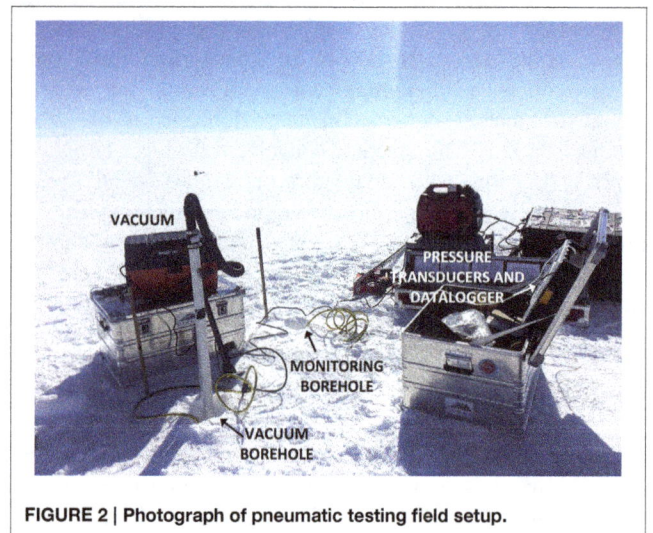

FIGURE 2 | Photograph of pneumatic testing field setup.

additional images sources/sinks. For each pair of images added, these imbalances persist; the solution by the method of images with two boundaries thus requires an infinite series of images (Bear, 1972; Shan et al., 1992). The resulting series solution is convergent because the later images in the series are increasingly farther away from the finite domain of interest. Designating the impermeable boundary at the depth of the first ice layer with thickness ~20 cm or greater (according to stratigraphy from cores drilled at each site), we found that this had no effect on the pressure solution in our test domain, suggesting that the shallowest extensive ice layer is too deep to influence the pressure response in the overlying firn and thinner refrozen melt layers. The solution for u can also be modified to treat the vacuum

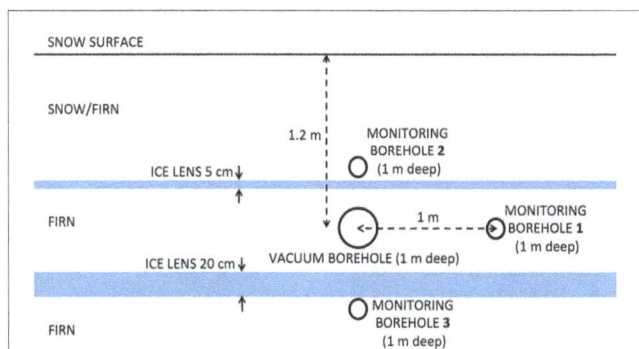

FIGURE 3 | Schematic of alternative pneumatic test setup, drilling boreholes horizontally into a firn pit wall at DYE-2. Vacuum was applied in a sealed borehole drilled 1 m horizontally into a layer of firn bounded above and below by ice layers as observed in the pit wall. Pressure response to the vacuum was measured in sealed boreholes at three locations: (1) in the same layer of firn as the vacuum, (2) above a 5 cm-thick ice layer, and (3) below a 20 cm-thick ice layer. The monitoring point located above the thinner (5 cm) ice layer from the vacuum exhibited a lower pressure response compared to the response in the same layer of firn as the vacuum, and the monitoring point across the thicker (20 cm) ice layer from the vacuum exhibited a negligible response.

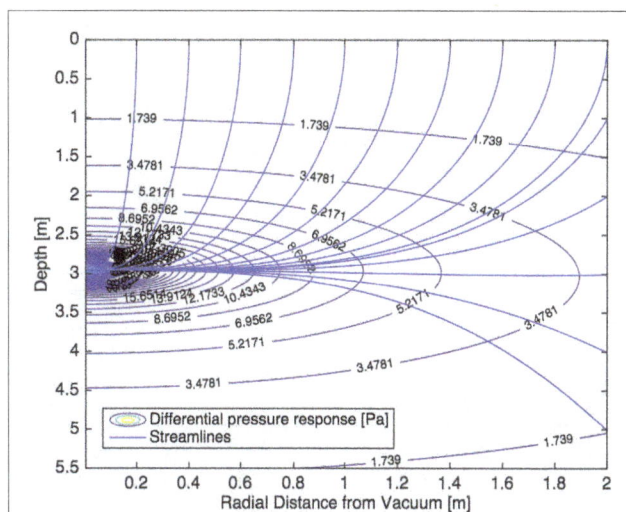

FIGURE 5 | Sample pressure field due to air flow induced by a vacuum in one of our pneumatic tests. In this example, the vacuum is applied at 3 m depth. Snow surface is at the top of the plot. Labeled contour lines show the differential pressure response. Blue lines indicate air flow streamlines.

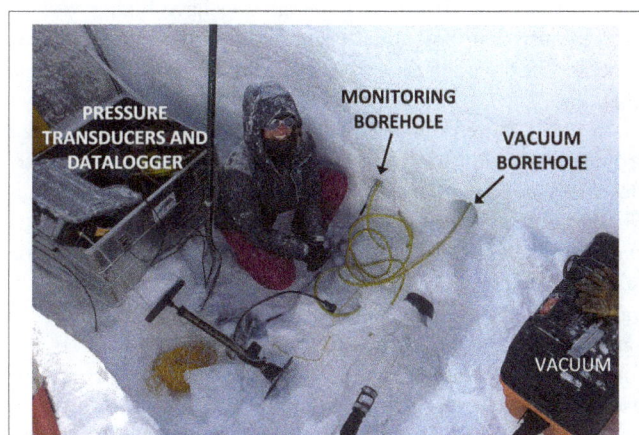

FIGURE 4 | Photograph of alternative pneumatic test setup, drilling boreholes horizontally into a firn pit wall at DYE-2.

as a line sink in the small section of open borehole below the plug, rather than as a point sink (Shan et al., 1992). The permeability values inferred using these modified analyses fell in the same range and exhibited the same anisotropy patterns as those obtained using the simpler solution (Equation 4) for a point-source in a semi-infinite domain.

RESULTS

Inferred Firn Permeability and Anisotropy at Six Sites on the Greenland Ice Sheet with Different Stratigraphy

We pilot tested this method for inferring firn permeability from pneumatic testing in April–May 2016 before the onset of melt at

six sites on the Greenland Ice Sheet (**Figure 6**). Site characteristics and specific test dates are listed in **Table 1**. Five of the sites (KAN-U, DYE-2, EKT, NASA-SE, and Saddle) are situated in south Greenland, near the Arctic Circle. These sites experience melt in some or most years; ice layers and lenses were observed in the firn cores drilled at each location and in snow pits. Ice layers are particularly abundant at KAN-U, which experiences the most melt of these five sites, and also at DYE-2. The sixth site, EastGRIP, is located farther north than the other sites and very rarely experiences melt, rendering it a useful control site for permeability of dry firn in the absence of refrozen lenses.

Using the analytical solution for air flow and optimization methods for parameter estimation described in Section Analytical Solution for air Flow through Firn, horizontal and vertical permeability were inferred for each test conducted at the field sites. **Figure 7** presents the horizontal (k_r, dashed line) and vertical (k_z, solid line) permeability profiles for five sites, along with the depth of ice layers and lenses at each site (indicated by horizontal dark blue lines, as recorded from stratigraphy analysis of a firn core drilled near the pneumatic test site at each location). The horizontal extent of ice layers and lenses is not known. Permeability at KAN-U was investigated separately because of the thick ice layer at that site, and will be described below in Section Permeability in a Thick Ice Layer. The colored shading in **Figure 7** represents possible ranges of k_r and k_z due to uncertainty in flow rate measurement. The depth range associated with each permeability estimate (vertical extent of the lines) corresponds to the depths of the two monitoring points where pressure response was measured for a given applied vacuum depth. In most cases, horizontal permeability is greater than vertical permeability. For tests that span several ice layers or lenses, this anisotropy is more pronounced, which intuitively makes sense: it is more difficult for air to flow vertically through ice layers, or around ice lenses, than to flow horizontally through

TABLE 1 | Characteristics for sites on the Greenland ice sheet where permeability tests were conducted.

Site	KAN-U	Dye-2	EKT	Saddle	NASA-SE	EastGRIP	Summit
Latitude (deg)	67.000	66.473	66.985	65.999	66.480	75.627	72.580
Longitude (deg)	−47.023	−46.283	−44.394	−44.500	−42.500	−35.942	−38.505
Elevation (m)	1812	2092	2343	2457	2389	2675	3199
Mean annual temperature (deg C)	−16.7	−18.4	−21.0	−21.0	−20.7	−30.6	−30.2
Test Date (2016)	28 April	9 May	2 May	6 May	4 May	16 May	N/A

Characteristics for Summit are also shown for comparison. Mean annual temperature is the mean annual air temperature for 1979–2014 from MAR 3.5.2 (Fettweis et al., 2013) forced by ERA Interim.

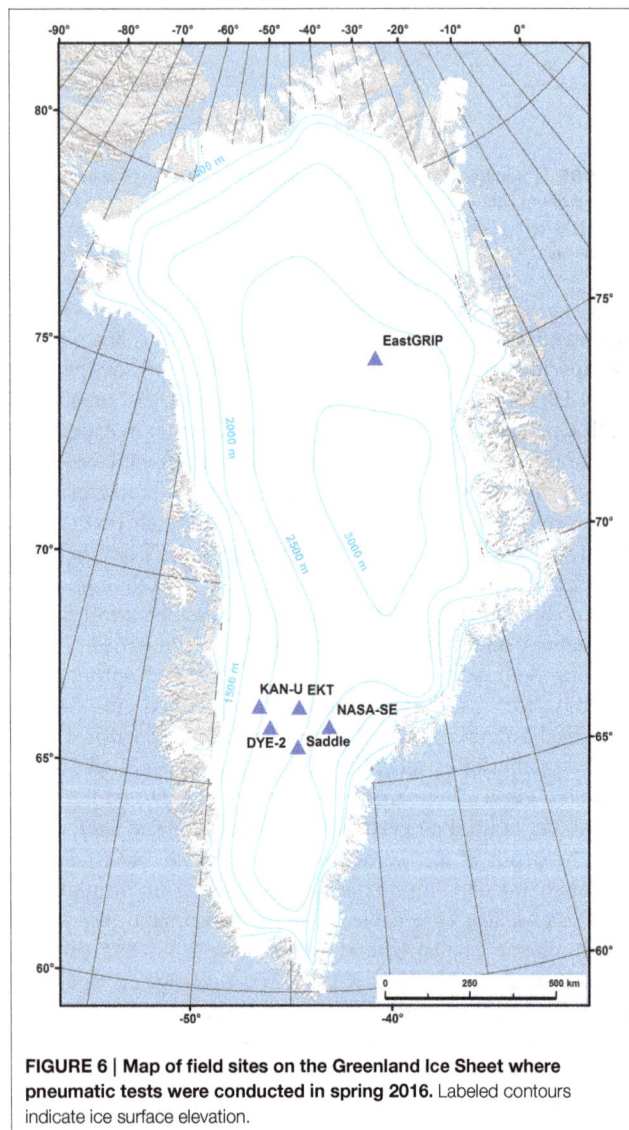

FIGURE 6 | **Map of field sites on the Greenland Ice Sheet where pneumatic tests were conducted in spring 2016.** Labeled contours indicate ice surface elevation.

example, the top test at DYE-2 and the top and middle tests at Saddle). This distinction may be related to the horizontal extent of individual ice lenses, which was not measured. In contrast to the high anisotropy of tests spanning several ice layers, horizontal and vertical permeability at EastGRIP are more similar in magnitude, with vertical permeability even being slightly higher than horizontal permeability in the deeper tests at that location. It is reasonable to find low anisotropy here, considering that EastGRIP has very few ice lenses; the deeper tests span dry firn without even thin refrozen ice lenses. In the absence of ice lenses, it is possible to have higher vertical permeability than horizontal permeability due to metamorphic processes at the grain scale (Luciano and Albert, 2002; Calonne et al., 2012).

In **Table 2**, we compare the anisotropy ratio k_r/k_z inferred from optimization with k_r/k_z from the direct solution of u_1/u_2 described in Section Analytical Solution for Air Flow through Firn. The results agree closely, with the exception of four tests that exhibit extremely high anisotropy across ice layers. For these tests that span several ice layers from which we infer very small k_z values from optimization (and correspondingly large anisotropy ratios), a direct solution for the anisotropy ratio k_r/k_z is not found due to its large magnitude (the equations for u_1 and u_2 each approach $1/\infty$, so u_1/u_2 becomes similar to dividing zero by zero). This implies that the inferred values for k_r and k_z from optimization are more uncertain in these cases, but the overall inference of very low vertical permeability and high anisotropy across ice layers remains valid.

Permeability in a Thick Ice Layer

Our lowest-elevation field site is KAN-U, which experiences the most melt each summer and, as a result, has the most prevalent refrozen ice layers. Firn cores drilled in spring 2016 at this site revealed a continuous solid ice layer from a depth of approximately 1.4–8 m. This configuration provides a unique field setup to illustrate the capabilities of our pneumatic test method for inferring permeability in the presence of thick sections of refrozen ice layers. At this site, differential pressure response was measured in a single borehole at the point of applied vacuum for incremental depths, from 0.2 m down to 2.0 m, shown in **Figure 8A**. An abrupt transition in the generated suction is apparent when the vacuum is applied in "solid" ice (below 1.4 m). The differential pressure measured at 1.8 m and 2.0 m is equal to that measured with the vacuum hose completely

the firn between layers. In some of the tests that span several ice layers or lenses (see the middle test at EKT, the middle and deepest tests at NASA-SE, the deepest test at Saddle, and the deepest test at DYE-2), the vertical permeability is very low (<1 × 10⁻¹⁰ m²). In other tests that span ice layers or lenses, however, the vertical permeability is not quite so low (for

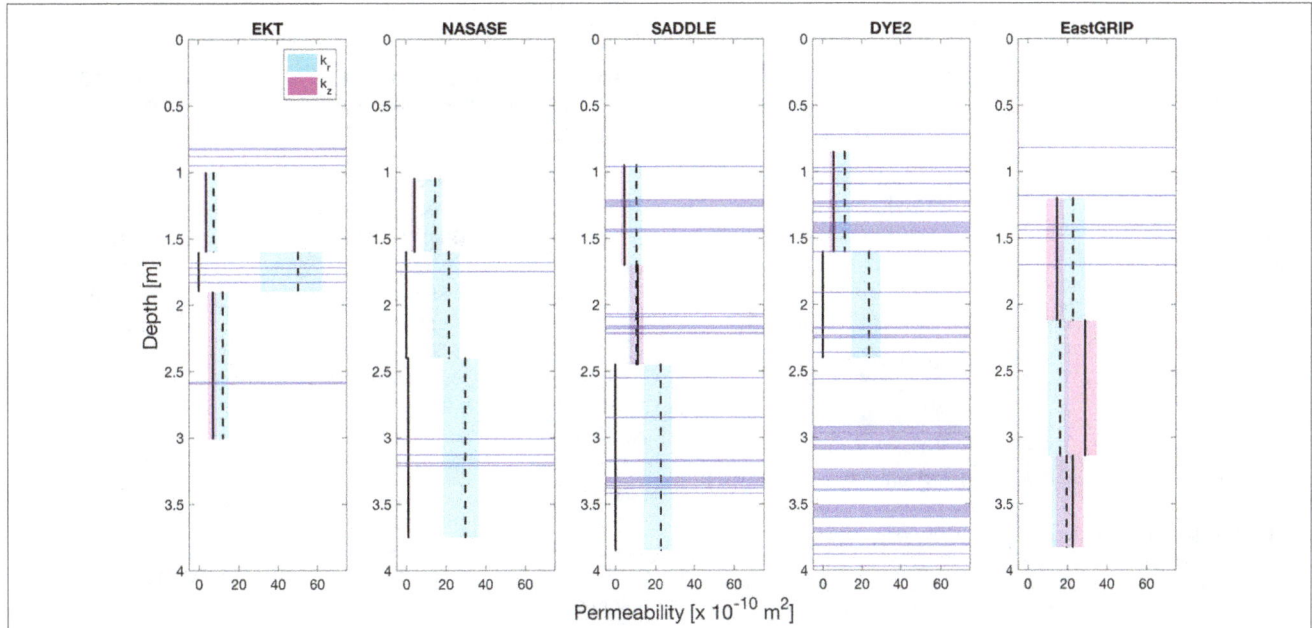

FIGURE 7 | Inferred horizontal (k_r) and vertical (k_z) firn permeability values at five sites on the Greenland Ice Sheet. The vertical depth range for each estimate corresponds to the depth between the two monitoring point depths where differential pressure was measured for each vacuum depth. Horizontal dark blue lines indicate the depths of ice layers and lenses observed in cores at each site. Cyan and magenta shading represents inferred permeability limits based on the range of airflow rates produced by the vacuum in the field. In some cases that span several ice layers, the inferred vertical permeability is very low ($k_z < 1 \times 10^{-10}$ m^2, essentially impermeable).

TABLE 2 | Anisotropy ratio (k_r/k_z) for each test inferred from optimization and by direct solution.

Test ID	Vacuum depth [m]	Monitoring depths [m]	k_r/k_z (optimization)	k_r/k_z (direct solution)
EKT 1	1.00	1.00, 1.60	2.0	2.0
EKT 2	2.10	1.60, 1.90	3.3×10^6	N/A
EKT 3	2.95	1.90, 3.01	1.7	1.7
NASA-SE 1	1.05	1.05, 1.60	3.5	3.4
NASA-SE 2	1.95	1.60, 2.40	620	N/A
NASA-SE 3	2.85	2.40, 3.75	31.1	31.9
SADDLE 1	1.00	0.95, 1.70	2.3	2.3
SADDLE 2	2.00	1.70, 2.45	0.9	0.9
SADDLE 3	3.10	2.45, 3.85	1.2×10^6	N/A
DYE-2 1	1.00	0.85, 1.60	2.1	2.0
DYE-2 2	2.05	1.60, 2.40	8.3×10^5	N/A
EastGRIP 1	0.90	1.20, 2.12	1.6	1.6
EastGRIP 2	2.01	2.12, 3.14	0.6	0.5
EastGRIP 3	2.90	3.14, 3.83	0.9	0.9

Test ID labels identify the site name and test number (numbered from shallowest to deepest). For the majority of the tests, k_r/k_z inferred by optimization agrees closely with k_r/k_z from the direct solution. Four tests that span several ice layers (EKT 2, NASA-SE 2, SADDLE 3, and DYE-2 2) exhibit extraordinarily high anisotropy, essentially implying negligible vertical permeability. For these cases, the direct solution for k_r/k_z breaks down numerically. This reflects a higher level of uncertainty for k_r and k_z inferred from optimization for these tests, but the general inference of high anisotropy and very low vertical permeability across the ice layers holds.

sealed, indicating that: (a) the thick refrozen ice layer is nearly impermeable, and (b) the inflatable plugs achieve a good seal against the borehole walls. Permeability values can be inferred from this test scenario using the analytical solution described in Section Analytical Solution for Air Flow through Firn. In this case, we consider the differential pressure measured just below the vacuum, and treat that pressure as uniform on a small sphere (with radius 5 cm) representing the open space in the vacuum below the borehole. For the two monitoring points needed for the analysis, we select two points on this sphere and employ the optimization methods for parameter estimation as described previously. **Figure 8B** presents horizontal and vertical permeability values inferred from the field tests at KAN-U. The results support what we might expect in this situation. In the top meter of snow and firn (with a few minor ice layers and lenses), horizontal permeability is greater than vertical permeability, but even the horizontal permeability is relatively low compared to our other sites ($<5 \times 10^{-10}$ m^2). Both horizontal and vertical permeability generally decrease further with depth, and both become very small ($<0.3 \times 10^{-10}$ m^2) in the thick ice below 1.5 m. This indicates that the meters-thick ice layer significantly decreases the permeability.

Pneumatic Tests into a Firn Pit Wall

We also conducted three additional tests at DYE-2 with the alternative test orientation involving boreholes drilled horizontally into a snow pit wall (shown in **Figures 3, 4**). First, the pressure response was observed in a borehole sealed 1 m deep, located 1 m radially away from the vacuum borehole, in the same layer of firn bounded by the ice layers (labeled Monitoring Borehole 1 in **Figure 3**); the measured differential pressure response here was 12.4 Pa. Next, the pressure was monitored in

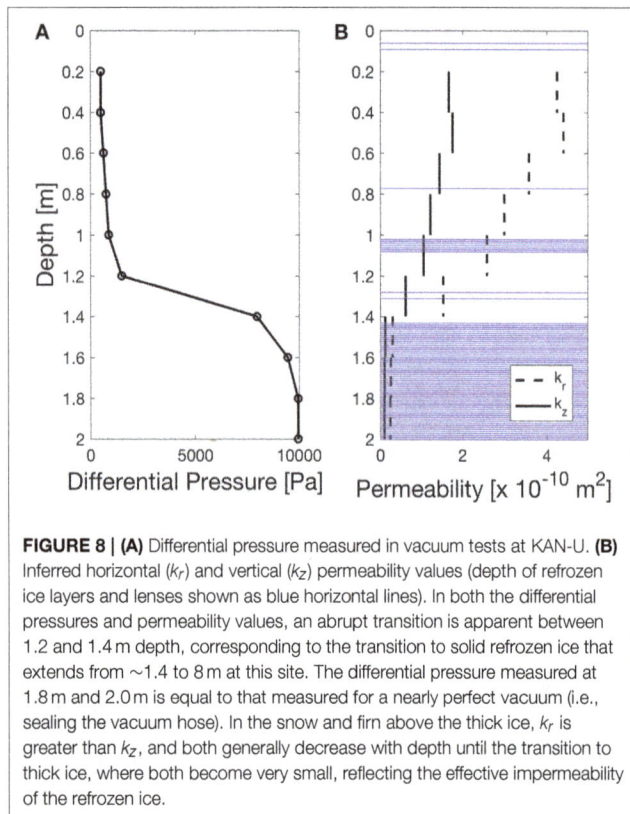

FIGURE 8 | (A) Differential pressure measured in vacuum tests at KAN-U. (B) Inferred horizontal (k_r) and vertical (k_z) permeability values (depth of refrozen ice layers and lenses shown as blue horizontal lines). In both the differential pressures and permeability values, an abrupt transition is apparent between 1.2 and 1.4 m depth, corresponding to the transition to solid refrozen ice that extends from ~1.4 to 8 m at this site. The differential pressure measured at 1.8 m and 2.0 m is equal to that measured for a nearly perfect vacuum (i.e., sealing the vacuum hose). In the snow and firn above the thick ice, k_r is greater than k_z, and both generally decrease with depth until the transition to thick ice, where both become very small, reflecting the effective impermeability of the refrozen ice.

TABLE 3 | Inferred permeability with 95% confidence intervals associated with the optimization method for parameter estimation.

Test ID	k_r [m^2 × 10^{-10}] (95% confidence)	k_z [m^2 × 10^{-10}] (95% confidence)
EKT 1	7.518 (7.498, 7.537)	3.723 (3.717, 3.730)
EKT 2	50.32 (33.79, 66.86)	3.055 × 10^{-5} (−1.321, 1.322)
EKT 3	12.00 (11.99, 12.02)	7.043 (7.036, 7.049)
NASA-SE 1	14.807 (14.805, 14.809)	4.268 (4.267, 4.268)
NASA-SE 2	21.68 (−9.251, 52.61)	0.035 (−7.754, 7.825)
NASA-SE 3	29.85 (29.48, 30.22)	0.961 (0.800, 1.123)
SADDLE 1	10.77 (10.73, 10.82)	4.724 (4.710, 4.738)
SADDLE 2	10.63 (10.45, 10.82)	11.30 (11.16, 11.44)
SADDLE 3	22.84 (−14.73, 60.40)	3.793 × 10^{-5} (−30.98, 30.98)
DYE-2 1	11.38 (11.35, 11.40)	5.518 (5.510, 5.526)
DYE-2 2	23.72 (−40.57, 88.02)	4.325 × 10^{-10} (−16.19, 16.19)
EastGRIP 1	22.86 (21.16, 24.57)	14.75 (13.95, 15.55)
EastGRIP 2	16.28 (15.50, 17.06)	29.03 (28.07, 29.99)
EastGRIP 3	19.67 (19.50, 19.83)	22.75 (22.56, 22.94)

In most cases, the confidence intervals are narrow, indicative of a true optimum found in the solution. Values shaded in gray correspond to tests that span several ice layers, with very low vertical permeability. The confidence intervals for these highlighted cases are large, without a unique solution. This suggests that the assumption of an effective homogeneous anisotropic material to represent the firn/ice combination in these cases is not fully appropriate, and it is apparent that the ice layers act as impermeable boundaries.

a borehole just above the 5 cm ice layer (Monitoring Borehole 2). In this case the differential pressure was 8.7 Pa, suggesting that this ice layer interferes with air flow but is not completely impermeable (the horizontal extent of the ice layer into the wall is unknown). For the third test (Monitoring Borehole 3), the pressure response was measured below the 20 cm-thick ice layer below the vacuum. The differential pressure response here was 1.2 Pa—a very small response, suggesting that the 20 cm-thick ice layer substantially diminishes vertical permeability.

DISCUSSION

Comparison to Previously Published Values

Permeability measurements for Greenland firn have been previously published for Summit (Albert et al., 1996; Albert and Shultz, 2002; Luciano and Albert, 2002; Adolph and Albert, 2014). We attempted to conduct pneumatic tests at Summit during our field campaign, but suffered equipment failure in the extreme cold during our only opportunity for these tests (−33°C, with freezing fog). Our northern field site at EastGRIP, however, can be reasonably compared to Summit with similar surface temperatures (see Table 1); both sites experience very little melt and so consist of mainly dry firn. When they have occasionally experienced melt in the last several years (for example, in the high melt year 2012), the refrozen melt layers are very thin, on the order of 1–2 mm. Average snow and firn density measured in the top 3 m at Summit and EastGRIP in 2016 indicate densities approximately 21% higher at EastGRIP, owing primarily to lower

accumulation rates. Adolph and Albert (2014) found, however, that there was not a direct relationship between permeability and density at Summit (for near-surface firn), so this difference in density does not necessarily undermine a comparison of permeabilities at the two sites.

The vertical permeability values measured for the top 3 m of snow and firn at Summit in 2000 typically fell between 2 and 100×10^{-10} m^2 (Albert and Shultz, 2002). Albert et al. (1996) published both horizontal and vertical permeability for the top meter of snow at Summit, with values for both ranging between 2 and 60×10^{-10} m^2. Horizontal and vertical permeability values were also reported at Summit by Luciano and Albert (2002), with a range of $10–65 \times 10^{-10}$ m^2. At EastGRIP, we infer vertical permeability (k_z) in the range $10–40 \times 10^{-10}$ m^2, and horizontal permeability (k_r) of $10–30 \times 10^{-10}$ m^2, which are consistent with the range reported for dry snow and firn at Summit. For our two deeper tests at EastGRIP (below 2 m), we infer vertical permeability to be slightly higher than horizontal permeability (k_r/k_z of 0.9 and 0.6 for the two depths, respectively); Luciano and

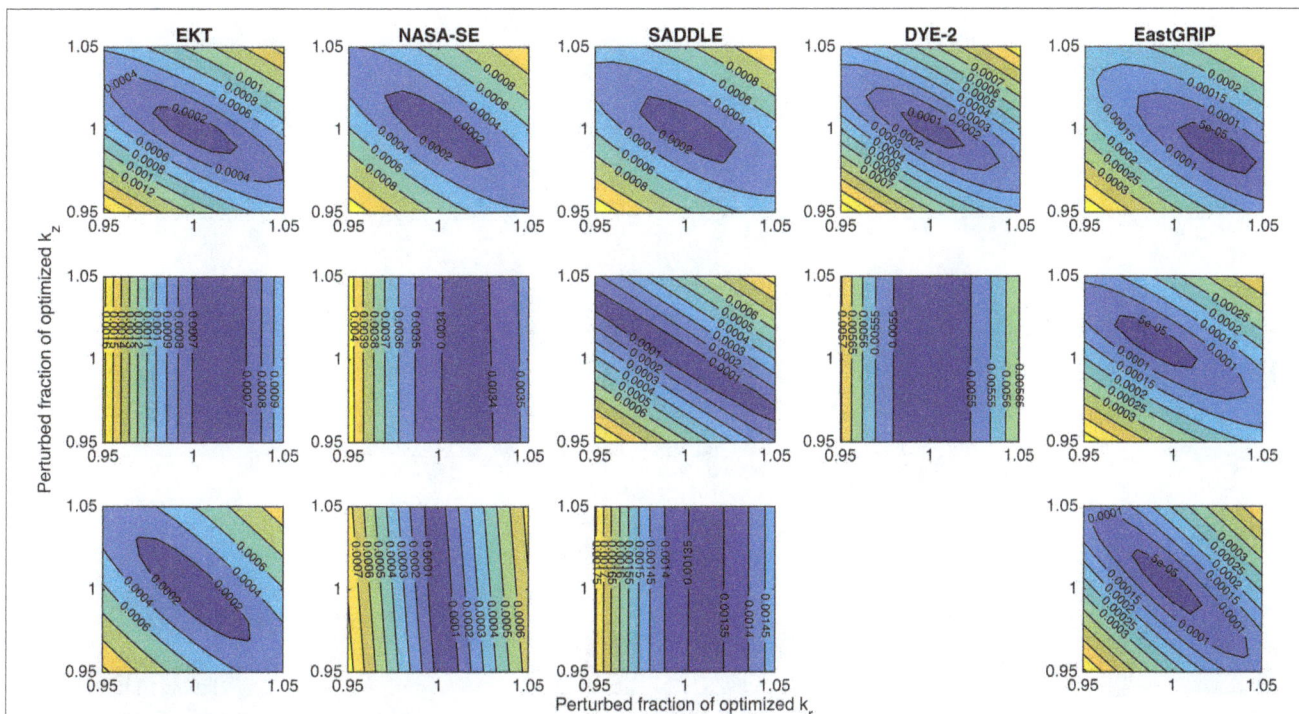

FIGURE 9 | Root-mean-square error (RMSE) in pressure response for each pneumatic test are obtained by perturbing about the optimal estimates of permeabilities. Tests are shown for each site in order of depth. In most cases, a clear optimum is apparent near the center of the plot. The tilt of the RMSE contour lines indicates a dependence in the solution between k_r and k_z. In several cases (EKT 2, NASA-SE 2, NASA-SE 3, SADDLE 3, DYE-2 2), there is not a clear RMSE optimum. These cases correspond to tests that span several ice layers, with very low vertical permeability. The assumption of an effective homogeneous anisotropic material in these cases does not adequately represent the behavior resulting from distinct impermeable layers and results in higher uncertainties for k_r estimates for these tests.

Albert (2002) also found slightly higher vertical than horizontal permeability for comparable depths in dry firn at Summit. This anisotropy is likely due to firn microstructure; for example, Calonne et al. (2012) measured higher vertical than horizontal permeability (k_r/k_z ~0.6) in evolved depth hoar.

Uncertainty in Permeability Estimates

We wish to clearly articulate sources of error and uncertainty in our measurements and analysis. In this section we present confidence intervals associated with optimization for the inferred permeability values, the uncertainty associated with measurement error, and discuss practical difficulties that were encountered during the field tests.

Uncertainty Associated with Optimization

In Section Inferred Firn Permeability and Anisotropy at Six Sites on the Greenland Ice Sheet with Different Stratigraphy, we inferred permeability based on optimal fitting of our measured pressure responses with an analytical solution for air flow through an anisotropic porous medium. From the optimization process (using the MATLAB nonlinear least-squares solver *lsqnonlin*, https://www.mathworks.com/help/optim/ug/lsqnonlin.html), we obtain 95% confidence intervals associated with each inferred permeability value using the MATLAB function *nlparci*

(https://www.mathworks.com/help/stats/nlparci.html, which uses the optimized parameter estimates, residuals, and Jacobian generated by *lsqnonlin*). **Table 3** presents k_r and k_z inferred for each pneumatic test along with these confidence intervals. In most cases, the confidence intervals are narrow (<2 × 10^{-10} m^2), indicating a clear optimum solution. In some cases, however, the confidence intervals are relatively large (up to 60 × 10^{-10} m^2), and may even include nonphysical negative permeability values. These nonconforming cases correspond to tests that span several ice layers or lenses within the firn. The broad confidence intervals suggest that the assumption of a semi-infinite effective/equivalent homogeneous medium is not physically representative in these cases, where sets of distinct ice layers may act like impermeable boundaries.

To help confirm the robustness of the k_r and k_z estimates, sensitivity analysis of the root-mean-squared error (RMSE) between predictions and observations was performed by perturbing each parameter estimate (k_r and k_z) by ±5% about the optimal estimate. The contours of RMSE in parameter space visualize how well-defined the optima are, and also determine whether there are interactions between the parameter estimates. **Figure 9** presents the resulting RMSE contours in measured pressure responses for each perturbation combination. In most cases, a clear optimum is apparent near the center of the contour plot. The RMSE contours are tilted, however, indicating

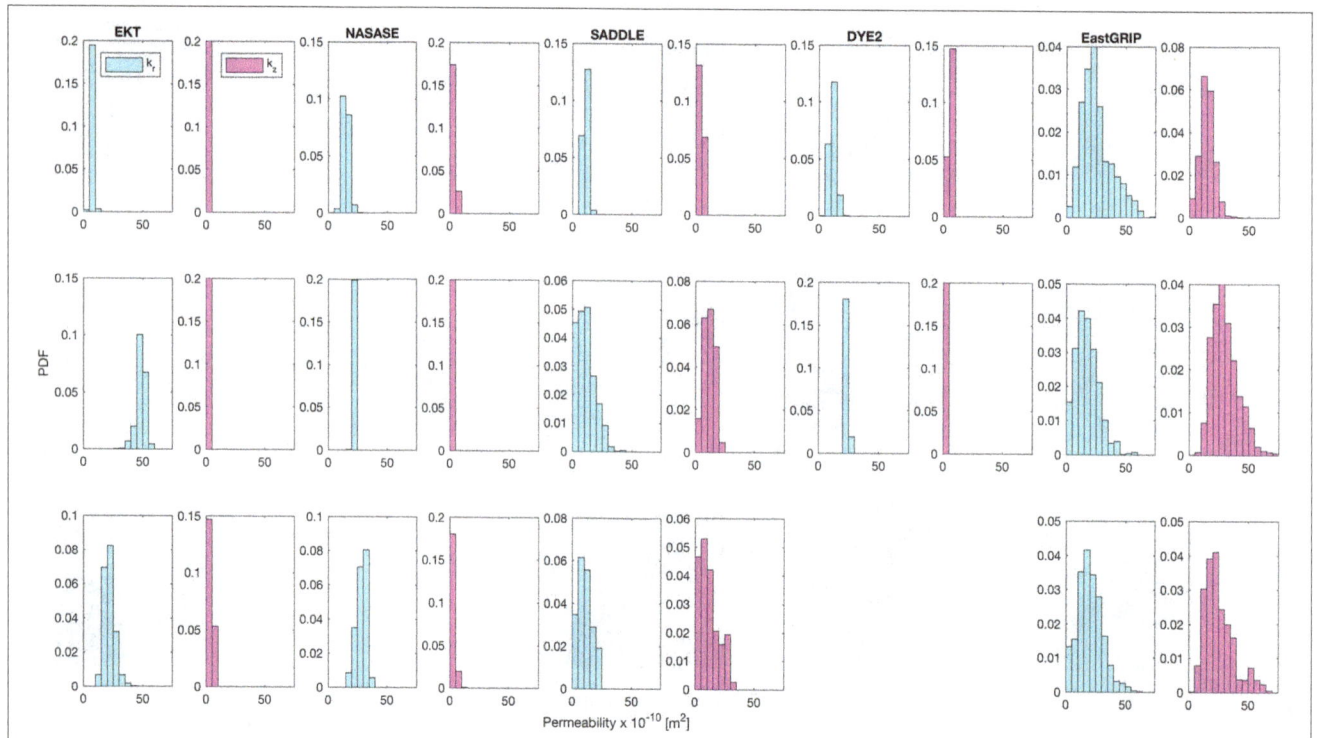

FIGURE 10 | Permeability inferred from pneumatic tests with perturbations of measured pressure responses to account for measurement error.
Assuming a normal distribution of measurement error with $\mu = 0$, $\sigma = 0.5$ Pa, we applied the optimization solution to find k_r and k_z for 1,000 randomly sampled errors for each pneumatic test. The resulting probability density function (PDF) of k_r and k_z for each test at each site is shown. Referencing the ice layer/lens depths shown in **Figure 7**, these distributions support the finding that in general, $k_r > k_z$ in the presence of ice layers or lenses. At EastGRIP, k_r and k_z are more similar in magnitude. Broader PDF distributions correspond to tests in which low pressure responses were observed, making these tests more sensitive to measurement error.

some dependence between the k_r and k_z estimates (for two independent variables, the contours would appear as circles or ellipses with major and minor axes aligned with the horizontal or vertical axes). In some cases, there is no clear center to the RMSE contours, but rather the contour lines are vertical or near vertical; these cases correspond to those tests that span several ice layers and are essentially impermeable to vertical flow. The shape of the pressure RMSE contours in these cases shows that the RMSE is relatively insensitive to estimates of k_z (note that k_z is very small in these cases, with correspondingly small perturbations). As discussed previously, the optimized permeability is less well constrained for these cases that span several ice layers.

Uncertainty Due to Measurement Error

During our pneumatic tests, steady-state pressure responses were read from analog differential pressure gauges and recorded by hand in a field notebook. Because pressure responses to the vacuum are small (in the range of 1–10 Pa), any fluctuations due to wind or a slight tilt of the gauge box can cause unintended errors. All efforts were made in the field to eliminate these errors. Where possible, manually recorded pressure responses were also verified and confirmed by the digitally recorded pressure responses on the datalogger. Some tests were not recorded digitally due to power supply challenges in cold conditions. A comparison of steady-state differential pressures for all tests with both a digital and manual record reveals close agreement

and only minor discrepancies between the two, with a median difference of 0.5 Pa. To estimate the uncertainty in inferred permeability associated with possible measurement error, we assume the error to be normally distributed with standard deviation $\sigma = 0.5$ Pa. We perturb the observed pressure responses for each pneumatic test with a random sample of 1,000 errors from this normal distribution and solve for the corresponding optimized k_r and k_z for each of the perturbations. **Figure 10** presents the probability density function (PDF) for k_r and k_z for each test depth at each site. These distributions fall within the same range as seen before (5–60 × 10^{-10} m^2 for k_r, and 0–30 × 10^{-10} m^2 for k_z), and support the general patterns discussed above in Section Inferred Firn Permeability and Anisotropy at Six Sites on the Greenland Ice Sheet with Different Stratigraphy. In the presence of ice layers and lenses, k_r is typically greater than k_z. At EastGRIP, k_r and k_z are more similar in magnitude. Tests with a wider distribution spread correspond to tests in which the measured pressure response was low, making them more sensitive to pressure measurement error.

Equipment Challenges and Flow Rate Uncertainty

The thermal anemometer we used for measuring flow rate was damaged and ceased to function partway through the field campaign after exposure to particularly cold conditions (below $-24°C$). The flow measurements that were recorded successfully prior to its damage, however, correspond to a fairly narrow range of 3.5–7.0 ms^{-1}. Because of this uncertainty (and also the general fluctuating nature of thermal anemometer flow rate measurement), shaded error bars on the inferred permeability values presented in **Figure 7** are included to reflect the range in flow rate generated by the portable wet/dry vacuum. Permeability values corresponding to the mean measured flow rate (5.1 ms^{-1}) are indicated as solid (k_z) or dashed (k_r) black lines on the figure.

The inflatable rubber plugs used to seal the boreholes at depth were inflated through tubes connected to a bicycle pump at the surface. The integrity of the seals against the borehole walls is important. On especially cold days (below $-20°$C), the tube connections to the pump contracted and required constant vigilance to ensure proper seals were maintained. Only tests in which solid hand-tight seals (tugging on the attached tubing yielded no movement) held in both the vacuum and monitoring boreholes for the duration of a test were recorded and used in the analysis above. If a seal was lost, the plug and equipment noticeably dropped to rest at the bottom of the borehole. In future applications, a portable air compressor could be used to maintain and ensure constant pressure in the plugs, and more expensive custom silicone plugs could be used, with temperature ratings for all connections and equipment down to $-40°$C. While it is certainly preferable to conduct the pneumatic tests in less extreme cold, logistics in the field do not always allow for such fair-weather flexibility.

Conclusions and Recommendations for Future Work

From these initial tests on the Greenland Ice Sheet, we have found *in-situ* pneumatic testing to be a promising method that can be used to infer horizontal and vertical firn permeability in the field. Accurate permeability estimates are necessary for meltwater infiltration, transport, and retention models that estimate runoff, a key component of surface mass balance. In general, we find higher anisotropy in firn permeability across refrozen ice layers and lenses than in dry firn. In the presence of ice layers and lenses, vertical permeability ranges from $0–30 \times 10^{-10}$ m^2, with horizontal permeability on the order of $5–60 \times 10^{-10}$ m^2. In dry firn without refrozen lenses or layers, both vertical and horizontal permeability range between approximately $10–30 \times 10^{-10}$ m^2.

Researchers using this method in future campaigns could employ more pressure transducers at various monitoring depths and distances from the vacuum for each test to generate a more complete picture of the permeability structure at a given site. To avoid being limited by temperature conditions during available test opportunities during a field campaign, all equipment should

be rated down to $-40°$C Additionally, a stronger vacuum motor could potentially be used to stimulate measureable pressure responses farther away from the vacuum (taking care not to destroy the firn structure itself). Finally, it would be particularly illuminating to examine in detail the change in permeability over multiple years at locations on the Greenland Ice Sheet with increasing meltwater production. Such a study has the potential to clarify changes accompanying increased ice layer and ice lens formation that results from a warming climate, with ramifications for meltwater percolation and surface mass balance of the entire ice sheet. This technique is also suitable for broad application in Antarctica and other glaciers and ice caps.

AUTHOR CONTRIBUTIONS

AS designed, built, and tested the equipment, performed field experiments, analyzed data, and wrote the majority of the manuscript. HR provided guidance in the experimental design and various analysis techniques, and assisted in writing the manuscript. EW was involved in the conceptual and technical design, helped build the field setup, conducted preliminary tests in Colorado, and provided valuable insight on soil-vapor-extraction system design. MM offered the opportunity to collaborate in Greenland to conduct the field tests, assisted in manuscript preparation, and provided encouragement and expertise on all firn-related matters. WC and CS assisted in field experiments in Greenland, helped design site-specific tests, and were involved in preparation of the manuscript.

FUNDING

This work was funded primarily by a Dean's Graduate Student Research Grant from the Graduate School at the University of Colorado. Additional support was provided by an American Geophysical Union Cryosphere Innovation Award for Students, a NASA Earth and Space Sciences Fellowship (NNX14AL24H), and an Innovative Seed Grant from the University of Colorado. Logistical support was provided in collaboration with the FirnCover project (NASA ROSES grant NNX15AC62G).

ACKNOWLEDGMENTS

Many thanks to the FirnCover 2016 team for support and camaraderie in the field (MM, CS, WC, Darren Hill, Baptiste Vandecrux, Shawn Marshall, Samira Samimi, Achim Heilig, Bastian Gerling, and Leander Gambal), and also to CH2MHill Polar Field Services.

REFERENCES

Adolph, A. C., and Albert, M. R. (2014). Gas diffusivity and permeability through the firn column at Summit, Greenland: measurements and comparison to microstructural properties. *Cryosphere* 8, 319–328. doi: 10.5194/tc-8-319-2014

Albert, M. R., Arons, E. M., and Davis, R. E. (1996). "Firn properties affecting gas exchange at Summit, Greenland: ventilation possibilities," in *Chemical Exchange between the Atmosphere and Polar Snow*, eds E. W. Wolff and R. C. Bales (Berlin; Heidelberg: Springer), 561–565.

Albert, M. R., and Perron, F. E. (2000). Ice layer and surface crust permeability in a seasonal snow pack. *Hydrol. Process.* 14, 3207–3214. doi: 10.1002/1099-1085(20001230)14:18%3C3207::AID-HYP196%3E3.3.CO;2-3

Albert, M. R., and Shultz, E. F. (2002). Snow and firn properties and air–snow transport processes at Summit, Greenland. *Atmos. Environ.* 36, 2789–2797. doi: 10.1016/S1352-2310(02)00119-X

Ambach, W., Blumthaler, M., and Kirchlechner, P. (1981). Application of the gravity flow theory to the percolation of melt water through firn. *J. Glaciol.* 27, 67–75.

Bader, H. (1939). "Mineralogische und strukturelle Charakterisierung des Schnees und seiner Metamorphose," in *Der Schnee und seine Metamorphose*, eds H. Bader, J. Neher, O. Eckel, C. Thams, R. Haefeli (Bern: Kummerly & Frey), 1–61. (Beiträge zur Geologie der Schweiz, Geotechnische Serie Hydrologie 3, Transl. Tienhaven, J. C. V., 1954: Snow and its metamorphism, SIPRE Translations, 14: 1–55).

Bear, J. (1972). *Dynamics of Fluids in Porous Materials*. New York, NY: Elsevier.

Bender, J. A. (1957). *Air Permeability of Snow*. US Army Snow Ice and Permafrost Research Establishment. Wilmette, IL: Corps of Engineers.

Calonne, N., Geindreau, C., Flin, F., Morin, S., Lesaffre, B., Du Roscoat, S. R., et al. (2012). 3-D image-based numerical computations of snow permeability: links to specific surface area, density, and microstructural anisotropy. *Cryosphere* 6, 939–951. doi: 10.5194/tc-6-939-2012

Charalampidis, C., van As, D., Colgan, W., Fausto, R., MacFerrin, M., and Machguth, H. (2016). Thermal tracing of retained meltwater in the lower accumulation area of the southwestern Greenland ice sheet. *Ann. Glaciol.* 57, 1–10. doi: 10.1017/aog.2016.2

Colbeck, S. C. (1975). A theory for water flow through a layered snowpack. *Water Res. Res.* 11, 261–266.

Collins, R. E. (1961). *Flow of Fluids through Porous Media*. New York, NY: Reinhold.

Courville, Z. R., Albert, M. R., Fahnestock, M. A., Cathles, L. M., and Shuman, C. A. (2007). Impacts of an accumulation hiatus on the physical properties of firn at a low-accumulation polar site. *J. Geophys. Res.* 112. doi: 10.1029/2005JF000429

Courville, Z., Hörhold, M., Hopkins, M., and Albert, M. (2010). Lattice-Boltzmann modeling of the air permeability of polar firn. *J. Geophys. Res.* 115:F04032. doi: 10.1029/2009JF001549

Fettweis, X., Franco, B., Tedesco, M., van Angelen, J. H., Lenaerts, J. T. M., van den Broeke, M. R., et al. (2013). Estimating the Greenland ice sheet surface mass balance contribution to future sea level rise using the regional atmospheric climate model MAR. *Cryosphere* 7, 469–489. doi: 10.5194/tc-7-469-2013

Fierz, C., Armstrong, R. L., Durand, Y., Etchevers, P., Greene, E., McClung, D. M., et al. (2009). "The International classification for seasonal snow on the ground," in *IHP-VII Technical Documents in Hydrology N° 83, IACS Contribution N° 1, UNESCO-IHP* (Paris).

Harper, J., Humphrey, N., Pfeffer, W. T., Brown, J., and Fettweis, X. (2012). Greenland ice-sheet contribution to sea-level rise buffered by meltwater storage in firn. *Nature* 491, 240–243. doi: 10.1038/nature11566

Hörhold, M. W., Albert, M. R., and Freitag, J. (2009). The impact of accumulation rate on anisotropy and air permeability of polar firn at a high-accumulation site. *J. Glaciol.* 55, 625–630. doi: 10.3189/002214309789471021

Jeans, J. H. (1908). *The Mathematical Theory of Electricity and Magnetism*. Cambridge: University Press.

Johnson, P. C., Stanley, C. C., Kemblowski, M. W., Byers, D. L., and Colthart, J. D. (1990). A practical approach to the design, operation, and monitoring of *in situ* soil-venting systems. *Groundwater Monit. Remediation* 10, 159–178.

Krabill, W., Hanna, E., Huybrechts, P., Abdalati, W., Cappelen, J., Csatho, B., et al. (2004). Greenland ice sheet: increased coastal thinning. *Geophys. Res. Lett.* 31:L24402. doi: 10.1029/2004GL021533

Luciano, G. L., and Albert, M. R. (2002). Bidirectional permeability measurements of polar firn. *Ann. Glaciol.* 35, 63–66. doi: 10.3189/172756402781817095

Machguth, H., MacFerrin, M., van As, D., Charalampidis, C., Colgan, W., Fausto, R. S., et al. (2016). Greenland meltwater storage in firn limited by near-surface ice formation. *Nat. Clim. Chang.* 6, 390–393. doi: 10.1038/nclimate2899

Marsh, P., and Woo, M. K. (1985). Meltwater movement in natural heterogeneous snow covers. *Water Resour. Res.* 21, 1710–1716.

Mikkelsen, A. B., Hubbard, A., MacFerrin, M., Box, J. E., Doyle, S. H., Fitzpatrick, A., et al. (2016). Extraordinary runoff from the Greenland ice sheet in 2012 amplified by hypsometry and depleted firn retention. *Cryosphere* 10, 1147–1159. doi: 10.5194/tc-10-1147-2016

Milne-Thomson, L. M. (1938). *Theoretical Hydrodynamics*. New York, NY: Macmillan.

Pfeffer, W. T., Meier, M. F., and Illangasekare, T. H. (1991). Retention of Greenland runoff by refreezing: implications for projected future sea level change. *J. Geophys. Res.* 96, 22117–22124. doi: 10.1029/91JC02502

Rignot, E., Velicogna, I., Van den Broeke, M. R., Monaghan, A., and Lenaerts, J. T. M. (2011). Acceleration of the contribution of the Greenland and Antarctic ice sheets to sea level rise. *Geophys. Res. Lett.* 38:L05503. doi: 10.1029/2011GL046583

Shan, C., Falta, R. W., and Javandel, I. (1992). Analytical solutions for steady state gas flow to a soil vapor extraction well. *Water Resour. Res.* 28, 1105–1120. doi: 10.1029/91WR02986

Shimizu, H. (1970). Air permeability of deposited snow. *Contrib. Inst. Low Temperature Sci. A* 22, 1–32.

Sommerfeld, R. A., and Rocchio, J. E. (1993). Permeability measurements on new and equitemperature snow. *Water Res. Res.* 29, 2485–2490. doi: 10.1029/93WR01071

Velicogna, I., and Wahr, J. (2006). Acceleration of Greenland ice mass loss in spring 2004. *Nature* 443, 329–331. doi: 10.1038/nature05168

Weinig, W. T. (1992). "Monitoring the vadose zone during pneumatic pumping tests," in *Proceedings of the Sixth National Outdoor Action Conference, National Ground Water Association* (Westerville, OH), 133–146.

Conflict of Interest Statement: The authors declare that the research was conducted in the absence of any commercial or financial relationships that could be construed as a potential conflict of interest.

3

Accumulation Rates during 1311–2011 CE in North-Central Greenland Derived from Air-Borne Radar Data

Nanna B. Karlsson[1]*[†], Olaf Eisen[2,3]*, Dorthe Dahl-Jensen[1], Johannes Freitag[2], Sepp Kipfstuhl[2], Cameron Lewis[4,5], Lisbeth T. Nielsen[1], John D. Paden[4], Anna Winter[2] and Frank Wilhelms[2,6]

[1] Centre for Ice and Climate, The Niels Bohr Institute, University of Copenhagen, Copenhagen, Denmark, [2] Alfred-Wegener-Institut Helmholtz-Zentrum für Polar- und Meeresforschung, Bremerhaven, Germany, [3] Department of Geosciences, University of Bremen, Bremen, Germany, [4] Center for Remote Sensing of Ice Sheets, University of Kansas, Lawrence, KS, USA, [5] Sandia National Laboratories, Albuquerque, NM, USA, [6] Department of Crystallography, Geoscience Centre, University of Göttingen, Göttingen, Germany

Edited by:
Felix Ng,
University of Sheffield, UK

Reviewed by:
Gordon Stuart Hamilton,
University of Maine, USA
Robert Hawley,
Dartmouth College, USA

*Correspondence:
Nanna B. Karlsson
nbkarlsson@nbi.ku.dk
Olaf Eisen
olaf.eisen@awi.de

[†] Present Address:
Nanna B. Karlsson,
Alfred-Wegener-Institut
Helmholtz-Zentrum für Polar- und
Meeresforschung, Bremerhaven,
Germany

Specialty section:
This article was submitted to
Cryospheric Sciences,
a section of the journal
Frontiers in Earth Science

Radar-detected internal layering contains information on past accumulation rates and patterns. In this study, we assume that the radar layers are isochrones, and use the layer stratigraphy in combination with ice-core measurements and numerical methods to retrieve accumulation information for the northern part of central Greenland. Measurements of the dielectric properties of an ice core from the NEEM (North Greenland Eemian Ice Drilling) site, allow for correlation of the radar layers with volcanic horizons to obtain an accurate age of the layers. We obtain 100 a averaged accumulation patterns for the period 1311–2011 for a 300 by 350 km area encompassing the two ice-core sites: NEEM and NGRIP (North Greenland Ice Core Project). Our results show a clear trend of high accumulation rates west of the ice divide and low accumulation rates east of the ice divide. At the NEEM site, this accumulation pattern persists throughout our study period with only minor temporal variations in the accumulation rate. In contrast, the accumulation rate shows more pronounced temporal variations (based on our centennial averages) from 170 km south of the NEEM site to the NGRIP site. We attribute this variation to shifts in the location of the high–low accumulation boundary away from the ice divide.

Keywords: surface mass balance, Greenland Ice Sheet, ice-penetrating radar, internal glacier stratigraphy, inverse methods

1. INTRODUCTION

Knowledge of past accumulation is essential for studying the evolution of ice sheets and their response to climate change. Estimates of past accumulation rates are often based on ice-core records that represent point measurements rather than spatially distributed observations. Unfortunately, ice core measurements of accumulation rates remain sparse, especially in remote areas such as central Greenland. Here, we present the accumulation pattern over the last seven centuries for an area encompassing two deep ice-core locations: the NEEM (North Greenland Eemian Ice Drilling, 77.45°N, 51.06°W) and the NGRIP (North Greenland Ice Core Project, 75.1°N, 42.32°W) ice-core drill sites.

The specific surface mass balance (SMB) and its spatial distribution is a key parameter for elucidating total ice sheet mass balance. In order to retrieve the SMB, studies have employed various approaches such as firn cores and/or snow pit studies at times combined with weather station data (e.g., Ohmura and Reeh, 1991; Bales et al., 2001, 2009), and climate models (e.g., Ettema et al., 2009; Burgess et al., 2010; Hanna et al., 2011; Box et al., 2013). Regardless of the approach, the estimated accumulation rates of the Greenland Ice Sheet (GrIS) display the same overall pattern: high rates on the coast and low rates in the interior. The location of the ice divide influences the SMB, especially in central northern Greenland, since it marks the highest points in the interior of the ice sheet (see **Figure 1**, inset). For this region, Ohmura and Reeh (1991) showed that the ice divide acts as a topographic barrier for water vapor transported from the west coast. Numerous studies have confirmed that this topographic barrier results in high accumulation rates on the west coast, decreasing rates as the elevation rises toward the interior, and very low accumulation rates east of the ice divide (e.g., Bales et al., 2001; Ettema et al., 2009; Burgess et al., 2010; Box et al., 2013). Similar gradients in accumulation rates have been observed across ice divides in Antarctica (Neumann et al., 2008; Koutnik et al., 2016). Another example of elevation-controlled SMB is the ice rises in coastal East Antarctica (Lenaerts et al., 2014), although they are of a substantially smaller spatial scale (a few tens of kilometers).

Sound interpretation of ice-core measurements for the history of regional accumulation rates requires knowledge of the SMB distribution around the core site. Otherwise, a shift in a spatially-uniform SMB pattern could be interpreted as an apparent change in accumulation rate history, in other words, a spatial signal can cloud or corrupt a temporal signal. A critical question is therefore whether the spatial pattern of accumulation has been stationary or varying in the past. This question is difficult to address by considering ice-core measurements alone since they only provide information at singular locations. Ice core sites positioned at ice divides that are transient sites or known to migrate, or positioned on the flank close to ice divides, are susceptible to this problem. This is a known issue for both the NEEM and NGRIP ice cores (North Greenland Ice Core Project members, 2004; Dahl-Jensen et al., 2013), which were retrieved from locations at the ice divide in the north-eastern and central parts of Greenland, respectively (**Figure 1**).

Our study employs the identification, tracing and modeling of internal radar reflectors, also termed internal reflection horizons or internal layers. In the following, we will use the term "(internal) layer" to emphasize that we are working with the radar reflectors that are visible to the human eye in the L1B product available on the CReSIS website (Center for Remote Sensing of Ice Sheets, University of Kansas: www.cresis.ku.edu). The L1B product contains geolocated radar echo strength profiles with corresponding information on time, longitude, latitude, elevation and flight path. We refer readers to the CReSIS data documentation (available on the CReSIS website) for more details. We note in a sense a layer is not an absolute horizon, but rather an average of the power reflected by the family of reflectors that occur within that radar resolution bin. In other

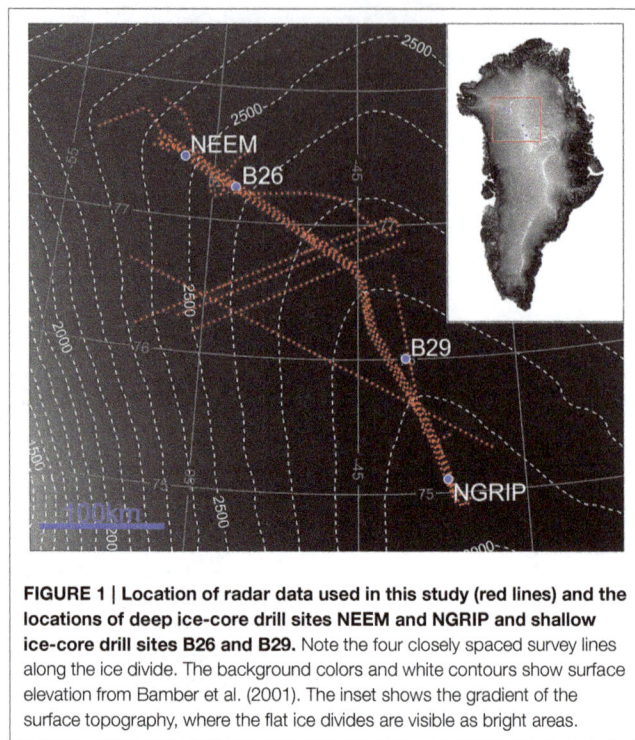

FIGURE 1 | Location of radar data used in this study (red lines) and the locations of deep ice-core drill sites NEEM and NGRIP and shallow ice-core drill sites B26 and B29. Note the four closely spaced survey lines along the ice divide. The background colors and white contours show surface elevation from Bamber et al. (2001). The inset shows the gradient of the surface topography, where the flat ice divides are visible as bright areas.

words, multiple layers that are closely spaced can appear as a single layer in the radar data. The frequency of the radar system determines how close the layers need to be in order to appear as one layer. Most internal layers detected by radar originate from physical properties imprinted at the past ice surface quasi-simultaneously over large areas. Consequently, such layers are commonly considered to be isochrones (Eisen et al., 2008) and represent spatially distributed time markers.

Approaches using radar observations to derive spatial patterns of accumulation have previously been applied at NGRIP (Dahl-Jensen et al., 1997; Steinhage et al., 2004), and at other ice-core sites such as the European Project for Ice Coring in Antarctica in Dronning Maud Land (Eisen et al., 2005), Siple Dome in Central West Antarctica (Nereson et al., 2000) and around the West Antarctic Ice Sheet Divide ice-core site (Neumann et al., 2008; Koutnik et al., 2016). Fine-resolution (i.e., ground-based) radars have been employed to derive accumulation patterns over larger distances along ice divides in both Antarctica (Richardson and Holmlund, 1999; Frezzotti et al., 2005; Fujita et al., 2011) and Greenland (Hawley et al., 2014). Such radars typically only penetrate a few hundred meters into the ice sheet. Layers at these depths are close enough to the surface that the dynamic influence of ice flow (such as layer thinning) is usually small, given typical ice thicknesses of more than 2000 m (e.g., Bamber et al., 2013). Air-borne radars operating at similar frequencies allow for even larger spatial coverage and air-borne radar datasets have recently been exploited to derive accumulation rates from 2009 to 2012 over parts of the GrIS (Koenig et al., 2016). In all cases, the depth distribution of shallow layers depends primarily on surface accumulation and densification rates. They in turn depend on the overall climate setting, e.g., temperature and radiation, additionally the densification also depends on accumulation rate, seasonal distribution of accumulation, and potentially impurity content (Freitag et al., 2013). Several studies have applied formal inverse approaches that take into account the spatio-temporal variability of accumulation and other factors when analysing radar layer stratigraphy to derive accumulation rates (e.g., Leysinger-Vieli et al., 2007; Waddington et al., 2007; Eisen, 2008; Simonsen et al., 2013; Koutnik et al., 2016). With the recent publication of an extensive radar layer data set for GrIS (MacGregor et al., 2015) such approaches will become increasingly important.

Here we apply the inverse method developed by Nielsen et al. (2015) to a radar dataset obtained by aircraft in the area between the NEEM drill site and the NGRIP drill site. We apply an electromagnetic wave propagation model of radar wave propagation to replicate reflection signatures in order to reliably convert age–depth functions from a firn core to an age–traveltime distribution and thereby assign ages to individual layers. The inverse approach returns centennially averaged accumulation rates and provides a robust estimate of uncertainties for the output accumulation fields.

2. DATA

This study uses multiple data sets. Our primary dataset is the internal layers imaged by radar. The layers provide the isochronous stratigraphy needed to infer accumulation rates. In addition, we rely on measurements made on an ice core retrieved at the NEEM drill site, specifically density and conductivity measurements, in order to date the internal radar layers. We first describe the acquisition and assembling of the necessary data sets. We present the methodology used for processing and interpreting the data sets in later sections.

2.1. Airborne Radar Data and Internal Layer Dataset

The airborne RES (radio-echo sounding) data (Gogineni, 2012) were collected using the most recent versions of the accumulation radar (Lewis et al., 2015) and are part of an extensive data set acquired by CReSIS during the last several decades (Kanagaratnam et al., 2004; Lewis et al., 2015). The accumulation radar has a bandwidth of 300 MHz or more, which provides 0.45 m resolution of internal layers in ice when accounting for the frequency domain windowing that is applied during data processing and an ice dielectric permittivity of 3.15 (we refer to Lewis et al., 2015 for more details on the data processing).

The data have been post-processed and time-synchronized to high precision GPS locations. Here, we use data collected in the campaigns conducted over Greenland in 2011 and 2012 (see **Figure 1**). In 2011, the bandwidth spans 565–885 MHz and in 2012, the bandwidth spans 600–900 MHz. We selected four flight lines that follow the ice divide with a distance of ~1 km between them and seven flight lines that intersect the ice divide at different angles. The data acquisition took place in March–May 2011 and April–May 2012 but in the following we assume that all data were acquired in spring 2011. The investigated time periods therefore

cover from spring to spring, i.e., the first period is spring 1311– spring 1411. This imposes an uncertainty in the age assignment of less than 1% (for the 2012 data, since the youngest layer is more than 100 year old) which is substantially less than other uncertainties in our method (see below for an in-depth treatment of uncertainty assignment).

The internal layers were traced manually, and all overlapping flight lines were checked at crossover points to ensure consistency in the layer data set. Prior to the tracing, the radar frames were combined into radargrams of an approximate horizontal length of 50–80 km using Matlab software developed by MacGregor et al. (2015).

2.2. Density Data

We use two types of firn density data: From 13.2 m to 184.8 m density is available from the shallow core NEEM11S1 (NEEM 2011 shallow core 1) in 1.1 m resolution from conventional core weighing. From 6.0 m to 70.4 m the density has been derived from the NEEM11S1 core by means of radioscopic imaging yielding density at sub-cm resolution (Freitag et al., 2013). We use the high resolution density where available. Below 184.8 m, we use a constant density of 905 kg m^{-3}, which corresponds to the average density in the sections above. Following Gfeller et al. (2014) we set the surface density to 340 kg m^{-3} and missing densities in the uppermost part have been replaced by linearly interpolated values between the surface density and the first measured density value from the firn core.

2.3. Conductivity Data

Dielectric profiling (DEP) was carried out along the NEEM11S1 core in the field using a conventional DEP bench as described by Wilhelms (1996) to obtain dielectric properties at a 5 mm sample interval from 6.0 m to 175.0 m depth. Although the bench records the complex conductance and capacitance, for the purpose investigated here, we use only the conductance measured by the DEP bench at 250 kHz, as a proxy for the conductivity as encountered by a radar wave. The conductivity is inversely scaled with the center frequency of radar operation of 750 MHz. To obtain a complete record, the DEP record is extended to the surface by assuming a constant conductivity between the surface and the first data point. Gaps in the record are linearly interpolated.

3. METHODS

Our study is based on the assumption that the depth of radar layers directly inform on accumulation rates if the age of the layers is known. The first step toward obtaining these two pieces of information is to link the observed layers with the NEEM ice core. This allows us to transfer the ice-core ages to our traced layers. We use the density and dielectric properties of the ice core as input for a model of electromagnetic wave propagation in ice ("emice," Section 3.1). The model converts the depth-scale of the ice core to a two-way travel time (TWT) scale, and, importantly it calculates a synthetic radargram that can be directly compared to the observed radargram (**Figure 2**). The traced layers can then be dated by matching them with the

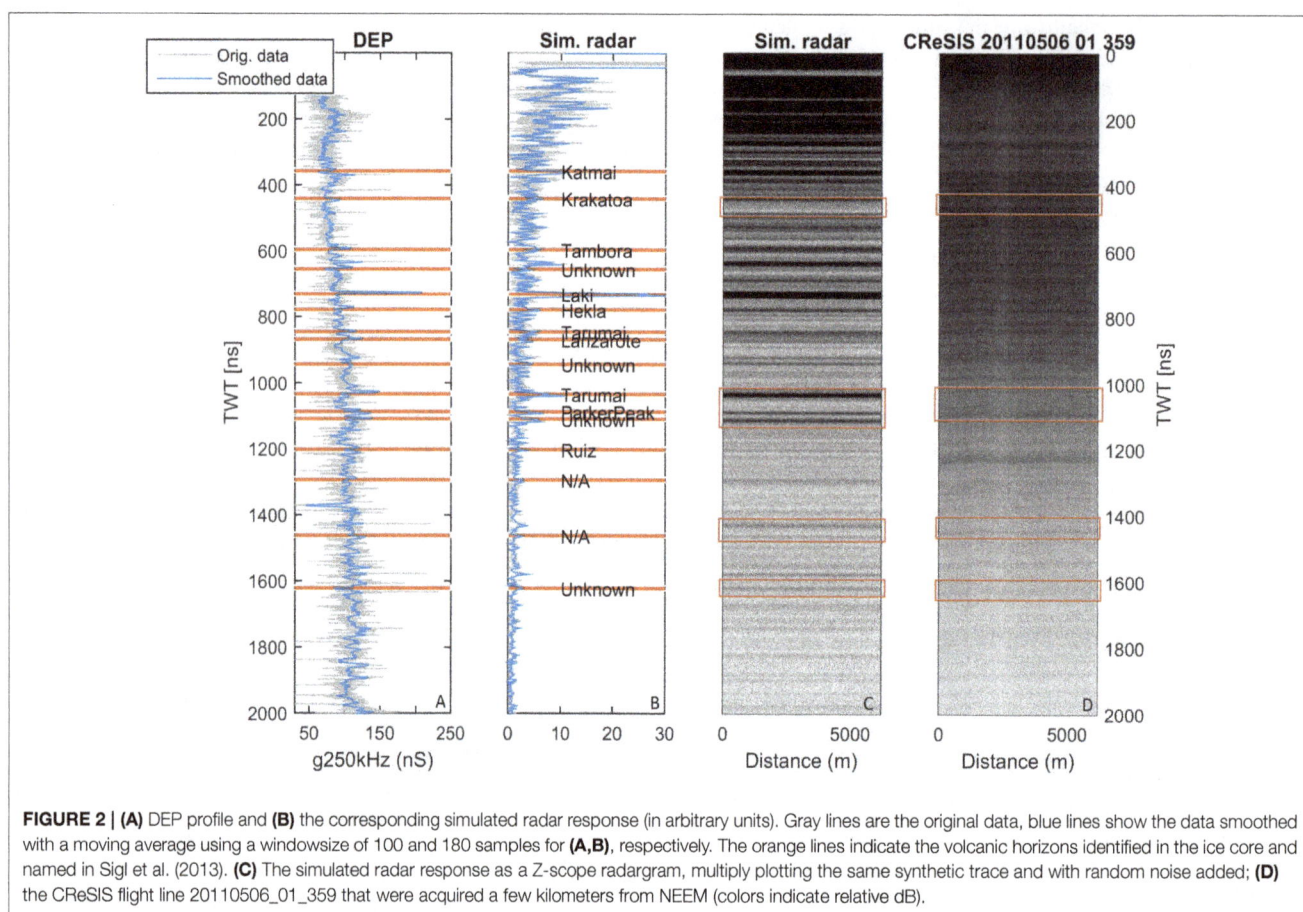

FIGURE 2 | (A) DEP profile and **(B)** the corresponding simulated radar response (in arbitrary units). Gray lines are the original data, blue lines show the data smoothed with a moving average using a windowsize of 100 and 180 samples for **(A,B)**, respectively. The orange lines indicate the volcanic horizons identified in the ice core and named in Sigl et al. (2013). **(C)** The simulated radar response as a Z-scope radargram, multiply plotting the same synthetic trace and with random noise added; **(D)** the CReSIS flight line 20110506_01_359 that were acquired a few kilometers from NEEM (colors indicate relative dB).

synthetic layers whose ages are known (Section 3.2). The next step is to convert the vertical positions of the layers in TWT to depth. In order to do this we need information on the firn density. Since direct observations of firn density are sparse in our study area, we use a 1D firn densification model that calculates the density at each data point (Section 3.3). This densification model contains a number of unknown parameters (including the accumulation rate) that vary spatially. We cannot therefore directly calculate the depth or the accumulation rate, but instead we construct an inverse method (Section 3.4) that uses our observations (dated and traced radar layers) to obtain the most likely range of each model parameter for every data point. In fact, the easiest approach is to let the inverse method operate in the TWT-domain and then convert back to depth once the best parameters are found. Evidently, several factors contribute to the uncertainty in the final age-TWT distribution, which we use to infer accumulation rates: the conversion of the DEP data on a depth-scale to a simulated radargram in TWT, the dating (i.e., the age–depth distribution) of the firn core, the identification and dating of the layers in the simulated radargram and the matching of the simulated radargram with the observed CReSIS radar data. Estimating the total uncertainty of the age of the layers is a fundamental input to the inverse approach. We will further discuss the uncertainties and their relevance for the results in detail below.

3.1. Electromagnetic Wave Propagation Model

The conversion of DEP data in the depth domain to radar data in the TWT–domain requires the ordinary relative permittivity of firn $\varepsilon'(z)$, which determines radar wave speed V as a function of depth z as

$$V(z) = c_0/\sqrt{\varepsilon'(z)}, \tag{1}$$

where c_0 is the speed of light in a vacuum. Permittivity of firn depends on the firn density and the permittivity of pure ice. The application of permittivity mixing formulae as well as the uncertainty in density data yield an effective uncertainty in the employed conversion from depth to TWT (Eisen et al., 2006). To reduce its impact, we employ radar-wave numerical modeling as established by Eisen et al. (2003) to optimize the uncertainty in the final conversion by matching synthetic radar signatures with the ones recorded by the radar data in the field. This can be considered as a calibration of the depth-TWT conversion.

Missing values of density and conductivity were linearly interpolated and the combined series resampled at a 5 mm sample interval. The DEP operations in the field did not provide sufficiently accurate calibration to determine absolute values of permittivity and conductivity to apply the complex-valued DECOMP equation (Eisen et al., 2006), which considers mixing of the real and imaginary parts of the permittivity in the two-phase firn system. The missing calibration of the DEP data is insignificant when obtaining reflections in synthetic radargrams, as these mainly depend on the relative changes in permittivity and conductivity. Ultimately, to obtain a reliable permittivity which determines the location of reflections at depth, we apply the real-valued Looyenga (1965) mixing model

$$\varepsilon' = (\frac{\rho}{\rho_i}[\sqrt[3]{\varepsilon'_i} - 1] + 1)^3, \tag{2}$$

using the available density records ρ with pure-ice values $\rho_i = 917$ kg m^{-3} and $\varepsilon'_i = 3.15$. The latter value has commonly been derived from field measurements (e.g., Eisen et al., 2006) and confirmed by laboratory studies (Bohleber et al., 2012).

Conductivity and density serve as input data in the electromagnetic wave propagation model "emice," a numerical representation of the Maxwell equations. We operated the model in one vertical dimension at 2 cm resolution, with conductivity accordingly averaged to that scale (cf., Eisen et al., 2006). In the upper part of the firn column, a number of synthetic radar reflections are generated by changes in the density data. As the footprint covered by the radar systems is more than an order of magnitude larger than the area covered by the firn core, considering the density at full resolution causes a higher level of ambiguities when connecting synthetic and real radar records. We therefore smoothed the density record with a running mean of 20 m. The smoothing preserved the average density but reduced the signal variability in the synthetic radar trace. The final synthetic trace is shown in **Figures 2B,C**.

3.2. Layer Matching

In order to assign an age to the observed layers in the CReSIS dataset, we construct a synthetic radargram using data from the NEEM11S1 shallow core as described above. This synthetic radargram is then compared to CReSIS data records that were acquired a few kilometers from NEEM. We calculated the correlation coefficient between the radargrams using offsets from −200 ns to 200 ns in increments of 5 ns. Correlation coefficients range from 0.42 to 0.6, with higher values (>0.54) for offsets larger than 10 ns. The best match, however, was found by manually comparing the two radargrams from observed and modeled data. **Figure 2** shows the DEP profile, the simulated radargram and the CReSIS radargram. The matching volcanic signals are included in **Table 1**.

We note that a strong signal in the DEP record (and consequently in the modeled data) is not necessarily reflected in the observed data as a distinct layer. For example, the Laki eruption of 1782–1784 is one of the five strongest DEP signals for the entire Holocene in the NEEM core (Sigl et al., 2013), but there is no obvious candidate for this single strong reflection in the observed radar data. This is likely because the radar layers represent the average of the power reflected by potentially several horizons. We therefore base our matching not only on the existence of a layer but also on layer sequences, i.e., distinctive patterns of layers, rather than using reflection magnitudes.

The best match between modeled and observed radar data is obtained by shifting the simulated radargram upwards by

TABLE 1 | Volcanoes used to construct the age-TWT relationship at NEEM.

Volcano	Assigned date	Depth at NEEM (m)	TWT at NEEM (ns)	Year b2k
Katmai	1914 AD	35	359	86 a
Krakatoa	1885 AD	43	442	115 a
Tambora	1817 AD	61	598	183 a
Unknown	1810 AD	63	657	190 a
Laki	1784 AD	69	733	216 a
Hekla	1767 AD	79	780	233 a
Tarumai	1740 AD	73	847	260 a
Lanzarote	1731 AD	82	870	269 a
Unknown	1696 AD	90	945	304 a
Tarumai	1668 AD	96	1035	332 a
Parker Peak	1642 AD	102	1088	358 a
Unknown	1637 AD	103	1110	363 a
Ruiz	1596 AD	112	1203	404 a
Unknown	1454 AD	146	1623	546 a

The dating is based on the NEEM dating scale (Rasmussen et al., 2013) and the work by Sigl et al. (2013).

approximately 55 ns which corresponds to a few meters of firn. This offset could be explained by several causes. One is probably due to the uncertainty in reconstructing the density and thereby the TWT in the upper few meters, since good density measurements are missing here. Another reason for the offset is undulations in the observed layers caused by the presence of the camp site, and we see a slight dip downwards in layer depths close to NEEM. This was also observed during the 2015 ground-based radar campaign. Finally, the shallow core was drilled on a small snow mound made by several years of camp activities, and such an offset is therefore not surprising. Thus, the identification of distinct layer patterns rather than individual layers form the basis for the matching. One example of such a pattern match is the double peak at approx. ~1100 ns (where the first peak is the Parker Peak eruption, 1641–1643 at 102 m) and the single, strongly-reflecting layer above (from the Tarumai eruption, 1667–1668 at 96 m). The pattern with a doublet and a single layer above is also present in the CReSIS data.

The identification of the layers with volcanic horizons leads to an age–TWT relationship that we use to date the traced layers. We use ages published by Sigl et al. (2013) and in the cases where the volcanic event spans several years, we use the average age. Because we are only considering the top few hundred meters of firn, the change in TWT with age is linear, which is in agreement with the age–depth relationship constructed from the NEEM core (Rasmussen et al., 2013). Note that two of the layers in **Figure 2** (annotated N/A in **Figure 2B**) do not match any identified volcanic horizons even though a strong reflection was visible in both the modeled and observed radar data. During the layer tracing, we picked the most strongly reflecting and easily identifiable layers, which in some cases were not the volcanic horizons themselves. The ages of the traced layers were therefore calculated from the age-TWT relationship, rather than directly dated from the ice-core record. An example of a traced radargram can be seen in the Supplementary Figure S2.

3.3. Firn Model

The propagation of radar waves in firn is highly dependent on the density of the firn column. Thus, in order to convert TWT to true depth, we need a firn densification model. Several different modeling approaches exist, but the most common approach is based on the theory set up by Herron and Langway (1980), that assumes an exponential increase of density with time. The Herron–Langway model can be described by

$$\frac{d\rho}{dt} = \begin{cases} c_0 (\rho_i - \rho), & \rho \le \rho_c \\ c_1 (\rho_i - \rho), & \rho_c < \rho < \rho_i \end{cases} \tag{3}$$

where ρ is the density as a function of time (and depth), ρ_i is the constant density of ice, $\rho_i = 917$ kg m^{-3}, ρ_c is the threshold density of 550 kg m^{-3}, and c_0, c_1 are compaction rate constants that depend on the SMB, surface density ρ_s and temperature. For a more complete introduction to firn models, we refer the reader to the original study by Herron and Langway (1980), or several new studies that have expanded on this model (e.g., Arthern et al., 2010; Ligtenberg et al., 2011; Simonsen et al., 2013).

Herron and Langway (1980) suggest values for the surface density and their compaction rate constants based on measurements of firn and ice cores from Greenland and Antarctica. To get a good estimate of ρ_s, c_0 and c_1 for our study area, we compare seven density records collected across central and northern Greenland. This comparison can be found in the Supplementary Table 1 and Supplementary Figure S1 along with full citation of the data (see also Bolzan and Strobel, 1994; Wilhelms, 1996; Schwager, 2000). The comparison clearly shows that the recommended parameter values in the Herron-Langway model overestimate the densification for all sites. We therefore fit an exponential function of the Herron-Langway type to the mean of the density records and obtain the best fitting values for ρ_s, c_0 and c_1. Equation (3) is then solved using a Crank-Nicolson finite-difference method (Crank and Nicolson, 1996). This density profile is used as a first guess in the inverse method (described below). The unknown precise values of the parameters c_0, c_1 and ρ_s are estimated by the inverse model.

3.4. Inverse Method

The depth of radar layers for a given (time-varying) accumulation rate can be approximated using a 1D flow model, where we assume that the layers thin with a constant thinning rate:

$$w(z) = -\frac{a}{H}(H - z) \tag{4}$$

where w is vertical velocity, a is accumulation rate in meter ice equivalent, H is the local ice thickness and z is depth, also in meter ice equivalent. This flow model is coupled to the densification model to obtain true depths, which can then be converted to an age-TWT scale for a given accumulation rate and given density profile. If all variables were known: layer depth, layer age and density profile, we could immediately calculate the accumulation since the deposition of the layer, as that would simply equal the amount of snow above the layer. In this study, we know the age of the layers, while the depth is only known in the TWT domain, which depends on the wave speed and

thus density, which in turn depends on the accumulation rate along with other parameters. Our problem is, in other words, underdetermined. The use of an inverse method together with a priori information is the most suitable way to obtain a reasonably accurate range of solutions to the problem.

All data are horizontally interpolated to a 5 km grid. For each data point, we use the depths and ages of the traced layers to infer centennially averaged accumulation rates for the last 700 years, and the three parameters of the Herron-Langway firn model, ρ_s, c_0, and c_1, along the flight lines. The inversion is performed using an iterative inverse method where the misfit between the observed and modeled layer depth is minimized using a gradient descent technique (cf., Waddington et al., 2007).

The misfit we seek to minimize consists of two terms. The first term is the data-model misfit, given by

$$J_d = \sum \left(\frac{t_d - t_m(\mathbf{p}\{n\})}{\sigma_d} \right)^2, \tag{5}$$

where the subscripts d and m indicate data and model respectively, σ is the standard deviation, t is layer depths in TWT, and $\mathbf{p}$ is the set of model parameters for n observations. The term J_d measures how well the solution of the forward model fits the traced layer depths converted to TWT t_d.

The second term is a set of regularization constraints on the model parameters, which are added to the cost function to prevent overfitting the data and to avoid unphysical model parameter values. For the three densification parameters, this term is estimated as the deviation from the expected value of the parameter by

$$J_{reg}^{(1)} = \sum \left(\frac{|c_i\{n\} - c_{ref}|}{\sigma_c} \right)^2. \tag{6}$$

For the accumulation parameters, the term measures the spatial consistency as the deviation of an accumulation rate from the surrounding accumulation rates, weighted according to the inverse distance r between the points, thus given as

$$J_{reg}^{(2)} = \sum_i \left(\frac{\Sigma_j(\dot{b}_i - \dot{b}_j)^2}{\nu r_{i,j}} \right)^2, \tag{7}$$

where $\dot{b}_i$ is the accumulation rate at point i, $\dot{b}_j$ is the accumulation rate at point j, r_{ij} is the distance between them, and ν is a normalization constant based on the expected spatial variability of the data. We set $\nu = 2.5 \times 10^{-4}$. The combined regularization constraint:

$$J_{reg} = J_{reg}^{(1)} + J_{reg}^{(2)}, \tag{8}$$

will tend to produce a smooth solution of the model parameters (Aster et al., 2013). If this regularization constraint were not included, we could have obtained a smaller data-model misfit but that would have led to unphysical values and/or unphysically large spatial variability for the densification parameters and accumulation rate parameters. We evaluate the optimal weight of the regularization constraints, determining the trade-off between data-model misfit and model "smoothness" by an L-curve analysis (Aster et al., 2013).

The L-curve is constructed by considering the total misfit $J = J_d + \omega J_{reg}$. Here, J_{reg} and the corresponding J_d is plotted against each other for different values of ω, showing how either contributes to the total misfit. The value of ω that is closest to where this curve bends is chosen as the optimal weight (e.g., Aster et al., 2013).

The Jacobian of the total misfit, J, is calculated by perturbing each model parameter in turn to obtain the corresponding change in the total misfit. The correction to the model parameters which should minimize this, $\Delta \mathbf{p}$, is then obtained following the approach of Waddington et al. (2007) (see also Nielsen et al., 2015). To avoid potential overshooting of the minimum in the iteration process, the model parameters are updated by $\alpha \Delta \mathbf{p}$ (Aster et al., 2013), where we use a value of $\alpha = 0.5$. The decrease in misfit with increasing iteration number is shown in the Supplementary Figure S3.

3.5. Uncertainty Assignment

Several uncertainties are associated with the methods presented here. We first address the uncertainties pertaining to the matching of the radar layers with the ice-core record and the corresponding age determination. When modeling the radar response to the measured changes in permittivity and conductivity (see Section 3.1), we make several assumptions. We linearly interpolate missing values of density and conductivity (in the upper part of the core), the permittivity of pure ice is assumed to be $\varepsilon_i' = 3.15$ and we smooth the density signal to a lower resolution to avoid high-frequency signals. The first leads to an unknown offset in the record, while the assumption of $\varepsilon_i' = 3.15$ was found to cause shifts on the order of 5–10 ns compared to using a value of $\varepsilon_i' = 3.20$ or $\varepsilon_i' = 3.10$. The impact of smoothing the density signal is not likely to cause shifts in the radargram but might mask some layers.

The uncertainties mentioned above impact the forward modeling of radar signals, but a larger contributing factor of uncertainty is probably our manual tracing of radar layers. Although utmost care was taken to ensure that the same layer was followed between different flight tracks, as the flight lines enter low accumulation areas, seemingly bright and distinct layers may merge and become difficult to trace. Likewise, when the accumulation increases, some times a layer splits into two or more layers. When tracing layers, we consistently checked crossover points between different flight lines. Where discrepancies were identified, the layer tracing was rectified for consistency. Typically, the errors were 20 ns or less.

The assignment of ages to the layers is also associated with an error. The best match between the simulated radargram and the observed radargram, established by experienced radar data analysts, relies on their subjective decision. It is difficult to assign a specific error range to this uncertainty and the validity of the age assignment is discussed in more detail in sections below. In order to test the impact of erroneous age assignment to the layers, we perturb the assigned ages randomly by ±15 a and assess the resulting difference in accumulation rates. We also run the model with a different age-TWT relationship to investigate the impact

of potential dating errors on the accumulation rate. The results of these tests will also be discussed in Section 5.1.

Based on the considerations above we assign an uncertainty of 50 ns to the observed layer depths, in order to ensure that we are not overfitting the layers. This uncertainty is used in Equation (5) when calculating the allowed misfit.

4. RESULTS

The inversion scheme returns the average accumulation rates during the period 1311–2011 CE (Common Era) in averages of 100 a. The inversion was performed with five iterations and an uncertainty of 50 ns. The solution converged after a few iterations (see Supplementary Figure S3).

Figure 3 shows the average accumulation rates for the period 1311–2011 CE. We retrieve accumulation rates between 110 and 250 kg m^{-2} a^{-1}, with a mean value of 187 kg m^{-2} a^{-1}. The NEEM ice core site is situated between the 200 and 210 kg m^{-2} a^{-1} contours, while NGRIP is between the 170 and 180 kg m^{-2} a^{-1} contours. The uncertainty in the 700 a average is estimated to 6%, which translates to 10–12 kg m^{-2} a^{-1} at the two ice cores sites. See below for a full treatment of uncertainty estimation.

Spatial variation in accumulation rates is clearly observed in **Figure 3** and as expected, the accumulation rate increases as the elevation decreases. This manifests itself as a decrease in accumulation rates in the north-south direction. South of 76°N the ice divide lies approximately along the 170 kg m^{-2} a^{-1} contour. North of 76°N the ice divide bends into a lower accumulation area (between the 150 and 170 kg m^{-2} a^{-1} contours) but farther north it reenters a higher accumulation area. To further investigate this spatial variation, **Figure 4** shows the results of the inversion on a section along the ice divide. The average accumulation rate (black line in **Figure 4**) decreases upstream from NEEM until approximately 170 km along the flight line, and after that the accumulation increases. From this point and upstream toward NGRIP, the accumulation rate is constant as the ice divide lies on an accumulation contour. We also recover a spatial gradient in the east-west direction: moving across the ice divide from west to east, there is a marked decrease in accumulation rates. The steep gradient in accumulation rates across the ice divide is evident from **Figure 5**, where the average accumulation rate for the period 1311–2011 CE drops from 270 kg m^{-2} a^{-1} to 120 kg m^{-2} a^{-1} over less than 200 km.

Finally, we consider the variation in the spatial pattern over time. **Figure 4** shows the centennial accumulation rates vs. distance from NEEM. To ensure the robustness of the result, we combine all four flight lines along the ice divide to form the figure, and bin the accumulation rate results into bins of 5 km width, using the distance from NEEM as a distance scale. The thick colored lines show the mean of the centennial accumulation rate from each bin and the shading indicates the range of values in each bin (results from each flight line can be seen in the Supplementary Figures S5–S8). The shading pinches in the right-hand end of the plot because there is only one flight line represented in the last ~20 km. In **Figure 4**, a marked temporal variation is evident along the ice divide. The northern part of

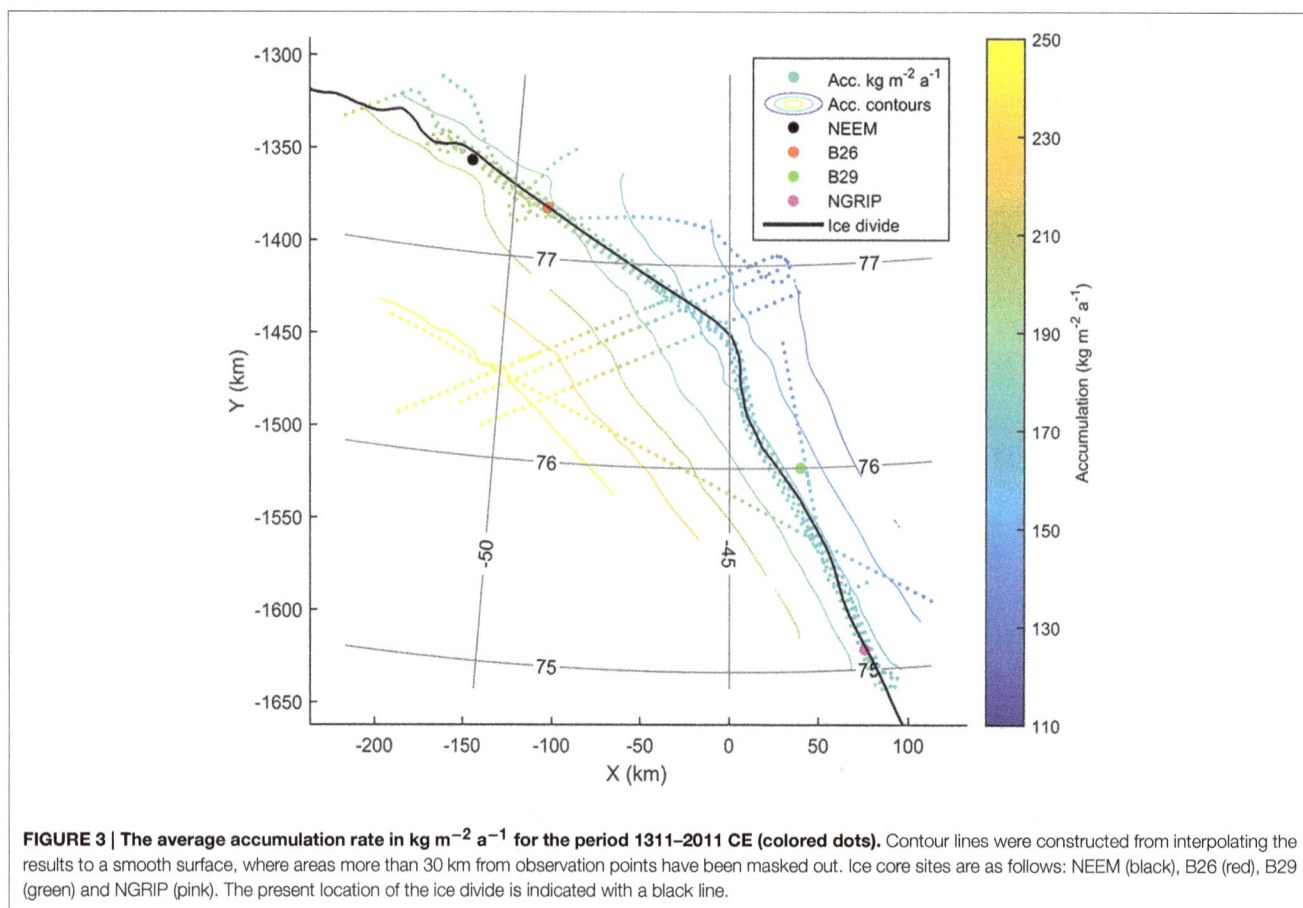

FIGURE 3 | The average accumulation rate in kg m⁻² a⁻¹ for the period 1311–2011 CE (colored dots). Contour lines were constructed from interpolating the results to a smooth surface, where areas more than 30 km from observation points have been masked out. Ice core sites are as follows: NEEM (black), B26 (red), B29 (green) and NGRIP (pink). The present location of the ice divide is indicated with a black line.

the section shows small variations in accumulation rate until approximately 170 km from NEEM. South of this point and to the NGRIP site the variation in accumulation is substantially larger, with the 1311–1411 CE average having the lowest accumulation rate and the 1911–2011 CE and 1511–1611 CE averages having the highest.

Figure 6 shows the accumulation rate anomalies for each 100 a period relative to the average accumulation rate for 1311–2011 CE (figures showing the absolute accumulation rates for the periods can be found in the Supplementary Figure S4). In the area around NEEM (all data points within 5 km of the camp) the accumulation rates for all periods are close to the 700 a average (within ±3%). This is in contrast to the southern half of the domain where anomalies occur. For example, the period 1511–1611 CE (**Figure 6E**) shows consistently high accumulation rates on both sides of the divide compared to the NEEM area, whereas the period 1711–1811 CE (**Figure 6C**) has higher accumulation in the southern part of the study area, but decreasing accumulation west of the ice divide. Most anomalies do not exceed ±10%.

For the centennial averages the inverse scheme is not so well constrained, since each centennial average relies on fewer layers than for the 1311–2011 CE average. Particularly the end points are not as well resolved in these cases. The average accumulation pattern for the period 1911–2011 CE is especially uncertain since

our youngest layer is 116 a old, and this accumulation pattern is therefore only constrained indirectly by the depth of the deeper layers.

5. DISCUSSION

The NEEM and NGRIP ice-core sites are both located on the boundary between high and low accumulation areas in West and East Greenland. Our analysis indicates that the accumulation pattern at NEEM has been relatively stable for the period 1311–2011 CE. In contrast, we retrieve temporal variations in accumulation rate upstream from NEEM and around the NGRIP drill site. Below, we discuss the uncertainties that influence the confidence in our results and quantify the likely uncertainty range. We compare our results with accumulation rates estimated from ice-core records and outputs from regional climate models. Finally, we make a case for careful selection and interpretation of accumulation records from ice cores when elucidating a climate signal.

5.1. Uncertainties

The uncertainties in our study can be split into two groups: (i) uncertainties inherent in our applied method (i.e., the fact that we use a one-dimensional model that disregards horizontal ice flow, and the ability of the firn densification model to

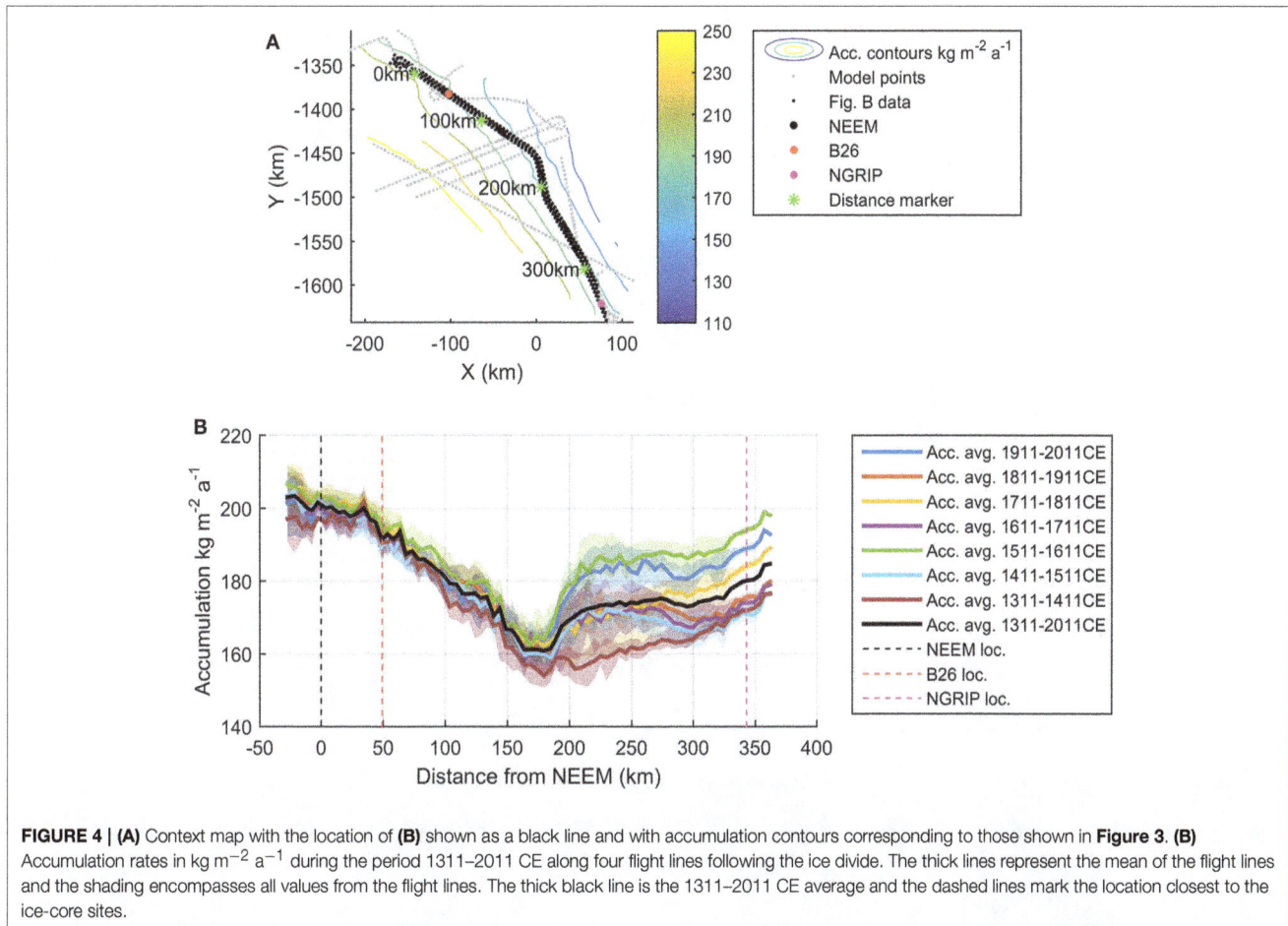

FIGURE 4 | (A) Context map with the location of **(B)** shown as a black line and with accumulation contours corresponding to those shown in **Figure 3**. **(B)** Accumulation rates in kg m^{-2} a^{-1} during the period 1311–2011 CE along four flight lines following the ice divide. The thick lines represent the mean of the flight lines and the shading encompasses all values from the flight lines. The thick black line is the 1311–2011 CE average and the dashed lines mark the location closest to the ice-core sites.

represent the density), and (ii) uncertainties introduced by errors in our procedure (for example, the errors associated with dating, matching, and tracing the layers).

We address the inherent uncertainties first. Using a one-dimensional model that only incorporates vertical advection (i.e., a local-layer approximation, Waddington et al., 2007; Nielsen et al., 2015) means that any horizontal transport of mass is not taken into account. West of the ice divide, a snow particle will travel through lower accumulation areas toward an area of high accumulation rates. In this case, our one-dimensional model underestimates the accumulation rate. The opposite is true east of the ice divide where particles travel from higher to lower accumulation areas. We therefore expect our findings to be a conservative estimate of spatial variations in accumulation rate. The application of a local layer approximation is justified partly because the study area is situated in a low ice-flow velocity region implying that the particles are not likely to have travelled far during the period of interest. Using balance velocities from the SeaRISE (Sea-level Response to Ice Sheet Evolution) project constructed by Johnson (2009), the estimated maximum distance that particles in our domain may have traveled in 700 a is 12 km. This is only a few grid points in our model resolution and not significant compared to the spatial variability of the expected accumulation pattern. By constructing the trajectories back in

time, we can compare accumulation rates at the model grid with the accumulation rate at a point where the 700 a snow would have originated. In this way we find the largest discrepancy of 8%, but most points have a discrepancy of less than 5%. We therefore conclude that neglecting the horizontal advection is justified and is not significant for our results. This is in line with the findings of Nielsen et al. (2015). However, since the inverse scheme also imposes a degree of smoothness, it is likely that the accumulation rate gradient (e.g., across the divide) is more pronounced than our results suggest.

The densification model provides the input to convert depth to TWT. We assess this model by comparing the modeled density profile with two density profiles from shallow cores B26 and B29 (**Figure 1**), using the parameters obtained from the inverse method. The upper 20 m have a discrepancy between the observed density and the modeled density upwards of 40 kg m^{-3} (B26) and 60 kg m^{-3} (B29). Below 20 m the difference between model and observation is much smaller and does not exceed 20 kg m^{-3} for either core. When the modeled and observed density profiles are used in the depth–to–TWT conversion, the resulting difference is less than 10 ns for either core. This is well within our assigned data uncertainty of 50 ns. Thus, it seems unlikely that our densification model introduces significant uncertainties.

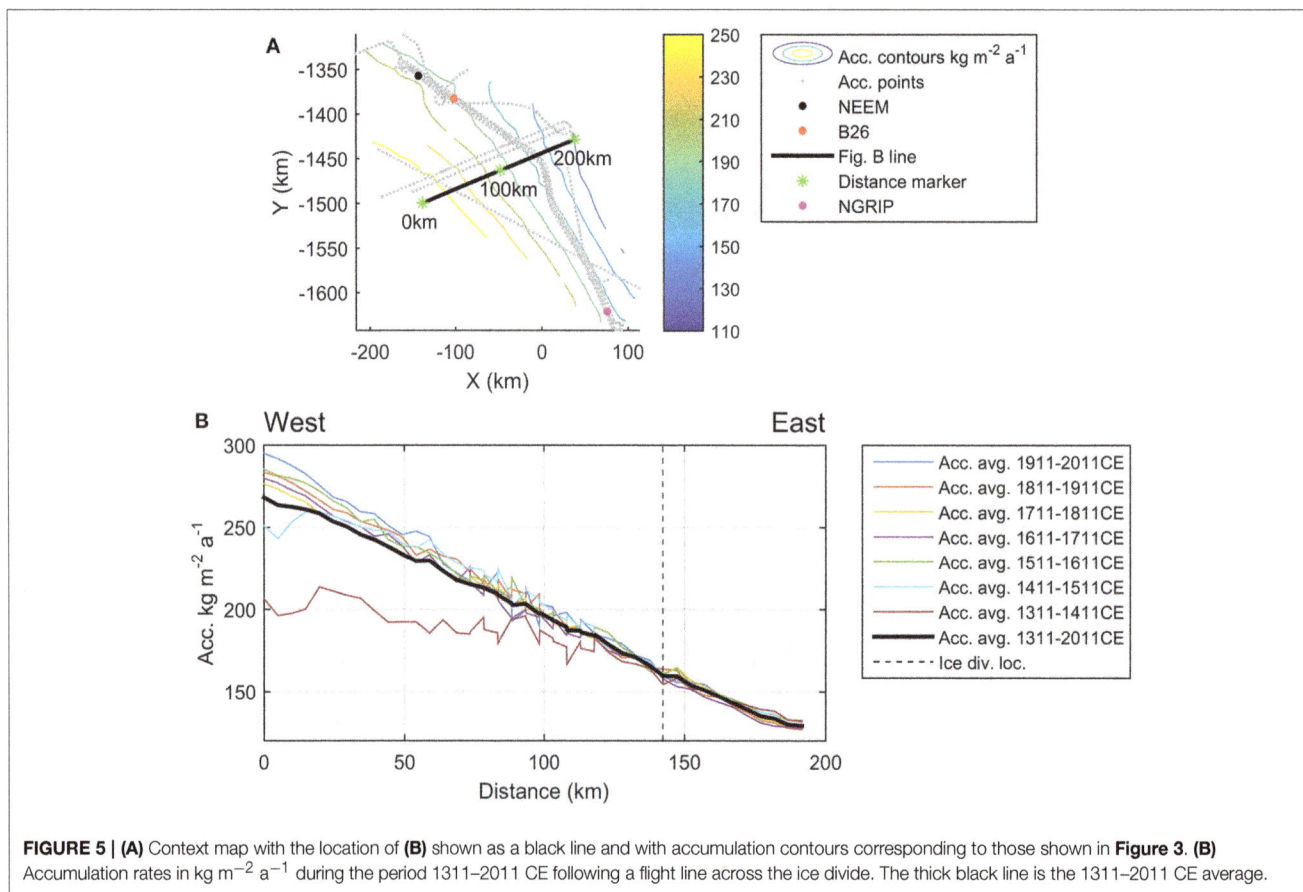

FIGURE 5 | (A) Context map with the location of **(B)** shown as a black line and with accumulation contours corresponding to those shown in **Figure 3. (B)** Accumulation rates in kg m^{-2} a^{-1} during the period 1311–2011 CE following a flight line across the ice divide. The thick black line is the 1311–2011 CE average.

The second group of uncertainties is more difficult to assess. One uncertainty is introduced by the errors associated with matching, dating and tracing the radar layers. A mistake in dating would mean a shift in the accumulation rate for the whole traced layer dataset, while a mistake in the tracing (e.g., an erroneous jump from one layer to another) would impact the resulting accumulation pattern. We obtain a less than 3% difference between the NEEM accumulation record and our accumulation reconstruction (this is discussed below), and we are confident that the internal layers have been dated correctly. To test the impact of an incorrect layer dating further, we have run the model with an offset in the age-TWT relationship. The age-TWT relationship and the result of the test are included in the Supplementary Figures S9, S10. The offset causes the 1911–2011 CE accumulation rate to be substantially larger, and the corresponding 1311–2011 CE average accumulation rate is therefore increased. The overall pattern of decreasing accumulation rates until approximately 170 km from NEEM and then increasing accumulation rates persists. We also recover the same temporal variability of the accumulation rates (although not as clearly as in **Figure 4**), where the centennial averages seem to be more variable starting 170 km from NEEM. Thus, the variability is not strongly dependent on the dating of the layers, especially, the results based on the deeper (older) layers seem to be relatively robust.

It is also possible that errors were introduced in the process of tracing the layers. The main reason, however, why very few errors were introduced in the layer tracing, can be observed in Supplementary Figure S6. Here, there has been an error in one (or more) layers belonging to the 1711–1811 CE time period. The yellow line can be observed to suddenly decrease by ~10 kg m^{-2} a^{-1} at 260 km and then increase the same amount at 340 km. During the jumps the line fluctuates back and forth because all layers were traced with overlaps between neighboring flight lines, and the inverse method therefore received two conflicting TWT-values for the same locations This is the only place where we observe a jump in accumulation rate. To further test the impact of such an error, we have run the model while randomly assigning ±15 a (<2% offset) to the layer dating— corresponding to a jump in approximately 20 ns. This leads to discrepancies in accumulation rates upwards of 18% compared to ice core measurements. In other words, a small error in the age assignment (which is equivalent to tracing an incorrect layer) leads to discernible deterioration in the agreement between model results and observations. Therefore, we consider it likely that all errors in the layer tracing have been detected and corrected. Furthermore, the age–depth scale at NEEM forms the basis of our age assignment, and given the low misfit between the model results and the measurements at NEEM, it is unlikely that there have been significant errors when the layers were dated.

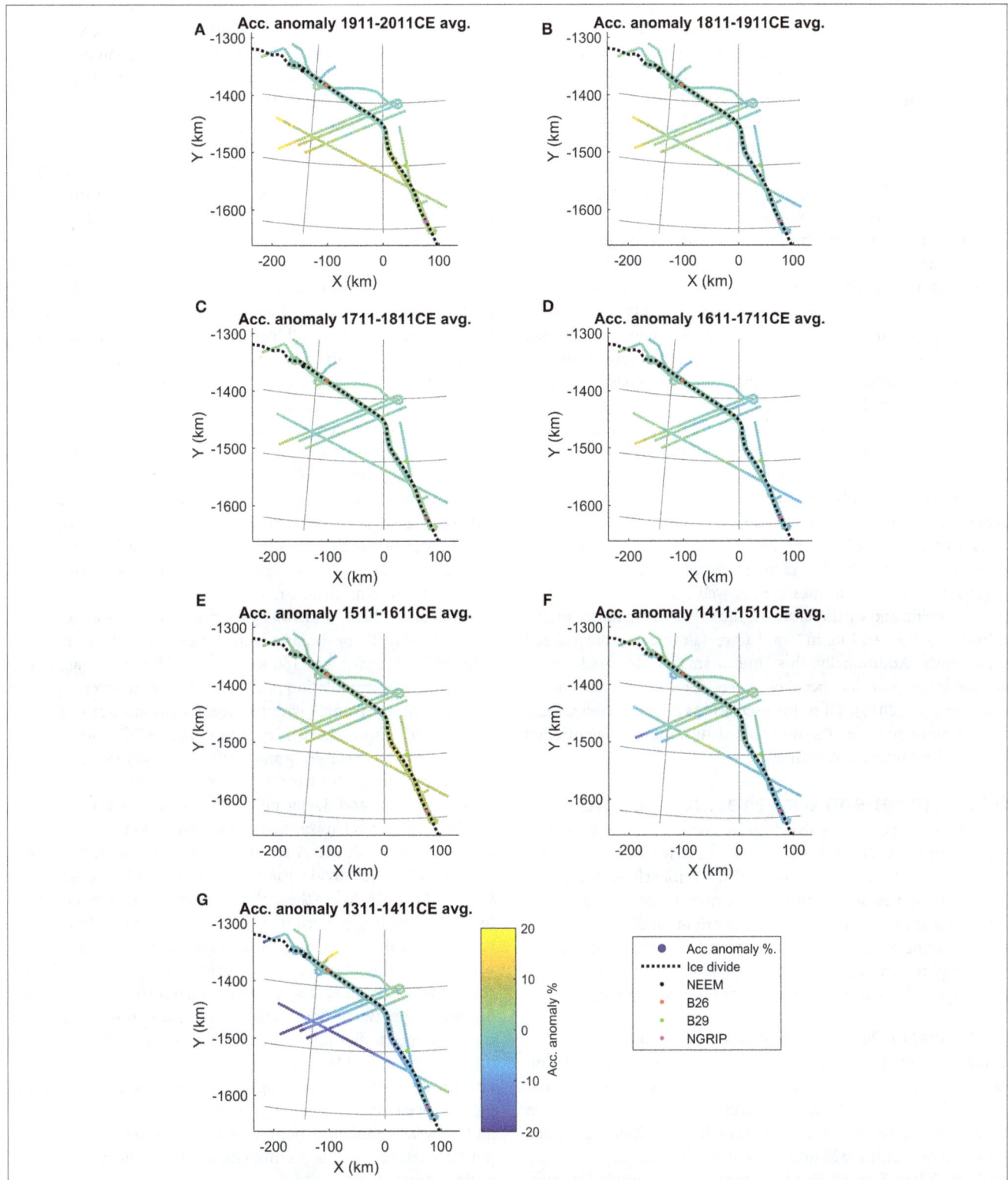

FIGURE 6 | (A–G) Centennially averaged accumulation rates that have been normalized with respect to the accumulation rates for the period 1311–2011 CE. The gray grid lines indicate longitude–latitude. The periods shown are **(A)** 1311–1411 CE, **(B)** 1411–1511 CE, **(C)** 1511–1611 CE, **(D)** 1611–1711 CE, **(E)** 1711–1811 CE, **(F)** 1811–1911 CE, **(G)** 1911–2011 CE.

Finally, we traced the layers starting at a high accumulation area and moving toward low accumulation areas. In this case, layers tend to pinch together thus picking the correct layer is less prone to errors than for areas where layers tend to split in two. Therefore, we argue that the spread in accumulation rates in **Figure 4** reflects real changes in accumulation pattern in time. However, the exact value of the accumulation rate for the periods could be slightly higher or lower, and especially for the 1611–1711 CE period it is possible that there is a slow drift in our results leading to an overestimation in accumulation rate.

Figure 4 offers some insights into the uncertainty of our results. The figure shows a general trend of decreasing and then increasing accumulation over spatial scales of $\gtrsim 10^2$ km along the ice divide. In order to quantify the uncertainty, we consider any oscillation over a spatial scale smaller than $\sim 10^2$ km to be a result of the uncertainties in our method. Based on this consideration, we estimate that the centennially averaged accumulation rates have an uncertainty of ~ 12 kg m^{-2} a^{-1} and the 1311–2011 CE accumulation rate has an uncertainty of ~ 8 kg m^{-2} a^{-1}. For the 1311–2011 CE accumulation rate, an uncertainty of ~ 8 kg m^{-2} a^{-1} corresponds at most to 6% of the reported accumulation rates. For the centennially averaged accumulation rates, an uncertainty of ~ 12 kg m^{-2} a^{-1} is translated into an 8% uncertainty based on the accumulation rates along the ice divide. As noted elsewhere, the centennially averaged accumulation rates are not well-behaved at end points in our domain and we therefore use the values along the ice divide to translate the ~ 12 kg m^{-2} a^{-1} uncertainty into a normalized uncertainty. Additionally, this kind of small scale variability in accumulation rate has been observed elsewhere in Greenland (Hawley et al., 2014). Thus, the oscillations might be correct and we therefore consider the normalized uncertainties mentioned above to be conservative estimates.

5.2. Comparison with Previous Records

Ice-core records, such as the deep ice cores NEEM and NGRIP (Andersen et al., 2006; Rasmussen et al., 2013) and the shallow cores B26 and B29 (Schwager, 2000; Weißbach et al., 2016), provide estimates of accumulation rates going back in time. We obtain a first insight into the spatial distribution of accumulation by comparing several cores with our model results. Additionally, we compare our findings with large-scale accumulation rates across Greenland estimated by regional climate models.

5.2.1. Spatial Distribution of Accumulation

Table 2 summarizes the comparisons between accumulation rates inferred from measurements of deep and shallow cores and our modeled results. With a few exceptions, differences between ice core measurements and model results fall within our stated uncertainty. For the B26 and B29 cores, the measurements only go back 500 a. Two studies have measured accumulation rates before; a study by Schwager (2000), and a more recent study by Weißbach et al. (2016). The latter constructed a new age-scale for the ice cores leading to slightly lower accumulation rates compared to the results by Schwager (2000). For the B26 core, the model results overestimate the accumulation rate compared to the measured rates. Potential reasons for this overestimation are

discussed below. The difference between measured and modeled accumulation rates is smaller for the B29 core, where only the accumulation rates from Weißbach et al. (2016) fall outside our uncertainty range. This discrepancy is likely introduced by the difference in accumulation rates for the period 1911-2011 CE, where our results indicate anomalously high accumulation rates (174 ± 14 kg m^{-2} a^{-1}), while the results from Weißbach et al. (2016) show low accumulation rates (138 kg m^{-2} a^{-1}). This is also the period where our method is not well constrained. If instead the periods 1511-1911 CE are compared, the discrepancy decreases to 4% which is within our stated uncertainty.

In **Figures 7A,B** we compare our findings with the output from two modeling studies by Ettema et al. (2009) and Burgess et al. (2010) respectively, which provide average SMB values for the period 1958–2007. Both studies use regional climate models (RCM) validated with *in situ* measurements from across the ice sheet. The models were run with a spatial resolution of 11 km (Ettema et al., 2009) and 24 km (Burgess et al., 2010) and the resulting accumulation rates are therefore smoother than our accumulation rates that are on a 5 km resolution. Note that we have regridded the RCM data to 5 km resolution for direct comparison with our results. In both studies, the accumulation pattern has a gradient across the ice divide, with contours approximately parallel to the ice divide. There is a notable difference in the value of the accumulation rate between the two studies. The accumulation rates at NEEM are 150 kg m^{-2} a^{-1} (Ettema et al., 2009) and 180 kg m^{-2} a^{-1} (Burgess et al., 2010), while at NGRIP the accumulation rates are 110 kg m^{-2} a^{-1} (Ettema et al., 2009) and 160 kg m^{-2} a^{-1} (Burgess et al., 2010). Our modeled accumulation rates display the same overall pattern but agree more strongly with the accumulation rates of Burgess et al. (2010). This is also evident from **Figures 7C,D** which show the accumulation rate along and across the ice divide. In contrast, the accumulation rates from Ettema et al. (2009) are on average 39 kg m^{-2} a^{-1} and 37 kg m^{-2} a^{-1} (along and across the ice divide, respectively) lower than our accumulation rates and those of Burgess et al. (2010). A study by Box et al. (2013) has shown that east of the ice divide, the model employed by Burgess et al. (2010) (Polar MM5) has the highest accumulation rates of the two RCMs. Interestingly, Vernon et al. (2013) found that Polar MM5 returns accumulation rates above observational values for areas with low accumulation rates. However, the Vernon et al. (2013) study does not include observations from our study region, so the discrepancy between observations and results might vary from region to region. For more in-depth discussions of SMB patterns from different RCMs, we refer the reader to the two studies referenced above. Here, we suggest that while the accumulation pattern presented in Ettema et al. (2009) is likely correct, the high–low accumulation boundary in Ettema et al. (2009) lies too far west, so that the accumulation rates at the ice divide are underestimated.

Our retrieved spatial pattern of accumulation can also be compared to results from studies of shallow ice-cores. In **Figure 7E** we show the results of Bales et al. (2001), who produced several short-term records of accumulation rates in our study area (see also Supplementary Table S1). The cores do not date further back than the 1930s but there is a clear agreement

TABLE 2 | Comparison between the modeled accumulation rates from this study and measured accumulation rates from ice and firn cores.

Ice core	Ice core period	References	Comparison period	Ice core acc. (kg m^{-2} a^{-1})	Modeled acc. (kg m^{-2} a^{-1})	% diff.
NEEM	1320–1980 CE	Rasmussen et al., 2013	1311–2011 CE	204	202 ± 12	<1
NGRIP	1301–1984 CE	Andersen et al., 2006	1311–2011 CE	177	179 ± 11	1.1
B26	1501–1983 CE	Schwager, 2000	1511–2011 CE	179	193 ± 12	6.8
	1512–1994 CE	Weißbach et al., 2016	1511–2011 CE	175	193 ± 12	10
B29	1501–1983 CE	Schwager, 2000	1511–2011 CE	152	161 ± 11	5.3
	1512–1994 CE	Weißbach et al., 2016	1511–2011 CE	149	161 ± 11	8.7
B-2-200	1942–1955 CE	Bales et al., 2001	1311–2011 CE	220	204 ± 12	7.3
B-2-225[a]	1939–1955 CE	Bales et al., 2001	1311–2011 CE	185	187 ± 11	1.1
B-2-250-4-0[a]	1939–1955 CE	Bales et al., 2001	1311–2011 CE	165	178 ± 11	7.7
B-4-100	1939–1955 CE	Bales et al., 2001	1311–2011 CE	188	193 ± 12	2.7
B-4-125	1940–1955 CE	Bales et al., 2001	1311–2011 CE	205	199 ± 12	2.8
B-4-25[b]	1939–1955 CE	Bales et al., 2001	1311–2011 CE	175	173 ± 10	1.3
B-4-50	1939–1955 CE	Bales et al., 2001	1311–2011 CE	175	179 ± 11	2.4
B-4-75	1940–1955 CE	Bales et al., 2001	1311–2011 CE	185	186 ± 11	<1
G(C)-N-SITE	1943–1973 CE	Bales et al., 2001	1311–2011 CE	151	148 ± 9	2.3
PATER-C	1952–1954 CE	Bales et al., 2001	1311–2011 CE	140	196 ± 12	40
PATER-C1[a]	1952–1954 CE	Bales et al., 2001	1311–2011 CE	170	175 ± 10	2.7
PATER-C2	1952–1954 CE	Bales et al., 2001	1311–2011 CE	150	148 ± 9	1.1

*"Ice core period" indicates the period that is compared to the modeled accumulation rates from the "Comparison period". Cores denoted [a] are shown in **Figure 7C** and the core denoted b is shown in **Figure 7D**. For geographical coordinates see Supplementary Table 1.*

with our results (cf. **Table 2**). This agreement also indicates that the accumulation rates in the 1930s–1950s were similar to the 700 a average accumulation rate. The difference between the accumulation rates of Bales et al. (2001) and our results is 8% or less, with one exception: the PATER-C site located west of the ice divide at 76.72°N, −47.33°E (difference of 40%, see also **Figure 7E**). We do not have an immediate explanation for the large discrepancy at this site but it should be noted that the core in question only covers the period 1952–1954 CE. Thus, the discrepancy could either be an error in the dating of the core, or a localized accumulation phenomenon (in space and/or time).

5.2.2. Temporal Variation of Accumulation

We next compare the temporal variation of our modeled accumulation rates with the estimated temporal variation of accumulation rates from the ice cores (**Figure 8**). The model results from the last century are not discussed in the comparison because they are not well constrained due to the lack of layers in that age bracket. A comparison with the NEEM record (**Figure 8A**) shows minor discrepancies that are all well within our expected uncertainty. For the NGRIP record (**Figure 8B**), the difference between modeled accumulation rates and ice core measurements are larger than at NEEM. The largest difference occurs for the 1511–1611 CE period where the our model results are higher than the accumulation rates from the ice core. In contrast, the modeled accumulation rates are lower than the NGRIP record for the 1411–1511 CE and 1311–1411 CE period. With the exception of the 1511–1611 CE period, where modeled accumulation rates differ by

9% compared to the ice core measurements, the difference between the model results and the NGRIP record is less than 6%.

For the B26 core (**Figure 8C**), our centennially averaged accumulation rates all overestimate the accumulation rates compared to the ice core. The period 1511–1611 CE has the highest discrepancy (14%) while the other periods are within our uncertainty range of 8%. This is in contrast to the NEEM and NGRIP ice cores where the discrepancies between modeled and measured accumulation rates do not display a clear bias. Closer inspection of **Figure 4B** reveals that the model indicates a local accumulation minimum at this site, but the dip is probably obscured by the imposed smoothness of the inverse method and, the fact that upstream effects are not included in our one-dimensional model. This implies that the accumulation rates are overestimated by our model at this site. One factor contributing to the overestimated accumulation rates could be the location directly above a "unit of disrupted radiostratigraphy" identified by Panton and Karlsson (2015) (see Figure 4 in that study). These units of disrupted radiostratigraphy, although often originating from processes acting at the ice-bedrock interface, may cause disturbances all the way up to the ice surface. Thus, the depth of the layers may be influenced by processes other than accumulation, that are not included in our model. Similar to the results for the B26 core, the accumulation rates for the B29 core have the largest difference (9%) between model results and observations for the 1511–1611 CE period. All other periods have differences of less than 5%. For this core there is no clear bias in the modeled accumulation rates.

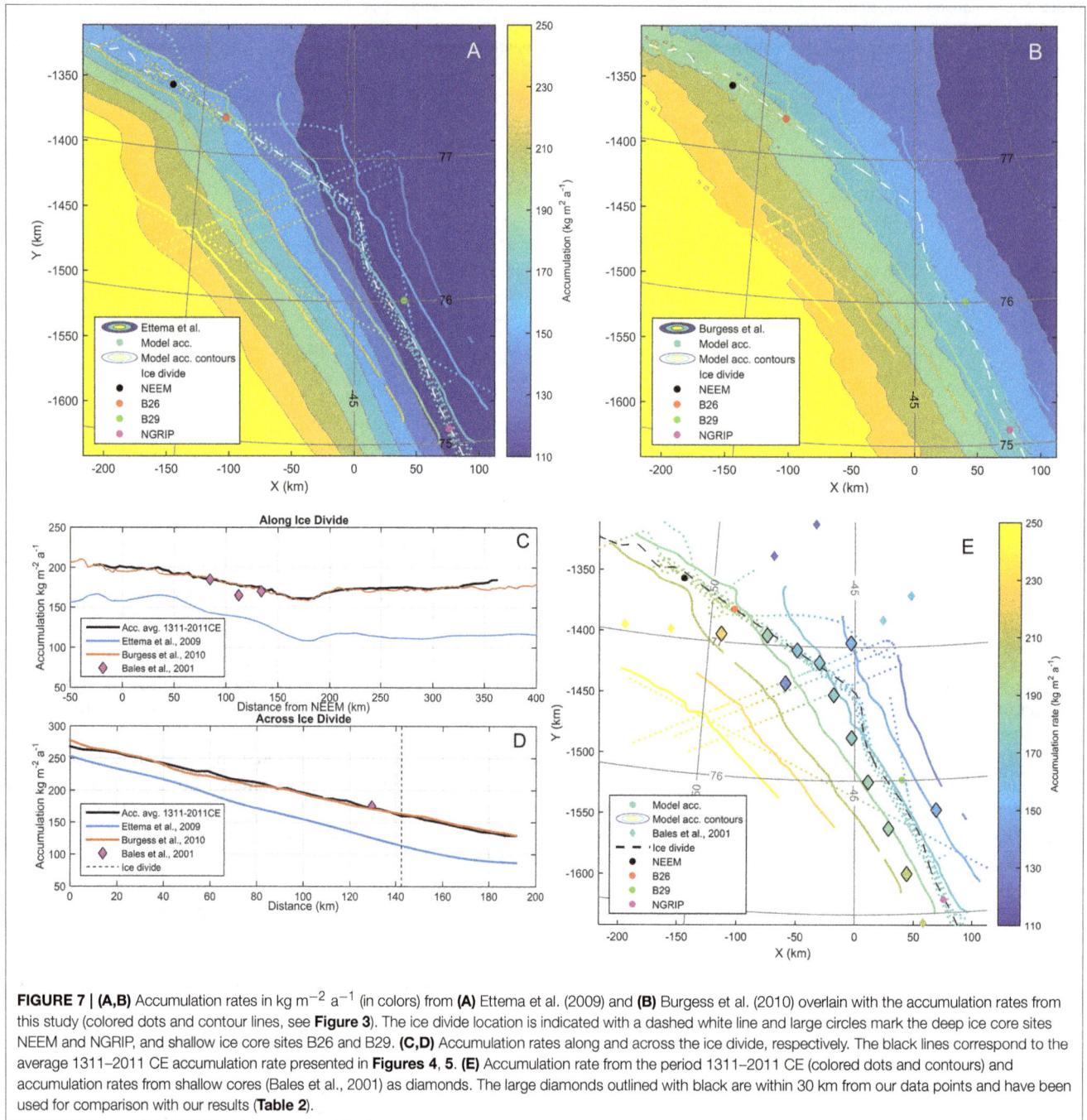

FIGURE 7 | (A,B) Accumulation rates in kg m^{-2} a^{-1} (in colors) from (A) Ettema et al. (2009) and (B) Burgess et al. (2010) overlain with the accumulation rates from this study (colored dots and contour lines, see Figure 3). The ice divide location is indicated with a dashed white line and large circles mark the deep ice core sites NEEM and NGRIP, and shallow ice core sites B26 and B29. (C,D) Accumulation rates along and across the ice divide, respectively. The black lines correspond to the average 1311–2011 CE accumulation rate presented in Figures 4, 5. (E) Accumulation rate from the period 1311–2011 CE (colored dots and contours) and accumulation rates from shallow cores (Bales et al., 2001) as diamonds. The large diamonds outlined with black are within 30 km from our data points and have been used for comparison with our results (Table 2).

To summarize, our results agree well with ice core measurements. The period with the largest discrepancy is the 1511–1611 CE period. Considering that three out of four ice core sites indicate that our results for this period overestimate the accumulation rate, it is possible that an error was introduced either during our dating of (one of) the layers, or during our layer tracing. Results from the remaining periods (with the exception of the 1911–2011 CE period that we exclude) are robust and show consistent accumulation rate in the northern part of the model domain, and temporal variability in the southern part (e.g., Figure 4).

5.3. Representativeness of Ice-Core Accumulation Records

Our analysis indicates that in the area around NEEM the accumulation pattern has been stable for the period 1311–2011 CE. Changes in accumulation rates derived from this ice core can thus be reliably interpreted as changes in amount of precipitation on a regional scale in this part of Greenland. In contrast, we find changes in accumulation pattern starting approximately 170 km upstream from NEEM and around the NGRIP drill site. This implies that changes in accumulation rates from the NGRIP ice core might be influenced by, or even

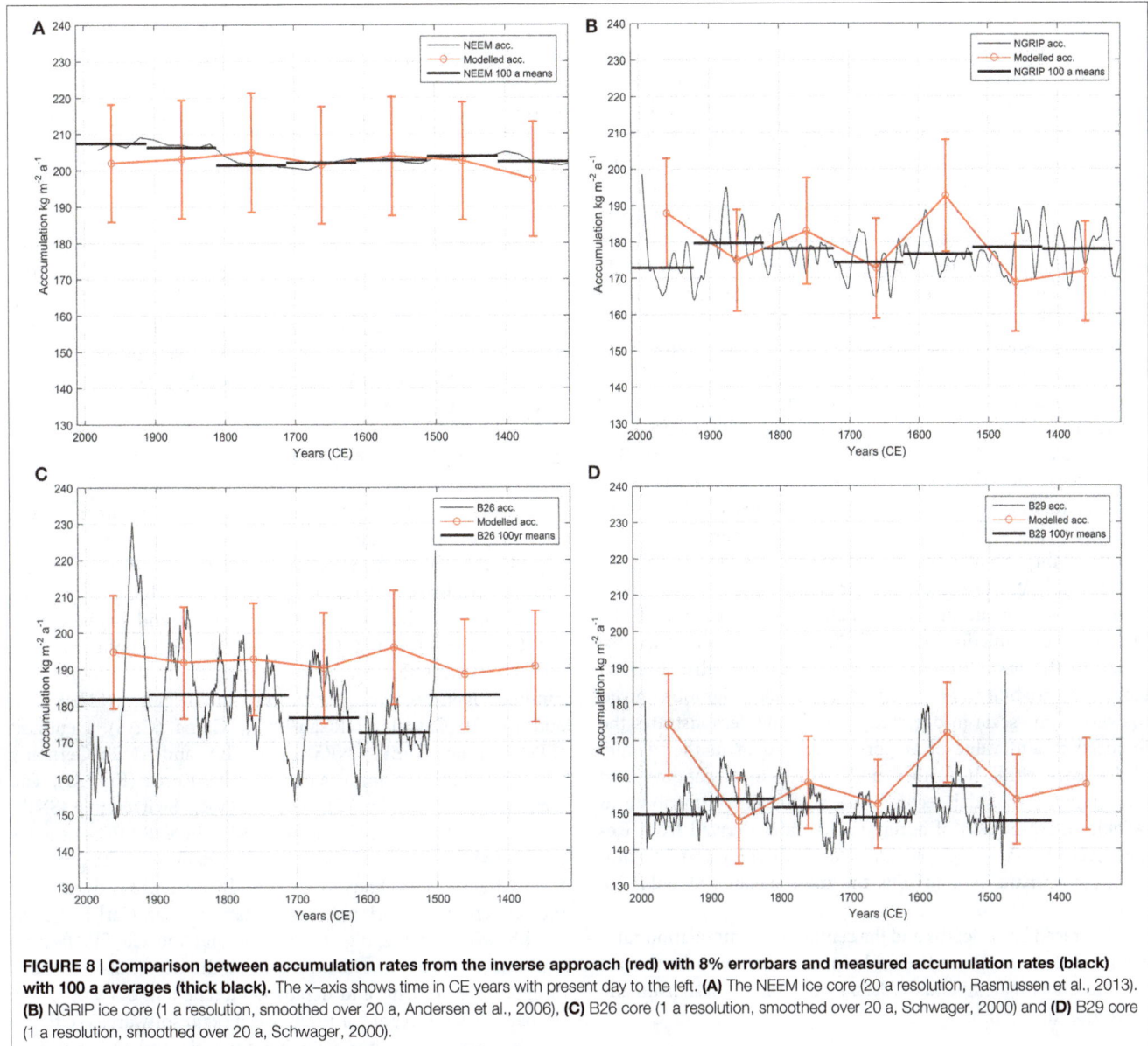

FIGURE 8 | Comparison between accumulation rates from the inverse approach (red) with 8% errorbars and measured accumulation rates (black) with 100 a averages (thick black). The x–axis shows time in CE years with present day to the left. **(A)** The NEEM ice core (20 a resolution, Rasmussen et al., 2013). **(B)** NGRIP ice core (1 a resolution, smoothed over 20 a, Andersen et al., 2006), **(C)** B26 core (1 a resolution, smoothed over 20 a, Schwager, 2000) and **(D)** B29 core (1 a resolution, smoothed over 20 a, Schwager, 2000).

obscured by, changes in the accumulation pattern. Such changes can be triggered by a shift in the position of the ice divide. However, studies suggest that the relaxation time of the divide-position in response to changes in accumulation is 500–1000 years for the Greenland Ice Sheet (see Hindmarsh, 1996). We therefore consider it unlikely that the ice divide could have changed position to an extent that it affected the accumulation pattern. Instead, we attribute the changes in accumulation rates to migration of the accumulation pattern across the ice divide, and not ice divide displacement. In conclusion, care should be taken when interpreting variations in accumulation rate from the NGRIP ice core as an indicator of global or regional change in precipitation.

To summarize, there is overall agreement between our model results and measured accumulation rates from NEEM and NGRIP, and the shallow cores B26 and B29. Differences between our model and the observations for all periods (except for the period 1511–1611 CE) are less than 3% for NEEM, less than 6% for NGRIP, less than 8% for the B26 core, and less than 5% for B29 core. Although some uncertainties might have been introduced during the manual tracing of the layers, we find it likely that the variation in accumulation rate and thereby the implied shifts in accumulation pattern represent real climate events. This ultimately means that during the period 1311–2011 CE the NEEM ice core is more likely than the NGRIP ice core to have preserved a history of the past regional weather patterns that affected northern Greenland. Our assertion is based on the fact that the temporal accumulation signal is stable in the area around NEEM. In contrast, changes in accumulation rates deduced from the NGRIP core could reflect local shifts in accumulation patterns rather than a climate signal.

6. CONCLUSION

We used internal radar layers as input for a numerical model in order to retrieve accumulation rates. The crucial process of assigning correct ages to the layers was supported by the use of a model of electromagnetic wave propagation in ice, in conjunction with dielectric measurements and density data from the NEEM11S1 core. Using 16 dated and manually traced layers over an area encompassing the NEEM and NGRIP ice–core sites, and expanding 300 km across the ice divide, we retrieved accumulation rates in 100 a means for the period 1311–2011 CE. We used a gradient–descent inverse method that includes a 1D firn–densification model and solves for the best accumulation rates and densification parameters. Our modeled accumulation rates agree with spatially distributed model results from regional climate models. There is also good agreement between our centennial accumulation rate averages and measurements from the deep ice cores NEEM and NGRIP and the shallow cores B26 and B29. We find that while the NEEM ice-core site has stable accumulation rates during the study period, the NGRIP ice core is situated in a region that exhibits more temporal variability. The variability initiates approximately mid-way between NEEM and NGRIP. We interpret this variability as an indication of a locally varying accumulation pattern, specifically that the high–low accumulation boundary has migrated across the ice divide in the past. We consider it unlikely that the observed temporal variability is a result of ice divide migration, given the short time-scale in our study. Our study demonstrates the importance and value of acquiring radar data in the vicinity of ice core sites. It further underscores that mapping and interpretation of radar records may significantly contribute to reliable interpretation of accumulation rates derived from ice-core records on a regional scale, and provides insights into the weather patterns and thus the mass balance of polar ice sheets.

The traced layer depths and the centennial accumulation rates are available from the Centre for Ice and Climate website (http://www.iceandclimate.nbi.ku.dk/data/) or upon request from the corresponding author.

AUTHOR CONTRIBUTIONS

NK and OE designed the study and wrote the manuscript. NK analyzed the radar data, traced the layers, constructed the firn model, did the layer matching and conducted the inverse model runs. OE constructed and calculated the synthetic radar trace. LN designed the inverse method. JF conducted the high resolution density measurements. SK did the DEP measurements and logged the NEEM11S1 core together with NK. JP and CL designed and operated the radar instrument and radar data processor. AW collected the 2015 radar data and contributed to the discussions. DD-J and FW secured the funding that made this study possible.

FUNDING

Funding for NK was provided by the European Research Council Advanced Grant no. 246815 WATERundertheICE. The Center for Ice and Climate is funded by the Danish National Research Foundation. Data and data products from CReSIS were generated with support from NSF grant ANT-0424589 and NASA grant NNX10AT68G, and NASA grant NNX10AT68G as a part of NASA Operation IceBridge.

ACKNOWLEDGMENTS

This work initially started at the Karthaus summer school on Glaciers and Ice Sheets in 2013. We thank the long-term organizer Hans Oerlemans for his sustained efforts to bring together students and teaching scientists, enabling such spin-off results. The NEEM project is directed and organized by the Center of Ice and Climate at the Niels Bohr Institute and US NSF, Office of Polar Programs. It is supported by funding agencies and institutions in Belgium (FNRS-CFB and FWO), Canada (NRCan/GSC), China (CAS), Denmark (FIST), France (IPEV, CNRS/INSU, CEA, and ANR), Germany (AWI), Iceland (RannIs), Japan (NIPR), Korea (KOPRI), The Netherlands (NWO/ALW), Sweden (VR), Switzerland (SNF), United Kingdom (NERC) and the USA (US NSF, Office of Polar Programs). We acknowledge the use of data and data products from CReSIS. We acknowledge Prasad Gogineni for developing the concept of the airborne accumulation radar, Carl Leuschen for his work on the digital systems, and the CReSIS faculty, staff and students who contributed to the accumulation radar design, development, and deployment. The authors would like to thank J. MacGregor for use of his *pickgui* software. We thank our reviewers and editor, Felix Ng, for insightful and thorough comments that improved the manuscript.

REFERENCES

Andersen, K. K., Ditlevsen, P. D., Rasmussen, S. O., Clausen, H. B., Vinther, B. M., Johnson, S. J., et al. (2006). Retrieving a common accumulation record from Greenland ice cores for the past 1800 years. *J. Geophys. Res.* 111:D15106. doi: 10.1029/2005JD006765

Arthern, R. J., Vaughan, D. G., Rankin, A. M., Mulvaney, R., and Thomas, E. R. (2010). *In situ* measurements of Antarctic snow compaction compared with predictions of models. *J. Geophys. Res.* 115:F03011. doi: 10.1029/2009JF001306

Aster, R. C., Borchers, B., and Thurber, C. H. (2013). *Parameter Estimation and Inverse Problems.* Cambridge, MA: Elsevier.

Bales, R. C., Guo, Q., Shen, D., McConnell, J. R., Du, G., Burkhart, J. F., et al. (2009). Annual accumulation for Greenland updated using ice core data developed during 2000–2006 and analysis of daily coastal meteorological data. *J. Geophys. Res.* 114:D06116. doi: 10.1029/2008JD011208

Bales, R. C., McConnell, J. R., Mosley-Thompson, E., and Csatho, B. (2001). Accumulation over the Greenland ice sheet from historical and recent records. *J. Geophys. Res. Atmos.* 106, 33813–33825. doi: 10.1029/2001JD900153

Bamber, J. L., Griggs, J. A., Hurkmans, R. T. W. L., Dowdeswell, J. A., Gogineni, S. P., I. Howat, J. M., et al. (2013). A new bed elevation dataset for Greenland. *Cryosphere* 7, 499–510. doi: 10.5194/tc-7-499-2013

Bamber, J. L., Layberry, R. L., and Gogineni, S. P. (2001). A new ice thickness and bed data set for the Greenland ice sheet: 1. measurement, data reduction, and errors. *J. Geophys. Res. Atmos.* 106, 33773–33780. doi: 10.1029/2001JD900054

Bohleber, P., Wagner, N., and Eisen, O. (2012). Permittivity of ice at radio frequencies: Part ii. artificial and natural polycrystalline ice. *Cold Reg. Sci. Technol.* 83,84, 13–19. doi: 10.1016/j.coldregions.2012.05.010

Bolzan, J. F., and Strobel, M. (1994). Accumulation-rate variations around summit, Greenland. *J. Glaciol.* 40, 56–66. doi: 10.3198/1994JoG40-134-56-66

Box, J. E., Cressie, N., Bromwich, D. H., Jung, J.-H., van den Broeke, M., van Angelen, J. H., et al. (2013). Greenland ice sheet mass balance reconstruction. Part I: net snow accumulation (1600Ǔ2009). *J. Clim.* 26, 3919–3934. doi: 10.1175/JCLI-D-12-00373.1

Burgess, E. W., Forster, R. R., Box, J. E., Mosley-Thompson, E., Bromwich, D. H., Bales, R. C., et al. (2010). A spatially calibrated model of annual accumulation rate on the Greenland Ice Sheet (1958-2007). *J. Geophys. Res.* 115:F02004. doi: 10.1029/2009JF001293

Crank, J., and Nicolson, P. (1996). A practical method for numerical evaluation of solutions of partial differential equations of the heat-conduction type. *Adv. Comput. Math.* 6, 207–226. doi: 10.1007/BF02127704

Dahl-Jensen, D., Albert, M. R., Aldahan, A., Azuma, N., Balslev-Clausen, D., Baumgartner, M., et al. (2013). Eemian interglacial reconstructed from a Greenland folded ice core. *Nature* 493, 489–494. doi: 10.1038/nature11789

Dahl-Jensen, D., Gundestrup, N. S., Keller, K., Johnsen, S. J., Gogineni, S. P., Allen, C. T., et al. (1997). A search in north Greenland for a new ice-core drill site. *J. Glaciol.* 43, 300–306.

Eisen, O. (2008). Inference of velocity pattern from isochronous layers in firn, using an inverse method. *J. Glaciol.* 54, 613–630. doi: 10.3189/002214308786570818

Eisen, O., Frezzotti, M., Genthon, C., van den Broeke, E., Dixon, D. A., Ekaykin, A., et al. (2008). Ground-based measurements of spatial and temporal variability of snow accumulation in East Antarctica. *Rev. Geophys.* 46:RG2001. doi: 10.1029/2006RG000218

Eisen, O., Rack, W., Nixdorf, U., and Wilhelms, F. (2005). Characteristics of accumulation rate in the vicinity of the EPICA deep-drilling site in Dronning Maud Land, Antarctica. *Ann. Glaciol.* 41, 41–46. doi: 10.3189/172756405781813276

Eisen, O., Wilhelms, F., Nixdorf, U., and Miller, H. (2003). Revealing the nature of radar reflections in ice: Dep-based fdtd forward modeling. *Geophys. Res. Lett.* 30:1218. doi: 10.1029/2002gl016403

Eisen, O., Wilhelms, F., Steinhage, D., and Schwander, J. (2006). Instruments and Methods: Improved method to determine radio-echo sounding reflector depths from ice-core profiles of permittivity and conductivity. *J. Glaciol.* 52, 299–310. doi: 10.3189/172756506781828674

Ettema, J., van den Broeke, M. R., van Meijgaard, E., van de Berg, W. J., Bamber, J. L., Box, J. E., et al. (2009). Higher surface mass balance of the Greenland ice sheet revealed by high-resolution climate modeling. *Geophys. Res. Lett.* 36:L12501. doi: 10.1029/2009GL038110

Freitag, J., Kipfstuhl, S., and Laepple, T. (2013). Core-scale radioscopic imaging: a new method reveals density-calcium link in antarctic firn. *J. Glaciol.* 59, 1009–1014. doi: 10.3189/2013JoG13J028

Frezzotti, M., Pourchet, M., Flora, O., Gandolfi, S., Gay, M., Urbini, S., et al. (2005). Spatial and temporal variability of snow accumulation in east Antarctica from traverse data. *J. Glaciol.* 51, 113–124. doi: 10.3189/172756505781829502

Fujita, S., Holmlund, P., Andersson, I., Brown, I., Enomoto, H., Fujii, Y., et al. (2011). Spatial and temporal variability of snow accumulation rate on the east antarctic ice divide between dome fuji and EPICA DML. *Cryosphere* 5, 1057–1081. doi: 10.5194/tc-5-1057-2011

Gfeller, G., Fischer, H., Bigler, M., Schüpbach, S., Leuenberger, D., and Mini, O. (2014). Representativeness and seasonality of major ion records derived from neem firn cores. *Cryosphere* 8, 1855–1870. doi: 10.5194/tc-8-1855-2014

Gogineni, P. (2012). *Cresis Accumulation Radar Data*. Lawrence: Digital Media.

Hanna, E., Huybrechts, P., Cappelen, J., Steffen, K., Bales, R. C., Burgess, E., et al. (2011). Greenland Ice Sheet surface mass balance 1870 to 2010 based on Twentieth Century Reanalysis, and links with global climate forcing. *J. Geophys. Res.* 116, 1–20. doi: 10.1029/2011JD016387

Hawley, R. L., Courville, Z. R., Kehrl, L. M., Lutz, E. R., Osterberg, E. C., Overly, T. B., et al. (2014). Recent accumulation variability in northwest Greenland from ground-penetrating radar and shallow cores along the Greenland inland traverse. *J. Glaciol.* 60, 375–382. doi: 10.3189/2014JoG13J141

Herron, M. M., and Langway, C. C. (1980). Firn densification: an empirical model. *J. Glaciol.* 25, 373–385.

Hindmarsh, R. C. A. (1996). Stochastic perturbation of divide position. *Ann. Glaciol.* 23, 94–104.

Johnson, J. (2009). *Searise Master Data Set for Greenland*. Missoula: University of Montana. Available online at: http://websrv.cs.umt.edu/isis/index.php/Present_Day_Greenland

Kanagaratnam, P., Gogineni, S. P., Ramasami, V., and Braaten, D. (2004). A wideband radar for high-resolution mapping of near-surface internal layers in glacial ice. *IEEE Trans. Geosci. Remote Sens.* 42, 483–490. doi: 10.1109/TGRS.2004.823451

Koenig, L. S., Ivanoff, A., Alexander, P. M., MacGregor, J. A., Fettweis, X., Panzer, B., et al. (2016). Annual Greenland accumulation rates (2009–2012) from airborne snow radar. *Cryosphere* 10, 1739–1752. doi: 10.5194/tc-10-1739-2016

Koutnik, M. R., Fudge, T. J., Conway, H., Waddington, E. D., Neumann, T. A., Cuffey, K. M., et al. (2016). Holocene accumulation and ice flow near the west antarctic ice sheet divide ice core site. *J. Geophys. Res. Earth Surf.* 121, 907–924. doi: 10.1002/2015jf003668

Lenaerts, J. T. M., Brown, J., Broeke, M. R. v., Matsuoka, K., Drews, R., Callens, D., et al. (2014). High variability of climate and surface mass balance induced by Antarctic ice rises. *J. Glaciol.* 60, 1101–1110. doi: 10.3189/2014JoG14J040

Lewis, C., Gogineni, S. P., Rodriguez-Morales, F., Panzer, B., Stumpf, T., and J. D. Paden, C. L. (2015). Airborne fine-resolution uhf radar: an approach to the study of englacial reflections, firn compaction and ice attenuation rates. *J. Glaciol.* 61, 89–100. doi: 10.3189/2015jog14j089

Leysinger-Vieli, G., Hindmarsh, R., and Siegert, M. (2007). Three-dimensional flow influences on radar layer stratigraphy. *Ann. Glaciol.* 46, 22–28. doi: 10.3189/172756407782871729

Ligtenberg, S. R. M., Helsen, M. M., and van den Broeke, M. R. (2011). An improved semi-empirical model for the densification of Antarctic firn. *Cryosphere* 5, 809–819. doi: 10.5194/tc-5-809-2011

Looyenga, H. (1965). Dielectric constant of heterogeneous mixtures. *Physica* 31, 401–406. doi: 10.1016/0031-8914(65)90045-5

MacGregor, J. A., Fahnestock, M. A., Catania, G. A., Paden, J. D., Gogineni, S. P., Young, S. K., et al. (2015). Radiostratigraphy and age structure of the Greenland Ice Sheet. *J. Geophys. Res. Earth Surf.* 120, 1–30. doi: 10.1002/2014JF003215

Nereson, N. A., Raymond, C. F., Jacobel, R. W., and Waddington, E. D. (2000). The accumulation pattern across siple dome, west antarctica, inferred from radar-detected internal layers. *J. Glaciol.* 46, 75–87. doi: 10.3189/172756500781833449

Neumann, T. A., Conway, H., Price, S. F., Waddington, E. D., Catania, G. A., and Morse, D. L. (2008). Holocene accumulation and ice sheet dynamics in central west antarctica. *J. Geophys. Res.* 113:F02018. doi: 10.1029/2007JF000764

Nielsen, L. T., Karlsson, N. B., and Hvidberg, C. S. (2015). Large-scale reconstruction of accumulation rates in northern Greenland from radar data. *Ann. Glaciol.* 56, 70–78. doi: 10.3189/2015AoG70A062

North Greenland Ice Core Project members (2004). High-resolution record of Northern Hemisphere climate extending into the last interglacial period. *Nature* 431, 147–151. doi: 10.1038/nature02805

Ohmura, A., and Reeh, N. (1991). New precipitation and accumulation maps for Greenland. *J. Glaciol.* 37, 140–148.

Panton, C., and Karlsson, N. B. (2015). Automated mapping of near bed radio-echo layer disruptions in the Greenland Ice Sheet. *Earth Planet. Sci. Lett.* 432, 323–331. doi: 10.1016/j.epsl.2015.10.024

Rasmussen, S. O., Abbott, P. M., Blunier, T., Bourne, A. J., Brook, E., Buchardt, S. L., Buizert, C., et al. (2013). A first chronology for the North Greenland Eemian Ice Drilling (NEEM) ice core. *Climate Past* 9, 2713–2730. doi: 10.5194/cp-9-2713-2013

Richardson, C., and Holmlund, P. (1999). Spatial variability at shallow snow-layer depths in central Dronning Maud Land, East Antarctica. *Ann. Glaciol.* 29, 10–16. doi: 10.3189/172756499781820905

Schwager, M. (2000). Ice core analysis on the spatial and temporal variability of temperature and precipitation during the late holocene in north Greenland. *Rep. Polar Res.* 362, 1–136. doi: 10.2312/BzP_0362_2000

Sigl, M., McConnell, J. R., Layman, L., Maselli, O., McGwire, K., Pasteris, D., et al. (2013). A new bipolar ice core record of volcanism from WAIS Divide and NEEM and implications for climate forcing of the last 2000 years. *J. Geophys. Res.* 118, 1151–1169. doi: 10.1029/2012JD018603

Simonsen, S. S., Stenseng, L., Adalgeirsdóttir, G., Fausto, R. S., Hvidberg, C. S., and Lucas-Picher, P. (2013). Assessing a multilayered dynamic firn-compaction model for Greenland with asiras radar measurements. *J. Glaciol.* 59, 545–558. doi: 10.3189/2013JoG12J158

Steinhage, D., Eisen, O., and Clausen, H. B. (2004). Regional and temporal variation of accumulation around North-GRIP derived from ground based ice-penetrating radar. *Ann. Glaciol.* 42, i326–i330. doi: 10.3189/172756405781812574

Vernon, C. L., Bamber, J. L., Box, J. E., Broeke, M. R. V. D., Fettweis, X., Hanna, E., et al. (2013). Surface mass balance model intercomparison for the Greenland ice sheet. *Cryosphere* 7, 599–614. doi: 10.5194/tc-7-599-2013

Waddington, E. D., Neumann, T. A., Koutnik, M. R., Marshall, H.-P., and Morse, D. L. (2007). Inference of accumulation-rate patterns from deep layers in glaciers and ice sheets. *J. Glaciol.* 53, 694–712. doi: 10.3189/002214307784409351

Weißbach, S., Wegner, A., Opel, T., Oerter, H., Vinther, B. M., and Kipfstuhl, S. (2016). Spatial and temporal oxygen isotope variability in northern Greenland – implications for a new climate record over the past millennium. *Clim Past* 12, 171–188. doi: 10.5194/cp-12-171-2016

Wilhelms, F. (1996). Measuring the conductivity and density of ice cores. *Rep. Polar Res.* 191. Available online at: http://epic.awi.de/3394/

Conflict of Interest Statement: The authors declare that the research was conducted in the absence of any commercial or financial relationships that could be construed as a potential conflict of interest.

4

The Changing Impact of Snow Conditions and Refreezing on the Mass Balance of an Idealized Svalbard Glacier

Ward J. J. van Pelt[1]*, Veijo A. Pohjola[1] and Carleen H. Reijmer[2]

[1] Department of Earth Sciences, Uppsala University, Uppsala, Sweden, [2] Institute for Marine and Atmospheric Research Utrecht, Utrecht University, Utrecht, Netherlands

Glacier surface melt and runoff depend strongly on seasonal and perennial snow (firn) conditions. Not only does the presence of snow and firn directly affect melt rates by reflecting solar radiation, it may also act as a buffer against mass loss by storing melt water in refrozen or liquid form. In Svalbard, ongoing and projected amplified climate change with respect to the global mean change has severe implications for the state of snow and firn and its impact on glacier mass loss. Model experiments with a coupled surface energy balance—firn model were done to investigate the climatic mass balance and the changing role of snow and firn conditions for an idealized Svalbard glacier. A climate forcing for the past, present and future (1984–2104) is constructed, based on observational data from Svalbard Airport and a seasonally dependent projection scenario. With this forcing we mimic conditions for a typical inland Svalbard glacier. Results illustrate ongoing and future firn degradation in response to an elevational retreat of the equilibrium line altitude (ELA) of 31 m decade^{-1}. The temperate firn zone is found to retreat and expand, while cold ice in the ablation zone warms considerably. In response to pronounced winter warming and an associated increase in winter rainfall, the current prevalence of refreezing during the melt season gradually shifts to the winter season in a future climate. Sensitivity tests reveal that in a present and future climate the density and thermodynamic structure of Svalbard glaciers are heavily influenced by refreezing. Refreezing acts as a net buffer against mass loss. However, the net mass balance change after refreezing is substantially smaller than the amount of refreezing itself, which can be ascribed to melt-enhancing effects after refreezing, which partly offset the primary mass-retaining effect of refreezing.

Keywords: Svalbard, mass balance, snow, future, climate change, refreezing

Edited by:
Horst Machguth,
University of Zurich, Switzerland

Reviewed by:
Koji Fujita,
Nagoya University, Japan
Kjetil Schanke Aas,
University of Oslo, Norway

*Correspondence:
Ward J. J. van Pelt
ward.van.pelt@geo.uu.se

Specialty section:
This article was submitted to
Cryospheric Sciences,
a section of the journal
Frontiers in Earth Science

1. INTRODUCTION

During the most recent decades, Arctic temperatures have increased at an amplified rate compared to the global mean (ACIA, 2005; IPCC, 2013), which has been ascribed to retreating sea-ice cover and resulting ice—atmosphere feedbacks (Serreze and Francis, 2006; Bintanja and van der Linden, 2013). Additionally, precipitation in the Arctic has increased in recent decades in response to atmospheric moistening (Min et al., 2008; Zhang et al., 2012). Increases in both precipitation and temperature in the Arctic are projected to accelerate during the remainder of the twenty-first century (Bengtsson et al., 2011; Bintanja and Selten, 2014).

The climate on the high-Arctic archipelago of Svalbard, located at the northern end of the Atlantic warm water current, is highly sensitive to variability and trends in winter-time sea-ice extent (Divine and Dick, 2006; Day et al., 2012), as well as changes in atmospheric circulation patterns (Hanssen-Bauer and Førland, 1998). Observed temperatures at Svalbard Airport, Longyearbyen, between 1898–2012 show a marked positive mean trend of 0.26°C decade^{-1} (Nordli et al., 2014). A strong seasonal contrast prevails in observed temperature trends with markedly more pronounced warming in winter/spring than in summer (Førland et al., 2011; Bintanja and van der Linden, 2013; Nordli et al., 2014). Observational precipitation records in Svalbard since the early and mid-twentieth century reveal weakly positive trends, with low significance due to substantial local-scale and instrumental uncertainty (Førland and Hanssen-Bauer, 2000). Projections for Svalbard, based on empirical downscaling, indicate a continued modest precipitation increase and a three times higher warming rate up to the year 2100 than observed over the last 100 years (Førland et al., 2011).

Changing temperature and precipitation conditions affect the energy and mass balance of Svalbard glaciers. Estimates of the multi-decadal mass balance of all Svalbard glaciers, disregarding frontal ablation, are close to zero with values ranging between -0.05 m w.e. a^{-1} for 1979–2013 (Lang et al., 2015) and 0.08 m w.e. a^{-1} for 1957–2014 (Østby et al., in review). Frontal calving of tide-water glaciers is a substantial mass loss term and has been estimated at -0.18 m w.e. a^{-1} for 2000–2006 (Blaszczyk et al., 2009). Altogether, recent warming has enhanced glacier thinning, which has been estimated at -0.59 m a^{-1} for 1961–2005, based on airborne/satellite altimetry for six glaciers around Svalbard (James et al., 2012). As a dynamic response to glacier thinning, glaciers in Svalbard have shrunk by 7% of their total area during the past 30 years for a sample of around 400 glaciers (Nuth et al., 2013).

The interaction between a glacier surface, the atmosphere above and the underlying snow/firn determines the climatic mass balance. The climatic mass balance is defined as the sum of the surface mass balance, due to surface accumulation and ablation, and the internal mass balance (Cogley et al., 2011). Rising temperature induces more surface melting, but the timing and rate of melt water discharge depends strongly on the presence and state of seasonal snow and firn. Snow acts as a buffer against mass loss due to its mirroring effect on solar radiation, which reduces heat absorption and surface melt, and because it can store percolating melt water in either solid (refrozen) or liquid form. Refreezing of percolating and stored water in snow and firn contributes significantly to the mass balance of glaciers and ice sheets and has a pronounced impact on subsurface temperature and density (Jania et al., 1996; Wright et al., 2007; Reijmer et al., 2012). It is generally most pronounced in accumulation zones and peaks in spring when first melt enters the cold snow pack and in fall when the cold wave induces refreezing of stored liquid water.

While there has been much focus on the direct mass-retaining effect of refreezing (e.g., Pfeffer et al., 1991; Reijmer et al., 2012), there has been considerably less attention for the role of indirect effects after refreezing. The primary mass balance effect of refreezing is to retain melt/rain water in refrozen form, thereby reducing the fraction of melt/rain water that runs off. At the same time heat release and solid mass retention by refreezing affect the thermodynamic state and stratigraphy of the snow pack, thereby affecting the conductive heat flux and potentially the surface albedo, which in turn influence surface melt and runoff. A detailed assessment of the impact of refreezing on the mass balance of glaciers requires a coupled treatment of surface and snow processes, and is highly relevant for modeling melt and discharge of Arctic glaciers.

In a warming climate, higher melt rates lead to firn densification and thinning, ultimately causing a retreat of the perennial snow cover to higher elevations. Rapid near-surface densification in lower accumulation zones in south-west Greenland (Machguth et al., 2016) and western Svalbard (Van Pelt and Kohler, 2015) are manifestations of this. Shrinking accumulation zones (1) amplify melt rates in summer due to increased bare-ice exposure (ice—albedo feedback), and (2) reduce the potential for melt water retention by refreezing. Both effects reduce the buffering role of snow and firn against mass loss in a warming climate (e.g., Van Angelen et al., 2013; Van Pelt and Kohler, 2015).

Our aim with this study is to investigate the changing state of snow and firn conditions in Svalbard and its impact on glacier mass loss in a past, present and future climate (1984–2104). Numerical simulations are done with a coupled surface energy balance—firn model (EBFM; Van Pelt et al., 2012; Van Pelt and Kohler, 2015), on a synthetic elevation-dependent grid. We attempt to simulate evolving surface and snow/firn conditions along the centerline of an idealized inland Svalbard glacier. Such a set-up facilitates an analysis with a focus on the process understanding and helps to identify the role of feedback mechanisms. The constructed climate forcing relies on variability from the Svalbard Airport meteorological station, a seasonally-dependent future projection scenario (Førland et al., 2011), and observation-based elevation lapse rates. The model, climate forcing and experimental setup are described in Section 2. Model results are analyzed and discussed in Section 3. Finally, conclusions and recommendations are given in Section 4.

2. METHODS

2.1. Model

A coupled energy balance—firn model (EBFM) is used to perform simulations on a synthetic elevation dependent grid. Here, we briefly summarize the main qualitative model aspects; for a more complete overview of the coupled model and its implementation, see Van Pelt et al. (2012) and Van Pelt and Kohler (2015) and references therein. The model has previously been used to study the mass balance and firn evolution of Nordenskiöldbreen in central Svalbard (Van Pelt et al., 2012, 2013, 2014; Vega et al., 2016), and Kongsvegen and

Holtedahlfonna in western Svalbard (Christianson et al., 2015; Van Pelt and Kohler, 2015).

Driven by meteorological fields, at every model time-step a surface energy balance model sums the heat fluxes at the surface in order to determine the surface temperature and the energy involved in melting:

$$Q_{melt} = Q_{sw} + Q_{lw} + Q_{turb} + Q_{ghf} + Q_{rain}, \qquad (1)$$

where Q_{melt} is the energy available for melt, Q_{sw} is the net shortwave radiation, Q_{lw} is the net longwave radiation, Q_{turb} is the turbulent heat exchange by the latent and sensible heat flux, Q_{ghf} is the conductive ground heat flux, and Q_{rain} is the energy supplied by rainfall. The surface temperature is limited to 0°C, any excess energy results in melting of the surface layer ($Q_{melt} > 0$). Formulations of the individual energy fluxes can be found in Klok and Oerlemans (2002); a qualitative description of the individual fluxes is given below.

In order to solve the surface energy balance climate input of air temperature, relative humidity, cloud cover and precipitation is required by this model. Cloud cover, temperature and relative humidity affect the incoming shortwave radiation, which depends on the top-of-atmosphere radiation, atmospheric transmissivity and the grid orientation (Klok and Oerlemans, 2002). Reflected shortwave radiation is determined by the albedo, following a snow depth and age dependent formulation (Oerlemans and Knap, 1998). Incoming longwave radiation is formulated as a function of cloud cover, relative humidity and air temperature (Konzelmann et al., 1994), while the Stefan-Boltzmann law for a black body describes emitted longwave radiation. The description of turbulent sensible and latent heat exchange uses bulk equations by Oerlemans and Grisogono (2002), requiring climate input of air temperature and relative humidity, respectively. The ground heat flux represents conduction between the surface skin layer and the underlying snow/ice and can take both signs depending on the near-surface temperature gradient. Finally, the heat supplied by rainfall depends on air temperature and is generally a small term. The model distinguishes between snow and rainfall by using a gradual (linear) transition from snow to rain between air temperatures of 0.5 and 2.5°C; hence, at a temperature of 1.5°C precipitation consists of 50% rain and snow (Van Pelt et al., 2012).

Energy and mass exchange at the surface provides an upper boundary condition for the subsurface model, tracking the evolution and vertical distribution of temperature, density and water content. The multi-layer subsurface model is based on Reijmer and Hock (2008) and later modified and extended by Van Pelt et al. (2012). The energy consumed in melting the surface may be released again as latent heat below the surface when percolating or stored water refreezes in snow or firn layers. The subsurface model solves the thermodynamic equation, accounting for heat conduction and refreezing:

$$\rho c_p(T) \frac{\partial T}{\partial t} = \frac{\partial}{\partial z}\left(\kappa(\rho)\frac{\partial T}{\partial z}\right) + \frac{RL}{d}, \qquad (2)$$

where T is the layer temperature, ρ the layer density, $c_p(T)$ the specific heat capacity, $\kappa(\rho)$ the effective conductivity, R the

refreezing rate (kg m^{-2} s^{-1}), L the latent heat of fusion and d the layer thickness. The subsurface model additionally solves the densification equation, considering refreezing and gravitational packing of the snow $G(\rho, T)$ (Ligtenberg et al., 2011):

$$\frac{\partial \rho}{\partial t} = G(\rho, T) + \frac{R}{d}. \qquad (3)$$

Vertical water transport and refreezing is simulated using a bucket scheme (Greuell and Konzelmann, 1994; Reijmer and Hock, 2008; Van Pelt et al., 2012). Storage of percolating water occurs in two forms: (1) on its way down a small amount of water is held in temperate layers against capillary forces (irreducible water), and (2) water accumulates on top of impermeable ice to form a layer of slush water, thereby filling the remaining pore space. Runoff occurs at the firn-ice transition and is instantaneous in the case of bare-ice exposure and according to a slope-dependent time-scale for standing slush water within the snow-pack (Reijmer and Hock, 2008).

The climatic mass balance is a direct measure for mass changes resulting from the interaction of atmosphere, glacier surface and the underlying snow/firn (Cogley et al., 2011). It is therefore effectively a measure for mass change at and below the surface, down to the base of the snow or firn pack (if present). The climatic mass balance is defined here as the sum of mass gain, through precipitation and condensation/riming, and mass loss, through runoff and evaporation/sublimation. Refreezing may act as a buffer against mass loss where a firn/snowpack is present, as it reduces the amount of melt water that runs off. In our experiments the climatic mass balance is assumed to be equal to the sum of mass changes above the base of the model domain at 15–20 m depth; any potential refreezing below this depth is regarded as negligible.

2.2. Climate Forcing
2.2.1. Temporal Variability
In order to run the coupled model, meteorological input of precipitation, air temperature, cloud cover and relative humidity is required at a sub-daily temporal resolution. Meteorological data of the above parameters, collected at the coastal meteorological station at Svalbard Airport at 28 m a.s.l., Longyearbyen (Norwegian Meteorological Institute, eKlima.no) are used as time-series for 1975–2014. Six-hourly observations of temperature, cloud cover and relative humidity were interpolated to the 3-h model resolution, whereas observed 12-h precipitation sums were split into 3-h bins. The average temperature between 1 September 1975 and 31 August 2014 was equal to −5.0 °C, precipitation summed to 191 mm a^{-1}, mean relative humidity was equal to 74%, and average cloud cover was 64%.

Time-series of winter (DJF), spring (MAM), summer (JJA) and autumn (SON) mean observed temperature and precipitation between 1975–2014 (first 39 years in **Figure 1**), show a clear inhomogeneity in observed seasonal temperature trends. Over the observation period 1975–2014, winter warming (2.2°C decade^{-1}) exceeds summer warming (0.5°C decade^{-1}) by a factor of four. Much stronger winter than summer warming, referred to as Arctic winter warming (AWW), is one of the key

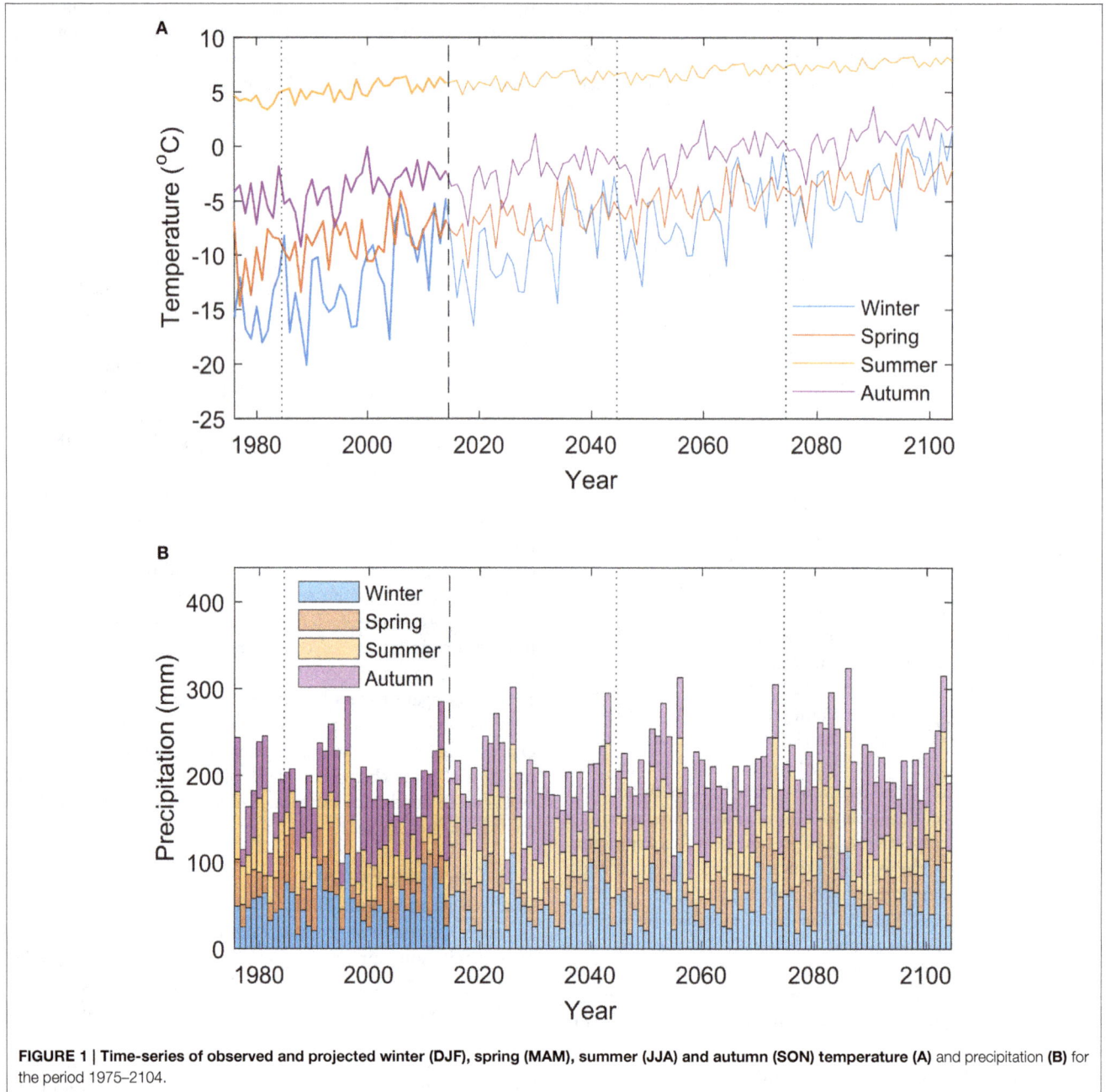

FIGURE 1 | Time-series of observed and projected winter (DJF), spring (MAM), summer (JJA) and autumn (SON) temperature (A) and precipitation **(B)** for the period 1975–2104.

features of Arctic Amplification and has been ascribed to the retreat of sea ice and the resulting ice—atmosphere feedbacks (Serreze and Francis, 2006; Bintanja and van der Linden, 2013).

Førland et al. (2011) used the NorACIA (Norwegian Arctic Climate Impact Assessment) regional climate model, with input from six atmosphere ocean general circulation model (AOGCM) simulations for a range of emission scenarios (A1B, A2, and B2), to simulate the past (1961–1990) and future (2071–2100) climate in the Svalbard region. The ensemble results present seasonal temperature and precipitation trends between 1961–1990 and 2071–2100, yielding linear seasonal temperature trends for Svalbard Airport of 1.0 (DJF), 0.6 (MAM), 0.3 (JJA) and

0.5°C decade^{-1} (SON) and precipitation trends of 0.4 (DJF), 2.6 (MAM), 1.0 (JJA), and 1.9 % decade^{-1} (SON) (Førland et al., 2011).

In addition to seasonal inhomogeneity in warming, at intra-seasonal time-scales (days to weeks) cold spells are found to warm faster than warm spells. This is apparent when comparing trends per season of above-average (upper 50%) and below-average (lower 50%) temperatures in the 1975–2014 Svalbard Airport record, as shown in **Figure 2**. Below-average temperatures were found to increase 1.5 (DJF), 2.1 (MAM), 2.1 (JJA), and 1.6 (SON) times faster than above-average temperatures, which indicates more rapid warming of cold spells for all seasons.

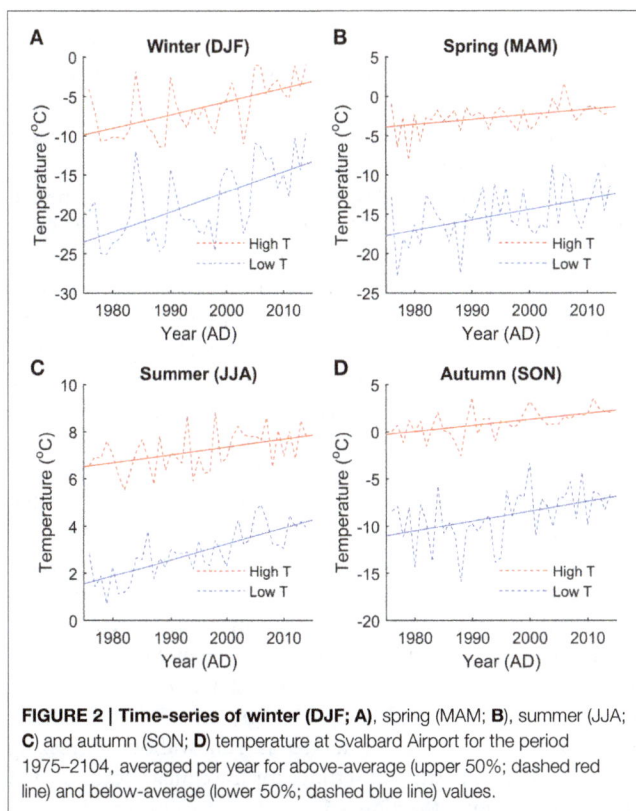

FIGURE 2 | Time-series of winter (DJF; A), spring (MAM; **B**), summer (JJA; **C**) and autumn (SON; **D**) temperature at Svalbard Airport for the period 1975–2104, averaged per year for above-average (upper 50%; dashed red line) and below-average (lower 50%; dashed blue line) values.

In order to extend the 1975–2014 climate forcing into the future, we first repeat variability of temperature, precipitation, cloud cover and relative humidity, observed at Svalbard Airport between 1984–2014, three times to cover the period 2014–2104. In a next step, we impose seasonal trends in temperature and precipitation, according to Førland et al. (2011). Finally, for the temperature projection we additionally account for the aforementioned inhomogeneity in warming of cold and warm spells, as observed in the Svalbard Airport record. More specifically, the following steps are applied to generate future time-series for temperature, precipitation, cloud cover and relative humidity:

1. Observed 3-h time-series of temperature, precipitation, cloud cover and relative humidity during the 30-year reference period (1984–2014), are repeated three times into the future to cover the 90-year period from 1 September 2014 to 31 August 2104.
2. Seasonal trends of temperature and precipitation in Førland et al. (2011) are imposed on the future time-series. Seasonally inhomogeneous trends are already present in the reference climate (1984–2014) and thus also within the repeated future 30-year blocks. Therefore, for every season corrections for seasonally inhomogeneous warming are applied per 30-year block, i.e., an entire 30-year block is shifted to conform with the change applicable for the midpoint of the 30-year period. After trend application, mean seasonal temperatures between 2015 and 2104 increase by 8.6 (DJF), 5.2 (MAM), 2.3 (JJA), and 4.6 °C (SON), and seasonal precipitation rates rise by 3 (DJF),

23 (MAM), 9 (JJA), and 17% (SON), in line with projections in Førland et al. (2011).
3. Inhomogeneity of warming of cold and warm spells within seasons during the observational period (**Figure 2**), is assumed to persist in a future climate. A time-dependent temperature correction is implemented by multiplying the seasonal trends applied in step 2 by a time-dependent function, which is <1 for above-average temperatures and >1 for below-average temperatures. The function is equal to $1 + \Delta T(t) * C_T$, where $\Delta T(t)$ is the temperature deviation at time-step t from the 30-year running seasonal mean, and C_T (in K^{-1}) is a seasonally-dependent coefficient, which represents the fractional change of the seasonal trend per degree temperature deviation from the 30-year running seasonal mean. With seasonal values of C_T of -0.034 (DJF), -0.057 (MAM), -0.160 (JJA), and -0.048 K^{-1} (SON) continuation of the observed inhomogeneity at Svalbard Airport between 1975–2014 (**Figure 2**) is assured in the future temperature scenario.

The resulting time-series of seasonally averaged temperature and precipitation for 1975–2104 are shown in **Figure 1**. The projected climate scenario includes changes in temperature and precipitation variability. Stronger winter than summer warming reduces the magnitude of the annual temperature cycle (**Figure 1A**). Additionally, more pronounced warming of cold than warm spells leads to a reduction of short-term (intra-seasonal) temperature variability. The latter is also responsible for the decrease of inter-annual variability in seasonal temperatures (**Figure 1A**), since relatively warm seasons (compared to the 30-year mean) will experience a weaker warming trend than relatively cold seasons. Finally, seasonal precipitation trends cause the magnitude of individual precipitation events to increase, while the frequency of precipitation events remains unchanged in the projected climate.

The above described approach to generate meteorological time-series provides a strong alternative to using climate output from a single future realization of a RCM, which is prone to substantial uncertainty given the generally large spread between individual RCM projection runs. The current approach also has some limitations. While we incorporate expected changes in the magnitude of temperature and precipitation variability, no potential long-term changes in the frequency and distribution of weather events are accounted for. For example, an increased frequency of heavy-precipitation events in the Northern hemisphere has been suggested (Min et al., 2011), although a recent analysis of extreme precipitation events in Svalbard reveals no significant trends over the most recent decades (Serreze et al., 2015). Furthermore, we do not account for potential trends in cloud cover and relative humidity, due to a current lack of constraints of long-term trends of these variables. This is believed to have a minor impact on the presented results, since (summer) temperature and precipitation are assumed to be the dominant factors influencing the climatic mass balance and snow/firn development. To test sensitivity of the model results to different climate scenarios, a set of climate sensitivity experiments is performed (Section 3.5).

2.2.2. Elevation Dependence

Experiments are performed on an elevation-dependent model grid, representing the centerline of a typical inland Svalbard glacier. Constructed time-series of temperature, precipitation, cloud cover and relative humidity (Section 2.2.1) apply at sea-level elevation. An elevation-dependent climate forcing is constructed by applying lapse rates for temperature and precipitation, using values typical for Svalbard glaciers. Based on previous estimates of temperature lapse rates in Svalbard which range between $-0.0040°C$ m^{-1} (Gardner et al., 2009) and $-0.0066°C$ m^{-1} (Nuth et al., 2012), we use a constant mean temperature lapse rate of $-0.0053°C$ m^{-1}. We apply an average precipitation lapse rate of 0.97 mm w.e. a^{-1} m^{-1}, based on GPR measurements in 1997–1999 in different regions on Svalbard (Sand et al., 2003; Winther et al., 2003). Winter stake balance measurements on Holtedahlfonna in western Svalbard (Baumberger, 2007; Nuth et al., 2012; Van Pelt and Kohler, 2015) and Nordenskiöldbreen in central Svalbard (Van Pelt et al., 2012, 2014) reveal that precipitation remains constant above an approximate elevation of 800–1000 m a.s.l. Based on this, we set the elevation above which the precipitation lapse rate reduces to zero at 900 m a.s.l. No elevation lapse rates were applied for relative humidity and cloud cover.

2.3. Model Setup and Experiments

Forced with 3-h climate input (Section 2.2), we perform time-dependent experiments with EBFM (Section 2.1), on a model grid covering an altitudinal range of 0–1500 m a.s.l. An altitudinal grid spacing of 25 m is used. The subsurface model uses a moving grid consisting of 200 vertical layers with a maximum layer thickness of 10 cm; depending on surface mass fluxes layers will be added or removed at the surface and base of the vertical model domain (Van Pelt and Kohler, 2015). In contrast to a fixed grid, a moving grid avoids averaging of layer properties and assures layer properties do not suffer from numerical diffusion. The model uses a 3-h time-step, except in the calculation of subsurface heat conduction, for which adaptive time-stepping is used to assure numerical stability. In the calculation of the incoming solar radiation, we assume the surface to be flat, we neglect any shading by surrounding topography, and set the geographic position to the approximate location of Longyearbyen in central Svalbard (78°12′ N, 15°26′ E). We adopt the same model parameter setup for simulating the energy balance and subsurface conditions as in Van Pelt and Kohler (2015). In Van Pelt and Kohler (2015) surface energy balance parameters have been calibrated to optimize model performance against stake and weather station data on glaciers around Kongsfjorden in western Svalbard. Calibrated constants include the aerosol transmissivity coefficient (affects incoming solar radiation), fresh snow and ice albedo (affects reflected solar radiation), emissivity exponents for clear-sky and overcast conditions (affects incoming longwave radiation), and the turbulent exchange coefficient (affecting turbulent heat exchange). Simulated subsurface density and temperature profiles have been validated against observational

data in Van Pelt et al. (2014) and Van Pelt and Kohler (2015). We do not account for any glacier dynamics, i.e., zero horizontal ice flow is assumed, which implies slight potential offsets (increasing with depth) of the presented vertical profiles. Given the shallow depth of the vertical model (up to 20 m), we believe the impact of ice flow on our results is likely to be small.

Experiments performed with EBFM cover the 120-year period between 1 September 1984 and 31 August 2104. Realistic simulation of subsurface conditions in the first years of the experiments requires initialized subsurface temperature, density and water content conditions at the start of the simulations (Van Pelt et al., 2012; Van Pelt and Kohler, 2015). To generate initial conditions, we perform a spin-up procedure in which the model is looped three times over the spin-up period 1975–1984, with meteorological input at sea-level from the Svalbard Airport record and lapse rates given in Section 2.2.2. Final output of an initialization run is used as input for the next iteration. During the 27-year spin-up procedure a snow and firn pack develops and deep density, temperature and liquid water storage approach a state consistent with the mean surface forcing between 1975–1984. Final output of the third initialization run provides initial conditions for the experiments starting 1 September 1984. Long-term runs done with EBFM include a standard run, using the above-described standard model settings, and a set of sensitivity experiments, in which the model is run without refreezing (Section 3.4) and perturbed settings for the climate scenario and elevation lapse rates (Section 3.5).

3. RESULTS AND DISCUSSION

In this section we will present results of the model experiments. We start with a discussion of the long-term subsurface evolution at a site in the lower accumulation zone (Section 3.1). Next, we present elevation profiles of the mass balance, refreezing and subsurface conditions (Section 3.2). Thereafter, the changing role of refreezing in a future climate is discussed (Section 3.3), followed by an assessment of the impact of refreezing on the mass balance and subsurface conditions (Section 3.4). Finally, we present the outcome of climate sensitivity experiments (Section 3.5).

3.1. Subsurface Evolution

Increasing melt and rain water percolation in snow and firn in a future climate impacts the subsurface thermodynamics, stratigraphy and water storage. To illustrate the consequences of ongoing and projected climate conditions on the state of snow/firn, we first focus on the simulated firn evolution for a site at an elevation of 800 m a.s.l. for the periods 2010–2014, 2040–2044, 2070–2074, and 2100–2014 (**Figure 3**).

In a present-day climate, the site at 800 m a.s.l. is located in the lower accumulation zone and a thick (>10 m) firn pack is present at this elevation (**Figure 3A**). Heat release by refreezing of percolating water quickly removes the winter cold content at the start of the melt season (**Figure 3B**). Melt-freeze cycles during the melt season cause rapid near-surface densification, leading to the formation of a summer surface with elevated density every melt season (**Figure 3A**). At this elevation, melt water percolates

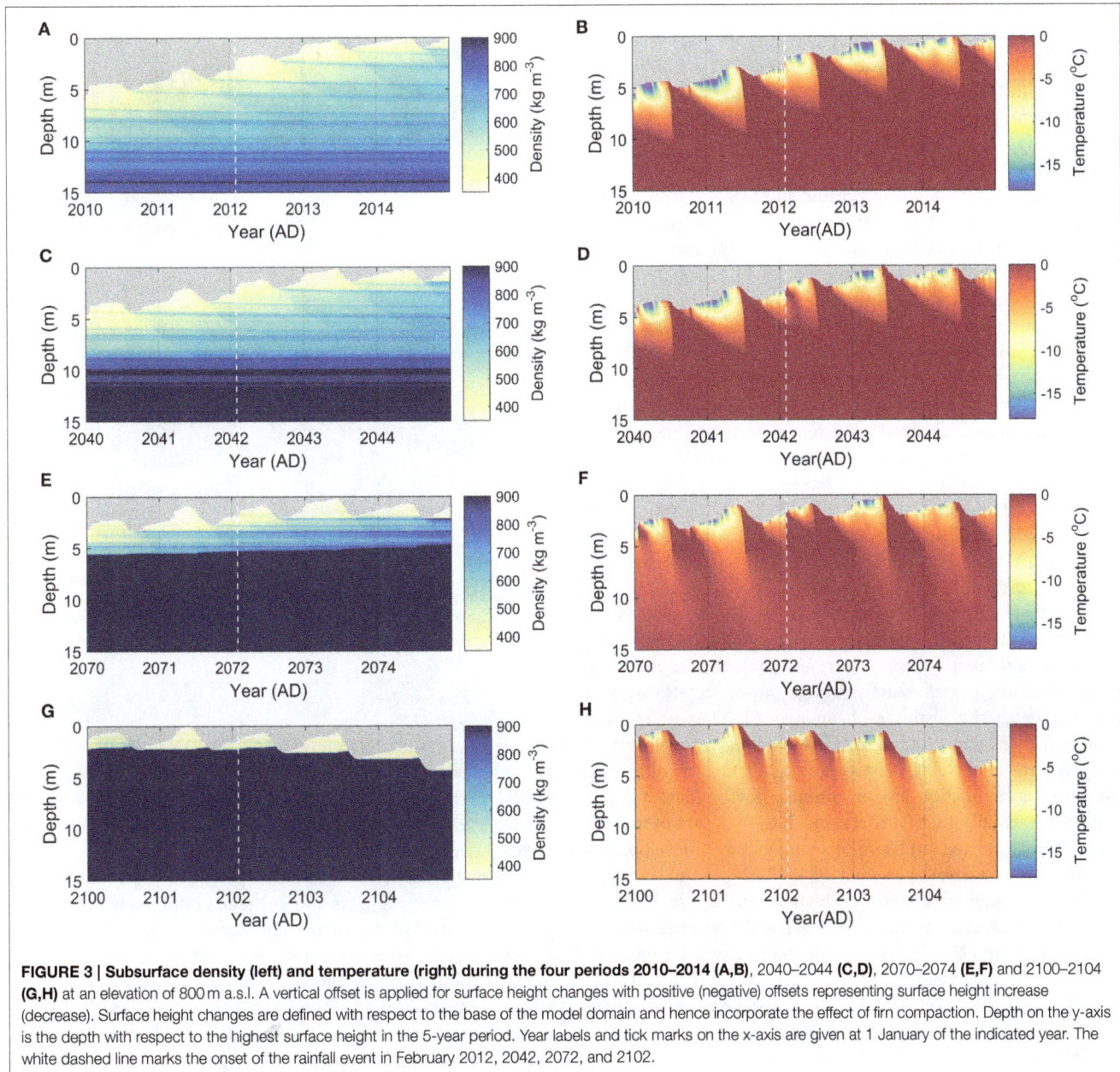

FIGURE 3 | Subsurface density (left) and temperature (right) during the four periods 2010–2014 (A,B), 2040–2044 **(C,D)**, 2070–2074 **(E,F)** and 2100–2104 **(G,H)** at an elevation of 800 m a.s.l. A vertical offset is applied for surface height changes with positive (negative) offsets representing surface height increase (decrease). Surface height changes are defined with respect to the base of the model domain and hence incorporate the effect of firn compaction. Depth on the y-axis is the depth with respect to the highest surface height in the 5-year period. Year labels and tick marks on the x-axis are given at 1 January of the indicated year. The white dashed line marks the onset of the rainfall event in February 2012, 2042, 2072, and 2102.

throughout the firn column every summer season (**Figure 3B**), filling a fraction of the pore space with irreducible water. During the winter season, refreezing occurs at increasing depth as stored irreducible water gradually freezes in response to surface and firn cooling. Winter-time cooling only removes stored water by refreezing in the first meters, which causes the deeper firn to remain temperate (**Figure 3B**). Firn densification in the first few meters is dominated by the effect of refreezing; deeper down, gravitational settling becomes a dominant factor (**Figure 3A**).

In a future climate, higher summer melt rates and continued densification by refreezing and settling cause the firn pack to rapidly thin in the first decades and to disappear by the end of the twenty-first century (**Figures 3C,E,G**). Toward the end of the twenty-first century, this site at 800 m a.s.l. turns into

ablation area, resulting in a net annual surface height lowering, absence of firn, and seasonal snow accumulation on top of the impermeable ice in winter (**Figure 3G**). The shallow firn—ice transition depth in the last decades limits deep water storage and refreezing, which causes deep temperatures to become non-temperate (**Figure 3H**). In addition to the lack of refreezing, the higher conductivity of ice compared to snow promotes more rapid winter-time cooling. Warm spells in winter and associated rainfall cause (additional) winter-time refreezing, removing part of the cold content in the upper meters. A clear example of this is seen in February 2012, 2042, 2072, and 2102; heavy winter rainfall causes deep percolation and refreezing, thereby substantially affecting subsurface temperatures (**Figures 3D,F,H**) and snow stratigraphy in subsequent months (**Figures 3C,E,G**).

With winter warming being more pronounced than summer warming, the frequency of winter rainfall events is projected to increase, having a major impact on the seasonal distribution of refreezing and the thermodynamic state of snow and firn, as will be further discussed in Section 3.3.

3.2. Elevation Profiles

Next, we discuss the mass balance and subsurface conditions for an elevation range of 0 to 1500 m a.s.l. Time-series of deep firn temperature (at 15-m depth) and bulk density (mean of first 10 m) are shown in **Figure 4**. Time-series of ELA, superposed in **Figure 4**, indicate a slightly non-linear increase from 500–600 m a.s.l. during the reference period (1984–2014) to 800–900 m a.s.l. during the final 30 years of the simulation (2074–2104). A mean linear trend of 31 m a.s.l. decade^{-1} was found, summing to a total ELA increase of 367 m a.s.l. over the simulation period. A comparison of ELA during the reference period to regional values presented in Hagen et al. (2003) and Möller et al. (2016) reveals that simulated ELA is more representative of the relatively dry inland regions in central Spitsbergen, than the wet coastal regions. This can in part be explained by the use of climate input from Svalbard Airport, located in central Spitsbergen, as a sea-level forcing.

The thermodynamic structure of Svalbard glaciers is strongly influenced by latent heat release after refreezing, which is dominant in the lower accumulation zone where deep percolation and water storage can occur ((Van Pelt and Kohler, 2015); Østby et al., in review). The current and projected thermodynamic structure with cold deep subsurface temperatures in the ablation area and temperate conditions in the lower accumulation area is common for Svalbard glaciers (Björnsson et al., 1996; Pettersson, 2004). It is noteworthy that the simulated subsurface temperature development during the reference period (1984–2014) resembles what has recently been observed at a high elevation site (1200 m a.s.l.) on Nordenskiöldbreen in central Spitsbergen. Observed temperatures at 12 m depth in a medium-length borehole in 1998 (Van de Wal et al., 2002) and in a shallow borehole in 2012 (Van Pelt et al., 2014) revealed a warming from cold to temperate conditions, which is in line with the simulated results shown in **Figure 4A**.

At high elevations in the accumulation zone the percolation depth of melt/rain water is limited and the winter cold wave is sufficiently strong to refreeze all stored water in the course of the winter season. This is not the case in the lower accumulation zone, where water stored deep in the firn (at more than a few meter depth) may survive the winter cooling. At these elevations temperate deep firn conditions remain and in case discharge through crevasses and moulins is low, deep firn water reservoirs (perennial firn aquifers) can potentially develop on top of the impermeable ice. As shown by Kuipers Munneke et al. (2014) the formation of deep perennial temperate firn requires moderate to high melt rates and high accumulation rates. After the discovery of perennial firn aquifers in Greenland (Forster et al., 2013), similar deep firn water reservoirs have recently been detected in ground-penetrating radar data in Svalbard in the accumulation zones on Lomonosovfonna (R.

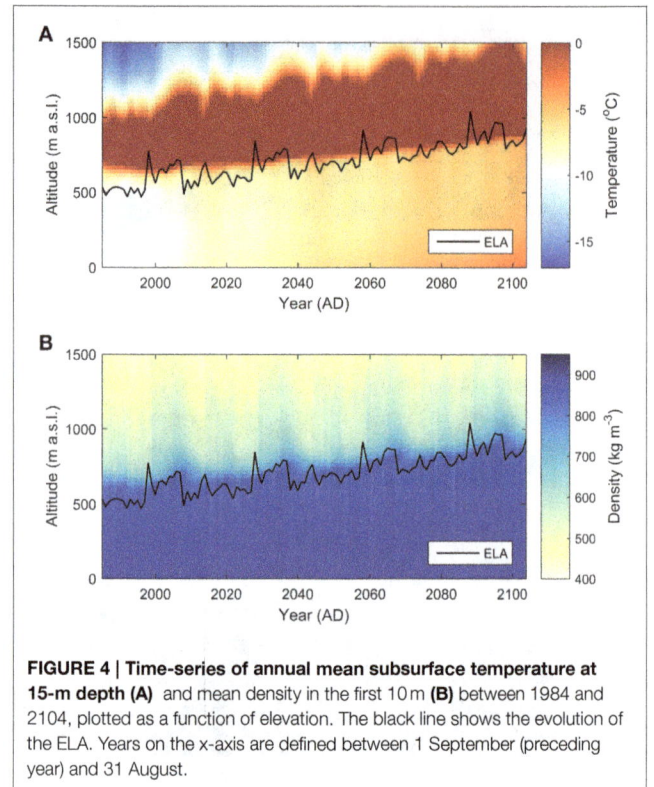

FIGURE 4 | Time-series of annual mean subsurface temperature at 15-m depth (A) and mean density in the first 10 m **(B)** between 1984 and 2104, plotted as a function of elevation. The black line shows the evolution of the ELA. Years on the x-axis are defined between 1 September (preceding year) and 31 August.

Pettersson, personal communication, 2016) and Holtedahlfonna (Christianson et al., 2015). Without appropriate physics in the model to simulate firn aquifer build-up and dynamics, we limit ourselves to identifying locations of potential firn aquifer development.

The projected subsurface development in **Figure 4A** reveals substantial warming of deep (ice) temperatures in the ablation area in response to increasing surface temperatures. The elevation band with temperate firn conditions in the lower accumulation zone is found to (1) migrate upwards in response to ELA retreat, and (2) to cover a larger elevation range in a warming climate (**Figure 4A**). Increased melt rates cause firn degradation and gradual depletion (**Figure 4B**), causing the firn line to migrate upwards with a delay of typically a few years relative to the rising ELA. The time-delay between ELA rise, firn line retreat and upward extension of the temperate firn zone is an indicator of the sensitivity of firn conditions to changing surface conditions. Widening of the temperate firn zone (**Figure 4A**) can be explained by an increased significance of rainfall at high elevations, as well as to preferential winter warming, causing a reduction of subsurface cold content at the start of the melt season.

Elevation profiles of mass balance for the periods 1984–2014, 2014–2044, 2044–2074, and 2074–2104 in **Figure 5A** reveal slightly accelerating upward migration of the ELA. The mean ELA equals 564 m a.s.l. for 1984–2014, 644 m a.s.l. for 2014–2044, 733 m a.s.l. for 2044–2074, and 831 m a.s.l. for 2075–2104. In line with the accelerated ELA rise, the mass balance sensitivity in the ablation area is found to increase in a future climate causing

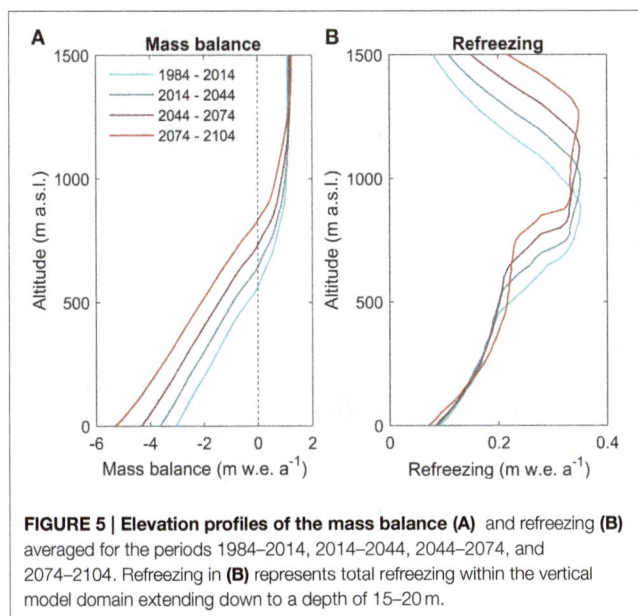

FIGURE 5 | Elevation profiles of the mass balance (A) and refreezing **(B)** averaged for the periods 1984–2014, 2014–2044, 2044–2074, and 2074–2104. Refreezing in **(B)** represents total refreezing within the vertical model domain extending down to a depth of 15–20 m.

accelerated mass loss toward the end of the simulation period (**Figure 5A**). Elevation profiles of refreezing (**Figure 5B**) show values ranging between 0.07 and 0.35 m w.e. a^{-1} for all periods, with patterns consistently showing a maximum of refreezing in the lower accumulation zone, in line with previous findings for all Svalbard glaciers (Østby et al., in review) and for individual glaciers in central Svalbard (Van Pelt et al., 2012) and western Svalbard (Van Pelt and Kohler, 2015). A similar distribution with a maximum of refreezing in the accumulation zone has also been shown in other regions, such as on the Tibetan Plateau (Fujita et al., 1996). Comparing height-profiles of refreezing between periods reveals an upward shift of the maximum in time in line with the projected firn line retreat. Widening in time of the elevation band with highest refreezing rates is apparent and explains the aforementioned extension of the temperate firn zone (**Figure 4A**).

3.3. Seasonality of Refreezing

Next, we discuss changes in the seasonal pattern of refreezing. The seasonal distribution of refreezing at elevations of 300, 600, and 900 m a.s.l., averaged over three different periods (1984–2014, 2029–2059, and 2074–2104), is shown in **Figure 6**. During the reference period (1984–2014, **Figure 6A**), refreezing of melt or rain water is typically most pronounced during the melt season when first melt enters cold seasonal snow or firn, and due to melt-freeze cycles following the daily cycle in temperature and insolation (Colbeck, 1987; Pfeffer and Humphrey, 1998). In the accumulation zone at 900 m a.s.l., a substantial fraction of the total refreezing occurs outside the melt season (**Figure 6A**) and can be ascribed to refreezing of stored irreducible water, continuing throughout the winter season. During the reference period, small amounts of winter-time refreezing occur at 300 m a.s.l. and can be attributed to rare rainfall events. During the melt season, refreezing at 300 m a.s.l. is mostly concentrated in June-July, when first melt enters the initially cold seasonal snow pack,

which disappears in the course of the melt season; higher up, refreezing continues throughout the melt season (**Figure 6A**).

In a future climate, refreezing is projected to become less substantial during the melt season and more significant during the winter season at all three elevations (**Figures 6B,C**). Pronounced winter warming leads to an increased frequency of warm spells with positive temperature excursions, causing increased rainfall occurrence in winter. Latent heat release after refreezing of rain water in winter snow causes snow and firn to warm considerably, which has major implications for the thermodynamic profile and substantially reduces the cold content at the start of the melt season. As a result, refreezing during the melt season will decrease at the expense of winter-time refreezing. An example of this is shown in **Figure 3** after a heavy rainfall event in February 2012, 2042, 2072, and 2102.

In contrast to rainfall, surface melting in a warming climate remains very small in winter, mostly due to low or absent solar radiation. For 2074–2104, the ratio of summed melt to summed rainfall between November and March ranges from 3% at 900 m a.s.l. to 20% at 300 m a.s.l., respectively, which confirms that increased future winter-time refreezing is mostly induced by rain events.

3.4. Impact of Refreezing on Subsurface Conditions and Mass Balance

Refreezing of melt and rain water in snow and firn is known to exert a major influence on subsurface temperature and density profiles, and it has been shown to provide a substantial buffer against mass loss of Arctic glaciers (e.g., Wright et al., 2007; Van Angelen et al., 2013). In order to investigate the impact of refreezing on the mass balance, as well as subsurface thermodynamics and stratigraphy, we performed an additional experiment (1984–2104) in which the effects of refreezing are ignored. More specifically, in this experiment we do not allow for any solid mass retention and heat release by refreezing. As a result, all available melt and rain water at the surface will eventually run off; liquid water storage as irreducible water or slush water may still occur and delays runoff. Results of the two runs, averaged over the whole simulation period, are compared in **Figure 7**.

Elevation profiles for the refreezing and no-refreezing runs show a major impact of refreezing on temperatures at 15-m depth, particularly in the lower accumulation zone (**Figure 7A**). Without refreezing subsurface temperature is only governed by heat conduction and the vertical mass flux and shows a nearly linear decrease with elevation as a result of atmospheric cooling with elevation. Differences of up to 10°C are apparent in the lower accumulation zone. In the ablation area, the impact of refreezing on 15-m subsurface temperature is negligible. Here, heat released by refreezing in seasonal snow is to a large extent removed through melting and discharge of the snow layer.

Without refreezing the vertical density profile in the accumulation zone is determined by gravitational settling and surface accumulation. Solid mass retention by refreezing increases long-term mean bulk density in the upper 10 m by up to 117 kg m^{-3} in the lower accumulation zone (**Figure 7B**).

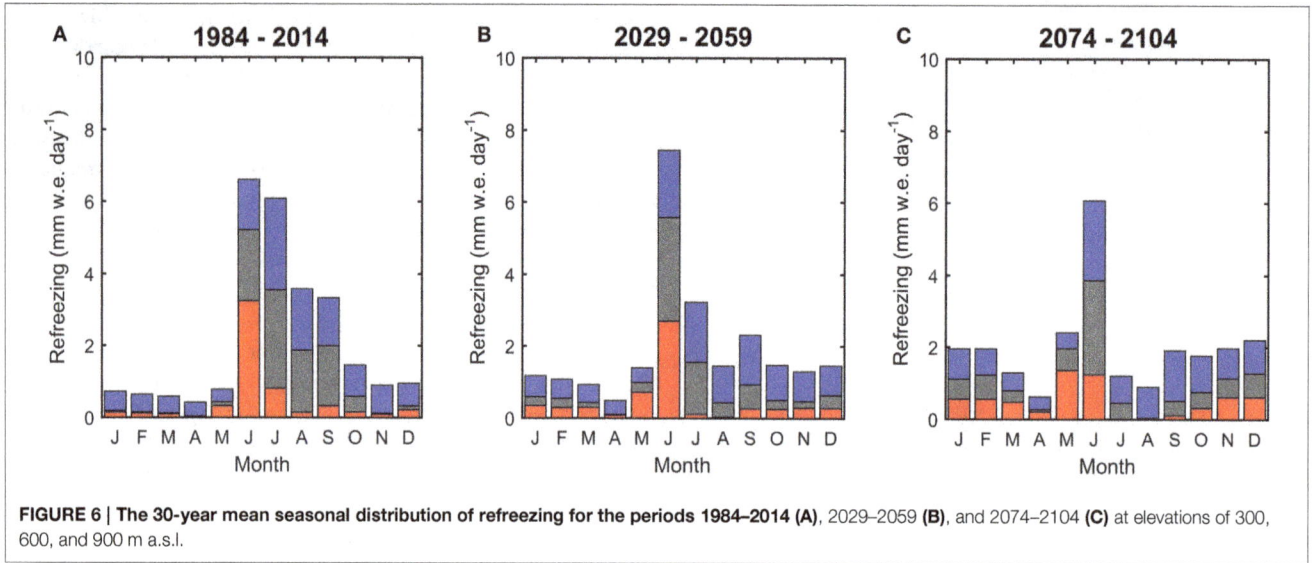

FIGURE 6 | The 30-year mean seasonal distribution of refreezing for the periods 1984–2014 (A), 2029–2059 **(B)**, and 2074–2104 **(C)** at elevations of 300, 600, and 900 m a.s.l.

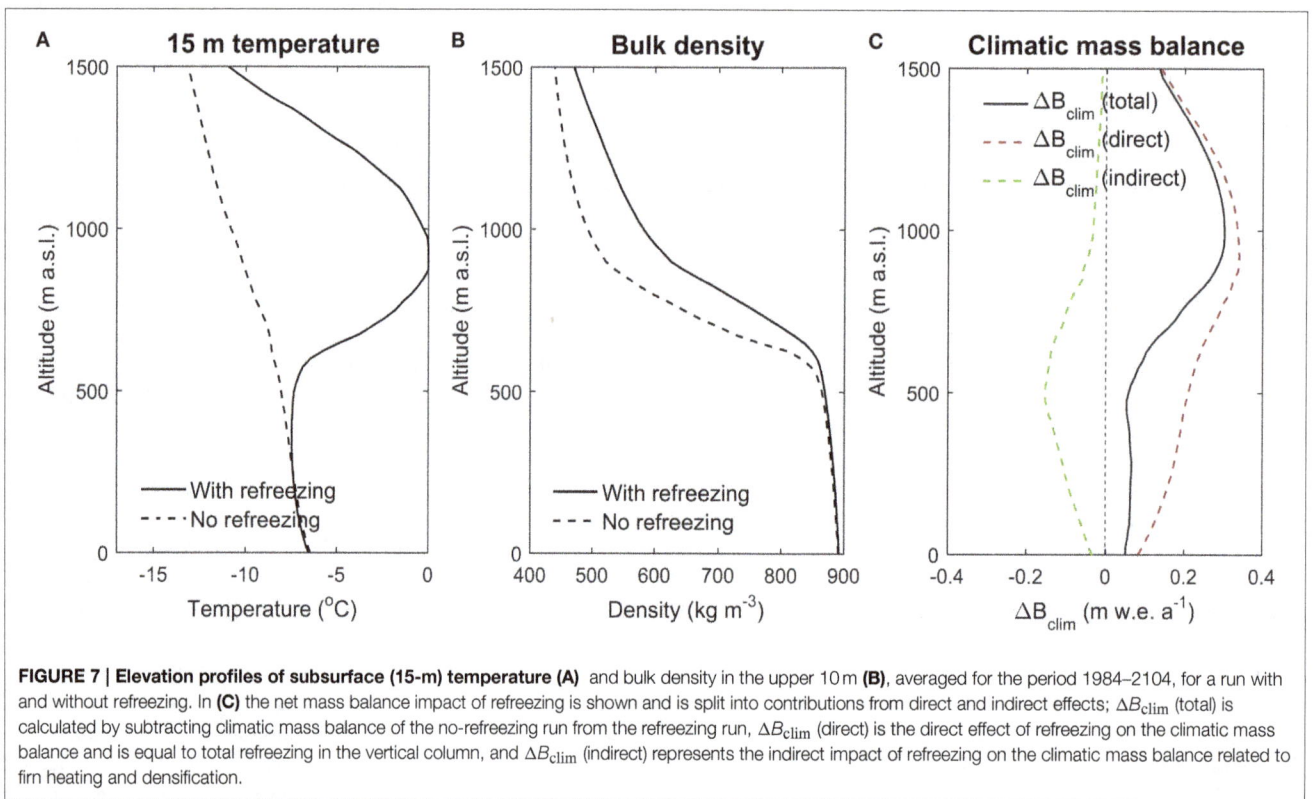

FIGURE 7 | Elevation profiles of subsurface (15-m) temperature (A) and bulk density in the upper 10 m **(B)**, averaged for the period 1984–2104, for a run with and without refreezing. In **(C)** the net mass balance impact of refreezing is shown and is split into contributions from direct and indirect effects; ΔB_{clim} (total) is calculated by subtracting climatic mass balance of the no-refreezing run from the refreezing run, ΔB_{clim} (direct) is the direct effect of refreezing on the climatic mass balance and is equal to total refreezing in the vertical column, and ΔB_{clim} (indirect) represents the indirect impact of refreezing on the climatic mass balance related to firn heating and densification.

Averaged over the simulation period and the elevation range 900–1500 m a.s.l., i.e., above the maximum firn line elevation, we find relative contributions to total densification of compaction and refreezing of 64 and 36%, respectively.

Refreezing of percolating or stored water reduces the amount of melt and rain water that runs off, which implies a positive direct impact on the mass balance. However, due to indirect effects after refreezing, as a result of firn heating and densification, the impact of refreezing on the climatic mass

balance is not necessarily equal to the amount of refreezing itself. Firstly, heat release during refreezing increases subsurface temperatures, inducing a more positive (or less negative) conductive heat flux at the surface, with the positive flux direction defined toward the surface. A more positive conductive heat flux at the surface may result in higher melt rates as surface temperatures are more readily raised to melting point. Secondly, refreezing adds solid mass to the snow pack and may create ice layers, which, when exposed at the surface, lowers the surface albedo and enhances melt rates. Thirdly, the mass addition by refreezing may prolong the lifetime of seasonal snow and thus reduces the period of bare-ice exposure in summer, thereby reducing surface melt. In order to quantify the net impact of refreezing on the climatic mass balance, we calculate the difference between the climatic mass balance (B_{clim}) in the refreezing and no-refreezing experiments (ΔB_{clim}; "refreezing" minus "no-refreezing"; black line in **Figure 7C**). The total ΔB_{clim} (black line) is positive at all elevations (**Figure 7C**), indicating a general positive impact of refreezing on the mass balance. The direct ΔB_{clim} (red line) refers to the direct mass retaining effect of refreezing, and is equal to the amount of refreezing itself (equivalent to the mean of the four lines in **Figure 5B**). By differencing the total and direct mass balance effect of refreezing, we estimate the net indirect effect of refreezing (green line in **Figure 7C**). The net indirect impact of refreezing is the summed effect of the three aforementioned indirect effects; this sensitivity experiment does not allow us to further distinguish individual contributions. The net impact of indirect effects on the mass balance is negative, which indicates that the positive indirect effect (potential prolongation of snow cover) is more than compensated for by negative indirect effects (melt enhancement due to snow pack heating and ice layer formation). A maximum net impact of indirect effects is found in the high ablation area at an approximate elevation of 500 m a.s.l. Averaged over the 0–1500 m a.s.l. elevation range, the mass balance effect of refreezing is 31% smaller than refreezing itself.

3.5. Climate Sensitivity

The above results are sensitive to the chosen future climate scenario, described in Section 2.2. To quantify this sensitivity we performed additional runs for the period 1984–2104 with perturbed precipitation and temperature trends. Sensitivity experiments include runs with zero long-term seasonal trends in temperature ("no ΔT" scenario) and precipitation ("no ΔP" scenario), and runs in which seasonal trends are doubled for temperature ("double ΔT" scenario) and precipitation ("double ΔP" scenario). We additionally perform an experiment in which inhomogeneity in warming of warm and cold spells is ignored ("no spell correction scenario"). For the "no ΔT," "double ΔT," "no ΔP" and "double ΔP" scenarios, sensitivity of the mass balance, deep temperature (15-m) and bulk density (0–10 m) is shown in **Figure 8**. For all experiments sensitivity values, averaged over the elevation range 0–1500 m a.s.l., are presented in **Table 1**.

Due to relatively weak trends for precipitation in the standard run, the mass balance, deep temperature and bulk density are all found to be much more sensitive to perturbations of the temperature scenario than perturbations of the precipitation scenario (**Table 1**; **Figure 8**). Averaged over the 0–1500 m a.s.l. elevation range the mass balance responds about twice as strong to doubling of future seasonal temperature trends than to a no-change scenario, which indicates a strong non-linearity of the mass balance sensitivity to temperature changes (**Table 1**). The mass balance sensitivity to temperature changes is highest in the ablation zone and decreases with elevation (**Figure 8A**), in line with previous findings in Van Pelt et al. (2012). In a

"double ΔT" scenario a mean future ELA for the period 2014–2104 of 1125 m a.s.l., and 1330 m a.s.l. for the period 2074–2104, would imply that nearly all glaciers in Svalbard would have maximum elevations below the ELA toward the end of the twenty-first century (Nuth et al., 2013). Finally, ignoring inhomogeneity in warming of cold and warm spells ("no spell correction scenario") has a modest negative impact on the simulated elevation-averaged mass balance (**Table 1**); in this sensitivity experiment, seasonal warm extremes are amplified leading to higher summer melt rates, thereby enhancing future mass loss.

Deep temperatures and bulk densities in a "double ΔT" scenario, averaged over the period 2014–2104, are strongly influenced by rapid upward migration of the ELA, which effectively smooths out the elevation-dependent variability seen in the "no ΔT" scenario, in which long-term trends in the ELA are absent (**Figure 8B**). Doubling of temperature trends causes pronounced subsurface warming with respect to the standard run (**Table 1**), yet cold deep temperature conditions remain in the ablation area even during the last 30 years of the simulation. Based on the above, it seems very unlikely that the general thermodynamic structure with cold deep temperatures in the ablation area and temperate conditions in the lower accumulation zone will change in the near-future. The impact of the "no spell correction scenario" on elevation-averaged deep temperatures is small, while the impact on density is substantial (**Table 1**). The latter can be ascribed to an increased significance of rainfall and subsequent refreezing in winter due to higher temperatures during warm spells.

As described in Section 2.2.2, we use fixed elevation lapse rates for temperature and precipitation. In practice, temperature and precipitation lapse rates may vary in space and we test the sensitivity of the modeled results to typical lapse rate perturbations representing regional variability across Svalbard. The magnitude of the chosen temperature lapse rate perturbations ($\pm0.0010°C$ m^{-1}), relative to the standard value of -0.0053°C m^{-1}, reflects variability in reported values across Svalbard, ranging from -0.0040 (Gardner et al., 2009) to -0.0066°C m^{-1} (Nuth et al., 2012). Sand et al. (2003) reported precipitation lapse rates for Svalbard ranging from 0 to 2.6 mm w.e. a^{-1} m^{-1}; based on this, we perform sensitivity experiments with the standard value (0.97 mm w.e. a^{-1} m^{-1}) perturbed by ±1.0 mm w.e. a^{-1} m^{-1}. Sensitivity results are averaged over the period 2014–2104 for the elevation range 0–1500 m a.s.l., and summarized in **Table 2**. The mass balance, deep temperature and bulk density are most sensitive to the applied perturbations of the precipitation lapse rate. The impact of temperature lapse rate perturbations is typically about half the impact of the precipitation lapse rate perturbations. **Table 2** further reveals that the mass balance is more sensitive to positive temperature and negative precipitation lapse rate changes than to perturbations of opposite sign. Furthermore, the high impact of negatively perturbing the precipitation lapse rate can be ascribed to low precipitation for the whole elevation range in this experiment, which has a severe impact on firn area extent, subsurface temperature and density. Altogether, the above suggests larger regional variability in mass balance and subsurface conditions

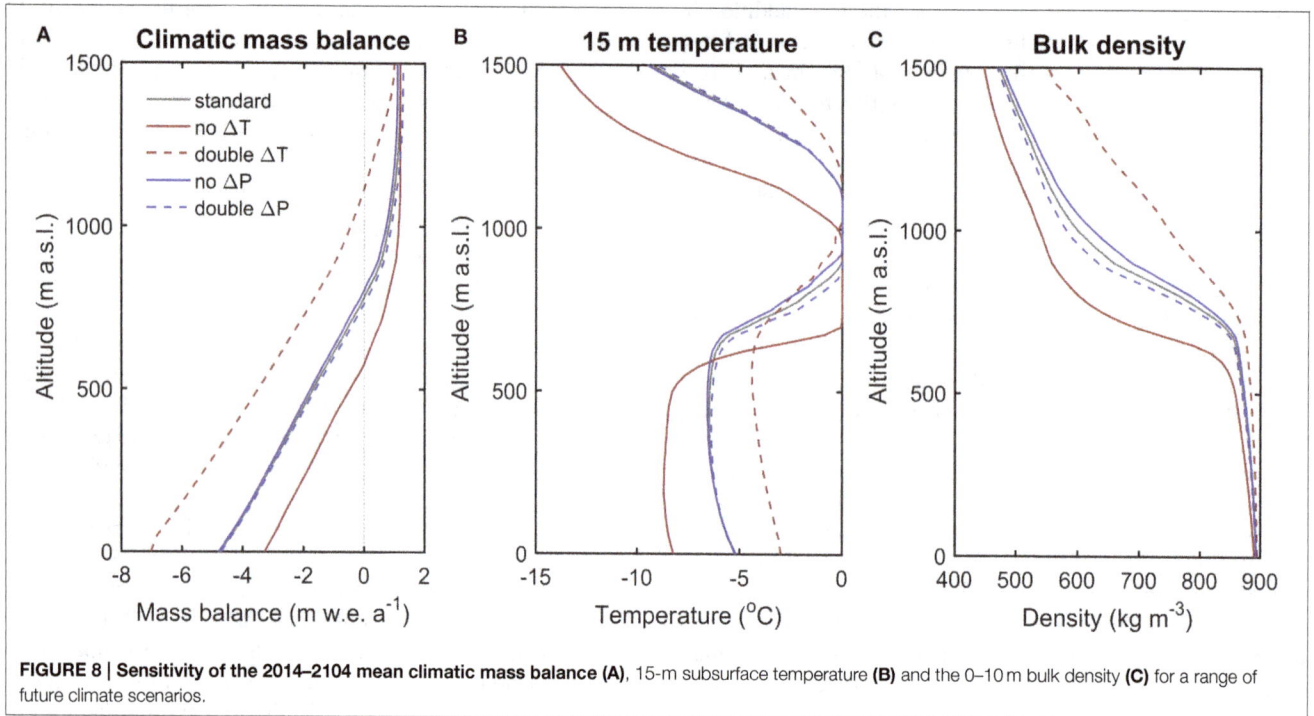

FIGURE 8 | Sensitivity of the 2014–2104 mean climatic mass balance (A), 15-m subsurface temperature **(B)** and the 0–10 m bulk density **(C)** for a range of future climate scenarios.

TABLE 1 | Sensitivity of the 2014–2104 mean climatic mass balance (B_{clim}), 15-m deep temperature (T_{deep}) and 0–10 m bulk density (ρ_{bulk}) to different temperature and precipitation scenarios.

Experiment	B_{clim} (m w.e. a^{-1})	ΔB_{clim} (m w.e. a^{-1})	T_{deep} (°C)	ΔT_{deep} (°C)	ρ_{bulk} (kg m^{-3})	$\Delta \rho_{bulk}$ (kg m^{-3})
Standard	−0.79	–	−4.18	–	729	–
No ΔT	−0.06	+0.73	−6.21	−2.03	681	−48
Double ΔT	−2.26	−1.47	−2.57	+1.61	792	+63
No ΔP	−0.88	−0.09	−4.31	−0.13	737	+9
Double ΔP	−0.71	+0.08	−4.03	+0.15	721	−7
No spell correction	−1.23	−0.44	−4.01	+0.17	763	+34

Calculated values are averaged values over the elevation range 0–1500 m a.s.l. ΔB_{clim}, ΔT_{deep}, and $\Delta \rho_{bulk}$ represent differences of B_{clim}, T_{deep}, and ρ_{bulk} relative to the standard run (perturbation run minus standard run).

TABLE 2 | Sensitivity of the 2014–2104 mean climatic mass balance (B_{clim}), 15-m deep temperature (T_{deep}) and 0–10 m bulk density (ρ_{bulk}) to different lapse rates for temperature (γ_T) and precipitation (γ_P).

Experiment	Standard lapse rate	Perturbation	B_{clim} (m w.e. a^{-1})	ΔB_{clim} (m w.e. a^{-1})	T_{deep} (°C)	ΔT_{deep} (°C)	ρ_{bulk} (kg m^{-3})	$\Delta \rho_{bulk}$ (kg m^{-3})
Standard			−0.79		−4.18		729	
γ_T	−0.0053 K m^{-1}	−0.0010	−0.55	+0.25	−5.59	−1.41	692	−36
		+0.0010	−1.15	−0.36	−3.54	+0.64	774	+45
γ_P	0.97 mm w.e. a^{-1} m^{-1}	−1	−1.86	−1.07	−8.21	−4.03	862	+133
		+1	−0.09	+0.70	−2.90	+1.28	668	−61

Calculated values are averaged values over the elevation range 0–1500 m a.s.l. ΔB_{clim}, ΔT_{deep}, and $\Delta \rho_{bulk}$ represent differences of B_{clim}, T_{deep}, and ρ_{bulk} relative to the standard run (perturbation run minus standard run).

can be expected from spatial variation of precipitation lapse rates rather than temperature lapse rates.

4. CONCLUSIONS

Model experiments with a coupled surface energy balance—firn model (EBFM) were done to investigate the climatic mass balance and the changing role of snow and firn conditions in an idealized Svalbard glacier setting. A climate forcing is generated on an elevation-dependent grid for the past, present and future (1984–2104), based on observational data from Svalbard Airport, observation-based elevation lapse rates, and a seasonally dependent projection scenario from an ensemble of future RCM runs. Experiments include a standard run and a set of sensitivity experiments to quantify the impact of refreezing and to assess sensitivity of the results to the chosen climate scenario and elevation lapse rates.

Results illustrate ongoing and future firn degradation and (delayed) firn line retreat in response to an upward ELA migration of 31 m a.s.l. decade^{-1} between 1984–2104. In a future climate, the elevation band with temperate deep firn in the lower accumulation area, caused by deep percolation, liquid water storage and year-round refreezing, is found to migrate upwards and expand. The cold ablation area warms considerably, yet remains non-temperate toward the end of the twenty-first century. In a future climate, refreezing remains of similar significance, with maximum refreezing rates occurring in the (retreating) lower accumulation zone. In response to pronounced winter warming and an associated increase in winter rainfall, the current prevalence of refreezing during the melt season gradually shifts to the winter season in a future climate. This can be ascribed to more frequent and substantial heat release in the winter snow pack after rainfall events, which in turn reduces the refreezing potential at the start of the melt season.

Comparison of the standard run and a no-refreezing experiment reveals that in a present and future climate the density and thermodynamic structure of Svalbard glaciers are heavily influenced by solid mass retention and heat release after refreezing. By computing the mass balance difference of the refreezing and no-refreezing experiments, we further quantified the mass balance impact of refreezing. The total mass balance effect is found to be positive at all elevations, indicating refreezing acts as a net buffer against mass loss. However, the mass balance effect of refreezing is found to be 31% smaller than mass retention by refreezing itself. This discrepancy is caused by indirect effects after refreezing, related to snow/firn heating, density changes and snow cover changes, which collectively have a negative mass balance impact that partly offsets the mass gain through melt water retention.

Climate sensitivity experiments with perturbed climate scenarios indicate a much higher sensitivity of mass balance, subsurface temperature and bulk density to the prescribed temperature than the precipitation scenario. Experiments with perturbed elevation lapse rates, based on observed regional variability across Svalbard, show a higher sensitivity of the results to precipitation lapse rate changes than to temperature lapse rate perturbations.

Our main aim with this work was to assess the role of snow/firn conditions and refreezing on the mass balance of glaciers in rapidly changing Svalbard climate. For this we chose an idealized elevation-dependent grid, which facilitated the presented analysis with a focus on process-understanding and a discussion of the role of feedbacks. Overall, the results indicate a decisive role for transient snow and firn conditions, and in particular refreezing, on glacier mass loss. This motivates the coupling of surface models to detailed snow models accounting for the subsurface thermodynamic, density and water content evolution in large-scale applications.

AUTHOR CONTRIBUTIONS

WvP performed the model experiments and wrote the manuscript. WvP, VP, and CR discussed the approach and model outcome. VP and CR provided thorough comments and feedback on the manuscript that helped to improve the manuscript quality.

ACKNOWLEDGMENTS

Meteorological data from the Svalbard Airport weather station were downloaded from the eKlima portal (www.eklima.no), provided by the Norwegian Meteorological Institute. We thank colleagues at Uppsala University, and in particular S. Marchenko, for discussion, comments and suggestions which have helped to improve the manuscript. Finally, we thank the reviewers and editor for their insightful feedback that has helped to improve the manuscript.

REFERENCES

ACIA (2005). *Arctic Climate Impact Assessment - Scientific Report*. New York, NY: Cambridge University Press.

Baumberger, A. (2007). *Massebalans på Kronebreen/Holtedahlfonna, Svalbard - kontrollerende faktorer*. Master thesis, University of Oslo.

Bengtsson, L., Hodges, K. I., Koumoutsaris, S., Zahn, M., and Keenlyside, N. (2011). The changing atmospheric water cycle in Polar Regions in a warmer climate. *Tellus A* 63A, 907–920. doi: 10.1111/j.1600-0870.2011.00534.x

Bintanja, R., and Selten, F. M. (2014). Future increases in Arctic precipitation linked to local evaporation and sea-ice retreat. *Nature* 509, 479–482. doi: 10.1038/nature13259

Bintanja, R., and van der Linden, E. C. (2013). The changing seasonal climate in the Arctic. *Sci. Rep.* 3, 1–8. doi: 10.1038/srep01556

Björnsson, H., Gjessing, Y., Hamran, S.-E., Hagen, J. O., Liestøl, O., Palsson, F., et al. (1996). The thermal regime of sub-polar glaciers mapped by multi-frequency radio-echo sounding. *J. Glaciol.* 42, 23–32.

Blaszczyk, M., Jania, J. A., and Hagen, J. O. (2009). Tidewater glaciers of Svalbard: recent changes and estimates of calving fluxes. *Polish Polar Res.* 30, 85–142.

Christianson, K., Kohler, J., Alley, R. B., Nuth, C., and Van Pelt, W. J. (2015). Dynamic perennial firn aquifer on an Arctic glacier. *Geophys. Res. Lett.* 42, 1418–1426. doi: 10.1002/2014GL062806

Cogley, J. G., Hock, R., Rasmussen, L. A., Arendt, A. A., Bauder, A., Braithwaite, R. J., et al. (2011). *Glossary of Glacier Mass Balance and Related Terms*.

Technical report, IHP-VII Technical Documents in Hydrology No. 86, IACS Contribution No. 2, UNESCO-IHP, Paris.

Colbeck, S. C. (1987). "A review of the metamorphism and classification of seasonal snow cover crystals," in *Avalanche Formation, Movement and Effects, Proceedings of the Davos Symposium, September 1986.*

Day, J. J., Bamber, J. L., Valdes, P. J., and Kohler, J. (2012). The impact of a seasonally ice free Arctic Ocean on the temperature, precipitation and surface mass balance of Svalbard. *Cryosphere* 6, 35–50. doi: 10.5194/tc-6-35-2012

Divine, D. V., and Dick, C. (2006). Historical variability of sea ice edge position in the Nordic Seas. *J. Geophys. Res.* 111:C01001. doi: 10.1029/2004JC002851

Førland, E. J., Benestad, R., Hanssen-Bauer, I., Haugen, J. E., and Skaugen, T. E. (2011). Temperature and precipitation development at svalbard 1900–2100. *Adv. Meteorol.* 2011, 1–14. doi: 10.1155/2011/893790

Førland, E. J., and Hanssen-Bauer, I. (2000). Increased precipitation in the Norwegian Arctic: true or false? *Clim. Change* 46, 485–509. doi: 10.1023/A:1005613304674

Forster, R. R., Box, J. E., Van den Broeke, M. R., Miège, C., Burgess, E. W., Van Angelen, J. H., et al. (2013). Extensive liquid meltwater storage in firn within the Greenland ice sheet. *Nat. Geosci.* 7, 95–98. doi: 10.1038/ngeo2043

Fujita, K., Seko, K., Ageta, Y., Jianchen, P., and Tandong, Y. (1996). Superimposed ice in glacier mass balance on the Tibetan Plateau. *J. Glaciol.* 42, 454–460.

Gardner, A. S., Sharp, M. J., Koerner, R. M., Labine, C., Boon, S., Marshall, S. J., et al. (2009). Near-surface temperature lapse rates over Arctic glaciers and their implications for temperature downscaling. *J. Climate* 22, 4281–4298. doi: 10.1175/2009JCLI2845.1

Greuell, W., and Konzelmann, T. (1994). Numerical modelling of the energy balance and the englacial temperature of the Greenland Ice Sheet. Calculations for the ETH-Camp location (West Greenland, 1155 m a.s.l.). *Glob. Planet. Change* 9, 91–114. doi: 10.1016/0921-8181(94)90010-8

Hagen, J. O., Kohler, J., Melvold, K., and Winther, J.-G. (2003). Glaciers in Svalbard: mass balance, runoff and freshwater flux. *Polar Res.* 22, 145–159. doi: 10.1111/j.1751-8369.2003.tb00104.x

Hanssen-Bauer, I., and Førland, E. (1998). Long-term trends in precipitation and temperature in the Norwegian Arctic: can they be explained by changes in atmospheric circulation patterns? *Clim. Res.* 10, 143–153. doi: 10.3354/cr010143

IPCC (2013). *Climate Change 2013: The Physical Science Basis. Contribution of Working Group I to the Fifth Assessment Report of the Intergovernmental Panel on Climate Change [Stocker, T.F., D. Qin, G. - K. Plattner, M. Tignor, S.K. Allen, J. Bosc hung, A. Nauels, Y. X.* Cambridge; New York: Cambridge University Press.

James, T. D., Murray, T., Barrand, N. E., Sykes, H. J., Fox, A. J., and King, M. A. (2012). Observations of enhanced thinning in the upper reaches of Svalbard glaciers. *Cryosphere* 6, 1369–1381. doi: 10.5194/tc-6-1369-2012

Jania, J., Mochnacki, D., and Gadek, B. (1996). The thermal structure of Hansbreen, a tidewater glacier in southern Spitsbergen, Svalbard. *Polar Res.* 15, 53–66. doi: 10.1111/j.1751-8369.1996.tb00458.x

Klok, E. L., and Oerlemans, J. (2002). Model study of the spatial distribution of the energy and mass balance of Morteratschgletscher, Switzerland. *J. Glaciol.* 48, 505–518. doi: 10.3189/172756502781831133

Konzelmann, T., van de Wal, R., Greuell, W., Bintanja, R., Henneken, E., and Abeouchi, A. (1994). Parameterization of global and longwave incoming radiation for the Greenland Ice Sheet. *Glob. Planet. Change* 9, 143–164. doi: 10.1016/0921-8181(94)90013-2

Kuipers Munneke, P., M. Ligtenberg, S. R., Van den Broeke, M. R., Van Angelen, J. H., and Forster, R. R. (2014). Explaining the presence of perennial liquid water bodies in the firn of the Greenland Ice Sheet. *Geophys. Res. Lett.* 41, 476–483. doi: 10.1002/2013GL058389

Lang, C., Fettweis, X., and Erpicum, M. (2015). Stable climate and surface mass balance in Svalbard over 1979-2013 despite the Arctic warming. *Cryosphere* 9, 83–101. doi: 10.5194/tc-9-83-2015

Ligtenberg, S. R. M., Helsen, M. M., and Van Den Broeke, M. R. (2011). An improved semi-empirical model for the densification of Antarctic firn. *Cryosphere* 5, 809–819. doi: 10.5194/tc-5-809-2011

Machguth, H., MacFerrin, M., Van As, D., Box, J. E., Charalampidis, C., Colgan, W., et al. (2016). Greenland meltwater storage in firn limited by near-surface ice formation. *Nat. Clim. Change* 6, 390–393. doi: 10.1038/nclimate2899

Min, S.-K., Zhang, X., and Zwiers, F. (2008). Human-induced Arctic moistening. *Science (New York, N.Y.)* 320, 518–520. doi: 10.1126/science.1153468

Min, S.-K., Zhang, X., Zwiers, F. W., and Hegerl, G. C. (2011). Human contribution to more-intense precipitation extremes. *Nature* 470, 378–381. doi: 10.1038/nature09763

Möller, M., Obleitner, F., Reijmer, C. H., Pohjola, V. A., Głowacki, P., and Kohler, J. (2016). Adjustment of regional climate model output for modeling the climatic mass balance of all glaciers on Svalbard. *J. Geophys. Res. Atmospheres.* 121, 5411–5429. doi: 10.1002/2015JD024380

Nordli, Ø., Przybylak, R., Ogilvie, A. E., and Isaksen, K. (2014). Long-term temperature trends and variability on Spitsbergen: the extended Svalbard Airport temperature series, 1898–2012. *Polar Res.* 33:21349. doi: 10.3402/polar.v33.21349

Nuth, C., Kohler, J., König, M., Von Deschwanden, A., Hagen, J. O., Kääb, A., et al. (2013). Decadal changes from a multi-temporal glacier inventory of Svalbard. *Cryosphere* 7, 1603–1621. doi: 10.5194/tc-7-1603-2013

Nuth, C., Schuler, T. V., Kohler, J., Altena, B., and Hagen, J. O. (2012). Estimating the long-term calving flux of Kronebreen, Svalbard, from geodetic elevation changes and mass-balance modelling. *J. Glaciol.* 58, 119–133. doi: 10.3189/2012JoG11J036

Oerlemans, J., and Grisogono, B. (2002). Glacier winds and parameterization of the related surface heat fluxes. *Tellus* 54A, 440–452. doi: 10.1034/j.1600-0870.2002.201398.x

Oerlemans, J., and Knap, W. H. (1998). A 1 year record of global radiation and albedo in the ablation zone of Morteratschgletscher, Switzerland. *J. Glaciol.* 44, 231–238.

Pettersson, R. (2004). *Dynamics of the Cold Surface Layer of Polythermal storglaciären, Sweden.* Ph.D. thesis, Stockholm University, Department of Physical Geography and Quaternary Geology. Stockholm University Dissertation Series.

Pfeffer, W. T., and Humphrey, N. (1998). Formation of ice layers by infiltration and refreezing of meltwater. *Ann. Glaciol.* 26, 83–91.

Pfeffer, W. T., Meier, M. F., and Illangasekare, T. H. (1991). Retention of Greenland runoff by refreezing: implications for projected future sea level change. *J. Geophys. Res.* 96, 117–122. doi: 10.1029/91JC02502

Reijmer, C. H., and Hock, R. (2008). Internal accumulation on Storglaciären, Sweden, in a multi-layer snow model coupled to a distributed energy- and mass-balance model. *J. Glaciol.* 54, 61–72. doi: 10.3189/002214308784409161

Reijmer, C. H., Van Den Broeke, M. R., Fettweis, X., Ettema, J., and Stap, L. B. (2012). Refreezing on the Greenland ice sheet: a comparison of parameterizations. *Cryosphere* 6, 743–762. doi: 10.5194/tc-6-743-2012

Sand, K., Winther, J.-G., Marechal, D., Bruland, O., and Melvold, K. (2003). Regional variations of snow accumulation on spitsbergen, svalbard, 1997-99. *Nordic Hydrol.* 34, 17–32.

Serreze, M. C., Crawford, A. D., and Barrett, A. P. (2015). Extreme daily precipitation events at Spitsbergen, an Arctic Island. *Int. J. Climatol.* 35, 4574–4588. doi: 10.1002/joc.4308

Serreze, M. C., and Francis, J. A. (2006). The Arctic amplification debate. *Clim. Change* 76, 241–264. doi: 10.1007/s10584-005-9017-y

Van Angelen, J. H., Lenaerts, J. T. M., Van den Broeke, M. R., Fettweis, X., and Van Meijgaard, E. (2013). Rapid loss of firn pore space accelerates 21st century Greenland mass loss. *Geophys. Res. Lett.* 40, 2109–2113. doi: 10.1002/grl.50490

Van de Wal, R. S. W., Mulvaney, R., Isaksson, E., Moore, J. C., Pinglot, J. F., Pohjola, V. A., et al. (2002). Reconstruction of the historical temperature trend from measurements in a medium-length borehole on the Lomonosovfonna plateau, Svalbard. *Ann. Glaciol.* 35, 371–378. doi: 10.3189/172756402781816979

Van Pelt, W. J. J., and Kohler, J. (2015). Modelling the long-term mass balance and firn evolution of glaciers around Kongsfjorden, Svalbard. *J. Glaciol.* 61, 731–744. doi: 10.3189/2015JoG14J223

Van Pelt, W. J. J., Oerlemans, J., Reijmer, C. H., Pettersson, R., Pohjola, V. A., Isaksson, E., et al. (2013). An iterative inverse method to estimate basal topography and initialize ice flow models. *Cryosphere* 7, 987–1006. doi: 10.5194/tc-7-987-2013

Van Pelt, W. J. J., Oerlemans, J., Reijmer, C. H., Pohjola, V. A., Pettersson, R., and Van Angelen, J. H. (2012). Simulating melt, runoff and refreezing on

Nordenskiöldbreen, Svalbard, using a coupled snow and energy balance model. *Cryosphere* 6, 641–659. doi: 10.5194/tc-6-641-2012

Van Pelt, W. J. J., Pettersson, R., Pohjola, V. A., Marchenko, S., Claremar, B., and Oerlemans, J. (2014). Inverse estimation of snow accumulation along a radar transect on Nordenskiöldbreen, Svalbard. *J. Geophys. Res. Earth Surf.* 119, 816–835. doi: 10.1002/2013JF003040

Vega, C. P., Pohjola, V. A., Beaudon, E., Claremar, B., Van Pelt, W. J. J., Pettersson, R., et al. (2016). A synthetic ice core approach to estimate ion relocation in an ice field site experiencing periodical melt: a case study on Lomonosovfonna, Svalbard. *Cryosphere* 10, 961–976. doi: 10.5194/tc-10-961-2016

Winther, J.-G., Bruland, O., Sand, K., Gerland, S., Marechal, D., Ivanov, B., et al. (2003). Snow research in Svalbard – an overview. *Polar Res.* 22, 125–144. doi: 10.1111/j.1751-8369.2003.tb00103.x

Wright, A. P., Wadham, J. L., Siegert, M. J., Luckman, A., Kohler, J., and Nuttall, A.-M. (2007). Modeling the refreezing of meltwater as superimposed ice on a high Arctic glacier: a comparison of approaches. *J. Geophys. Res.* 112:F04016. doi: 10.1029/2007jf000818

Zhang, X., He, J., Zhang, J., Polyakov, I., Gerdes, R., Inoue, J., et al. (2012). Enhanced poleward moisture transport and amplified northern high-latitude wetting trend. *Nat. Clim. Change* 3, 47–51. doi: 10.1038/nclimate1631

Conflict of Interest Statement: The authors declare that the research was conducted in the absence of any commercial or financial relationships that could be construed as a potential conflict of interest.

5

Parameterizing Deep Water Percolation Improves Subsurface Temperature Simulations by a Multilayer Firn Model

Sergey Marchenko [1,2]*, Ward J. J. van Pelt [1], Björn Claremar [1], Veijo Pohjola [1], Rickard Pettersson [1], Horst Machguth [3] and Carleen Reijmer [4]

[1] Department of Earth Sciences, Uppsala University, Uppsala, Sweden, [2] Department of Geophysics, The University Centre in Svalbard, Longyearbyen, Norway, [3] Department of Geography, University of Zurich, Zurich, Switzerland, [4] Institute for Marine and Atmosphere Research, Utrecht University, Utrecht, Netherlands

Edited by:
Michael Lehning,
École Polytechnique Fédérale de
Lausanne (EPFL), Switzerland

Reviewed by:
Hiroyuki Hirashima,
National Research Institute for Earth
Science and Disaster Resilience,
Japan
Nander Wever,
École Polytechnique Fédérale de
Lausanne (EPFL), Switzerland

*Correspondence:
Sergey Marchenko
sergey.marchenko@geo.uu.se

Specialty section:
This article was submitted to
Cryospheric Sciences,
a section of the journal
Frontiers in Earth Science

Deep preferential percolation of melt water in snow and firn brings water lower along the vertical profile than a laterally homogeneous wetting front. This widely recognized process is an important source of uncertainty in simulations of subsurface temperature, density, and water content in seasonal snow and in firn packs on glaciers and ice sheets. However, observation and quantification of preferential flow is challenging and therefore it is not accounted for by most of the contemporary snow/firn models. Here we use temperature measurements in the accumulation zone of Lomonosovfonna, Svalbard, done in April 2012–2015 using multiple thermistor strings to describe the process of water percolation in snow and firn. Effects of water flow through the snow and firn profile are further explored using a coupled surface energy balance - firn model forced by the output of the regional climate model WRF. *In situ* air temperature, radiation, and surface height change measurements are used to constrain the surface energy and mass fluxes. To account for the effects of preferential water flow in snow and firn we test a set of depth-dependent functions allocating a certain fraction of the melt water available at the surface to each snow/firn layer. Experiments are performed for a range of characteristic percolation depths and results indicate a reduction in root mean square difference between the modeled and measured temperature by up to a factor of two compared to the results from the default water infiltration scheme. This illustrates the significance of accounting for preferential water percolation to simulate subsurface conditions. The suggested approach to parameterization of the preferential water flow requires low additional computational cost and can be implemented in layered snow/firn models applied both at local and regional scales, for distributed domains with multiple mesh points.

Keywords: firn, firn modeling, preferential flow, internal accumulation, Lomonosovfonna, Svalbard, firn water content

1. INTRODUCTION

Glaciers and ice sheets occupy a substantial fraction of mountain ranges and polar areas and are important components in the system of feedbacks linking the atmosphere, land, and ocean. Glaciers are sensitive to environmental perturbations and are often used as indicators of the past and present climate changes (e.g., Meese et al., 1994; Haeberli et al., 2007). The observed global air temperature rise during the twentieth and twenty-first century (Vaughan et al., 2013) leads to elevated rates of melt water production on glaciers. According to recent estimates one third of the observed rate of the sea level rise is explained by the accelerated rate of mass loss from the Greenland and Antarctic ice sheets, another third is coming from the glaciers and ice caps (Church et al., 2011).

One of the uncertainty sources in the estimates of the contribution of glaciers and ice sheets to the sea-level rise stems from the fact that the melt water generated at the glacier surface does not necessarily lead to runoff. Liquid water can refreeze (e.g., Pfeffer et al., 1991; Machguth et al., 2016) or be stored in the firn pores forming perennial firn aquifers (e.g., Forster et al., 2014; Christianson et al., 2015). Refreezing effectively reduces the amount of melt water that runs off and is most significant in the accumulation zone (e.g., Van Pelt et al., 2012). At the same time, the release of latent heat accompanying refreezing effectively warms the snow and firn column (e.g., Zdanowicz et al., 2012). Once the snow and firn profile is temperate and water cannot be accommodated in the pores, excess water contributes to runoff and eventually to the sea level rise. Firn line retreat to higher elevation and consecutive shrinking of accumulation zones in a warming climate lead to an overall reduction of refreezing, which implies an acceleration of glacier runoff (e.g., van Angelen et al., 2013; Van Pelt et al., 2016). Thus it is important to understand the processes involved in the mass and energy exchange at and below the glacier surface to explain the observed changes in the Earth's ice cover and predict its role in a changing climate.

Multi-layer snow and firn models have been used for description of the mass and energy fluxes in the uppermost tens of meters below the glacier surface. Depending on their exact implementation they may include descriptions of snow precipitation and wind drift, metamorphism, gravitational settling, conductive heat transport, water infiltration, refreezing, and runoff. These processes determine the evolution of the state variables like temperature, grain size and structure, density and water content. The models are typically forced either by observational data or by the surface schemes describing the energy and mass fluxes at the surface to which they are coupled through albedo, ground heat flux, and other feedbacks. For instance, the firn model initially suggested by Greuell and Oerlemans (1986) and developed further by several studies (e.g., Greuell and Konzelmann, 1994; Reijmer and Hock, 2008; Van Pelt et al., 2012) was successfully used in energy/mass balance studies on glaciers and ice sheets (e.g., Bassford et al., 2006; Reijmer et al., 2012; Van Pelt and Kohler, 2015). The CROCUS (Brun et al., 1989) and SNOWPACK (Bartelt and Lehning, 2002) models were originally designed for seasonal snow studies but were also successfully utilized for thicker snow, firn and ice packs

(Obleitner and Lehning, 2004; Fettweis, 2007; Gascon et al., 2014; Lang et al., 2015).

Most models assume that the vertical water percolation is laterally uniform and is regulated by two properties of each layer: refreezing capacity and the potential of capillary and adhesive forces to hold water against gravity. However, field and laboratory observations suggest considerable horizontal gradients in the rate of vertical water flow. At certain points concentration of water flow occurs and water infiltrates deeper than the background wetting front forming preferential flow paths. Occurrence of the latter in porous soils infiltrated by water is widely acknowledged (e.g., Hill and Parlange, 1972; Hendrickx and Flury, 2001). Existence of the preferential flow paths in snow and firn has been shown by numerous stratigraphical studies using dye tracing experiments and thick sections (McGurk and Marsh, 1995; Schneebeli, 1995; Bøggild, 2000; Waldner et al., 2004; Campbell et al., 2006; Williams et al., 2010). Further evidences have been provided by temperature tracking of water refreezing events (Sturm and Holmgren, 1993; Conway and Benedict, 1994; Pfeffer and Humphrey, 1996; Humphrey et al., 2012; Cox et al., 2015), radar surveys (Albert et al., 1999; Williams et al., 2000), and observations at the melting snow surface (Williams et al., 1999).

It is important to account for the preferential water flow in estimates of the energy and mass fluxes at glaciers as the mechanism can bring water to the deep snow and firn layers. Deep water percolation serves as an effective way to increase the temperature in deep snow and firn layers through the release of latent heat. A model assuming no preferential flow is likely to overestimate the density gradient with depth and produce a colder subsurface temperature profile than what exists in reality (Gascon et al., 2014), especially at the onset of the melt season (Steger et al., 2017). These effects may results in biased estimates of energy and mass fluxes at and below the surface.

A number of attempts have been undertaken to parameterize preferential flow in snow models. An approach used in one dimensional models is to divide the water available at the surface between two wetting fronts corresponding to the laterally uniform and preferential water flow. For constraining that division, Marsh and Woo (1984b, 1985) used data from multiple lysimeters. Katsushima et al. (2009) and Wever et al. (2016) applied thresholds based on the water content of snow to allocate water to the matrix and preferential flow domains. An alternative approach was suggested by Bøggild (2000) who allowed melt water to affect only a fraction of the profile's refreezing and retention capacity rising from 0.22 (Marsh and Woo, 1984a) to 1 over time. To induce preferential water flow in multidimensional snow models Illangasekare et al. (1990) and Hirashima et al. (2014) perturbed the spatial distribution of subsurface density and grain size.

It has been shown that using the above-mentioned approaches allows simulating preferential flow paths that are intuitively expected in a cold snow pack with a melting surface and qualitatively comply with the empirical evidences. However, constraining the parameters used in the models for characterizing snow and firn (e.g., hydraulic conductivity, water entry pressure, size and shape of snow grains and pores, scale of their spatial heterogeneities) is challenging as reliable high resolution datasets on the spatial variability of snow properties are complicated to retrieve and hence scarce. The fine scale of the processes involved in preferential water flow currently limits the applicability of 2D and 3D snow models with detailed physics in snow/firn simulations at larger scales. However, for distributed regional scale modeling of the mass and energy fluxes at glaciers the details of water infiltration processes might be of limited importance, while accurate estimation of the effect of water percolation on the evolution of the bulk density and temperature is more crucial.

The purpose of the present article is to find an efficient way to describe preferential water flow in snow and firn and by this to increase model performance in reproducing observed subsurface temperature evolution. It is based on the field data on temperature evolution in the upper 12 m of the snow and firn pack measured at Lomononosovfonna, Svalbard, during 2012-2014 by nine thermistor strings horizontally separated by 3-8.5 m. The results of measurements are compared with the temperature simulations using a multilayer snow and firn model (Van Pelt et al., 2012; Van Pelt and Kohler, 2015). A simple routine for describing the effects of preferential water percolation on snow and firn conditions is implemented and tuned to minimize the error between the simulated and observed temperature evolutions. Sensitivity of the model to different infiltration schemes, depths of water percolation and surface melt rate is tested and discussed. Finally, we discuss the potential development of deep percolation schemes and their implementation in firn models, which will help to reduce the uncertainty in simulated subsurface conditions.

2. STUDY SITE

Lomonosovfonna is a >500 km^2 ice field in the central part of the Spitsbergen island, Svalbard, nourishing several outlet glaciers including Nordenskiöldbreen (**Figure 1**). During the twentieth to twenty-first century the archipelago has seen a pronounced rise in air temperature (Førland et al., 2011; Nordli et al., 2014), glacier retreat (Nuth et al., 2013), and thinning (Moholdt et al., 2010; Nuth et al., 2010; James et al., 2012). At the summit of Lomonosovfonna the surface temperature experienced a rise of ca 2-3°C during the twentieth century (Van de Wal et al., 2002). Recent decades at Lomonosovfonna and Nordenskiöldbreen are dominated by an amplified and seasonally-dependent warming and the mean annual air temperature increased by more than 1°C during 1989-2010. The resulting glacier-averaged mass balance during this period was calculated to be −0.39 m w.e. yr^{-1} (Van Pelt et al., 2012).

The study site is in the accumulation zone of the Lomonosovfonna ice field on a flat spot at 78.824°N, 17.432°E, 1,200 m a.s.l., well above the equilibrium line, which for Nordenskiöldbreen was estimated to be at 719 m a.s.l. (Van Pelt et al., 2012). The local glacier thickness derived from a GPR survey is 192 ± 5.1 m (Pettersson, 2009, unpublished data) and the thickness of the firn layer according to coring results was around 20 m in 2009 (Wendl, 2014). Recent modeling results in the area showed that at the elevation of our study site the melt rate during 1989-2010 was on average ≈0.34 m w.e. yr^{-1} (Van Pelt et al., 2012). Accumulation at Lomonosovfonna is affected by the wind drift (Pälli et al., 2002; Van Pelt et al.,

2014) and measured rates range between 0.58±0.13 and ≈0.75 m w.e. yr^{-1} (Van Pelt et al., 2014; Wendl et al., 2015). Subsurface temperature measurements at Lomonosovfonna during summer season were carried out in 1965 (Zinger et al., 1966), 1980 (Zagorodnov and Zotikov, 1980), and more recently by Marchenko et al. (2017). While in Marchenko et al. (2017) the focus was on subsurface density and stratigraphy, in this study the analysis concentrates on the subsurface temperature evolution.

3. DATA
3.1. AWS and Stake Measurements
For monitoring of the surface energy and mass fluxes an automatic weather station (AWS) was installed at Lomonosovfonna in April 2013 and reinstalled in April 2014 and 2015. Air temperature was measured every 3 h during 23 April-17 August in 2013 and 29 April 2014-1 May 2015. Downwelling and upwelling fluxes of short- and long-wave radiation (SW↓, SW↑, LW↓, LW↑, respectively) were measured every 30 min and averaged values were recorded every 3 h during 23 April-14 July 2013 and 28 April-26 August 2014. Details on the used instruments are given in the **Table 1**.

The AWS data were filtered to exclude potentially erroneous values from the analysis. Firstly, we discarded the values collected at battery voltage below the limits provided by the manufacturers of the corresponding instruments. Secondly, values outside of the feasible range were rejected: for air temperature this was <−40 and >10°C, for SW↓ and SW↑ this was <0 and >800 W m^{-2} and for LW↓ <0 and >350 W m^{-2}. Additional constraints were also applied to LW↑ assuming that the glacier surface cannot be colder than −50°C and warmer than 0°C. The minimal (140 W m^{-2}) and maximum (316 W m^{-2}) values for the fluxes were calculated using the Stefan-Boltzmann law assuming a surface emissivity of 1, as:

$$LW \uparrow = \sigma \cdot T^4,$$

where $\sigma = 5.67 \cdot 10^{-8}$ Wm^{-2}K^{-4} and T is the surface temperature in Kelvin. Radiation measurements are prone to significant biases due to deposition of rime on the sensors. Values possibly measured during periods dominated by riming were interpreted from occasional decreases in short wave radiative fluxes accompanied by a small difference between LW↓ and LW↑ and excluded from the analysis.

Measurements of snow depth and snow surface height at stake S11 during field campaigns in April 2012-2015 (**Figure 1**) are also used for validation of the simulated surface energy and mass fluxes. Although, the stake is situated at a distance of 1.5 km from our site, results from radar surveys (Van Pelt et al., 2014) suggest that there is no significant gradient in net annual accumulation rate between the two sites. Measurements at the stake yield estimates of accumulation during the period from August (when the summer surface is formed) to April and the net surface mass flux between two field campaigns. It has to be noted, that the measurements also include the effect of snow and firn settling. Winter accumulation rate measurements include the effect of snow settling that occurs above the previous summer surface and

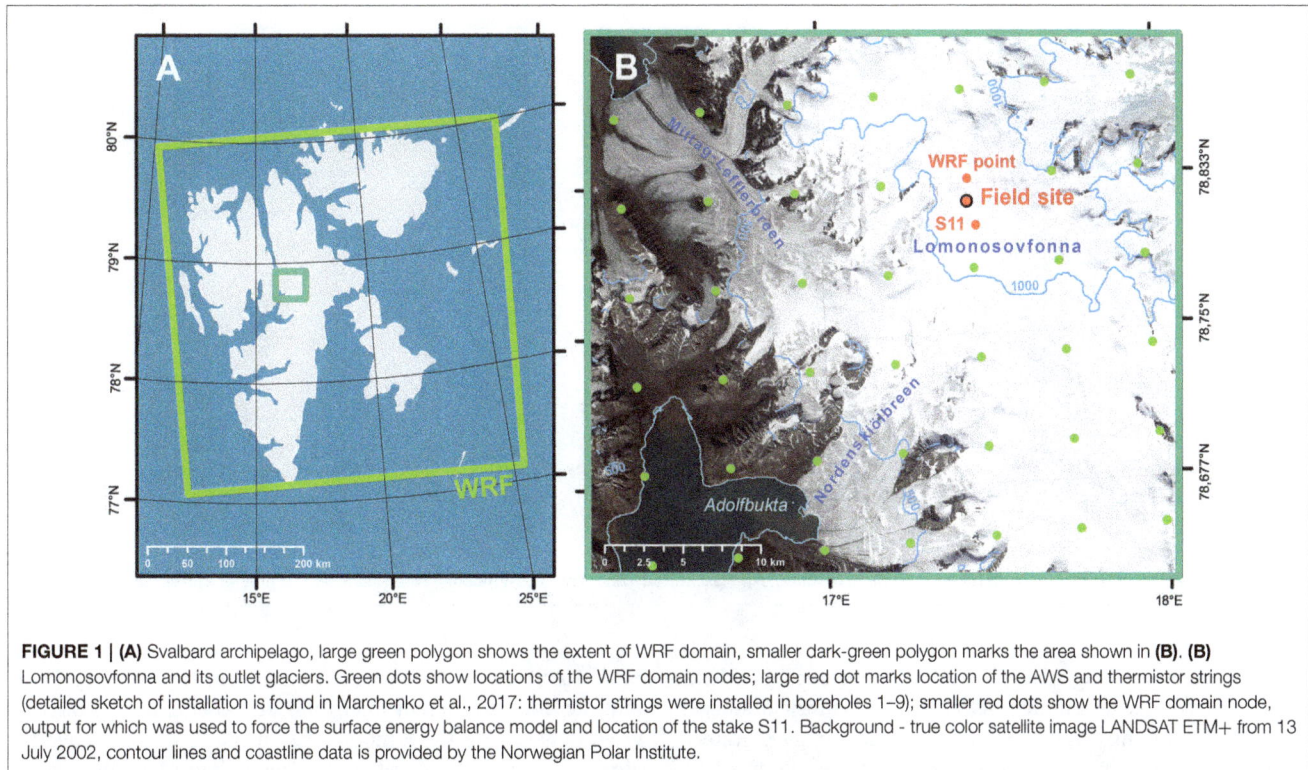

FIGURE 1 | (A) Svalbard archipelago, large green polygon shows the extent of WRF domain, smaller dark-green polygon marks the area shown in **(B)**. **(B)** Lomonosovfonna and its outlet glaciers. Green dots show locations of the WRF domain nodes; large red dot marks location of the AWS and thermistor strings (detailed sketch of installation is found in Marchenko et al., 2017: thermistor strings were installed in boreholes 1–9); smaller red dots show the WRF domain node, output for which was used to force the surface energy balance model and location of the stake S11. Background - true color satellite image LANDSAT ETM+ from 13 July 2002, contour lines and coastline data is provided by the Norwegian Polar Institute.

TABLE 1 | Instruments used for measurements of the air temperature, radiative energy fluxes, and subsurface temperature.

Parameters		22 April 2012	22 April 2013	17 April 2014
Data logger			Campbell Scientific CR10X	
Air temperature		–	Vaisala HMP45A	Rotronic HC2S3
Radiative energy fluxes		–	Kipp and Zonen CNR1	Kipp and Zonen CNR4
Snow and firn temperature	Thermistors	RTI-electronics ACC-003,5 kOhm at 25°C	Betatherm 100K6A1i, 100 kOhm at 25°C	
	Reference resistors	Rho Point 8G16D, 10 kOhm	Rho Point 8G16D, 100 kOhm	
	Excitation voltage, mV	280	900	

net annual accumulation measurements can be expected to be influenced by the gravitational densification occurring above the stake bottom. The latter can reach significant values as the in 2012 the stake bottom was at 8 m below the surface.

3.2. Snow and Firn Density

Firn density was measured in four cores drilled in April 2012–2014 by a Kovacs corer. The lengths of the cores were 9.9, 11.3 and 13.6 m. Details on the routines applied for extraction and processing of the cores were presented by Marchenko et al. (2017). For the further analysis the density profiles originally measured with a high vertical resolution were smoothed using a 1 m running average filter to exclude spikes caused by thin and likely not laterally consistent ice layers (Marchenko et al., 2017) and to ensure mass conservation.

3.3. Snow and Firn Temperature

Multiple thermistor strings were used to continuously record the evolution of the snow and firn temperature. In April 2012, 2013, and 2014 nine custom manufactured thermistor strings

were installed in holes drilled by Kovacs auger with a 5.5 cm diameter. The depths of sensors on the strings are presented in **Figure 2**. The strings were installed in a 3 × 3 square grid pattern with 3 m separation between neighboring holes. To minimize preferential percolation of melt water along the cables, the holes were backfilled with drill chips and surface snow.

Each thermistor was connected in series with a reference temperature stable resistor to form a resistive half bridge. Precisely measured excitation voltage was applied to the circuit and the voltage drop over reference resistor was measured using a data logger. To increase the number of channels scanned by the logger several relay multiplexors were connected to it. Details on the applied electrical schemes are listed in the **Table 1**. During post-processing the voltages measured by the data logger were first converted to resistances and then to temperature values using the manufacturer guidelines.

The resulting datasets span the periods 22 April–9 October in 2012, 23 April–12 July in 2013, and 17 April 2014–11 April 2015. Time gaps between the periods covered by subsurface

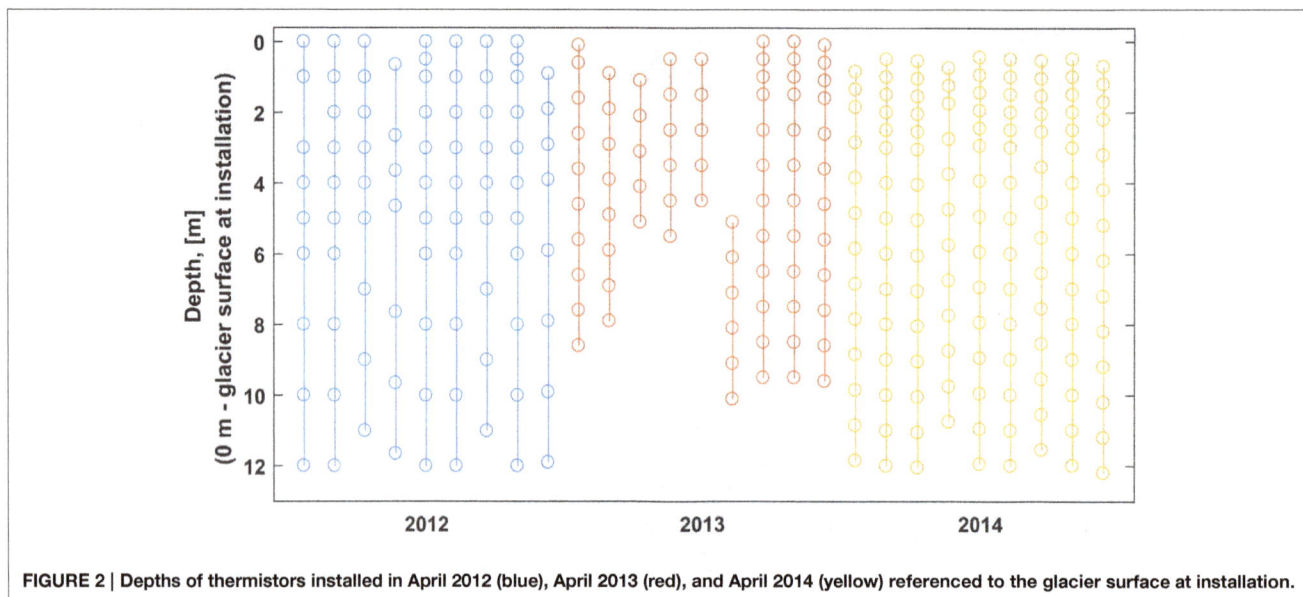

FIGURE 2 | Depths of thermistors installed in April 2012 (blue), April 2013 (red), and April 2014 (yellow) referenced to the glacier surface at installation.

temperature monitoring are explained by the technical problems that occurred during summer 2012 (rapid battery drainage) and 2013 (flooding of the equipment case by melt water). To facilitate the comparison of firn temperature measured by multiple thermistor strings the data from individual strings were first interpolated to a common depth grid using shape-preserving piecewise cubic interpolation. In the next step, the resulting datasets were horizontally averaged to obtain a more spatially representative dataset for each year. We also define three time intervals with rapid subsurface temperature changes subjectively estimated from the results of measurements and interpreted as being driven by the release of latent heat from refreezing water: 2 June–25 September in 2012, 3 June–12 July in 2013, and 4 July–21 September in 2014. These time intervals are used in the further analysis of field data and for its comparison with the model outputs.

4. MODELING

4.1. Regional Climate Model WRF

The Weather Research and Forecast (WRF) regional climate model (Skamarock et al., 2008) has been previously shown to successfully reproduce the evolution of meteorological parameters measured by weather stations in Svalbard (Claremar et al., 2012; Aas et al., 2015, 2016). WRF was forced by ERA Interim weather re-analysis data (Dee et al., 2011) with the spatial resolution of 0.75°, the sea surface temperature and sea ice data from the Operational Sea Surface Temperature and Sea Ice Analysis dataset (Donlon et al., 2012) with 0.05° resolution. We used the dataset on land terrain elevation from the US Geological Survey, while glacier mask was provided by the Norwegian Polar Institute (Nuth et al., 2010). The following modules were used in WRF: land surface scheme NOAH-MP (Niu et al., 2011), turbulence scheme MYNN-2.5 (Nakanishi and Niino, 2006), radiation parameterization RRTMG (Iacono

et al., 2008), cumulus convection from Grell and Dévényi (2002) and the double-moment micro-physics scheme from Morrison et al. (2005). The WRF model was run for a domain covering almost the entire Svalbard archipelago (**Figure 1A**) on a spatial resolution of 5.5 km (**Figure 1B**). For the further analysis we used four WRF output variables for the domain point that is closest to our study site and lies at 1,053 m a.s.l. (**Figure 1B**). Air temperature and relative humidity, precipitation and fractional cloud cover accounting for the presence of clouds in all atmospheric levels produced by WRF at 3 h temporal resolution serve as input for the coupled surface energy balance–firn model described below. Despite an elevation difference between the observation site and the model grid point of around 150 m a negative air temperature bias with respect to AWS measurements was found. However, no lapse rate based correction is applied as further discussed in Section 5.1.

4.2. Coupled Surface Energy Balance - Firn Model

To simulate the evolution of the surface and subsurface energy and mass fluxes we apply a model previously used for glaciers on Svalbard (Van Pelt et al., 2012, 2014; Van Pelt and Kohler, 2015). The model consists of two parts describing processes at and below the surface. The surface scheme developed along the lines of Klok and Oerlemans (2002) describes the energy and mass fluxes at the glacier surface and solves for the skin temperature, melt rate, and accumulation. The one dimensional subsurface scheme is based on the model presented by Greuell and Konzelmann (1994) modified by Reijmer and Hock (2008) and Van Pelt et al. (2012) to incorporate updated physics of liquid water storage, gravitational compaction, and heat conduction.

4.2.1. Surface Energy and Mass Fluxes

The surface scheme accounts for the following mass fluxes at the surface: accumulation due to solid precipitation and riming and mass loss due to melt and sublimation. The energy fluxes (in $W\,m^{-2}$) at the surface are described by the equation:

$$Q_m = SW_{net} + LW_{net} + Q_s + Q_{lat} + Q_{rain} + Q_{sub},$$

where Q_m is the energy available for melt, SW_{net} and LW_{net} are the short and long wave radiation balances, respectively, Q_s and Q_{lat} are the sensible and latent turbulent heat fluxes, Q_{rain} is the energy supplied by rainfall, Q_{sub} is the subsurface conductive heat flux. Detailed information on the formulations for individual energy fluxes can be found in Klok and Oerlemans (2002) and Van Pelt et al. (2012) and references therein.

On each time step the surface energy balance scheme estimates the above listed fluxes and solves for the surface temperature. In case the resulting value exceeds $0°C$, it is reset to ice melt temperature and the fluxes dependent on surface temperature (LW_{net}, Q_s, Q_{lat}, and Q_{sub}) are recomputed to produce the Q_m.

4.2.2. Firn Model

At each time step the subsurface model simulates temperature (T, $°C$), density (ρ, $kg\,m^{-3}$) and "wet" gravimetric water content ($\theta_w = m_w/m_{tot}$, grav.%, where m_w is the mass of water and m_{tot} is the total mass) of multiple snow, firn and ice layers. The model uses a moving grid in which the boundaries between layers are kept constant over time (Lagrangian scheme). That allows to take into account the effects of vertical advection of snow and firn layers with respect to the surface induced by mass fluxes at the surface. Gravitational settling of the layers is described following Ligtenberg et al. (2011): density of the layers increases with time as a function of the accumulation rate and local subsurface temperature. Conductive heat exchange in the profile is described by the Fourier's law with the conductivity of individual layers parameterized as a function of their density following Calonne et al. (2011). It is assumed that at the lower boundary of the model domain the heat flux equals zero. Although, the simulations are performed for a *ca.* 25 m thick domain, the output only for the upper 12 m is analyzed in detail, as the vertical extent of the empirical dataset on subsurface temperature and density is limited.

Refreezing of surface liquid water strongly influences the subsurface conditions. The routines applied to describe the processes at each time step of the model are described in **Figure 3**. The original subsurface scheme is built following the widely applied "bucket" approach and is referred to below as the "default water infiltration scheme." According to it all water is first allocated to the topmost layer after which its downwards propagation is estimated. Water is allowed to advance to the underlying layer when the refreezing capacity (*RC* in **Figure 3**) of the current one is eliminated and the gravimetric water content exceeds the maximum gravimetric water content (or water holding capacity—*WHC* in **Figure 3**) — potential of capillary and adhesive forces to retain water against gravity. Water holding capacity of snow and firn is parameterized as a function of density following Schneider and Jansson (2004) and decreases from 11 to 2 grav.% with the density rise from 350 to 900 $kg\,m^{-3}$. Temperature and density of subsurface layers changes in response to the refreezing of water. The

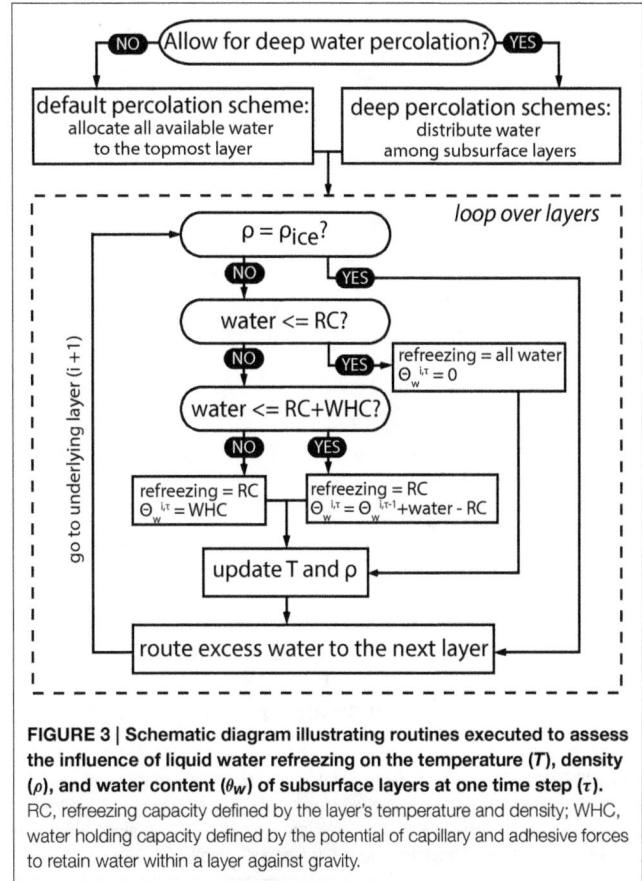

FIGURE 3 | Schematic diagram illustrating routines executed to assess the influence of liquid water refreezing on the temperature (*T*), density (*ρ*), and water content (*θ_w*) of subsurface layers at one time step (*τ*). RC, refreezing capacity defined by the layer's temperature and density; WHC, water holding capacity defined by the potential of capillary and adhesive forces to retain water within a layer against gravity.

model also describes buildup of slush water layer on top of the lowermost solid ice layer that is assumed to be impermeable and superimposed ice formation as a result of freezing of the slush layer. However, in our case no slush water is generated in the upper 15 m as density never reaches the density of ice there. Runoff is possible either from the bottom of the domain or from the slush layer as a function of time (Zuo and Oerlemans, 1996).

4.2.3. Deep Water Infiltration Schemes

Motivated by previous studies pointing to the highly inhomogeneous pattern of water flow in snow (e.g., Williams et al., 2010) and associated uncertainties in simulating the subsurface mass and energy fluxes (e.g., Gascon et al., 2014; Langen et al., 2017; Steger et al., 2017) we implement different parameterizations of the deep water percolation in snow and firn. Preferential water flow essentially concentrates water flow at certain parts of the snow and firn profile even if water supply at the surface exhibits no lateral variability (e.g., Katsushima et al., 2013). This effect is schematically illustrated by **Figure 4A**, which, however, is not based on field or simulation data. In a horizontally averaged 1-D case the phenomena can be expressed by allowing fractions of surface water to penetrate to different depths.

In each model time step the mass of liquid water available at the surface is distributed among the subsurface layers before the effect of refreezing is assessed (**Figure 3**). The distribution is

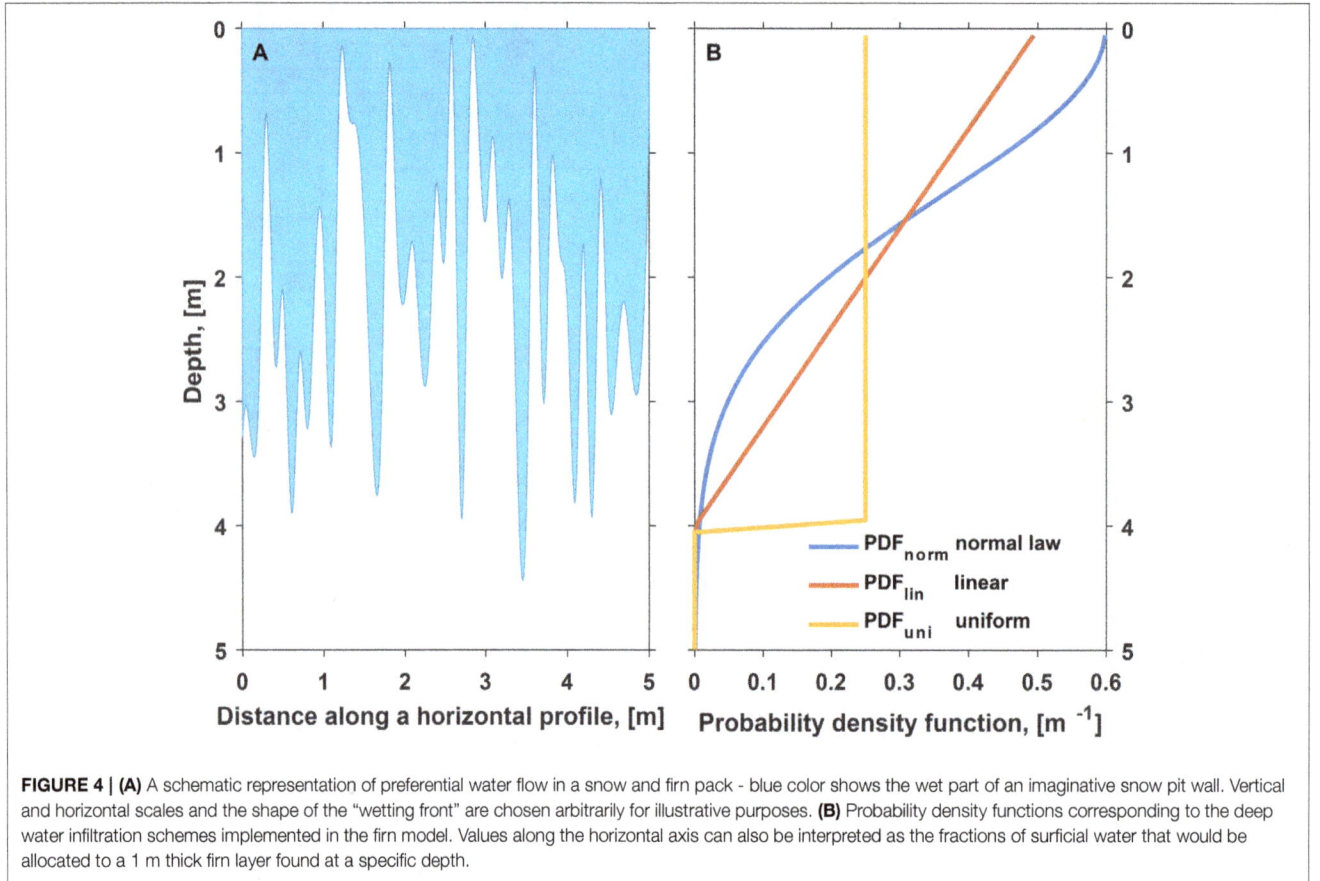

FIGURE 4 | (A) A schematic representation of preferential water flow in a snow and firn pack - blue color shows the wet part of an imaginative snow pit wall. Vertical and horizontal scales and the shape of the "wetting front" are chosen arbitrarily for illustrative purposes. (B) Probability density functions corresponding to the deep water infiltration schemes implemented in the firn model. Values along the horizontal axis can also be interpreted as the fractions of surficial water that would be allocated to a 1 m thick firn layer found at a specific depth.

governed by a probability density function $PDF(z, z_{lim})$, where argument z (m) is depth and z_{lim} (m) is the single tuning parameter—the depth below which no (or only a small fraction) liquid water is allocated. Since liquid water is generated at the surface it was assumed that deeper layers generally receive less water and the $PDF(z, z_{lim})$ function decreases with depth. We test three different implementations of this probabilistic approach with different shapes of the $PDF(z, z_{lim})$ function. Corresponding curves are presented in **Figure 4B**. In contrast to the default infiltration scheme, the tested deep percolation routines allow for instantaneous percolation below 0°C level, thereby mimicking the effect of piping at the lateral scale of several meters.

Uniform water infiltration scheme (yellow curve in **Figure 4B**) assumes that water is equally distributed above z_{lim} and no water is allowed to infiltrate below this level:

$$PDF_{uni}(z, z_{lim}) = \begin{cases} \frac{1}{z_{lim}}, & \text{if } z < z_{lim} \\ 0, & \text{otherwise} \end{cases}$$

According to the linear water infiltration scheme (red curve in **Figure 4B**) the $PDF(z, z_{lim})$ function decreases with depth at a constant rate and reaches 0 at z_{lim}:

$$PDF_{lin}(z, z_{lim}) = \begin{cases} \frac{2(z_{lim}-z)}{z_{lim}^2}, & \text{if } z < z_{lim} \\ 0, & \text{otherwise} \end{cases}$$

The normal law water infiltration scheme (blue curve in **Figure 4B**) distributes the water mass according to the corresponding probability density function with the standard deviation $\sigma = z_{lim}/3$, implying that 99.7% of the water stays above z_{lim}:

$$PDF_{norm}(z, z_{lim}) = 2\frac{\exp(-\frac{z^2}{2\sigma^2})}{\sigma\sqrt{2\pi}},$$

Thus with the same z_{lim} value the relative amount of surficial water allocated to the firn layers just above z_{lim} is maximum for the uniform infiltration scheme and minimal for the normal-law percolation. Using the above defined $PDF(z, z_{lim})$ functions we estimate the fraction of the surface water that is added to subsurface layer i with thickness dz_i as:

$$k^i = PDF(z_i, z_{lim}) \cdot dz_i.$$

The suggested schemes describing deep water percolation do not incorporate the physical details of the water transport processes occurring in snow and firn but are rather a statistical description of the horizontally-averaged effect of deep percolation on firn conditions at the scale of several meters. Although, we optimize the model to minimize the mismatch with the averaged temperature evolution measured on a 6 × 6 m plot, the results may also be useful for larger scale applications.

4.3. Model Setup

We set the values of the tuning parameters in the surface energy balance scheme such as the aerosol transmissivity exponent, emissivity exponent at clear-sky and overcast conditions, turbulent exchange coefficient following Van Pelt et al. (2012), who carried extensive calibration experiments at Nordenskiöldbreen and Lomonosovfonna using the data from an AWS at 600 m a.s.l. and readings at multiple stakes. In line with Van Pelt et al. (2014) the value of the fresh snow albedo was set to 0.89, which also resulted in best fit between the simulated and measured LW↑.

The coupled energy balance-firn model was run for a single point at the study site (**Figure 1B**) with a 3 h temporal resolution during 1 April 2012–1 May 2015. The subsurface scheme is based on a one-dimensional vertical domain comprising 250 layers with an initially uniform thickness of 0.1 m. The firn model is initiated by the temperature and density profiles measured in April 2012. On the 23^d and 18th April 2013 and 2014, respectively, the simulated density and temperature distributions are reset to the values measured on the corresponding dates to minimize cumulative errors. The gravimetric liquid water content in the profile is reset accordingly to ensure that no liquid water is present in layers at subfreezing temperature.

To explore the effect of deep water percolation on the simulated snow and firn temperature evolution the infiltration schemes described in Section 4.2.3 were implemented and tested for a range of z_{lim} values: 0.1 m and 0.5–25 m with a step of 0.5 m. We also investigate the model response to perturbation of the surface energy balance by an additional term E_+ set to −5 and 5 W m^{-2}, which corresponds to a *ca.* 22% decrease and a *ca.* 27% increase in surface melt rate, respectively. For the default infiltration scheme additional experiments were also performed for a range of E_+ values up to 50 W m^{-2}.

5. RESULTS

5.1. Surface Energy Fluxes

The simulated surface conditions at Lomonosovfonna are validated by comparing the output of the WRF model and of

the surface energy balance scheme with the AWS and stake measurements. The measured and simulated evolution of air temperature is presented in **Figure 5**. The modeled mean annual temperature at the study site during April 2012–April 2015 is −11.7°C and shows a pronounced seasonal cycle. In summer months the temperature reaches a few degrees above zero while during cold spells in winter it can drop down to −30°C (**Figure 5A**). The WRF model reproduces well the observed air temperature (**Figure 5B**, **Table 2**) with a root mean square difference (RMSD) of just below 2°C and an underestimation by −0.44°C averaged over the period of observation.

Agreement between the simulated and measured radiative energy fluxes is illustrated by the scatter plots in **Figure 6**. The mean deviation and RMSD values between the measured and modeled LW↓ values are 9 and 39 W m^{-2}, respectively (**Table 2**). The cloud of points in **Figure 6A** describing the model performance in reproducing the LW↓ can visually be divided into three parts. The points lying along the diagonal line correspond to a good match between the measured and simulated values, while the points occurring above and below the diagonal line indicate respectively over- and under-estimation of the energy flux by the model. Since, the LW↓ flux is heavily dependent on the cloud cover (e.g., Klok and Oerlemans, 2002; Stephens, 2005), we suggest that the three clusters of points are respectively

TABLE 2 | Model performance in reproducing the air temperature (*T$_a$*) and radiative fluxes measured at the AWS (see Figures 5, 6).

Parameter	k	b	r	RMSD	MD
T_a	0.94	−0.9	0.97	1.9	−0.4
LW↓	0.63	105.0	0.65	38.9	9.2
LW↑	0.89	31.7	0.89	11.4	1.7
SW↓	0.64	60.0	0.78	83.6	−18.7
SW↑	0.65	57.6	0.75	81.9	−9.9

k and b are the parameters in the linear fit: S = k · M + b (where S and M are the simulated and measured values), r is the Pearson correlation coefficient, RMSD is the root mean square difference and MD is the mean difference (bias) between the simulated and measured values.

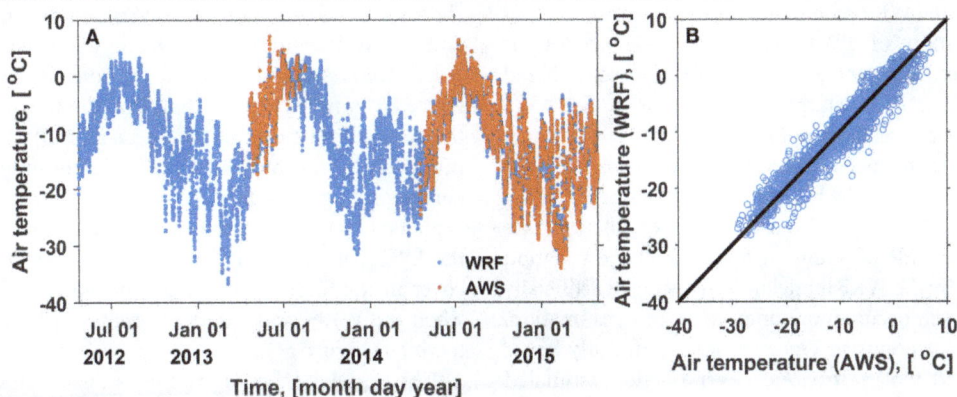

FIGURE 5 | Air temperature at the Lomonosovfonna ice field in April 2012–2015. (A) Values simulated using the regional climate model WRF (blue), AWS measurements (red). **(B)** Scatter plot illustrating the agreement between the simulated and measured values.

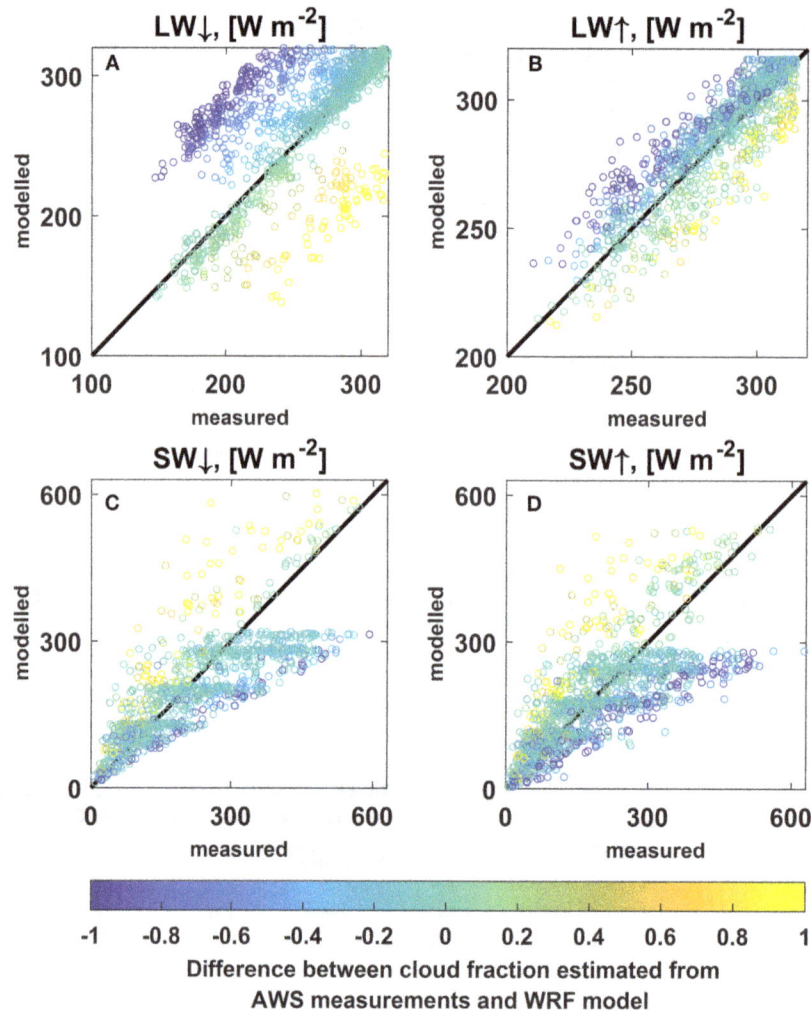

FIGURE 6 | Scatter plots illustrating the agreement between the measured and simulated intensities of the radiative energy fluxes at the Lomonosovfonna ice field during 23 April–14 July 2013 and 28 April–26 August 2014. (A) Long wave downwelling radiation. **(B)** Long wave upwelling radiation. **(C)** Short wave downwelling radiation. **(D)** Short wave upwelling radiation.

explained by: small errors, over- and underestimation of the cloud cover (n) by the WRF model.

This interpretation of LW↓ scatter plot (**Figure 6A**) is supported by an independent estimation of the n values based on the LW↓ and air temperature measurements at the AWS following the approach of Kuipers Munneke et al. (2011) which assumes a linear dependence between these two parameters for any given air temperature. The resulting differences between the AWS-estimated and WRF-predicted cloud cover fractions are shown as the color code of points in **Figure 6**. Overestimation of the cloud cover by the WRF model indeed results in artificially high LW↓ values and occurs more often than underestimation. The upper cluster of points in **Figure 6A** is significantly larger than the lower and the mean cloud cover fraction estimated from the LW↓ and air temperature measurements is 0.65, while the average of WRF derived values for the same period is 0.85.

The mean differences between the simulated and measured short wave radiative flux values (**Figures 6C,D; Table 2**) are −19 and −10 W m^{-2} for the downwelling and upwelling components, respectively, although RMSD are considerably larger and are above 80 W m^{-2}. We also find that the mean daily albedo (i.e., the averaged relation between mean daily SW↑ and SW↓ values) produced by the model (0.88) is significantly higher than the corresponding measured value (0.80), which is most probably explained by the measurement uncertainties related to the tilt of the AWS mast and riming of sensors. Overestimation of the cloud cover in the forcing data results in reduced SW↓ values, while when the WRF model underestimates the cloud cover, SW↓ is too high (**Figure 6C**).

The negative effect of the overestimated cloud cover on SW↓ is more than compensated by the positive effect on the LW↓. **Figure 7** illustrates the cumulative melt produced by the model during the 3 year period. It is apparent that the model forced by

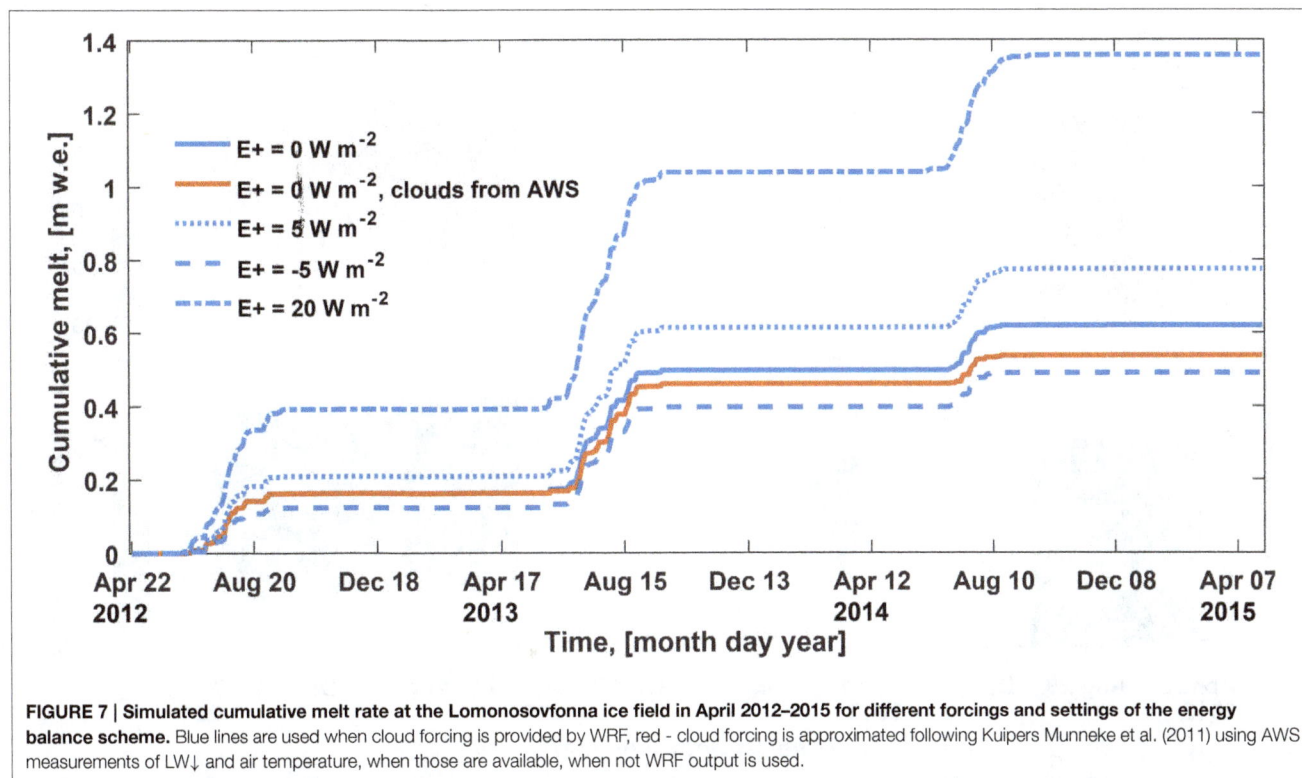

FIGURE 7 | Simulated cumulative melt rate at the Lomonosovfonna ice field in April 2012–2015 for different forcings and settings of the energy balance scheme. Blue lines are used when cloud forcing is provided by WRF, red - cloud forcing is approximated following Kuipers Munneke et al. (2011) using AWS measurements of LW↓ and air temperature, when those are available, when not WRF output is used.

overall lower n values based on AWS measurements, when those are available, produces significantly less melt water compared to the reference run done using the cloud cover from WRF (solid red and blue curves on **Figure 7** correspondingly). The effect is also seen in **Figure 6B** illustrating the agreement between measured and simulated LW↑: overestimation of cloud cover by WRF also results in higher simulated LW↑ values than measured by the radiometer (blue points).

The upwelling long wave radiative energy flux appears to be in close agreement with the measurements (**Figure 6B**, **Table 2**). LW↑ is a function of the surface temperature and is defined by the sum of all energy fluxes at the surface. High correlation (0.89) between the modeled and measured LW↑ (**Table 2**) hence suggests the high accuracy of simulated surface temperature, timing of melt events and also of the melt rate, assuming that the physics of energy fluxes above subfreezing and melting surfaces are similar. In absence of data to directly validate melt rate at the site this provides an important indirect source of validation of simulated melt rates. The absence of a clear bias in simulated surface temperature can be ascribed to compensating effects of the uncertainties in simulated cloud cover (overall positive), air temperature and surface albedo (negative), and turbulent fluxes.

The average winter (*ca.* August to April) and annual surface height change measured at stake S11 in April 2012–2015 is 1.8 and 1.3 m correspondingly. The average simulated surface height change during the period starting at the surface's lowermost position in July–September (12, 8, and 26 August 2012, 2013, and 2014) and ending at the day when the stake is revisited in April next year is 1.7 m. Estimation of snow gravitational

settling during the time following Ligtenberg et al. (2011) yields *ca.* 0.1 m and thus the surface height change is reduced to 1.6 m. The average annual increment in the surface height during April 2012–April 2015 is 1.7 m. If snow and firn settling above the layer that in April 2012 is at the depth of 8 m (stake length below the surface) is taken into account following Ligtenberg et al. (2011) the value decreases to 1.3 m. The close agreement between modeled and observed dynamics of the surface height suggests decent performance of the surface scheme forced by the WRF model output in reproducing the energy and mass fluxes at our field site on Lomonosovfonna.

The amount of melt water produced by the reference run of the model (blue solid curve in **Figure 7**) in 2013 (0.34 m w. e.) is significantly larger than in 2012 (0.16 m w. e.) and in 2014 (0.12 m w. e.). This finding agrees with glacier-averaged stake mass balance data for Nordenskiöldbreen (unpublished data described in Van Pelt et al., 2012) according to which the summer mass balance in 2012, 2013, and 2014 was −0.37, −0.79, and −0.35 m w.e., respectively.

5.2. Observed Snow and Firn Temperature Evolution

The measured snow and firn temperature evolution averaged over all thermistor strings installed in each of the field campaigns is presented in **Figure 8A**. The vertical depth scale is corrected using the modeled change in the surface height and thickness of subsurface layers to account for the accumulation and ablation at the surface and gravitational settling below it.

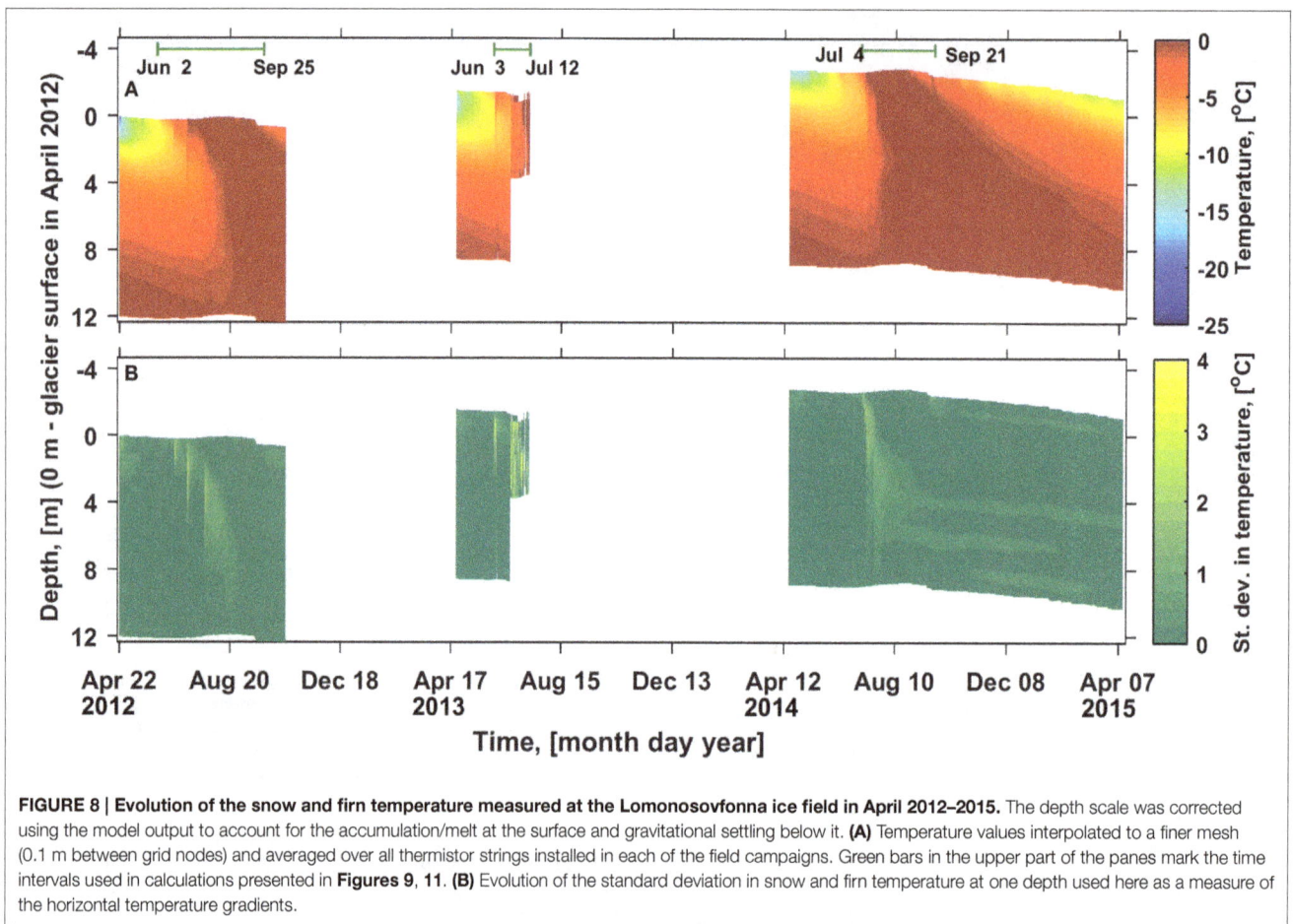

FIGURE 8 | Evolution of the snow and firn temperature measured at the Lomonosovfonna ice field in April 2012–2015. The depth scale was corrected using the model output to account for the accumulation/melt at the surface and gravitational settling below it. **(A)** Temperature values interpolated to a finer mesh (0.1 m between grid nodes) and averaged over all thermistor strings installed in each of the field campaigns. Green bars in the upper part of the panes mark the time intervals used in calculations presented in **Figures 9, 11. (B)** Evolution of the standard deviation in snow and firn temperature at one depth used here as a measure of the horizontal temperature gradients.

In April of each year temperature increases from *ca* −15°C at the surface to 0°C at *ca* 10 m depth (**Figure 8A**). Data from June and July 2012 and 2013 shows that warming of the profile occurred in several steps associated with events of intensive melt at the surface (**Figure 7**). The entire profile was temperate by August 24 in 2012 and by August 9 in 2014. Data for the later part of the summer season in 2013 is not available, however, considering the high melt rates simulated for that year (**Figure 7**), it can be expected that the upper 12 m of the firn profile became temperate even earlier than in the other two years. Snow and firn temperature evolution during the autumn and winter months was only measured in 2014. It is apparent that in the absence of surface melt starting from the first days of September, the layer of temperate snow and firn is eliminated due to the conductive heat flux toward the cooling surface.

To characterize the lateral variability of the snow and firn temperature measured simultaneously at the same depth standard deviations for corresponding sets of values from multiple thermistor strings were calculated (**Figure 8B**). The standard deviations reach up to 4°C and during all three melt seasons the areas with increased lateral variability in snow and firn temperature descend to greater depth over time following the 0°C isotherm, where intensive refreezing can be expected. The observation can be related to the horizontal gradients in the rate

of downward water flux and associated migration of maxima in lateral temperature variability. It is particularly obvious during the ablation season in 2012, when the melt water generated at the surface during the first warm spell on the 20th June reached no deeper than 2 m, while the consecutive ablation events on the 4th and 23rd July likely brought the water deeper: to 5 and 9 m at maximum, respectively. In 2014 increased lateral variability in snow and firn temperature is also observed during the autumn and winter months at 4.2, 6.3, and 8.5 m (**Figure 8**). Referenced to the glacier surface in April 2014 these depths correspond to 7, 9.1, and 11.3 m. The phenomena can be related to perched horizons with increased water content that have a horizontal extent less than the plot covered by our thermistor measurements (6×6 m). In April 2014 Marchenko et al. (2017) found multiple ice layers at 7, 8, 9, and 10.8 m depth in three boreholes where the thermistor strings used in present study were installed afterwards. These ice layers could have acted as impermeable horizons and retained water which locally inhibited the cold wave penetration.

The occasional high lateral temperature gradients during the 3 years of observations at Lomonosovfonna indicate a possible uncertainty in subsurface temperature measurements performed at single locations in accumulation zones of glaciers. Higher values of the standard deviations (**Figure 8B**) during the summer period suggest that the largest uncertainty can be expected during

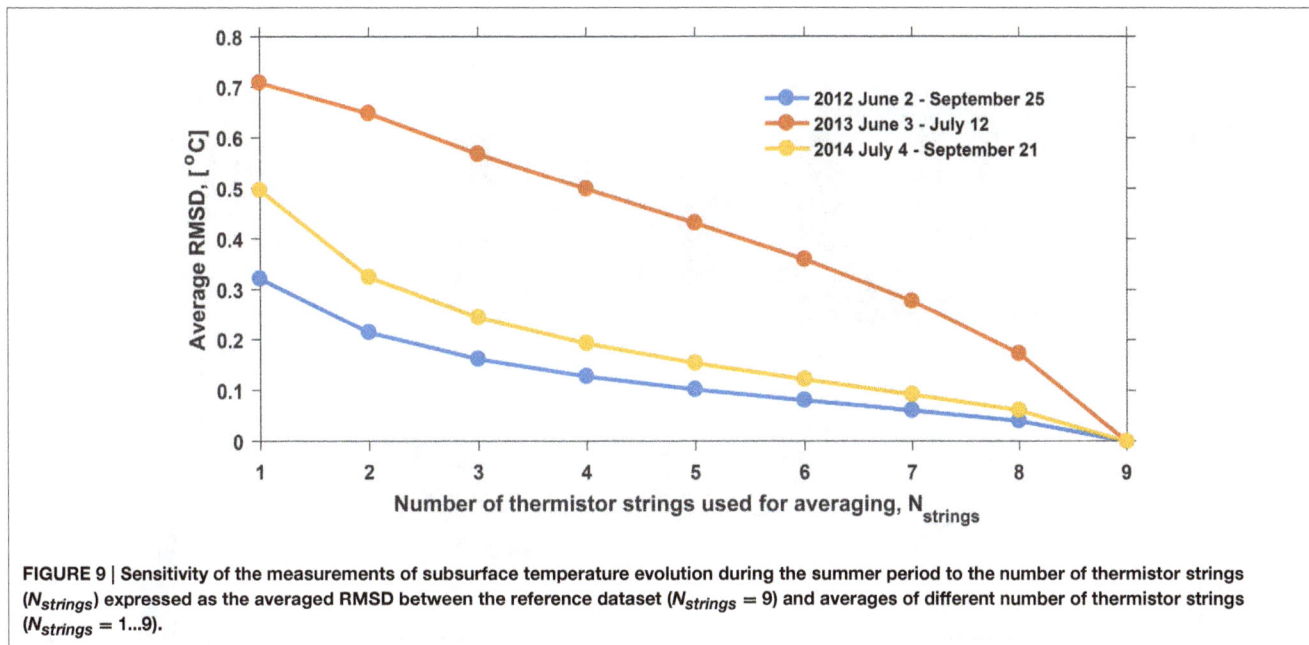

FIGURE 9 | Sensitivity of the measurements of subsurface temperature evolution during the summer period to the number of thermistor strings ($N_{strings}$) expressed as the averaged RMSD between the reference dataset ($N_{strings} = 9$) and averages of different number of thermistor strings ($N_{strings} = 1...9$).

liquid water infiltration in the snow and firn. Data from multiple thermistor strings installed in close vicinity of each other and monitored simultaneously during three melt seasons allows to investigate the effect of averaging the measured temperature evolution over a varying number of thermistor strings.

Firstly for each of the three periods with rapid subsurface temperature changes (see the green bars in **Figure 8A**) a reference dataset was defined as an average of all nine thermistor strings available ($N_{strings} = 9$). Secondly, multiple sample datasets were constructed by averaging data from $N_{strings} = 1...9$ thermistor strings. The resulting average RMSD between the reference and the samples are presented in **Figure 9**. For the 2012 and 2014 datasets the RMSD value decreases from $\approx 0.4°$C for single thermistor string to $\approx 0.25°$C for 3 strings, while the RMSD for larger $N_{strings}$ values decrease at a slower rate. For the dataset collected in 2013 the RMSD values also show a decrease with the number of strings but the values are generally higher and the concave pattern in lacking. We relate these observations to the shorter period of measurements and a lesser number of thermistors scanned during the season (see **Figure 2**) and note that the subsurface temperature dataset collected during 2013 is the most uncertain. Thus it becomes apparent that monitoring the summer evolution of subsurface temperature using multiple thermistor strings is beneficial as the resulting laterally averaged dataset is more representative for the study site.

5.3. Modeling Snow and Firn Temperature Evolution

Figure 10 presents the evolution of snow and firn temperature simulated with $E_+ = 0$ W m^{-2} and using the conventional water percolation scheme not allowing for deep water infiltration (**Figure 10A**) along with the corresponding deviations from the measurements (**Figure 10B**) and simulated gravimetric water content (**Figure 10C**). The general pattern of subsurface

temperature evolution agrees with the results of measurements. In May and June a gradual warming occurs due to the increase of air temperature. Between mid-June and July infiltrating water causes a fast downward expansion of the temperate snow and firn layer. Evolution of the water content profile follows the firn temperature changes: water saturates the potential of the capillary and adhesive forces immediately after the layer in question becomes temperate. In September, after surface melt ceases, the subsurface profile gradually cools, and the water suspended in pores is refrozen. The temperature below ≈ 10–12 m remains always at $0°$C, it is set to the observed state every April and the conductive cold wave does not reach that deep during 1 year.

Despite the general agreement between the simulated and measured subsurface temperature, the vertical extent and strength of perturbations in the snow and firn temperature deviate considerably from the observations (**Figure 10B**) with the simulated profile being generally colder. The difference between the simulated and measured temperature reaches up to 12.5 and $-8°$C. While only 0.3% of the RMSD values are $>1°$C, 42.3% of the deviations are below $-1°$C. Biases are the largest in July, when high surface melt rate results in fast and deep water infiltration, which is not captured by the firn model in its default configuration. Positive errors in simulated temperature occur very seldom and are restricted to the upper 1 m of the profile. We note that the depth of elevated negative anomalies increases over time as the deeper firn layers become temperate under the influence of infiltrating water. By the end of all three simulated ablation seasons the subsurface profile did not reach the temperate state. Cold firn was still present below 2 m depth in 2012 and 2014, and below 6.5 m in 2013, when the surface melt rate was higher and the temperate near-surface firn layer almost connected with the deep temperate firn.

Next we explore the sensitivity of the firn model results to different water infiltration schemes and the value of the

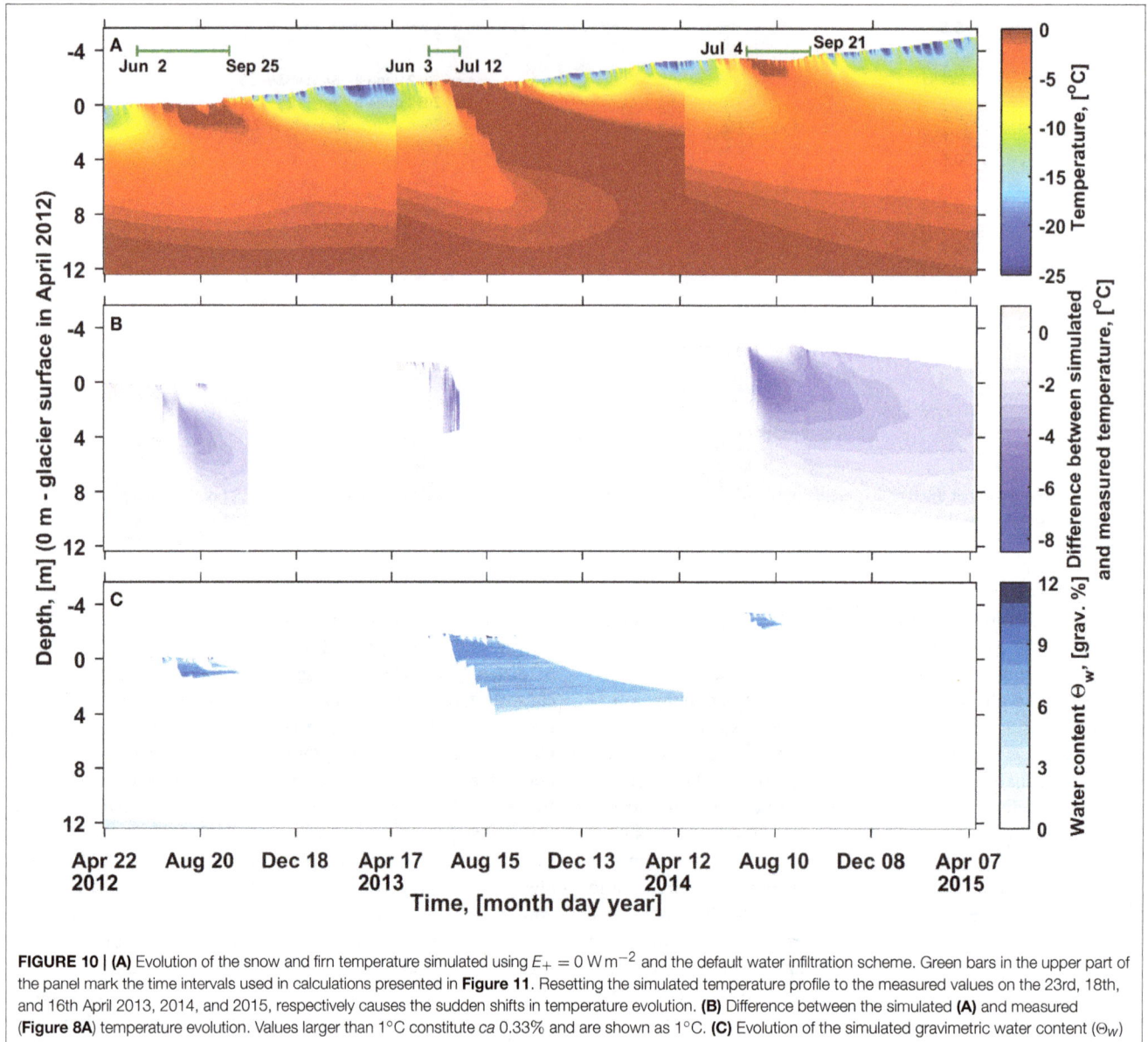

FIGURE 10 | (A) Evolution of the snow and firn temperature simulated using $E_+ = 0\,\mathrm{W\,m^{-2}}$ and the default water infiltration scheme. Green bars in the upper part of the panel mark the time intervals used in calculations presented in **Figure 11**. Resetting the simulated temperature profile to the measured values on the 23rd, 18th, and 16th April 2013, 2014, and 2015, respectively causes the sudden shifts in temperature evolution. **(B)** Difference between the simulated **(A)** and measured **(Figure 8A)** temperature evolution. Values larger than 1°C constitute *ca* 0.33% and are shown as 1°C. **(C)** Evolution of the simulated gravimetric water content (Θ_w)

percolation depth z_{lim}. The RMSD between simulated and measured subsurface temperature during the periods of rapid changes (**Figure 11**) is 1.5–2.8°C for the default infiltration scheme and all three deep infiltrations schemes with z_{lim} set to 0.1 m. With the increase of z_{lim} to 25 m the RMSD values in all cases exhibit a gradual decrease approximately by a factor of 2 and a subsequent increase almost to the initial values. The presence of a minimum in each case suggests that implementation of the deep water percolation scheme significantly improves the firn model performance in reproducing the measured subsurface temperature evolution. The z_{lim} values corresponding to the minimal RMSD between the simulated and measured temperature evolution lie in a wide range from 4.5 to 12 m for different years and parameterizations, with a consistent pattern of the uniform percolation exhibiting

the smallest values (4.5–6 m) and the normal law percolation the largest (7.5–12 m). At the same time, the minimal RMSD values reached for the different infiltration schemes vary only slightly.

Sensitivity experiments for the default configuration of the firn model revealed that decreasing the amount of melt ($E_+ = -5\,\mathrm{W\,m^{-2}}$) results in slightly higher values of the RMSD between the simulated and measured temperature evolution. Higher melt rates ($E_+ > 0\,\mathrm{W\,m^{-2}}$) improve the model performance (**Figure 11**). While a moderate perturbation of $E_+ = 5\,\mathrm{W\,m^{-2}}$ results only in a slight reduction of the RMSD, with significantly higher E_+ values it is possible to reach a similar effect as by implementation of the deep percolation schemes. However, that is only possible with $E_+ = 20\,\mathrm{W\,m^{-2}}$ in 2012, $E_+ = 22\,\mathrm{W\,m^{-2}}$ in 2013, and $E_+ = 32\,\mathrm{W\,m^{-2}}$ in 2014, which corresponds to increase in the annual melt amount by a factor 2.2, 1.9, and 4.2

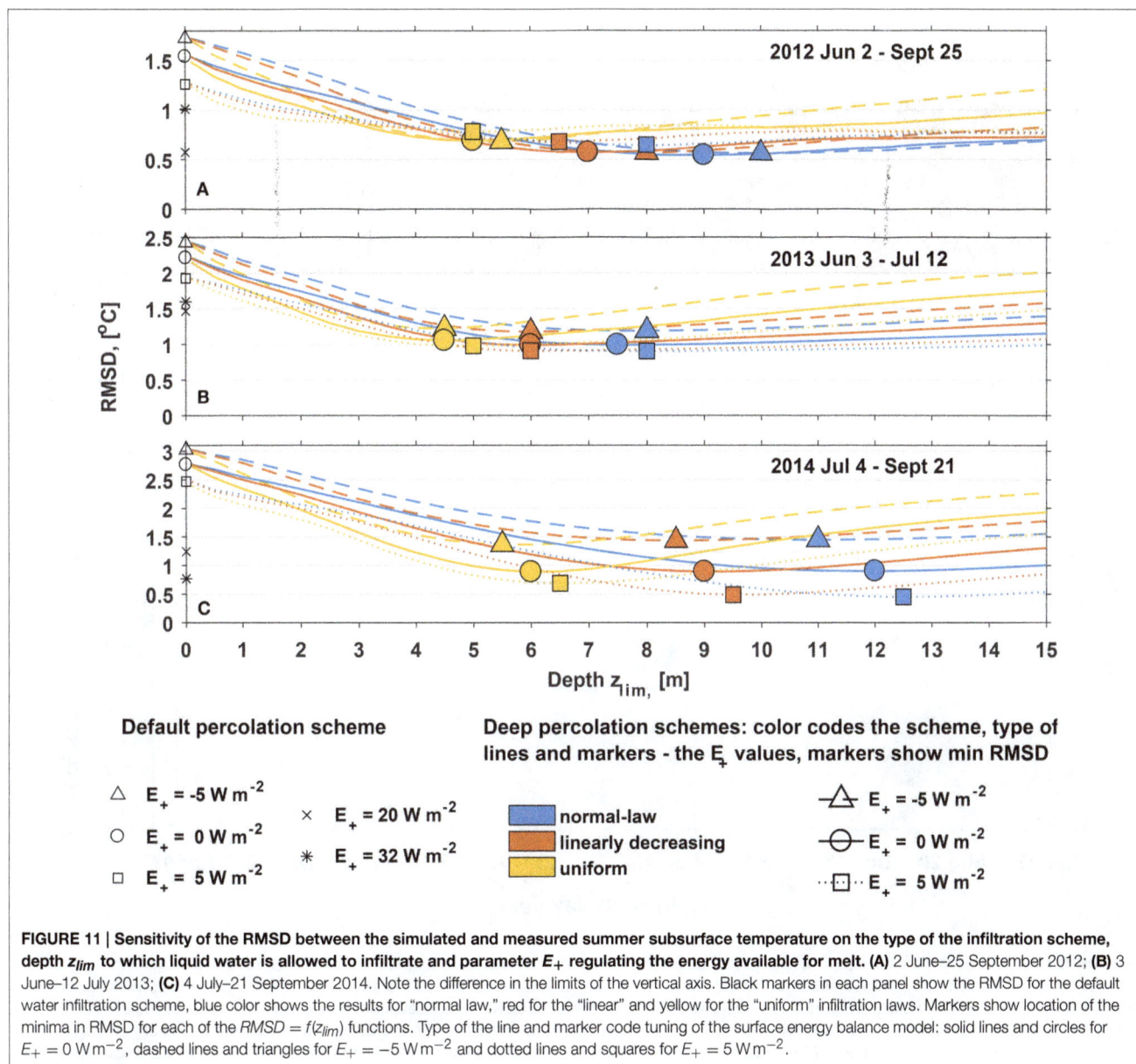

FIGURE 11 | Sensitivity of the RMSD between the simulated and measured summer subsurface temperature on the type of the infiltration scheme, depth z_{lim} to which liquid water is allowed to infiltrate and parameter E_+ regulating the energy available for melt. **(A)** 2 June–25 September 2012; **(B)** 3 June–12 July 2013; **(C)** 4 July–21 September 2014. Note the difference in the limits of the vertical axis. Black markers in each panel show the RMSD for the default water infiltration scheme, blue color shows the results for "normal law," red for the "linear" and yellow for the "uniform" infiltration laws. Markers show location of the minima in RMSD for each of the $RMSD = f(z_{lim})$ functions. Type of the line and marker code tuning of the surface energy balance model: solid lines and circles for $E_+ = 0\,\mathrm{W\,m^{-2}}$, dashed lines and triangles for $E_+ = -5\,\mathrm{W\,m^{-2}}$ and dotted lines and squares for $E_+ = 5\,\mathrm{W\,m^{-2}}$.

for the 3 years, respectively. Given the above described model performance in reproducing the observed energy and mass fluxes at the surface the major underestimation of melt rates seems unrealistic.

In most cases for any given z_{lim} value in the deep infiltration schemes setting E_+ to $-5\,\mathrm{W\,m^{-2}}$ (dashed lines) results in the highest RMSD values, while $E_+ = 5\,\mathrm{W\,m^{-2}}$ (dotted lines) leads to the minimal errors. However, the minimal RMSD values obtained with different settings of E_+ for the deep percolation schemes exhibit this dependence only in 2014, while in 2012 and 2013 the performance of the firn model depends only slightly on the settings of the surface melt rate.

Figure 12 presents the snow and firn temperature evolution simulated using $E_+ = 0\,\mathrm{W\,m^{-2}}$ and the uniform law percolation scheme with z_{lim}=5 m, which is the average of the three z_{lim}

values resulting in minimal RMSD for that scheme in 2012, 2013 and 2014. The misfit between the simulated and measured temperature evolution (**Figure 12B**) still reaches -8 and $12.5°C$, but the corresponding time and depth ranges are rather limited. Only 2.3% of the values are $>1°C$ and 13.6% are below $-1°C$, which is much lower than what was produced by the firn model in its default configuration (**Figure 10B**). A notable overestimation of the firn temperature occurs in July 2012, while in the other two seasons firn temperature is slightly underestimated. The firn profile reaches the temperate conditions by the end of ablation seasons in 2012 and in 2013 in line with the observations. However, in summer 2014 increase of the deep firn temperature remains underestimated. A better match is found with E_+ set to $5\,\mathrm{W\,m^{-2}}$ (blue square in **Figure 11C**) but a considerable cold firn layer is still present by the end of ablation season, suggesting

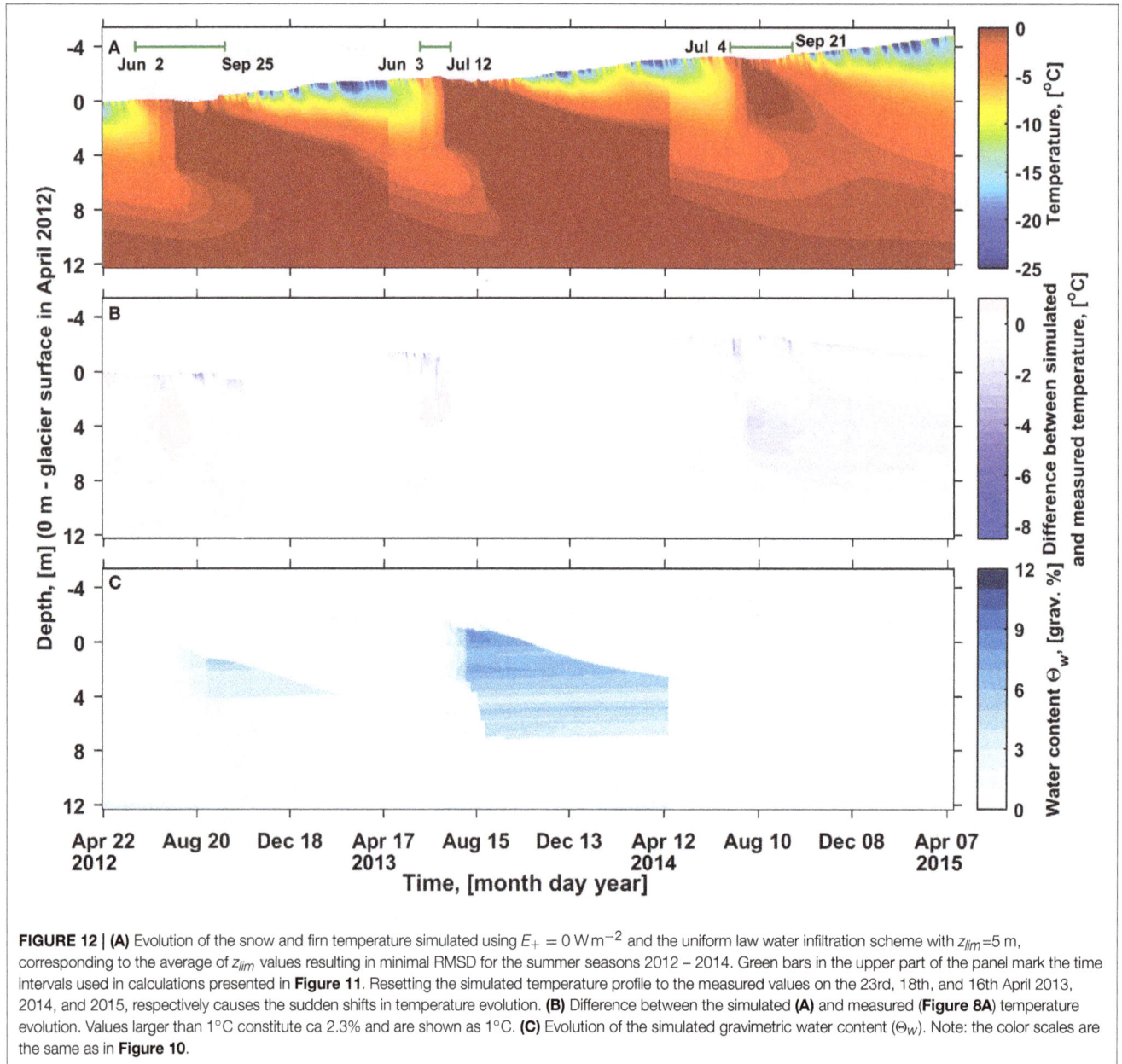

FIGURE 12 | (A) Evolution of the snow and firn temperature simulated using $E_+ = 0\,\mathrm{W\,m^{-2}}$ and the uniform law water infiltration scheme with z_{lim}=5 m, corresponding to the average of z_{lim} values resulting in minimal RMSD for the summer seasons 2012 – 2014. Green bars in the upper part of the panel mark the time intervals used in calculations presented in **Figure 11**. Resetting the simulated temperature profile to the measured values on the 23rd, 18th, and 16th April 2013, 2014, and 2015, respectively causes the sudden shifts in temperature evolution. **(B)** Difference between the simulated **(A)** and measured **(Figure 8A)** temperature evolution. Values larger than 1°C constitute ca 2.3% and are shown as 1°C. **(C)** Evolution of the simulated gravimetric water content (Θ_w). Note: the color scales are the same as in **Figure 10**.

that simulated melt rates are likely underestimated during this ablation period. The firn gravimetric water content (**Figure 12C**) follows the downwards propagation of the 0°C level but is considerably lower than in the case of default water percolations scheme (**Figure 10C**). The density profile is similar in both model realizations, suggesting no difference in the maximum gravimetric water content, and, consequently, significantly lower saturation values in case deep water percolation is assumed.

6. DISCUSSION

The air temperature simulated by WRF for the point at 1053 m a.s.l. is on average 0.4°C lower than the values measured by the AWS at 1,200 m a.s.l. even though an opposite tendency can be expected from the elevation difference of ≈150 m and

published lapse rate values for Svalbard of 4–6.6°C km⁻¹ (Wright et al., 2005; Nuth et al., 2012). Given the high correlation coefficient (0.97, see **Table 2**) between the measured and modeled values we conclude that in accordance with earlier results from Svalbard (Claremar et al., 2012; Aas et al., 2015, 2016) the WRF model demonstrated a prominent skill in reproducing the observed air temperature with a tendency for its underestimation.

The cloud cover generated by WRF clearly exceeds the approximations based on the AWS measurements of LW↓ and air temperature following Kuipers Munneke et al. (2011). In contrast with our findings, Aas et al. (2015) reported underestimation of the cloud optical depth at Ny Ålesund (20 m a.s.l.) and Etonbreen (370 m a.s.l., Austfonna). That inconsistency may be explained by the fact that our study site is at a considerably higher elevation.

We find that monitoring the evolution of subsurface temperature using multiple thermistor strings scanned simultaneously and installed in close vicinity from each other is advantageous. It allows to derive a more spatially representative dataset by averaging data from several strings. The latter is of particular relevance during the period of melt water infiltration occurring not homogeneously and resulting in the highest standard deviations in temperature measured at the same depth (**Figure 8B**). The shape of the $RMSD = f(N_{strings})$ function (**Figure 9**) suggests that representativity can be significantly improved by increasing the number of strings to 3 or 4. The typical values of RMSD between the reference and averages of different number of thermistor strings (**Figure 9**) are consistently and significantly lower than the RMSD between the simulated and measured subsurface temperature (**Figure 11**). This justifies utilization of the empirical data for constraining the suggested schemes describing deep water percolation and validating the modeling results. In a broader context it also proves the potential of temperature tracking of melt water in snow and firn packs (Conway and Benedict, 1994; Pfeffer and Humphrey, 1996; Humphrey et al., 2012; Cox et al., 2015; Charalampidis et al., 2016).

Subsurface temperature measurements at Lomonosovfonna provide evidence of warming that occurred during the last decades. The firn temperature measured at 10 m depth in August 1965 (Zinger et al., 1966, 1,050 m a.s.l.) and 1976 (Zagorodnov and Zotikov, 1980, 1,120 m a.s.l.) was in the range of -3 to $-2°C$ and the seasonal temperate surface layer reached the maximum depth of 2-3 m. This suggests that the upper reaches of the ice field, at that time, belonged to the cold firn zone according to the classification of Shumskii (1964). However, our results show that by the end of the ablation seasons 2012-2014 the firn at Lomonosovfonna was temperate down to 10 m meaning that the study site is now in the warm firn zone and runoff may occur.

Reproducing these observations appears to be challenging for the firn model with the default configuration of the water infiltration scheme (**Figure 10**). By the end of all three ablation seasons the simulated subsurface profile still contained thick layers of firn at subfreezing temperature, which eliminates the possibility of runoff. Underestimation of the snow and firn temperature also results in errors in the simulated conductive heat flux below the surface and consequently biased estimations of other components of the surface energy balance. Additionally, a higher viscosity of too cold firn and ice (Cuffey and Paterson, 2010) may induce biases in long-term simulations of glacier dynamics.

Increasing the surface melt rate by 27% by setting E_+ to 5 $W\,m^{-2}$ resulted only in moderate decrease of the RMSD between the simulated and measured firn temperature and did not produce a warm firn pack by the end of ablation seasons 2012-2014. Larger perturbations (E_+ = 20, 22, and 32 $W\,m^{-2}$ in 2012-2014) significantly reducing the RMSD between the simulated and measured subsurface temperature evolution seem unlikely. Serious changes in the settings of the surface energy balance scheme result in reduction of the performance in validation of the model output against data from

stake S11 and the AWS, particularly the LW↑ measurements. The spatial and temporal dynamics of the errors in simulated subsurface temperature evolution (**Figure 10B**) had the same principal characteristics as evolution of the standard deviations in temperature measured at the same depth (**Figure 8B**). Altogether this suggests that in conditions of short ablation period at Lomonosovfonna the latent heat of infiltrating and refreezing water heats up the subsurface profile more effectively than it is possible within the default infiltration scheme.

The tested deep percolation schemes allowed to significantly reduce the RMSD between the simulated and measured temperature evolution (**Figures 11, 12B**). The deep infiltration schemes result in more effective heating of the subsurface profile by the latent heat of refreezing water due to two mechanisms. Firstly, water going deeper releases latent heat at greater distance from the surface, where temperature gradients are low, and conductive heat exchange is not effective. At the same time, the upper part of the profile is allowed to maintain subfreezing temperature longer and by this be more readily warmed up by conduction from the melting surface. Secondly, by allowing a fraction of water to go past the upper firn horizons the potential of capillary and adhesive forces to retain some of the water in the upper layers is underused.

Two dimensional empirical data on the volumetric liquid water content of the seasonal snow packs presented by Techel and Pielmeier (2011) evidences that even though the source of liquid water is at the surface and the dominant flow is directed downwards, perched pockets with increased liquid water content can be present within snow pack along with pockets with reduced liquid water content. Thus the estimates of snow water holding capacity obtained by measuring the liquid mass retained by the sample after being entirely soaked and then drained (Coléou and Lesaffre, 1998; Schneider and Jansson, 2004) may be not immediately applicable in 1D snow and firn models at the onset of the water infiltration. Measurements of the volumetric water content in seasonal snow using upward directed radar conducted by Heilig et al. (2015) yielded the values of <2 vol. % at the onset of outflow from the snow pack registered by lysimeter. Converted to the relative gravimetric water content this corresponds to 6 grav. % assuming the density of 350 $kg\,m^{-3}$, which is significantly lower than the 10.7 grav. % resulting from the Schneider and Jansson (2004) parameterization. Earlier Reijmer et al. (2012) used a low water holding capacity value (2 grav. %) within the SOMARS model similar to the one used in present study to mimic the effect of preferential flow on subsurface conditions. However, such approach has the drawback of producing too high runoff rates once the water refreezing and retention capacity of the profile is exhausted.

The minimal RMSDs between the simulated and measured subsurface temperature (**Figure 11**) are consistently found with the largest z_{lim} values for the normal law infiltration (7.5-12 m) and with the smallest z_{lim} for the uniform scheme (4.5-6 m). This behavior of the firn model is explained by the properties of the probability density functions (**Figure 4B**). With the same settings of the z_{lim} parameter the uniform infiltration scheme allocates most water between $z_{lim}/2$ and z_{lim}, while the normal infiltration implies that the smallest amount of water reaches that deep. It can be noted that the high horizontal gradients in subsurface temperature (**Figure 8B**) were observed in the depth range that is most close to the z_{lim} values resulting in minimal RMSD derived using the uniform infiltration scheme (**Figure 11**). That is particularly apparent for the data from 2012. Therefore, we suggest that this pattern of water percolation most closely reproduces the subsurface water flow.

It has to be mentioned that constraining of the shape of the $PDF(z, z_{lim})$ function and of the z_{lim} values largely relies on the extensive subsurface temperature dataset collected at Lomonosovfonna. Implementation of the suggested approach to parameterization of the deep water percolation for other areas or time periods will require additional temperature measurements for validation of the model output. That might be rather challenging in case the firn model is run on a distributed grid.

In the same time the logic of the deep water infiltration parameterization is simple to implement and allows for further development. Potentially additional parameters can be incorporated in the suggested deep percolation functions for a more realistic description of the effect of preferential water flow. Firstly, the utilized assumption of a constant in time z_{lim} value is in contrast with the results of temperature measurements. From **Figure 8B** it is apparent that the maximum depth of water percolation increases over the course of a melt season, which calls for implementation of a time dependence in the model. Secondly, the rate of water infiltration is known to be strongly dependent on the water supply. This is the result of the positive feedbacks: between the relative water conductivity and the water saturation of snow for unsaturated flow (Colbeck, 1972) and between the hydraulic conductivity and grain size of snow for saturated flow (Shimizu, 1970). It is, however, not obvious whether preferential water flow is more pronounced under conditions of high surface melt. While it was found to be the case in soils (Glass et al., 1989), according to the experimental studies, sieved snow was not more prone to preferential water infiltration under high water supply rates (Katsushima et al., 2013). According to our results high water supply rates at the surface are associated with a less pronounced concentration of the flow in the snow and firn pack: in 2013 when most melt water was produced during the ablation period (**Figure 7**), the optimal z_{lim} values are minimal (**Figure 11**) and in 2014 the opposite situation is observed. Finally, subsurface density distribution (ice layers, snow/firn grain-size contrasts, etc.) greatly influences the pattern of water infiltration (Marsh and Woo, 1984a; Fierz et al., 2009), which could be taken into account by incorporating additional arguments in the model.

7. CONCLUSIONS

Evolution of subsurface temperature at 1200 m a.s.l. on Lomonosovfonna, Svalbard, was studied during April 2012–2015 using field measurements by multiple thermistor strings and simulations using a coupled surface energy balance–multilayer firn model. Climate forcing for the model is provided by output of the regional scale climate model WRF and the initial subsurface temperature and density distributions are approximated from field data. Simulated mass and energy fluxes at the surface are validated against stake and AWS data.

The snow and firn pack is found to be heavily influenced by liquid water refreezing during the summer period and associated release of the latent heat. By the end of ablation seasons 2012–2014 the upper 12 m of the snow and firn pack were isothermal at 0°C. Compared to the earlier measurements at the ice field, our results provide an evidence of the subsurface warming that occurred during the last decades. The horizontal gradients in subsurface temperature were largest during the period of active melt water infiltration in July and migrated downwards over time. They are interpreted as being associated with the preferential flow paths in firn below the background wetting front. As a consequence of inhomogeneous water flow in snow and firn the measurements from single thermistor strings may be not representative for the area. On the basis of measurements done simultaneously in nine closely placed thermistor strings during two melt seasons we suggest that averaging data from 3 to 4 thermistor strings can significantly increase reliability of the summer temperature data.

Firn model with the default configuration of the water infiltration scheme consistently produced an overall too cold firn profile during the summer seasons 2012–2014. Performance of the subsurface temperature simulations is significantly improved by introduction of deep water infiltration schemes. We describe the effects of preferential water flow in snow and firn using a probability density function that distributes surficial water among subsurface layers relying on a single tuning parameter—depth z_{lim} to which liquid water is allowed to reach. Three different implementations of the approach were tested.

Introduction of the deep infiltration schemes reduced the RMSD between the simulated and measured temperature by about a factor of 2, more effectively than increasing the melt rate by 27% and in line with a melt increase of *ca* 200–400% for different years. The minimal RMSD values were reached with z_{lim} set to 4.5–12 m with the largest and smallest values consistently corresponding to the normal law and uniform infiltration schemes.

The mechanism behind the suggested approach to description of preferential water flow in one dimensional snow and firn models relies on the under-use of the potential of the upper layers to refreeze and retain water against gravity due to the capillary and adhesive forces. Given the simple logic and dependence on the sole parameter (z_{lim}), suggested deep infiltration schemes can be implemented in firn models applied at a wide range of spatial scales and allow further development by introduction of additional arguments.

AUTHOR CONTRIBUTIONS

VP, RP, CR, and SM designed the study. SM designed the field experiments, modeling strategy, was responsible for preparation, and updating of the manuscript. CR and WvP provided the coupled surface energy balance-firn model and facilitated its utilization and implementation of the deep infiltration schemes.

BC ran the WRF model and provided the forcing data for the coupled surface energy balance-firn model. SM, WvP, HM, RP, VP contributed to collection of the empirical data in the field campaigns. All authors contributed to the preparation of the manuscript and its critical revision.

ACKNOWLEDGMENTS

This publication is contribution 85 of the Nordic Centre of Excellence SVALI funded by the Nordic Top-level Research Initiative. Authors appreciate the constructive feedback provided by the Editor and two Reviewers, their efforts helped to significantly improve the manuscript. Funding was also provided by the Vetenskapsrådet grant 621-2014-3735 (VP). Authors acknowledge the Swedish Polar Research Secretariat, the Norwegian Polar Institute and the University Centre in Svalbard for logistical support of the field campaigns. Additional funding of the field operations was provided by the Ymer-80 foundation, Arctic Field Grant of the Research Council of Norway, Margit Althins stipend of the Royal Swedish Academy of Sciences.

REFERENCES

Aas, K. S., Berntsen, T. K., Boike, J., Etzelmüller, B., Kristjánsson, J. E., Maturilli, M., et al. (2015). A comparison between simulated and observed surface energy balance at the Svalbard archipelago. *J. Appl. Meteorol. Climatol.* 54, 1102–1119. doi: 10.1175/JAMC-D-14-0080.1

Aas, K. S., Dunse, T., Collier, E., Schuler, T. V., Berntsen, T. K., Kohler, J., et al. (2016). The climatic mass balance of Svalbard glaciers: a 10-year simulation with a coupled atmosphere–glacier mass balance model. *Cryosphere* 10, 1089–1104. doi: 10.5194/tc-10-1089-2016

Albert, M., Koh, G., and Perron, F. (1999). Radar investigations of melt pathways in a natural snowpack. *Hydrol. Process.* 13, 2991–3000. doi: 10.1002/(SICI)1099-1085(19991230)13:18<2991::AID-HYP10>3.0.CO;2-5

Bartelt, P., and Lehning, M. (2002). A physical SNOWPACK model for the Swiss avalanche warning: Part I: numerical model. *Cold Regions Sci. Technol.* 35, 123–145. doi: 10.1016/S0165-232X(02)00074-5

Bassford, R. P., Siegert, M. J., Dowdeswell, J. A., Oerlemans, J., Glazovsky, A. F., and Macheret, Y. Y. (2006). Quantifying the mass balance of ice caps on Severnaya Zemlya, Russian high Arctic I: climate and mass balance of the Vavilov ice cap. *Arctic Antarct. Alp. Res.* 38, 1–12. doi: 10.1657/1523-0430(2006)038[0001:QTMBOI]2.0.CO;2

Bøggild, C. E. (2000). Preferential flow and melt water retention in cold snow packs in West-Greenland. *Nordic Hydrol.* 31, 287–300. Available online at: http://hr.iwaponline.com/content/31/4-5/287

Brun, E., Martin, E., Simon, V., Gendre, C., and Coleou, C. (1989). An energy and mass model of snow cover suitable for operational avalanche forecasting. *J. Glaciol.* 35, 333–342. doi: 10.1017/S0022143000009254

Calonne, N., Flin, F., Morin, S., Lesaffre, B., du Roscoat, S. R., and Geindreau, C. (2011). Numerical and experimental investigations of the effective thermal conductivity of snow. *Geophys. Res. Lett.* 38, L23501. doi: 10.1029/2011GL049234

Campbell, F. M. A., Nienow, P. W., and Purves, R. S. (2006). Role of the supraglacial snowpack in mediating meltwater delivery to the glacier system as inferred from dye tracer investigations. *Hydrol. Process.* 20, 969–985. doi: 10.1002/hyp.6115

Charalampidis, C., van As, D., Colgan, W. T., Fausto, R. S., Macferrin, M., and Machguth, H. (2016). Thermal tracing of retained meltwater in the lower accumulation area of the southwestern Greenland ice sheet. *Ann. Glaciol.* 57, 1–10. doi: 10.1017/aog.2016.2

Christianson, K., Kohler, J., Alley, R. B., Nuth, C., and van Pelt, W. J. J. (2015). Dynamic perennial firn aquifer on an Arctic glacier. *Geophys. Res. Lett.* 42, 1418–1426. doi: 10.1002/2014GL062806

Church, J. A., White, N. J., Konikow, L. F., Domingues, C. M., Cogley, J. G., Rignot, E., et al. (2011). Revisiting the Earth's sea-level and energy budgets from 1961 to 2008. *Geophys. Res. Lett.* 38:L18601. doi: 10.1029/2011GL048794

Claremar, B., Obleitner, F., Reijmer, C., Pohjola, V., Waxegård, A., Karner, F., et al. (2012). Applying a mesoscale atmospheric model to Svalbard glaciers. *Adv. Meteorol.* 2012:22. doi: 10.1155/2012/321649

Colbeck, S. C. (1972). A theory of water percolation in snow. *J. Glaciol.* 11, 369–385. doi: 10.1017/S0022143000022346

Coléou, C., and Lesaffre, B. (1998). Irreducible water saturation in snow: experimental results in a cold laboratory. *Ann. Glaciol.* 26, 64–68. doi: 10.1017/S0260305500014579

Conway, H., and Benedict, R. (1994). Infiltration of water into snow. *Water Resour. Res.* 30, 641–649. doi: 10.1029/93WR03247

Cox, C., Humphrey, N., and Harper, J. (2015). Quantifying meltwater refreezing along a transect of sites on the Greenland ice sheet. *Cryosphere* 9, 691–701. doi: 10.5194/tc-9-691-2015

Cuffey, K., and Paterson, W. S. B. (2010). *The Physics of Glaciers.* Academic Press. Available online at: https://www.elsevier.com/books/the-physics-of-glaciers/cuffey/978-0-12-369461-4

Dee, D. P., Uppala, S. M., Simmons, A. J., Berrisford, P., Poli, P., Kobayashi, S., et al. (2011). The ERA-Interim reanalysis: configuration and performance of the data assimilation system. *Q. J. R. Meteorol. Soc.* 137, 553–597. doi: 10.1002/qj.828

Donlon, C. J., Martin, M., Stark, J., Roberts-Jones, J., Fiedler, E., and Wimmer, W. (2012). The Operational Sea Surface Temperature and Sea Ice Analysis (OSTIA) system. *Remote Sens. Environ.* 116, 140–158. Advanced Along Track Scanning Radiometer(AATSR) Special Issue. doi: 10.1016/j.rse.2010.10.017

Fettweis, X. (2007). Reconstruction of the 1979–2006 Greenland ice sheet surface mass balance using the regional climate model MAR. *Cryosphere* 1, 21–40. doi: 10.5194/tc-1-21-2007

Fierz, C., Armstrong, R., Durand, Y., Etchevers, P., Greene, E., McClung, D., et al. (2009). *The International Classification for Seasonal Snow on the Ground.* IHP-VII Technical Documents in Hydrology, No. 83, IACS contribution No. 1, UNESCO-IHP, Paris. Available online at: http://unesdoc.unesco.org/images/0018/001864/186462e.pdf

Førland, E. J., Benestad, R., Hanssen-Bauer, I., Erik, H. J., and Skaugen, T. E. (2011). Temperature and precipitation development at Svalbard 1900–2100. *Adv. Meteorol.* 2011:893790. doi: 10.1155/2011/893790

Forster, R. R., Box, J. E., van den Broeke, M. R., Miege, C., Burgess, E. W., van Angelen, J. H., et al. (2014). Extensive liquid meltwater storage in firn within the Greenland ice sheet. *Nat. Geosci.* 7, 95–98. doi: 10.1038/ngeo2043

Gascon, G., Sharp, M., Burgess, D., Bezeau, P., Bush, A. B., Morin, S., et al. (2014). How well is firn densification represented by a physically based multilayer model? Model evaluation for Devon Ice Cap, Nunavut, Canada. *J. Glaciol.* 60, 694–704. doi: 10.3189/2014JoG13J209

Glass, R. J., Steenhuis, T. S., and Parlange, J.-Y. (1989). Wetting front instability: 2. Experimental determination of relationships between system parameters and two-dimensional unstable flow field behavior in initially dry porous media. *Water Resour. Res.* 25, 1195–1207. doi: 10.1029/WR025i006p01195

Grell, G. A., and Dévényi, D. (2002). A generalized approach to parameterizing convection combining ensemble and data assimilation techniques. *Geophys. Res. Lett.* 29, 38.1–38.4. doi: 10.1029/2002GL015311

Greuell, W., and Konzelmann, T. (1994). Numerical modelling of the energy balance and the englacial temperature of the Greenland Ice Sheet. Calculations for the ETH-Camp location (West Greenland, 1155 m a.s.l.). *Global Planet. Change* 9, 91–114. doi: 10.1016/0921-8181(94)90010-8

Greuell, W., and Oerlemans, J. (1986). Sensitivity studies with a mass balance model including temperature profile calculations inside the glacier. *Z. Gletsch.kd. Glazialgeol.* 22, 101–124. Available online at: https://dspace.library.uu.nl/handle/1874/21035

Haeberli, W., Hoelzle, M., Paul, F., and Zemp, M. (2007). Integrated monitoring of mountain glaciers as key indicators of global climate change: the European Alps. *Ann. Glaciol.* 46, 150–160. doi: 10.3189/172756407782871512

Heilig, A., Mitterer, C., Schmid, L., Wever, N., Schweizer, J., Marshall, H.-P., et al. (2015). Seasonal and diurnal cycles of liquid water in snow measurements and modeling. *J. Geophys. Res.* 120, 2139–2154. doi: 10.1002/2015JF003593

Hendrickx, J. M., and Flury, M. (2001). "Uniform and preferential flow mechanisms in the vadose zone," in *Conceptual Models of Flow and Transport in the Fractured Vadose Zone* (Washington, DC: The National Academies Press), 149–188. Available online at: https://www.nap.edu/catalog/10102/conceptual-models-of-flow-and-transport-in-the-fractured-vadose-zone

Hill, D. E., and Parlange, J.-Y. (1972). Wetting front instability in layered soils. *Soil Sci. Soc. Am. J.* 36, 697–702. doi: 10.2136/sssaj1972.03615995003600050010x

Hirashima, H., Yamaguchi, S., and Katsushima, T. (2014). A multi-dimensional water transport model to reproduce preferential flow in the snowpack. *Cold Regions Sci. Technol.* 108, 80–90. doi: 10.1016/j.coldregions.2014.09.004

Humphrey, N. F., Harper, J. T., and Pfeffer, W. T. (2012). Thermal tracking of meltwater retention in Greenland's accumulation area. *J. Geophys. Res. Earth Surf.* 117:F01010. doi: 10.1029/2011JF002083

Iacono, M. J., Delamere, J. S., Mlawer, E. J., Shephard, M. W., Clough, S. A., and Collins, W. D. (2008). Radiative forcing by long-lived greenhouse gases: calculations with the AER radiative transfer models. *J. Geophys. Res.* 113:D13103. doi: 10.1029/2008JD009944

Illangasekare, T. H., Walter, R. J., Meier, M. F., and Pfeffer, W. T. (1990). Modeling of meltwater infiltration in subfreezing snow. *Water Resour. Res.* 26, 1001–1012. doi: 10.1029/WR026i005p01001

James, T. D., Murray, T., Barrand, N. E., Sykes, H. J., Fox, A. J., and King, M. A. (2012). Observations of enhanced thinning in the upper reaches of Svalbard glaciers. *Cryosphere* 6, 1369–1381. doi: 10.5194/tc-6-1369-2012

Katsushima, T., Kumakura, T., and Takeuchi, Y. (2009). A multiple snow layer model including a parameterization of vertical water channel process in snowpack. *Cold Regions Sci. Technol.* 59, 143–151. doi: 10.1016/j.coldregions.2009.09.002

Katsushima, T., Yamaguchi, S., Kumakura, T., and Sato, A. (2013). Experimental analysis of preferential flow in dry snowpack. *Cold Regions Sci. Technol.* 85, 206–216. doi: 10.1016/j.coldregions.2012.09.012

Klok, E. L., and Oerlemans, J. (2002). Model study of the spatial distribution of the energy and mass balance of Morteratschgletscher, Switzerland. *J. Glaciol.* 48, 505–518. doi: 10.3189/172756502781831133

Kuipers Munneke, P., Reijmer, C. H., and van den Broeke, M. R. (2011). Assessing the retrieval of cloud properties from radiation measurements over snow and ice. *Int. J. Climatol.* 31, 756–769. doi: 10.1002/joc.2114

Lang, C., Fettweis, X., and Erpicum, M. (2015). Stable climate and surface mass balance in Svalbard over 1979 - 2013 despite the Arctic warming. *Cryosphere* 9, 83–101. doi: 10.5194/tc-9-83-2015

Langen, P. L., Fausto, R. S., Vandecrux, B., Mottram, R. H., and Box, J. E. (2017). Liquid water flow and retention on the Greenland ice sheet in the regional climate model HIRHAM5: local and large-scale impacts. *Front. Earth Sci.* 4:110. doi: 10.3389/feart.2016.00110

Ligtenberg, S. R. M., Helsen, M. M., and van den Broeke, M. R. (2011). An improved semi-empirical model for the densification of Antarctic firn. *Cryosphere* 5, 809–819. doi: 10.5194/tc-5-809-2011

Machguth, H., MacFerrin, M., van As, D., Box, J. E., Charalampidis, C., Colgan, W. T., et al. (2016). Greenland meltwater storage in firn limited by near-surface ice formation. *Nat. Clim. Change* 6, 390–393. doi: 10.1038/nclimate2899

Marchenko, S., Pohjola, V. A., Pettersson, R., van Pelt, W. J. J., Vega, C. P., Machguth, H., et al. (2017). A plot-scale study of firn stratigraphy at Lomonosovfonna, Svalbard, using ice cores, borehole video and GPR surveys in 2012–14. *J. Glaciol.* 63, 67–78. doi: 10.1017/jog.2016.118

Marsh, P., and Woo, M.-K. (1984a). Wetting front advance and freezing of meltwater within a snow cover: 1. Observations in the Canadian Arctic. *Water Resour. Res.* 20, 1853–1864. doi: 10.1029/WR020i012p01853

Marsh, P., and Woo, M.-K. (1984b). Wetting front advance and freezing of meltwater within a snow cover: 2. A simulation model. *Water Resour. Res.* 20, 1865–1874. doi: 10.1029/WR020i012p01865

Marsh, P., and Woo, M.-K. (1985). Meltwater movement in natural heterogeneous snow covers. *Water Resour. Res.* 21, 1710–1716. doi: 10.1029/WR021i011p01710

McGurk, B. J., and Marsh, P. (1995). "Flow-finger continuity in serial thick-sections in a melting sierran snowpack," in *Biogeochemistry of Seasonally Snow-Covered Catchments (Proceedings of a Boulder Symposium, July 1995) IAHS*

publ. no. 228. Available online at: http://iahs.info/uploads/dms/iahs_228_0081.pdf

Meese, D. A., Gow, A. J., Grootes, P., Stuiver, M., Mayewski, P. A., Zielinski, G. A., et al. (1994). The accumulation record from the GISP2 core as an indicator of climate change throughout the Holocene. *Science* 266, 1680–1682. doi: 10.1126/science.266.5191.1680

Moholdt, G., Nuth, C., Hagen, J. O., and Kohler, J. (2010). Recent elevation changes of Svalbard glaciers derived from ICESat laser altimetry. *Remote Sens. Environ.* 114, 2756–2767. doi: 10.1016/j.rse.2010.06.008

Morrison, H., Curry, J. A., and Khvorostyanov, V. I. (2005). A new double-moment microphysics parameterization for application in cloud and climate models. Part I: description. *J. Atmos. Sci.* 62, 1665–1677. doi: 10.1175/JAS3446.1

Nakanishi, M., and Niino, H. (2006). An improved Mellor–Yamada level-3 model: its numerical stability and application to a regional prediction of advection fog. *Bound. Layer Meteorol.* 119, 397–407. doi: 10.1007/s10546-005-9030-8

Niu, G.-Y., Yang, Z.-L., Mitchell, K. E., Chen, F., Ek, M. B., Barlage, M., et al. (2011). The community Noah land surface model with multiparameterization options (Noah-MP): 1. model description and evaluation with local-scale measurements. *J. Geophys. Res.* 116:D12109. doi: 10.1029/2010JD015139

Nordli, Ø., Przybylak, R., Ogilvie, A., and Isaksen, K. (2014). Long-term temperature trends and variability on Spitsbergen: the extended Svalbard Airport temperature series, 1898–2012. *Polar Res.* 33:21349. doi: 10.3402/polar.v33.21349

Nuth, C., Kohler, J., König, M., von Deschwanden, A., Hagen, J. O., Kääb, A., et al. (2013). Decadal changes from a multi-temporal glacier inventory of Svalbard. *Cryosphere* 7, 1603–1621. doi: 10.5194/tc-7-1603-2013

Nuth, C., Moholdt, G., Kohler, J., Hagen, J. O., and Kääb, A. (2010). Svalbard glacier elevation changes and contribution to sea level rise. *J. Geophys. Res.* 115, F01008. doi: 10.1029/2008JF001223

Nuth, C., Schuler, T. V., Kohler, J., Altena, B., and Hagen, J. O. (2012). Estimating the long-term calving flux of Kronebreen, Svalbard, from geodetic elevation changes and mass-balance modelling. *J. Glaciol.* 58, 119–133. doi: 10.3189/2012JoG11J036

Obleitner, F., and Lehning, M. (2004). Measurement and simulation of snow and superimposed ice at the Kongsvegen glacier, Svalbard (Spitzbergen). *J. Geophys. Res.* 109:D04106. doi: 10.1029/2003JD003945

Pälli, A., Kohler, J. C., Isaksson, E., Moore, J. C., Pinglot, J. F., Pohjola, V. A., et al. (2002). Spatial and temporal variability of snow accumulation using ground-penetrating radar and ice cores on a Svalbard glacier. *J. Glaciol.* 48, 417–424. doi: 10.3189/172756502781831205

Pfeffer, W. T., and Humphrey, N. F. (1996). Determination of timing and location of water movement and ice-layer formation by temperature measurements in sub-freezing snow. *J. Glaciol.* 42, 292–304. doi: 10.1017/S0022143000004159

Pfeffer, W. T., Meier, M. F., and Illangasekare, T. H. (1991). Retention of Greenland runoff by refreezing: Implications for projected future sea level change. *J. Geophys. Res.* 96, 22117–22124. doi: 10.1029/91JC02502

Reijmer, C. H., and Hock, R. (2008). Internal accumulation on Storglaciären, Sweden, in a multi-layer snow model coupled to a distributed energy- and mass-balance model. *J. Glaciol.* 54, 61–72. doi: 10.3189/002214308784409161

Reijmer, C. H., van den Broeke, M. R., Fettweis, X., Ettema, J., and Stap, L. B. (2012). Refreezing on the Greenland ice sheet: a comparison of parameterizations. *Cryosphere* 6, 743–762. doi: 10.5194/tc-6-743-2012

Schneebeli, M. (1995). "Development and stability of preferential flow paths in a layered snowpack," in *Biogeochemistry of Seasonally Snow-Covered Catchments (Proceedings of a Boulder Symposium July 1995) IAHS Publ. no. 228*, 89–95. Available online at: http://iahs.info/uploads/dms/iahs_228_0089.pdf

Schneider, T., and Jansson, P. (2004). Internal accumulation in firn and its significance for the mass balance of Storglaciären, Sweden. *J. Glaciol.* 50, 25–34. doi: 10.3189/172756504781830277

Shimizu, H. (1970). "Air permeability of deposited snow," in *Contributions from the Institute of Low Temperature Science* A22, 1–32. Available online at: http://hdl.handle.net/2115/20234

Shumskii, P. A. (1964). "Zones of Ice Formation," in *Principles of structural glaciology*, ed P. F. Shvetsov (New York, NY: Dover Publications, Inc.), 407–440.

Skamarock, W., Klemp, J., Dudhia, J., Gill, D., Barker, D., Wang, W., et al. (2008). *A Description of the Advanced Research WRF Version 3*. Technical Report, NCAR Technical Note NCAR/TN-475+STR. Available online at: http://www2.mmm.ucar.edu/wrf/users/docs/arw_v3.pdf

Steger, C., Reijmer, C., Van Den Broeke, M., Wever, N., Forster, R., Koenig, L., et al. (2017). Firn meltwater retention on the Greenland ice sheet: a model comparison. *Front. Earth Sci.* 5:3. doi: 10.3389/feart.2017.00003

Stephens, G. L. (2005). Cloud feedbacks in the climate system: a critical review. *J. Clim.* 18, 237–273. doi: 10.1175/JCLI-3243.1

Sturm, M., and Holmgren, J. (1993). Rain-induced water percolation in snow as detected using heat flux transducers. *Water Resour. Res.* 29, 2323–2334. doi: 10.1029/93WR00609

Techel, F., and Pielmeier, C. (2011). Point observations of liquid water content in wet snow – investigating methodical, spatial and temporal aspects. *Cryosphere* 5, 405–418. doi: 10.5194/tc-5-405-2011

van Angelen, J. H., Lenaerts, J. T. M., van den Broeke, M. R., Fettweis, X., and van Meijgaard, E. (2013). Rapid loss of firn pore space accelerates 21st century Greenland mass loss. *Geophys. Res. Lett.* 40, 21092113. doi: 10.1002/grl.50490

Van de Wal, R. S. W., Mulvaney, R., Isaksson, E., Moore, J. C., Pinglot, J. F., Pohjola, V. A., et al. (2002). Reconstruction of the historical temperature trend from measurements in a medium-length borehole on the Lomonosovfonna plateau, Svalbard. *Ann. Glaciol.* 35, 371–378. doi: 10.3189/172756402781816979

Van Pelt, W., and Kohler, J. (2015). Modelling the long-term mass balance and firn evolution of Glaciers around Kongsfjorden, Svalbard. *J. Glaciol.* 61, 731–744. doi: 10.3189/2015JoG14J223

Van Pelt, W. J., Pettersson, R., Pohjola, V. A., Marchenko, S., Claremar, B., and Oerlemans, J. (2014). Inverse estimation of snow accumulation along a radar transect on Nordenskiöldbreen, Svalbard. *J. Geophys. Res.* 119, 816–835. doi: 10.1002/2013JF003040

Van Pelt, W. J. J., Oerlemans, J., Reijmer, C. H., Pohjola, V. A., Pettersson, R., and van Angelen, J. H. (2012). Simulating melt, runoff and refreezing on Nordenskiöldbreen, Svalbard, using a coupled snow and energy balance model. *Cryosphere* 6, 641–659. doi: 10.5194/tc-6-641-2012

Van Pelt, W. J. J., Pohjola, V. A., and Reijmer, C. H. (2016). The changing impact of snow conditions and refreezing on the mass balance of an idealized Svalbard glacier. *Front. Earth Sci.* 4:102. doi: 10.3389/feart.2016.00102

Vaughan, D. G., Comiso, J. C., Allison, I., Carrasco, J., Kaser, G., Kwok, R., et al. (2013). "Observations: Cryosphere, Chapter 4," in *Climate Change 2013: The Physical Science Basis. Contribution of Working Group I to the Fifth Assessment Report of the Intergovernmental Panel on Climate Change,* eds T. F. Stocker, D. Qin, G.-K. Plattner, M. Tignor, S. K. Allen, J. Boschung, A. Nauels, Y. Xia, V. Bex, and P. M. Midgley (Cambridge, UK; New York, NY: Cambridge University Press), 317–382. Available online at: http://ipcc.ch/pdf/assessment-report/ar5/wg1/WG1AR5_Chapter04_FINAL.pdf

Waldner, P. A., Schneebeli, M., Schultze-Zimmermann, U., and Flühler, H. (2004). Effect of snow structure on water flow and solute transport. *Hydrol. Process.* 18, 1271–1290. doi: 10.1002/hyp.1401

Wendl, I. A. (2014). *High resolution records of black carbon and other aerosol constituents from the Lomonosovfonna 2009 ice core.* Ph.D. thesis, Departement für Chemie und Biochemie der Universität Bern. Available online at: http://csold.unibe.ch/students/theses/phd/76.pdf

Wendl, I. A., Eichler, A., Isaksson, E., Martma, T., and Schwikowski, M. (2015). 800-year ice-core record of nitrogen deposition in Svalbard linked to ocean productivity and biogenic emissions. *Atmos. Chem. Phys.* 15, 7287–7300. doi: 10.5194/acp-15-7287-2015

Wever, N., Würzer, S., Fierz, C., and Lehning, M. (2016). Simulating ice layer formation under the presence of preferential flow in layered snowpacks. *Cryosphere* 10, 2731–2744. doi: 10.5194/tc-10-27 31-2016

Williams, M., Pfeffer, W., and Knoll, M. (2000). *Collaborative Experiment for Pulsed Radar Visualization of Water Flow Paths in Snow.* Boulder, CO: Technical Report, University of Colorado.

Williams, M. W., Erickson, T. A., and Petrzelka, J. L. (2010). Visualizing meltwater flow through snow at the centimetre-to-metre scale using a snow guillotine. *Hydrol. Process.* 24, 2098–2110. doi: 10.1002/hyp.7630

Williams, M. W., Sommerfeld, R., Massman, S., and Rikkers, M. (1999). Correlation lengths of meltwater flow through ripe snowpacks, colorado front range, USA. *Hydrol. Process.* 13, 1807–1826. doi: 10.1002/(SICI)1099-1085(199909)13:12/13<1807::AID-HYP891>3.0.CO;2-U

Wright, A., Wadham, J., Siegert, M., Luckman, A., and Kohler, J. (2005). Modelling the impact of superimposed ice on the mass balance of an Arctic glacier under scenarios of future climate change. *Ann. Glaciol.* 42, 277–283. doi: 10.3189/172756405781813104

Zagorodnov, V. S., and Zotikov, I. A. (1980). Core drilling at Spitsbergen (in Russian). *Materialy Glatsiologicheskih Issledovanii: Khronika, Obsujdeniya (Data of Glaciological Studies)* 40, 157–163.

Zdanowicz, C., Smetny-Sowa, A., Fisher, D., Schaffer, N., Copland, L., Eley, J., et al. (2012). Summer melt rates on Penny Ice Cap, Baffin Island: past and recent trends and implications for regional climate. *J. Geophys. Res.* 117:F02006. doi: 10.1029/2011JF002248

Zinger, Y. M., Koryakin, V. S., Lavrushin, Y. A., Markin, V. A., Mihalev, B. I., and Troitskiy, L. C. (1966). Study of glaciers at Spitsbergen by a Soviet expedition during summer 1965 (in Russian). *Materialy Glatsiologicheskih Issledovanii: Khronika, Obsujdeniya (Data of Glaciological Studies)* 12, 59–72.

Zuo, Z., and Oerlemans, J. (1996). Modelling albedo and specific balance of the Greenland ice sheet: calculations for the Søndre Strømfjord transect. *J. Glaciol.* 42, 305–317. doi: 10.1017/S0022143000004160

Conflict of Interest Statement: The authors declare that the research was conducted in the absence of any commercial or financial relationships that could be construed as a potential conflict of interest.

The reviewer NW and handling Editor declared their shared affiliation, and the handling Editor states that the process nevertheless met the standards of a fair and objective review.

6

Rock Magnetism of the Offshore Sediments of Lake Qinghai in the Western China

Peng Zhang[1]*, Shan Lin[1,2], Hong Ao[1], Lijuan Wang[1], Xiaoyan Sun[1,3] and Zhisheng An[1]

[1] State Key Laboratory of Loess and Quaternary Geology, Institute of Earth Environment, Chinese Academy of Sciences, Xi'an, China, [2] College of Earth Sciences, University of Chinese Academy of Sciences, Beijing, China, [3] College of Urban and Environment Sciences, Shanxi Normal University, Linfen, China

Edited by:
Juan Cruz Larrasoaña,
Instituto Geológico y Minero de
España, Spain

Reviewed by:
Luigi Jovane,
Instituto Oceanográfico da
Universidade de São Paulo, Brazil
Junsheng Nie,
Lanzhou University, China
Neli Jordanova,
National Institute in Geophysics,
Geodesy and Geography-Bulgarian
Academy of Science, Bulgaria

*Correspondence:
Peng Zhang
zhangpeng@ieecas.cn

Specialty section:
This article was submitted to
Geomagnetism and Paleomagnetism,
a section of the journal
Frontiers in Earth Science

Lake Qinghai is the largest lake in China and situated in an important climate-sensitive zone on the northeastern margin of the Tibetan Plateau, making it an ideal place to study the environmental evolution of the northwest China as well as the interplay between the Asian monsoon and the westerlies in the late Quaternary. In this study, detailed rock magnetic measurements were carried out on the offshore soils of Lake Qinghai. The dry grassland samples have higher magnetic susceptibility than that of the wet grassland samples, which suggests a higher concentration of magnetic minerals in the dry grassland and lower concentration of magnetic minerals in the wet grassland near the lake edge. The high concentration of the superparamagnetic (SP) magnetic minerals related to pedogenesis may also contribute to the high magnetic susceptibility of the dry grassland. The low magnetic susceptibility of the wet grassland may result from the conversion of strongly to weakly magnetic minerals and/or the dissolution of magnetic minerals. In addition, the Hm/(Gt+Hm) value has a positive correlation with the water content, thus can be taken as an effective proxy for the soil moisture.

Keywords: Lake Qinghai, rock magnetism, pedogenesis, goethite, hematite

INTRODUCTION

The iron-bearing minerals in rocks and sediments are sensitive to a range of environmental processes, which makes it possible to associate magnetic signals with environmental processes. Environmental magnetism is an interdisciplinary subject involving the application of rock and mineral magnetic techniques to situations in which the transportation, deposition, and transformation of magnetic grains are influenced by environmental processes in the atmosphere, hydrosphere, and lithosphere (Thompson and Oldfield, 1986; Oldfield, 1991; Verosub and Roberts, 1995; Liu et al., 2012). In recent decades, with the rapid development of the techniques for identifying magnetic minerals, environmental magnetism has been employed as an effective tool in the researches of sedimentation processes and environmental evolution recorded in marine and lacustrine sediments, loess-paleosol sequences, and soils (Kämpf and Schwertmann, 1983; Heller and Evans, 1995; Dekkers, 1997; Maher et al., 2002; Evans and Heller, 2003; van der Zee et al., 2003; Deng et al., 2007; Liu et al., 2007; Zhang et al., 2007; Ao et al., 2010).

Lake Qinghai, situated in the sensitive semi-arid zone between the Asian summer monsoon controlled (humid) and the westerlies influenced (arid) areas of Asia, is an ideal site to study the competing influence of two climate system in the late Quaternary (An et al., 2012). Previous

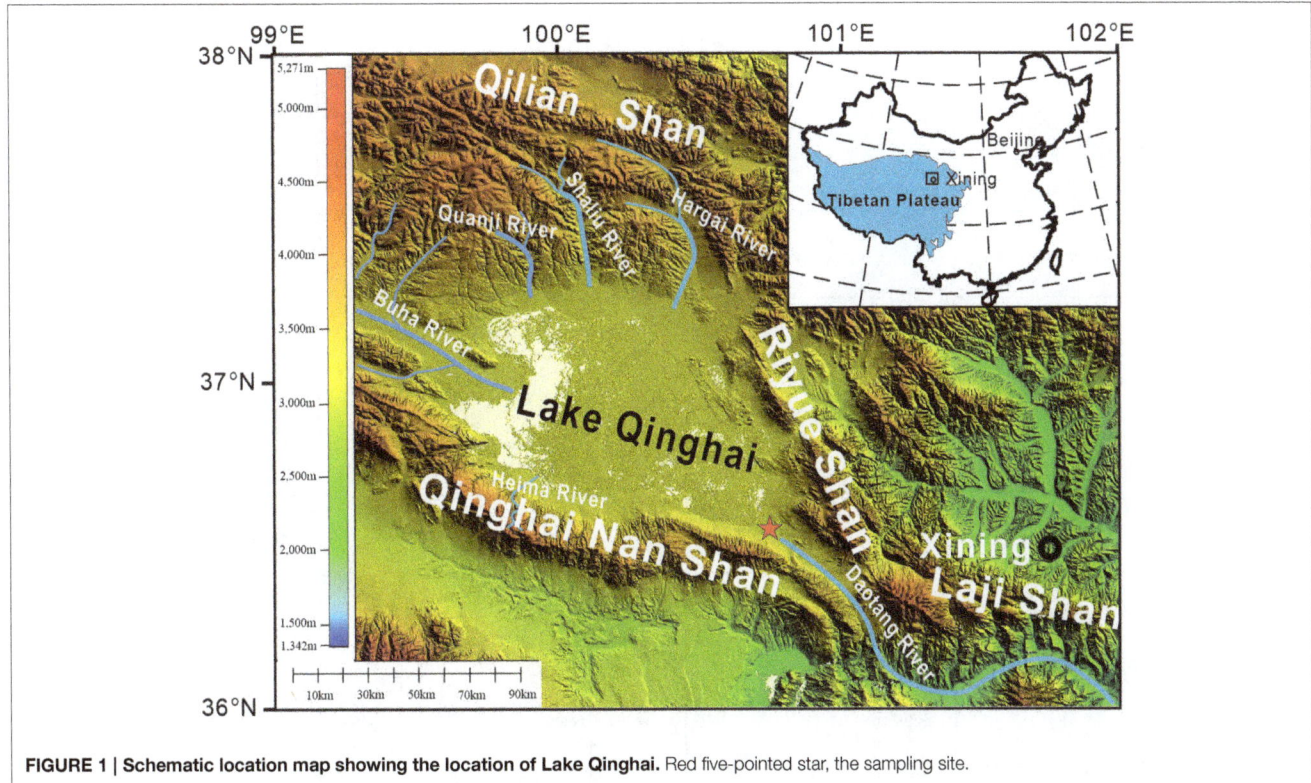

FIGURE 1 | Schematic location map showing the location of Lake Qinghai. Red five-pointed star, the sampling site.

studies have provided valuable information about the climatic changes and their responses to the interplay between the westerlies and the monsoons during the last 36 ka. For example, geochemical and palynological results from sediment cores QH85-14 (Kelts et al., 1989; Lister et al., 1991) and QH2000 (Liu et al., 2003; Shen et al., 2005) suggested a transition towards warm climate in Lake Qinghai at around 14.5 ka. The magnetic susceptibility of the bottom sediments of Lake Qinghai also has been employed as an important climatic proxy in paleoclimate researches (e.g., Wu, 1993; Shen et al., 2001; Zhang et al., 2002). The rock magnetism research was, however, rarely taken in the sediments of Lake Qinghai, which hindered the explanation of environmental magnetism in paleoclimatic studies of Lake Qinghai. The magnetic mineral assemblage in sediments is first influenced by weathering and transport process in the catchment areas. After deposition, it will also be altered by chemical and biogenic processes at the water/sediment interface and during diagenesis (Snowball, 1993; Verosub and Roberts, 1995; Demory et al., 2005). The catchment provides a significant amount of erodible material to the bottom sediment in the lake. As a result, the investigation of modern magnetic minerals in the catchment will enrich our understanding of the migration and alteration processes of magnetic minerals, which is necessary for further application of environmental magnetism in paleoclimatic studies.

In this study, we carried out detailed rock magnetic measurements on the offshore sediments of Lake Qinghai to improve our understanding of the different magnetic properties related to different sedimentary processes. We further discussed the validity of Hm/(Gt+Hm) as a proxy for the climate in the offshore sediments.

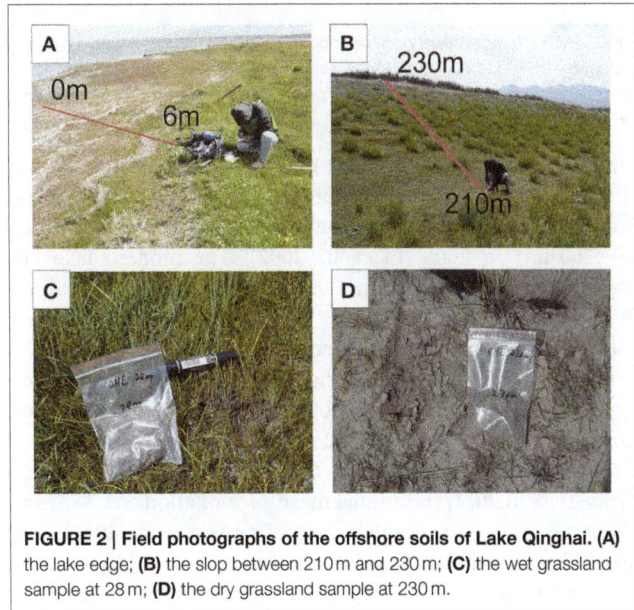

FIGURE 2 | Field photographs of the offshore soils of Lake Qinghai. (A) the lake edge; **(B)** the slop between 210 m and 230 m; **(C)** the wet grassland sample at 28 m; **(D)** the dry grassland sample at 230 m.

GEOLOGICAL SETTING AND SAMPLING

Geological Setting

Lake Qinghai is located on the northeastern margin of the Tibetan Plateau and west of the Chinese Loess Plateau (**Figure 1**), with an altitude of 3194 m above current sea level. As the largest saline lake in China, the onset of today's permanent Lake

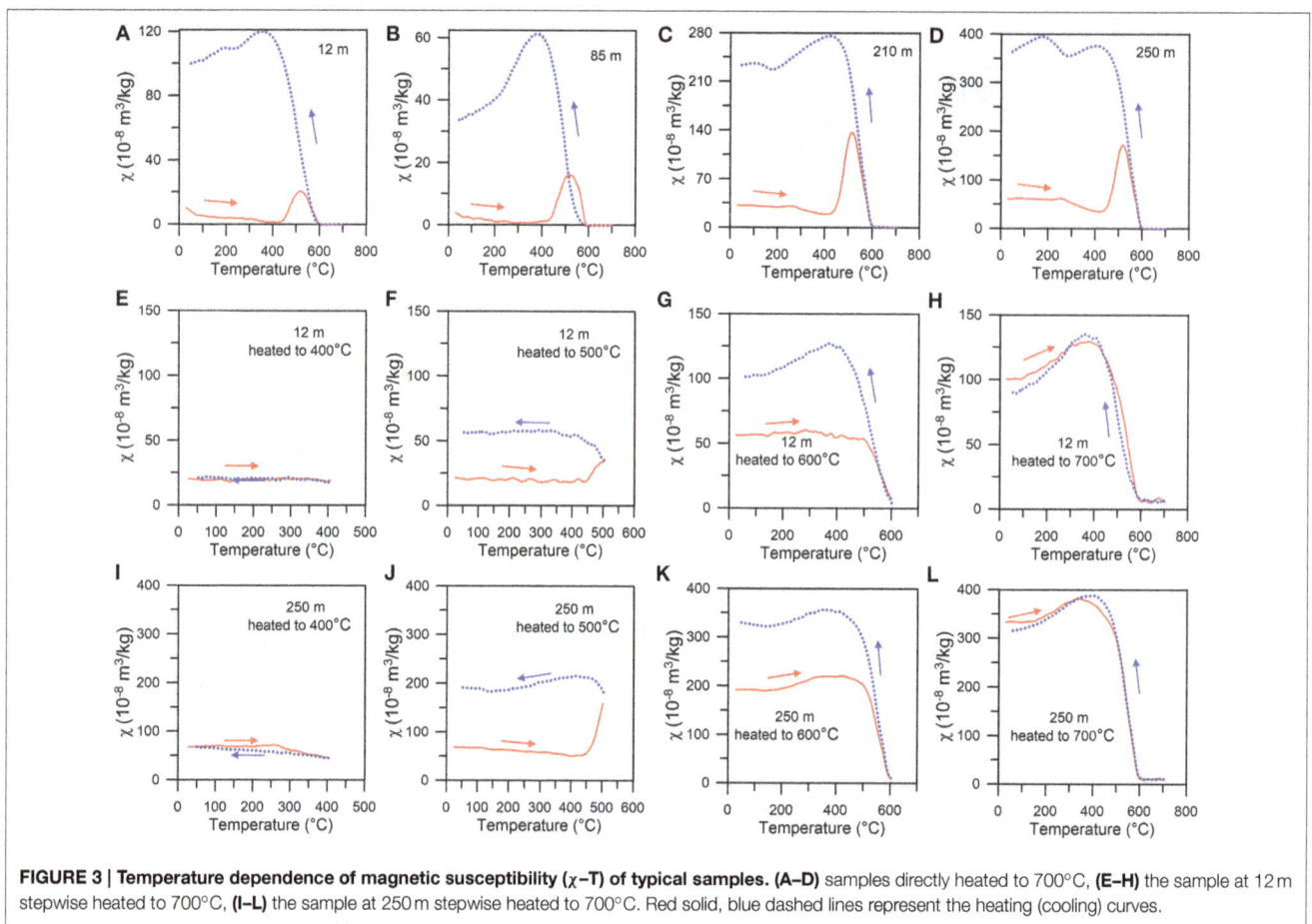

FIGURE 3 | Temperature dependence of magnetic susceptibility (χ–T) of typical samples. (A–D) samples directly heated to 700°C, **(E–H)** the sample at 12 m stepwise heated to 700°C, **(I–L)** the sample at 250 m stepwise heated to 700°C. Red solid, blue dashed lines represent the heating (cooling) curves.

Qinghai may occur at 11.5 ka, due to a major shift to humid climate throughout the Holocene (Jin et al., 2015). During the last century, the surface of the lake changed from 4980 km² to 4260 km² in 2006 (Li et al., 2007). The modern lake has a water volume of 71.6 km³ and a catchment area of about 29,660 km². The lake is currently fed by 5 major rivers, including Buha, Shaliu, Hargai, Quanji, and Heima Rivers (**Figure 1**). The primary runoff supply is from the Buha River in the west, which supplies annually ~50% of the total runoff and ~70% of the total sand loading to the lake. From 1951 to 2005, the average annual temperature was ~1.2°C within the catchment of Lake Qinghai. From 1959 to 2011, the annual mean precipitation was 383 mm, about 1/3–1/4 of the annual evaporation (Jin et al., 2011). The lake is frozen from late October to April.

The catchment of Lake Qinghai comprises of predominantly late Paleozoic marine limestone and sandstones, Triassic granites, Mesozoic diorite and granodiorite with minor late Cambrian phyllite and gneiss LIGCAS (1979). Two terraces formed since the last glaciation. The first terrace is mainly wet grassland which is exposed due to the retreat of the lake level thousands of years ago. The soils on the second terrace are developed on the sediments of the Lake Qinghai between 70 and 90 ka (Liu and Lai, 2010), but the sediments have been evolved into soils.

Field Sampling

We took samples at the end of August when the lake level was the highest. The sampling sites (36°32′37.75″ N, 100°42′38.18″) were located to the southeast of the Lake (**Figure 1**). Samples were collected along a 330-meter-long transection from the lake edge to the dry grassland (**Figure 2**). From 0 to 210 m, the ancient offshore alluvial/lacustrine sediments consist of brown-black sapropel with clay or silt (**Figures 2A,C**). There is a 10-meter-high slope from 210 to 230 m (**Figure 2B**). The samples of surface soils were collected from 230 to 330 m from the dry gray grassland on the second terrace (**Figure 2D**). Altogether, 41 samples were collected from the offshore sediments and soils in the southeast of Lake Qinghai.

METHODOLOGY

Magnetic Methods

The raw samples were oven dried at 45°C and grinded into powder for magnetic measurements. The low-frequency magnetic susceptibility (χ_{lf}) and high-frequency magnetic susceptibility (χ_{hf}) was measured with a Bartington MS2 meter at a frequency of 470 and 4700 Hz, respectively. Further, the frequency-dependent susceptibility χ_{fd} percent was calculated:

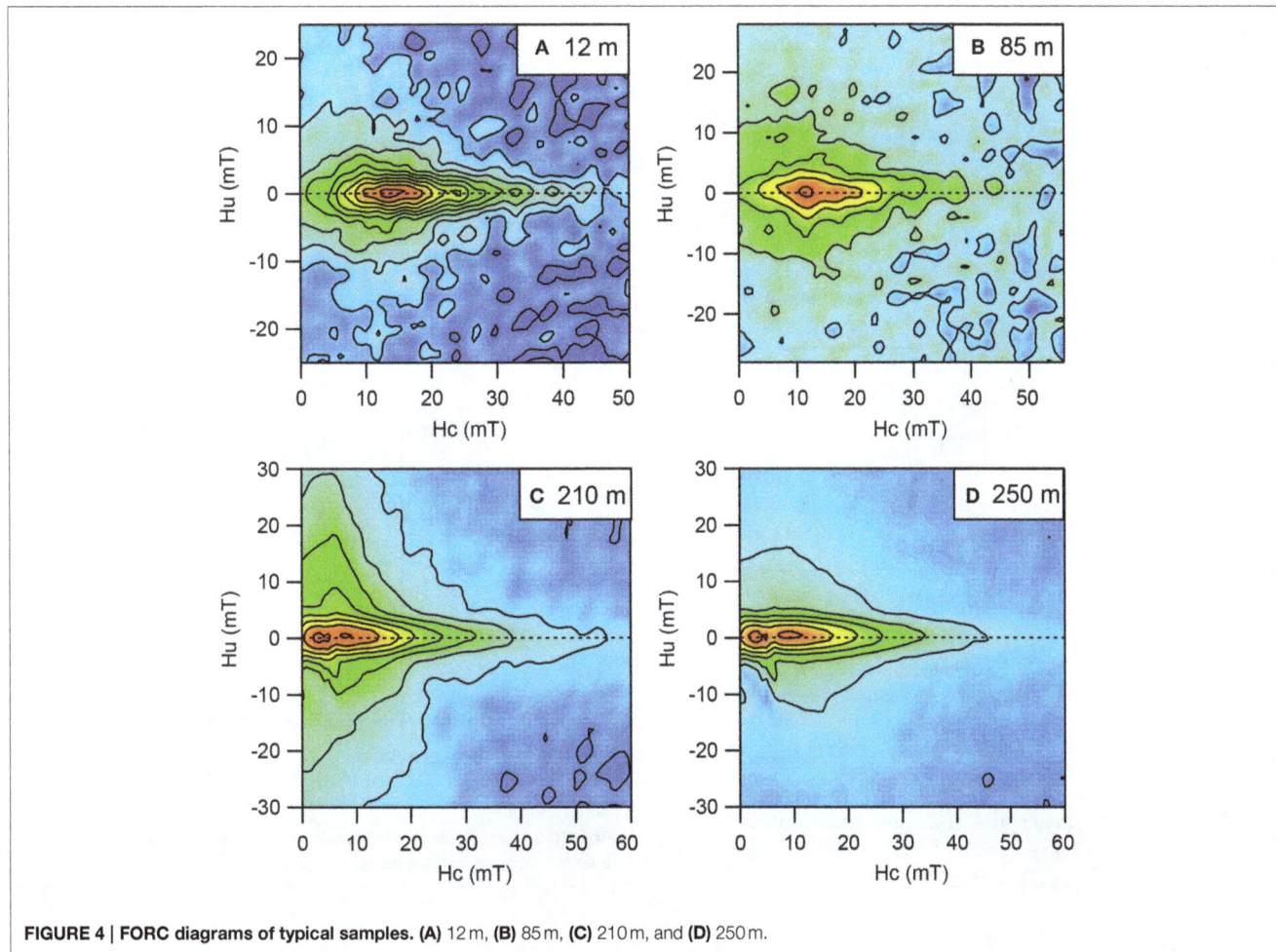

FIGURE 4 | FORC diagrams of typical samples. (A) 12 m, **(B)** 85 m, **(C)** 210 m, and **(D)** 250 m.

χ_{fd} percent $= (\chi_{lf}-\chi_{hf})/\chi_{lf} \times 100\%$. About 300 mg powdered samples were used for measuring temperature-dependent susceptibility (χ–T) curves with a MFK1 FA susceptometer equipped with a CS-3 high-temperature furnace (AGICO, Brno, Czech Republic). Measurements were done in an argon atmosphere from room temperature up to 700°C and back to room temperature (heating and cooling rate of ~6.5°C/min). The magnetic field during measurement was 300 A/m (peak-to-peak). The susceptibility of each sample was corrected for the background χ (furnace tube correction) using the CUREVAL 8.0 program (AGICO, Brno, Czech Republic). Isothermal remanent magnetization (IRM) acquisition curves were measured with an AGICO JR-6A dual speed spinner magnetometer in a magnetically shielded room (residual field <150 nT). IRMs were imparted with an impulse magnetizer (ASC, model IM-10-30). IRM acquisition curves consist of 32 field steps with a maximum field of 2.0 T. We define SIRM as IRM acquired at 2000 mT, IRM$_{-300}$ as IRM acquired at back-field 300 mT after being saturated at 2000 mT, and S-ratio as -IRM$_{-300}$/SIRM. The anhysteretic remnant magnetization (ARM) was imparted in a 100 mT alternating field with a superimposed 0.05 mT direct bias field using a D2000T Alternating Field Demagnetizer, and

measured with JR-6A dual speed spinner magnetometer. The χ_{ARM} was calculated by dividing the ARM intensity by the DC field strength (0.05 mT).

First-order reversal curve (FORC) diagrams were measured by a Vibrating Sample Magnetometer System (VSM 3900) with a maximum field of 1 T or 1.5 T. 120 FORCs were measured for each sample following the method of Roberts et al. (2000). The FORC diagrams are processed by the FORCinel software (Harrison and Feinberg, 2008) with a smoothing factor (SF) of 7.

The χ-T curves were measured at the Institute of Tibet Plateau Research, Chinese Academy of Sciences, CAS, Beijing, and FORC diagrams at the Institute of Geology and Geophysics, CAS, Beijing. The rest of the measurements were performed at the Institute of Earth Environment, CAS, Xi'an.

Non-magnetic Methods

To obtain the gravimetric water content, powdered samples were oven dried at 105°C for 2 days. We define water content as: the weight of lost water/ the weight of raw sample × 100%. The diffuse reflectance spectroscopy was carried out using Cary 4000

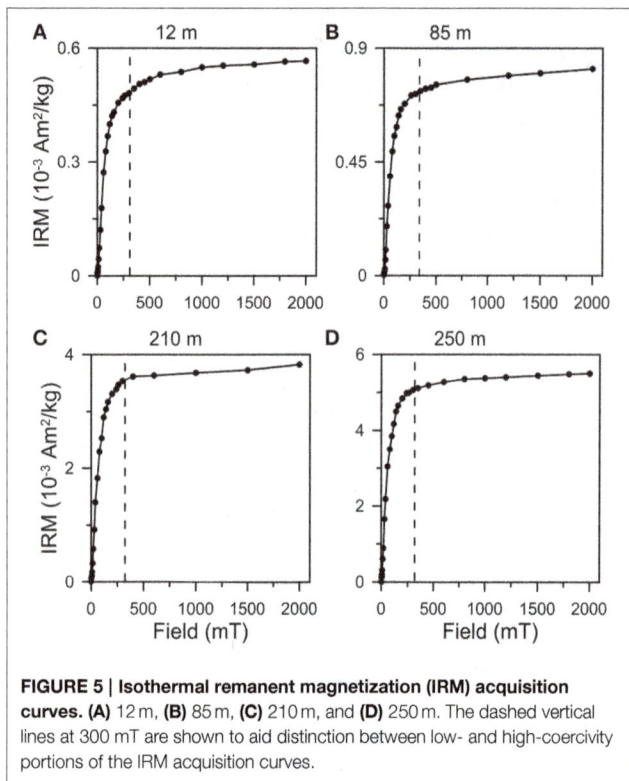

FIGURE 5 | Isothermal remanent magnetization (IRM) acquisition curves. **(A)** 12 m, **(B)** 85 m, **(C)** 210 m, and **(D)** 250 m. The dashed vertical lines at 300 mT are shown to aid distinction between low- and high-coercivity portions of the IRM acquisition curves.

UV-Vis spectrophotometer at a scan rate of 30 nm/min from 350 to 750 nm in 0.5 nm steps and the second derivative was calculated with the Cary UV software.

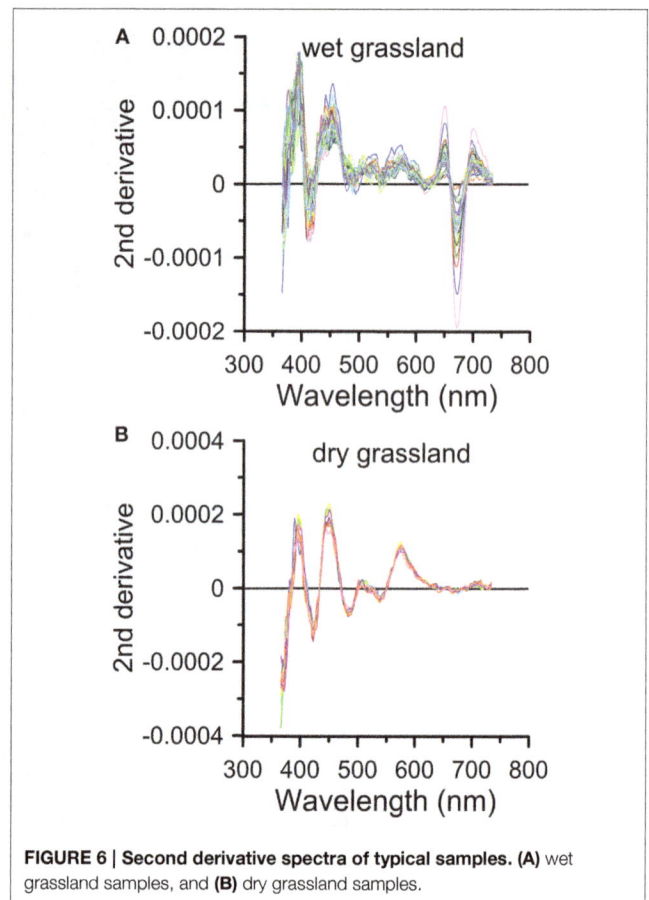

FIGURE 6 | Second derivative spectra of typical samples. **(A)** wet grassland samples, and **(B)** dry grassland samples.

RESULTS

χ-T Curves

The decrease of χ between 250 and 450°C in the heating curves of dry grassland samples (**Figures 3C,D**) may result from the inversion of ferrimagnetic maghemite to weakly magnetic hematite (Sun et al., 1995; Deng et al., 2006), suggesting the presence of a small amount of maghemite. The χ–T curves of all samples are characterized by a major peak at 450–600°C (**Figures 3A–D**), which is due to the appearance of new strongly magnetic minerals during heating. Step-wise heating of the samples suggests that the peak is related to the thermal transformations during laboratory heating (**Figures 3E–L**). The major drop in magnetic susceptibility at 520–600°C indicates the presence of magnetite in both the samples and the newly formed minerals (**Figure 3**). The transformation of iron oxyhydroxides (e.g., goethite) to a strongly magnetic phase with the presence of organic carbon occurred at 450–600°C according to the stepwise heating curves (Hanesch et al., 2006).

FORC Diagrams

FORC diagrams can be used to identify and discriminate between the different components in a mixed magnetic mineral assemblage (Roberts et al., 2000, 2014). FORC diagrams of samples at 12 and 85 m show closed contours that diverge along the H_c axis, suggesting the character of interacting SD

behavior (**Figures 4A,B**). The FORC diagrams of the sample at 210 m show wide vertical distribution along H_u axis with open contours (**Figure 4C**), which may indicate a PSD behavior. In addition, the low coercivity peak near the origin of the FORC diagram indicates the signals of superparamagnetic (SP) grains. The FORC diagrams of the sample at 250 m show two evident contour peaks of both SD and SP grains (**Figure 4D**), which may imply the presence of both SD and SP grains (Roberts et al., 2000).

IRM Analyses

All IRM acquisition curves undergo a major increase at low field and the acquired IRM reaches more than 85% of the SIRM at 300 mT (**Figure 5**), suggesting the dominance of low-coercivity magnetic minerals. But the IRM is not totally saturated up to 2.0 T (**Figure 5**), suggesting the presence of low concentration of high-coercivity magnetic minerals (e.g., hematite or goethite).

DRS Results

The difference in ordinate between the trough and the next peak at a longer wavelength, the band intensity, has been used as a proxy for the true band amplitude (Scheinost et al., 1998). In the second derivative of the reflectance spectrum, the band intensity at 424 nm (I_{424}) and 535 nm (I_{535}) is proportional to the

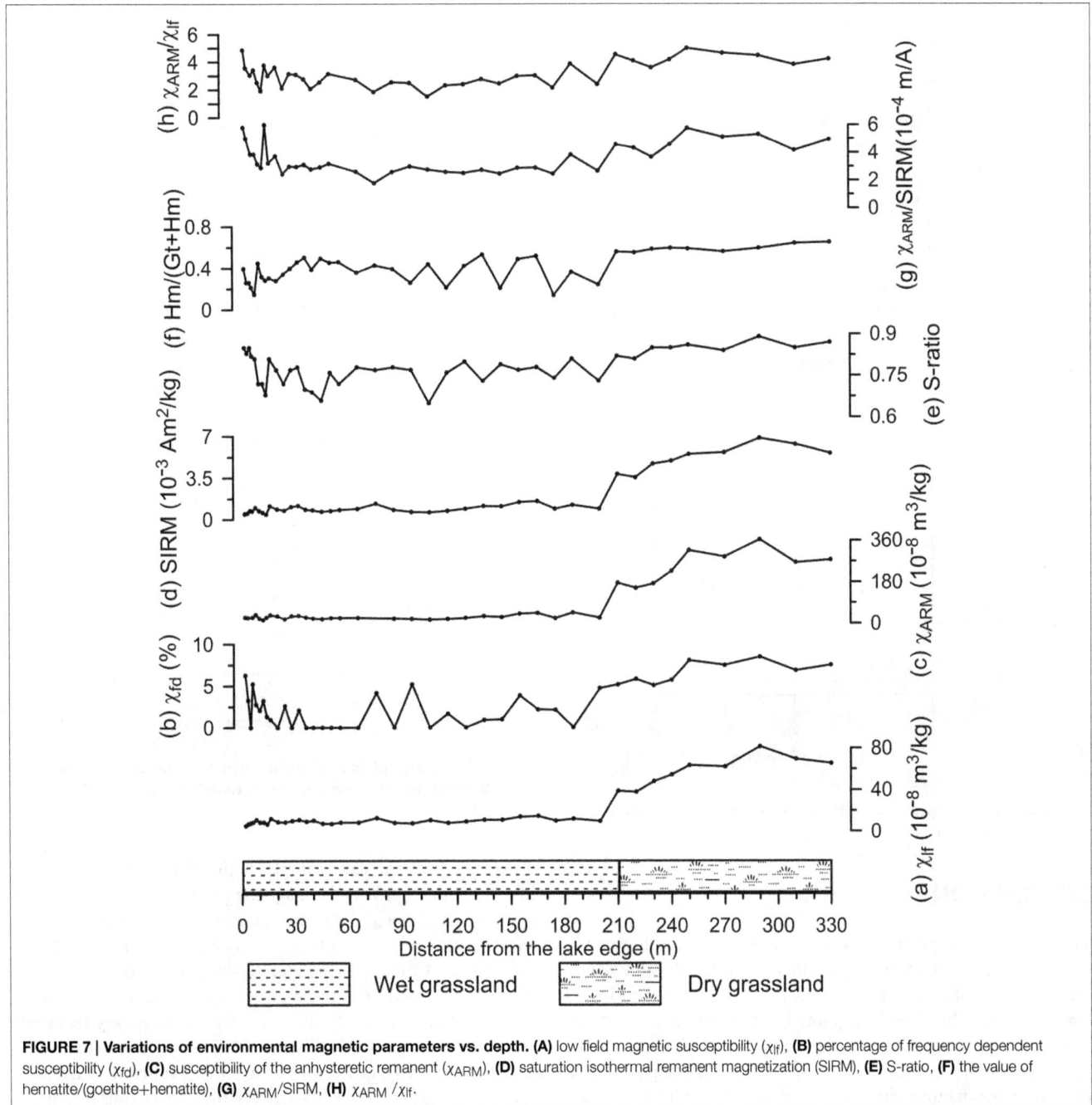

FIGURE 7 | Variations of environmental magnetic parameters vs. depth. (A) low field magnetic susceptibility (χ_{lf}), **(B)** percentage of frequency dependent susceptibility (χ_{fd}), **(C)** susceptibility of the anhysteretic remanent (χ_{ARM}), **(D)** saturation isothermal remanent magnetization (SIRM), **(E)** S-ratio, **(F)** the value of hematite/(goethite+hematite), **(G)** χ_{ARM}/SIRM, **(H)** χ_{ARM} /χ_{lf}.

concentration of goethite and hematite, respectively (Scheinost et al., 1998; Torrent et al., 2007). For most samples in this study, the I_{425} is more evident than I_{535} (**Figure 6**). Nonetheless, the band intensity is not the real concentration of the goethite or hematite, because the reflectance can be affected by other admixed minerals (the matrix effect) (Deaton and Balsam, 1991; Ji et al., 2002). The I_{535} of the wet grassland is not evident compared to that of the dry grassland, which may be related to the low concentration of the hematite in the wet grassland. In this study, we calculated the ratio Hm/(Gt+Hm) along the

transection following the regression functions of Torrent et al. (2007):

$$Y = -0.133 + 2.871* X - 1.709* X^2,$$
Where, Y = Hematite / (Hematite + Goethite),
$X = I_{535}/ (I_{425} + I_{535}).$

We only calculated the value of Y, which is not the absolute content of hematite or goethite, because the calculation of the citrate-bicarbonate-dithionite (CBD) extractable Fe remains controversial (Torrent et al., 2007; Buggle et al., 2014).

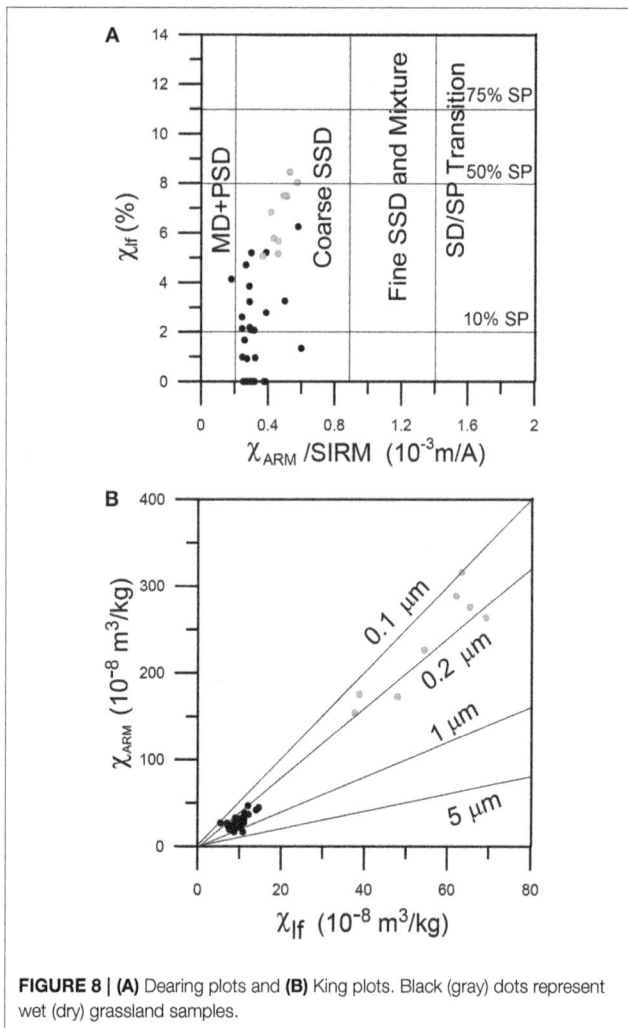

FIGURE 8 | (A) Dearing plots and (B) King plots. Black (gray) dots represent wet (dry) grassland samples.

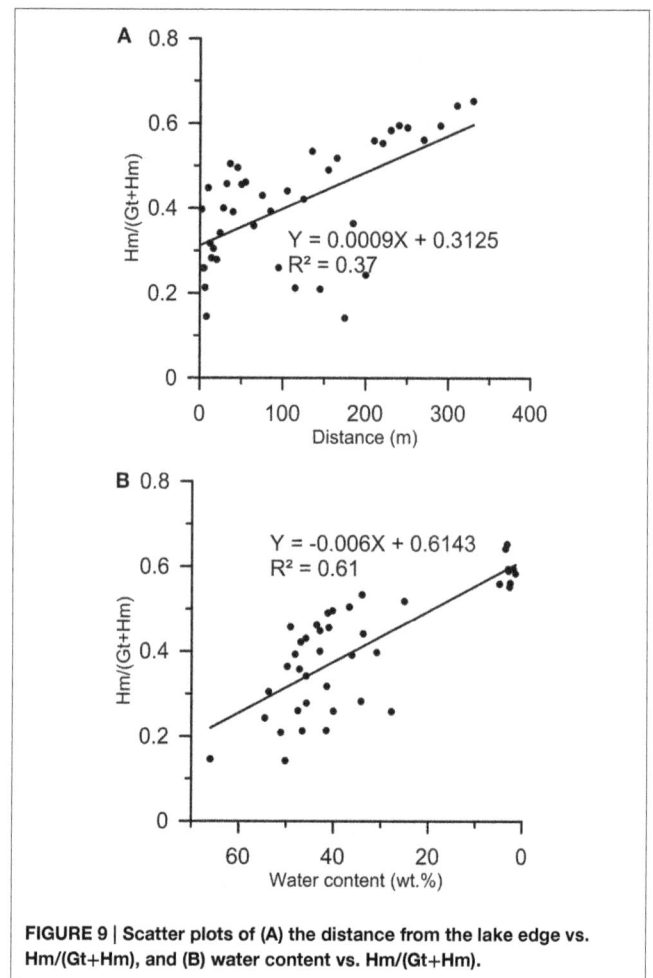

FIGURE 9 | Scatter plots of (A) the distance from the lake edge vs. Hm/(Gt+Hm), and (B) water content vs. Hm/(Gt+Hm).

DISCUSSION

The χ-T curves imply the presence of magnetite in both wet and dry grassland, but that of maghemite only in the dry grassland. The high value of χ_{fd} also suggests the presence of SP grains in the dry grassland. The FORC diagrams illustrate the dominant SD behavior in the wet grassland and the presence of SP, SD, and PSD grains in the dry grassland. The IRM and S-ratio suggests that the low-coercivity magnetic minerals dominate in all the samples (**Figures 5**, **7E**). Nonetheless, the IRM is not saturated at 2000 mT due to the presence of high-coercivity magnetic minerals (e.g., hematite and goethite). Additionally, the band intensity in the second derivative of the reflectance spectrum also provides evidences for the presence of goethite and hematite in the dry grassland and only goethite in the wet grassland (**Figure 6**).

When magnetite dominates the magnetic properties of the sediments, the Dearing plot and King plot can be used to estimate the grain size of the ferrimagnets (King et al., 1982; Dearing et al., 1997). Both the wet grassland and dry grassland samples are located in the coarse stable SD (SSD) region (**Figure 8A**), but the dry grassland samples show more SP contribution. The dry grassland samples are within 0.1–0.2 μm size range while

most of the wet grassland samples within 0.2–1 μm (**Figure 8B**). Inter-parameter proxies for magnetic grain-size variation, such as χ_{ARM}/χ_{lf} and $\chi_{ARM}/SIRM$, are sensitive to climate (Bloemendal and Liu, 2005). For example, χ_{ARM}/χ_{lf} and $\chi_{arm}/SIRM$ of surface soils on the Chinese Loess Plateau are proportional to annual mean temperature and annual mean precipitation (Nie et al., 2014). But in this study, this two values increased from wet grassland to dry grassland (**Figures 7G,H**), opposite to the results of Nie et al. (2014), which may be due to the different conversion process of magnetic minerals between the surface soils of Chinese Loess Plateau and the offshore sediments of Lake Qinghai.

The χ_{lf}, χ_{fd} percent, χ_{ARM}, and SIRM show similar trend along the transection from the lake edge to the dry grassland. All the parameters are relatively low in the wet grassland, but increase sharply in the dry grassland (**Figures 7A–D**). Consequently, we speculate that the wet grassland has lower concentration of total magnetic minerals than the dry grassland. The χ_{lf} of wet grassland is usually less than 15×10^{-8} m³/kg. The value is of the same magnitude as the χ of typical lacustrine sediments in northwest China, such as Pliocene and Holocene sediments of Lake Qinghai (Jun and Kelts, 2002; Fu et al., 2015), the Oligocene-Miocene lacustrine sediments in the Xining basin (Fang et al., 2015) and Oligocene lacustrine sediments in the

Lanzhou basin (Zhang et al., 2015). The χ_{lf} of dry grassland is usually over 35×10^{-8} m^3/kg, the same as the loess deposited in the Lake Qinghai region (Hunt et al., 1995; Lu et al., 2004; Wang et al., 2015).

The different magnetic properties of the wet and dry grassland may be linked to different conversion process of magnetic minerals, pedogenesis in the dry grassland, and/or the dissolution of magnetic minerals in the wet grassland. Magnetic minerals could be converted to goethite or other weakly magnetic minerals owing to the high water content of wet grassland, but converted to ferrimagnetic minerals in dry grassland, a hypothesis raised by Nie et al. (2016) based on the magnetic properties of red-clay on the Chines Loess Plateau. However, the different mineral transform route needs detailed study in the future.

The pedogenic fine-grained (SP+SD) maghemite is responsible for the enhanced magnetic susceptibility in the paleosols on the Chinese Loess Plateau (Zhou et al., 1990; Deng et al., 2000; Liu et al., 2007). In this study, the high magnetic susceptibility, the presence of SP and SD grans and maghemite may point to pedogenesis in the dry grassland. Besides, the dissolution of the magnetic minerals in the wet grassland may also contribute to the low magnetic susceptibility. The unusually low χ_{lf}, χ_{ARM}, and SIRM compared to the dry grassland soils may be due to the removal of most magnetite particles. What is more, the soils of dry grassland have a much wider magnetic grain-size distribution (SP, SD, and PSD particles), while the soils of wet grassland have a narrow grain-size distribution (SD particles). The several lines of evidence suggest that the pedogenesis and magnetic mineral dissolution are potential causes of different magnetic properties between the wet and dry grassland.

Of a variety of magnetic minerals, the hematite (α-Fe$_2$O$_3$) and goethite (α-FeOOH) which commonly occur in soils and sediments, are especially sensitive to the environmental variation. The transformations from ferrihydrite to hematite and goethite are favored by opposite climate conditions. Hematite is interpreted as being indicative of warm and dry conditions, while goethite indicative of cold and wet conditions (Kämpf and Schwertmann, 1983; Ji et al., 2004). Consequently, the ratio of hematite to goethite (Hm/Gt) or hematite to (goethite+hematite) (Hm/(Gt+Hm)) has been used as a climatic proxy in the paleoclimate studies (Ji et al., 2002; Zhang et al., 2007, 2009; Hao et al., 2009).

In this study, the Hm/(Gt+Hm) value has an indistinctive positive correlation ($R^2 = 0.37$, **Figure 9A**) with the distance from the lake edge (**Figure 7F**), which may be due to the abnormal water content of four samples at 115, 145, 175, and 200 m. The I_{535} of the wet grassland is not evident compared to that of the dry grassland, which may be related to the low concentration

of the hematite in the wet grassland. However, the Hm/(Gt+Hm) value has a much more positive correlation with the water content ($R^2 = 0.61$, **Figure 9B**), despite the low concentration of the hematite in the wet grassland. High value of Hm/(Gt+Hm) corresponds to low soil moisture, and vice versa. This suggests that the Hm/(Gt+Hm) value can be taken as an effective proxy for soil moisture, in line with the results from a 600 km E-W transect in south Brazil (Kämpf and Schwertmann, 1983). In paleoclimate studies, the Hm/(Gt+Hm) value (or Hm/Gt) was mostly regarded as an indicator of precipitation (Harris and Mix, 2002; Ji et al., 2004; Zhang et al., 2007). For example, the hematite/goethite (Hm/Gt) ratios are lower in paleosols (formed during wet period) than in loess (formed during dry period) (Ji et al., 2004). Modern process [this study and the results of Kämpf and Schwertmann (1983)] proves the validity for the use of Hm/Gt ratios in the paleoclimate researches.

CONCLUSIONS

In this study, detailed rock magnetic measurements were carried out in the offshore soils of Lake Qinghai. The results show a higher magnetic susceptibility of dry grassland samples than that of wet grassland, suggesting a higher concentration of magnetic minerals in the dry grassland and lower concentration of magnetic minerals in the wet grassland near the lake edge. The different magnetic properties of the wet and dry grassland may be linked to different conversion process of magnetic minerals, pedogenesis in the dry grassland, and/or the dissolution of magnetic mirerals in the wet grassland. Besides, the Hm/(Gt+Hm) value can be taken as an effective proxy for soil moisture.

AUTHOR CONTRIBUTIONS

PZ, HA, and ZA designed the study and contributed to discussion and interpretation of the results. PZ performed the fieldwork and led the writing of the paper. SL, LW, and XS contributed to the laboratory work and data representation.

ACKNOWLEDGMENTS

We thank Dawen Zhang and Suzhen Liu for their help in laboratory work. We thank Prof. Weiguo Liu, Huanye Wang, and Yan Yan for their useful suggestions. This study was financially supported by the National Basic Research Program of China (2013CB956402), National Natural Science Foundation of China (41174057, 41290253), and open fund of State Key Laboratory of Loess and Quaternary Geology (SKLLQG1535).

REFERENCES

An, Z. S., Colman, S. M., Zhou, W. J., Li, X. Q., Brown, E. T., Jull, A. J. T., et al. (2012). Interplay between the Westerlies and Asian monsoon recorded in Lake Qinghai sediments since 32 ka. *Sci. Rep.* 2:619. doi: 10.1038/srep00619

Ao, H., Deng, C. L., Dekkers, M. J., and Liu, Q. S. (2010). Magnetic mineral dissolution in Pleistocene fluvio-lacustrine sediments, nihewan basin (North China). *Earth Planet. Sci. Lett.* 292, 191–200. doi: 10.1016/j.epsl.2010.01.035

Bloemendal, J., and Liu, X. M. (2005). Rock magnetism and geochemistry of two plio–pleistocene chinese loess–palaeosol sequences—implications for quantitative palaeoprecipitation reconstruction. *Palaeogeogr. Palaeoclimatol. Palaeoecol.* 226, 149–166. doi: 10.1016/j.palaeo.2005.05.008

Buggle, B., Hambach, U., Müller, K., Zöller, L., Marković, S. B., and Glaser, B. (2014). Iron mineralogical proxies and Quaternary climate change in SE-European loess–paleosol sequences. *CATENA* 117, 4–22. doi: 10.1016/j.catena.2013.06.012

Dearing, J., Bird, P., Dann, R., and Benjamin, S. (1997). Secondary ferrimagnetic minerals in Welsh soils: a comparison of mineral magnetic detection methods and implications for mineral formation. *Geophys. J. Int.* 130, 727–736. doi: 10.1111/j.1365-246X.1997.tb01867.x

Deaton, B. C., and Balsam, W. L. (1991). Visible spectroscopy–a rapid method for determining hematite and goethite concentration in geological materials. *J. Sediment. Res.* 61, 628–632. doi: 10.1306/D4267794-2B26-11D7-8648000102C1865D

Dekkers, M. J. (1997). Environmental magnetism: an introduction. *Geol. Mijnb.* 76, 163–182. doi: 10.1023/A:1003122305503

Demory, F., Oberhänsli, H., Nowaczyk, N. R., Gottschalk, M., Wirth, R., and Naumann, R. (2005). Detrital input and early diagenesis in sediments from Lake Baikal revealed by rock magnetism. *Glob. Planet. Change* 46, 145–166. doi: 10.1016/j.gloplacha.2004.11.010

Deng, C. L., Liu, Q. S., Pan, Y. X., and Zhu, R. X. (2007). Environmental magnetism of chinese loess-paleosol sequences. *Quat. Sci.* 27, 193–209.

Deng, C. L., Shaw, J., Liu, Q. S., Pan, Y. X., and Zhu, R. X. (2006). Mineral magnetic variation of the Jingbian loess/paleosol sequence in the northern loess plateau of china: implications for quaternary development of asian aridification and cooling. *Earth Planet. Sci. Lett.* 241, 248–259. doi: 10.1016/j.epsl.2005.10.020

Deng, C. L., Zhu, R. X., Verosub, K. L., Singer, M. J., and Yuan, B. Y. (2000). Paleoclimatic significance of the temperature-dependent susceptibility of holocene loess along a NW-SE transect in the chinese loess plateau. *Geophys. Res. Lett.* 27, 3715–3718. doi: 10.1029/2000GL008462

Evans, M. E., and Heller, F. (2003). *Environmental Magnetism: Principles and Applications of Enviromagnetics.* Amsterdam: Academic Press.

Fang, X. M., Zan, J. B., Appel, E., Lu, Y., Song, C. H., Dai, S., et al. (2015). An Eocene–Miocene continuous rock magnetic record from the sediments in the Xining Basin, NW China: indication for Cenozoic persistent drying driven by global cooling and Tibetan Plateau uplift. *Geophys. J. Int.* 201, 78–89. doi: 10.1093/gji/ggv002

Fu, C. F., Bloemendal, J., Qiang, X. K., Hill, M. J., and An, Z. S. (2015). Occurrence of greigite in the Pliocene sediments of Lake Qinghai, China, and its paleoenvironmental and paleomagnetic implications. *Geochem. Geophys. Geosyst.* 16, 1293–1306. doi: 10.1002/2014GC005677

Hanesch, M., Stanjek, H., and Petersen, N. (2006). Thermomagnetic measurements of soil iron minerals: the role of organic carbon. *Geophys. J. Int.* 165, 53–61. doi: 10.1111/j.1365-246X.2006.02933.x

Hao, Q. Z., Oldfield, F., Bloemendal, J., Torrent, J., and Guo, Z. T. (2009). The record of changing hematite and goethite accumulation over the past 22 myr on the chinese loess plateau from magnetic measurements and diffuse reflectance spectroscopy. *J. Geophys. Res. Solid Earth* 114:B12101. doi: 10.1029/2009jb006604

Harris, S. E., and Mix, A. C. (2002). Climate and tectonic influences on continental erosion of tropical South America, 0–13 Ma. *Geology* 30, 447–450. doi: 10.1130/0091-7613(2002)

Harrison, R. J., and Feinberg, J. M. (2008). FORCinel: An improved algorithm for calculating first-order reversal curve distributions using locally weighted regression smoothing. *Geochem. Geophys. Geosyst.* 9, Q05016. doi: 10.1029/2008GC001987

Heller, F., and Evans, M. E. (1995). Loess magnetism. *Rev. Geophys.* 33, 211–240. doi: 10.1029/95RG00579

Hunt, C. P., Banerjee, S. K., Han, J., Solheid, P. A., Oches, E., Sun, W., et al. (1995). Rock-magnetic proxies of climate change in the loess-palaeosol sequences of the western Loess Plateau of China. *Geophys. J. Int.* 123, 232–244. doi: 10.1111/j.1365-246X.1995.tb06672.x

Ji, J. F., Balsam, W., Chen, J., and Liu, L. W. (2002). Rapid and quantitative measurement of hematite and goethite in the chinese loess-paleosol sequence by diffuse reflectance spectroscopy. *Clays Clay Miner.* 50, 208–216. doi: 10.1346/000986002760832801

Ji, J. F., Chen, J., Balsam, W., Lu, H. Y., Sun, Y. B., and Xu, H. F. (2004). High resolution hematite/goethite records from Chinese loess sequences for the last

glacial-interglacial cycle: rapid climatic response of the East Asian Monsoon to the tropical Pacific. *Geophys. Res. Lett.* 31, L03207. doi: 10.1029/2003gl018975

Jin, Z. D., An, Z. S., Yu, J. M., Li, F. C., and Zhang, F. (2015). Lake Qinghai sediment geochemistry linked to hydroclimate variability since the last glacial. *Quat. Sci. Rev.* 122, 63–73. doi: 10.1016/j.quascirev.2015.05.015

Jin, Z. D., You, C. F., Yu, J. M., Wu, L. L., Zhang, F., and Liu, H. C. (2011). Seasonal contributions of catchment weathering and eolian dust to river water chemistry, northeastern tibetan plateau: chemical and Sr isotopic constraints. *J. Geophys. Res. Earth Surface (2003-2012)* 116:F04006. doi: 10.1029/2011JF002002

Jun, Q. Y., and Kelts, K. R. (2002). Abrupt changes in climatic conditions across the late-glacial/Holocene transition on the NE TIBET-Qinghai Plateau: evidence from Lake Qinghai, China. *J. Paleolimnol.* 28, 195–206. doi: 10.1023/A:1021635715857

Kämpf, N., and Schwertmann, U. (1983). Goethite and hematite in a climosequence in southern Brazil and their application in classification of kaolinitic soils. *Geoderma* 29, 27–39. doi: 10.1016/0016-7061(83)90028-9

Kelts, K., Zao, C. K., Lister, G. S., Yu, J. Q., Zhang, G. H., Niessen, F., et al. (1989). Geological fingerprints of climate history: a cooperative study of Qinghai lake, China. *Eclogae Geol. Helv.* 82, 167–182.

King, J., Banerjee, S. K., Marvin, J., and Özdemir, Ö. (1982). A comparison of different magnetic methods for determining the relative grain size of magnetite in natural materials: some results from lake sediments. *Earth Planet. Sci. Lett.* 59, 404–419. doi: 10.1016/0012-821X(82)90142-X

Li, X. Y., Xu, H. Y., Sun, Y. L., Zhang, D. S., and Yang, Z. P. (2007). Lake-level change and water balance analysis at Lake Qinghai, west China during recent decades. *Water Resour. Manage* 21, 1505–1516. doi: 10.1007/s11269-0 06-9096-1

LIGCAS (Lanzhou Institute of Geology of Chinese Academy of Sciences) (1979). *A Synthetically Investigation Report on Qinghai Lake.* Beijing: Science Press.

Lister, G. S., Kelts, K., Zao, C. K., Yu, J. Q., and Niessen, F. (1991). Lake Qinghai, China: closed-basin like levels and the oxygen isotope record for ostracoda since the latest Pleistocene. *Palaeogeogr. Palaeoclimatol. Palaeoecol.* 84, 141–162. doi: 10.1016/0031-0182(91)90041-O

Liu, Q. S., Deng, C. L., Torrent, J., and Zhu, R. X. (2007). Review of recent developments in mineral magnetism of the Chinese loess. *Quat. Sci. Rev.* 26, 368–385. doi: 10.1016/j.quascirev.2006.08.004

Liu, Q. S., Roberts, A. P., Larrasoaña, J. C., Banerjee, S. K., Guyodo, Y., Tauxe, L., et al. (2012). Environmental magnetism: principles and applications. *Rev. Geophys.* 50, RG4002. doi: 10.1029/2012RG000393

Liu, X. J., and Lai, Z. P. (2010). Lake level fluctuations in Qinghai Lake in the Qinghai-Tibetan Plateau since the last interglaciation: abrief review and new data. *J. Earth. Environ.* 1, 79–89.

Liu, X. Q., Shen, J., Wang, S. M., Zhang, E. L., and Cai, Y. F. (2003). A 16 000-year paleoclimatic record derived from authigenetic carbonate of lacustrine sediment in Qinghai Lake. *Geol. J. Chin. Univ.* 9, 38–46.

Lu, H., Wang, X., Ma, H., Tan, H., Vandenberghe, J., Miao, X., et al. (2004). The Plateau Monsoon variation during the past 130 kyr revealed by loess deposit at northeast Qinghai–Tibet (China). *Global Planet. Change* 41, 207–214. doi: 10.1016/j.gloplacha.2004.01.006

Maher, B. A., Alekseev, A., and Alekseeva, T. (2002). Variation of soil magnetism across the Russian steppe: its significance for use of soil magnetism as a palaeorainfall proxy. *Quat. Sci. Rev.* 21, 1571–1576. doi: 10.1016/S0277-3791(02)00022-7

Nie, J., Stevens, T., Song, Y., King, J. W., Zhang, R., Ji, S., et al. (2014). Pacific freshening drives Pliocene cooling and Asian monsoon intensification. *Sci. Rep.* 4:5474. doi: 10.1038/srep05474

Nie, J. S., Song, Y. G., and King, J. W. (2016). A review of recent advances in red-clay environmental magnetism and paleoclimate history on the Chinese Loess Plateau. *Front. Earth Sci.* 4:27. doi: 10.3389/feart.2016.00027

Oldfield, F. (1991). Environmental magnetism–A personal perspective. *Quat. Sci. Rev.* 10, 73–85. doi: 10.1016/0277-3791(91)90031-O

Roberts, A. P., Heslop, D., Zhao, X., and Pike, C. R. (2014). Understanding fine magnetic particle systems through use of first-order reversal curve diagrams. *Rev. Geophys.* 52, 557–602. doi: 10.1002/2014RG000462

Roberts, A. P., Pike, C. R., and Verosub, K. L. (2000). First-order reversal curve diagrams: a new tool for characterizing the magnetic properties of natural samples. *J. Geophys. Res.* 105, 28461–28475. doi: 10.1029/2000JB900326

Scheinost, A. C., Chavernas, A., Barron, V., and Torrent, J. (1998). Use and limitations of second-derivative diffuse reflectance spectroscopy in the visible to near-infrared range to identify and quantity fe oxide minerals in soils. *Clays Clay Miner.* 46, 528–536. doi: 10.1346/CCMN.1998.0460506

Shen, J., Liu, X. Q., Wang, S. M., and Matsumoto, R. (2005). Palaeoclimatic changes in the Qinghai Lake area during the last 18,000 years. *Quat. Int.* 136, 131–140. doi: 10.1016/j.quaint.2004.11.014

Shen, J., Zhang, E. L., and Xia, W. L. (2001). Records from lake sediments of the Qinghai Lake to mirror climatic and environmental changes of the past about 1000 years. *Quat. Sci.* 21, 508–513.

Snowball, I. (1993). Mineral magnetic properties of Holocene lake sediments and soils from the Kårsa valley, Lappland, Sweden, and their relevance to palaeoenvironmental reconstruction. *Terra Nova* 5, 258–270. doi: 10.1111/j.1365-3121.1993.tb00257.x

Sun, W. W., Banerjee, S. K., and Hunt, C. P. (1995). The role of maghemite in the enhancement of magnetic signal in the Chinese loess-paleosol sequence: An extensive rock magnetic study combined with citrate-bicarbonate-dithionite treatment. *Earth Planet. Sci. Lett.* 133, 493–505. doi: 10.1016/0012-821X(95)00082-N

Thompson, R., and Oldfield, F. (1986). *Environmental Magnetism. London: Allen and Unwin.*

Torrent, J., Liu, Q. S., Bloemendal, J., and Barrón, V. (2007). Magnetic enhancement and iron oxides in the upper Luochuan loess–paleosol sequence, Chinese Loess Plateau. *Soil Sci. Soc. Am. J.* 71, 1570–1578. doi: 10.2136/sssaj2006.0328

van der Zee, C., Roberts, D. R., Rancourt, D. G., and Slomp, C. P. (2003). Nanogoethite is the dominant reactive oxyhydroxide phase in lake and marine sediments. *Geology* 31, 993–996. doi: 10.1130/G19924.1

Verosub, K. L., and Roberts, A. P. (1995). Environmental magnetism: Past, present, and future. *J. Geophys. Res.* 100, 2175–2192. doi: 10.1029/94JB02713

Wang, X. Y., Yi, S., Lu, H., Vandenberghe, J., and Han, Z. (2015). Aeolian process and climatic changes in loess records from the northeastern Tibetan Plateau: response to global temperature forcing since 30 ka. *Paleoceanography* 30, 612–620. doi: 10.1002/2014PA002731

Wu, R. J. (1993). Magnetic susceptibility (χ) and frequency dependent susceptibility (χ_{fd}) of lake sediments and their paleocilatic implication—The case of recent sediments of Qinghai Lake and Daihai Lake. *J. Lake Sci.* 5, 128–135. doi: 10.18307/1993.0204

Zhang, E. L., Shen, J., Wang, S. M., Xia, W. L., and Jin, Z. D. (2002). Climate and environment change during the past 900 years in Qinghai Lake. *J. Lake Sci.* 14, 30–38.

Zhang, P., Ao, H., An, Z. S., and Wang, Q. S. (2015). Rock magnetism properties of Oligocene sediments in the Lanzhou Baisn. *Chin. J. Geophys. Chin. Ed.* 58, 2445–2459. doi: 10.6038/cjg20150721

Zhang, Y. G., Ji, J. F., Balsam, W., Liu, L. W., and Chen, J. (2009). Mid-Pliocene Asian monsoon intensification and the onset of Northern Hemisphere glaciation. *Geology* 37, 599–602. doi: 10.1130/G25670A.1

Zhang, Y. G., Ji, J. F., Balsam, W. L., Liu, L. W., and Chen, J. (2007). High resolution hematite and goethite records from ODP 1143, South China Sea: Co-evolution of monsoonal precipitation and El Niño over the past 600,000 years. *Earth Planet. Sci. Lett.* 264, 136–150. doi: 10.1016/j.epsl.2007.09.022

Zhou, L. P., Oldfield, F., Wintle, A. G., Robinson, S. G., and Wang, J. T. (1990). Partly pedogenic origin of magnetic variations in Chinese loess. *Nature* 346, 737–739. doi: 10.1038/346737a0

Conflict of Interest Statement: The authors declare that the research was conducted in the absence of any commercial or financial relationships that could be construed as a potential conflict of interest.

Dynamic Changes at Yahtse Glacier, the Most Rapidly Advancing Tidewater Glacier in Alaska

William J. Durkin[1]*, Timothy C. Bartholomaus[2], Michael J. Willis[1, 3, 4] and Matthew E. Pritchard[1]

[1] Department of Earth and Atmospheric Sciences, Cornell University, Ithaca, NY, USA, [2] Department of Geological Sciences, University of Idaho, Moscow, ID, USA, [3] Geological Sciences, University of North Carolina at Chapel Hill, Chapel Hill, NC, USA, [4] Cooperative Institute for Research in Environmental Sciences (CIRES), University of Colorado, Boulder, CO, USA

Edited by:
*Alun Hubbard,
Aberystwyth University, UK*

Reviewed by:
*Qiao Liu,
Institute of Mountain Hazards and
Environment (CAS), China
Martin Rückamp,
Alfred-Wegener-Institut für Polar-und
Meeresforschung, Germany
Samuel Huckerby Doyle,
Aberystwyth University, UK*

*Correspondence:
*William J. Durkin
wjd73@cornell.edu.edu*

Specialty section:
*This article was submitted to
Cryospheric Sciences,
a section of the journal
Frontiers in Earth Science*

Since 1990, Yahtse Glacier in southern Alaska has advanced at an average rate of $\sim$100 m year^{-1} despite a negative mass balance, widespread thinning in its accumulation area, and a low accumulation-area ratio. To better understand the interannual and seasonal changes at Yahtse and the processes driving these changes, we construct velocity and ice surface elevation time series spanning the years 1985–2016 and 2000–2014, respectively, using satellite optical and synthetic aperture radar (SAR) observations. We find contrasting seasonal dynamics above and below a steep (up to 35% slope) icefall located approximately 6 km from the terminus. Above the icefall, velocities peak in May and reach their minima in October synchronous with the development of a small embayment at the calving terminus. The up-glacier minimum speeds, embayment, and plume of turbid water that emerges from the embayment are consistent with an efficient, channelized subglacial drainage system that lowers basal water pressures and leads to focused submarine melt in the calving embayment. However, velocities near the terminus are fastest in the winter, following terminus retreat, possibly off of a terminal moraine that results in decreased backstress. Between 1996 and 2016 the terminus decelerated by $\sim$40% at an average rate of $\sim$0.4 m day^{-1} year^{-1}, transitioned from tensile to compressive longitudinal strain rates, and dynamically thickened at rates of 1-6 m year^{-1}, which we hypothesize is in response to the development and advance of a terminal moraine. The described interannual changes decay significantly upstream of the icefall, indicating that the icefall may inhibit the upstream transmission of stress perturbations. We suggest that diminished stress transmission across the icefall could allow moraine-enabled terminus advance despite mass loss in Yahtse's upper basin. Our work highlights the importance of glacier geometry in controlling tidewater glacier re-advance, particularly in a climate favoring increasing equilibrium line altitudes.

Keywords: tidewater glacier dynamics, glacier advance, morainal bank, remote sensing, elevation time series, velocity time series

1. INTRODUCTION

The rapid retreat and thinning of tidewater glaciers is governed by processes that can be substantially decoupled from climate (e.g., Post et al., 2011). The contributions to sea level rise from tidewater glaciers are highly variable and contribute to large uncertainties in sea level rise projections (Pachauri et al., 2014). Tidewater glaciers lose mass through a combination of surface ablation and frontal ablation (itself the sum of losses from submarine melt and iceberg calving). Mass loss is enhanced during tidewater glacier retreat due to dynamic thinning, in which accelerated flow leads to thinning of upstream ice followed by further acceleration (Meier and Post, 1987; Pfeffer, 2007). Dynamic thinning ends once a tidewater glacier terminus restabilizes, greatly reducing mass loss through frontal and surface ablation. While the prevalence and urgency of tidewater glacier retreat has resulted in comparatively well-studied retreat processes, the processes by which tidewater glaciers transition from retreat into a stable or advance phase are poorly understood (Post et al., 2011) despite their importance for developing long-term projections of sea level rise.

In the tidewater glacier advance and retreat cycle, advance is driven by a positive mass balance gained over a high accumulation-area ratio. For example, Hubbard Glacier, the largest nonpolar tidewater glacier in the world, is advancing at a rate of 35 m year^{-1}, has a mass balance of +0.32 Gt year^{-1}, and an accumulation-area ratio of 0.95 (Motyka and Truffer, 2007; Larsen et al., 2015; Mcnabb et al., 2015; Stearns et al., 2015). An important component of tidewater glacier advance is the presence, growth, and migration of a moraine shoal that protects the terminus from submarine melting and buoyancy driven instabilities (Powell, 1991). Repeat bathymetric studies at Hubbard glacier reveal the presence of a moraine shoal that advances at ~32 m year^{-1}, roughly the same rate as terminus advance (Goff et al., 2012). This supports the hypothesis that the rate of terminus advance is limited by the rate at which erosion and sedimentation can build and prograde the moraine shoal (Motyka et al., 2006). In a study modeling tidewater glacier advance, Nick et al. (2007) found that while positive mass balance and a high accumulation-area ratio can initiate the advance phase, the glacier could not advance into deep water without the presence of a moraine shoal. However, the development of the moraine shoal is generally considered to be a second-order process in tidewater glacier advance (Powell, 1991).

Yahtse Glacier, located in Icy Bay, southern Alaska (**Figure 1**), is currently advancing at ~100 m year^{-1}, making it the state's fastest-advancing tidewater glacier (Mcnabb and Hock, 2014). Yahtse formed in 1961 from a tributary that separated from the retreating Guyot Glacier and retreated until 1985, at which point Yahtse terminated at the foot of a steep (~35%) icefall (Porter, 1989). In 1990 Yahtse entered a phase of sustained advance at an average rate of ~100 m year^{-1} (Mcnabb and Hock, 2014), which it maintains up to the time of writing. Repeat bathymetric soundings made 1.5 km from Yahtse's terminus show a 50 m shallowing between 1981 and 2011, evidence of sedimentation and the development of a moraine

shoal (Post, 1983; Bartholomaus et al., 2013). However, Yahtse is unusual among advancing tidewater glaciers because it has been characterized by a negative mass balance and thinning in its accumulation zone–both before and after the initiation of its advance phase (Muskett et al., 2008; Larsen et al., 2015). Between 1972 and 2000, the portion of the upper basin located above 1220 m had an area-averaged thinning rate of -0.9 ± 0.1 m year^{-1} (Muskett et al., 2008). The large upper basin continued to thin from 2006 to 2014, and despite thickening near the terminus yielded a -0.16 Gt year^{-1} mass balance (Larsen et al., 2015). This large upper basin is connected to the narrow (~2.5 km wide) terminating fjord by the steep icefall. This icefall appears to have had a significant role in characterizing Yahtse's dynamic behavior. Following the separation from Guyot Glacier in 1961 (Porter, 1989), aerial photographs show that Yahtse's near-terminus region was much thicker and featured slopes that were more shallow than they are at present (Molnia, 2008), to such an extent that the icefall was mostly concealed. As Yahtse retreated, the constriction at the icefall likely limited the delivery of ice to the terminus, causing the terminus region to thin and steepen until it retreated to the base of the icefall. Similar stretching and thinning occurred near the terminus of Columbia Glacier as it approached a smaller icefall in 2001 although, unlike Yahtse, Columbia Glacier's retreat continued past its icefall (O'Neel et al., 2005). McNabb et al. (2012) suggested that the sharp change in ice thickness over a short horizontal distance at Columbia Glacier could keep regions on opposing sides dynamically decoupled by inhibiting the upstream transmission of stresses. In this paper, we investigate the major kinematic changes occurring at Yahtse on the decadal and seasonal scales to better understand the role of Yahtse's geometry in facilitating its rapid advance.

2. DATA AND METHODS
2.1. Field Site
Our focus is on the lower section of Yahtse Glacier (**Figure 1**), specifically focusing on three points (A, B, and C) along the central flow line (**Figure 2A**). When referring to distances, we use a coordinate system in which the positive x-direction is along the central flow line and points in the upstream direction (**Figure 2A**). For analyses of velocity and ice elevations changes spanning multiple years, we use an Eulerian coordinate system in which the origin is at the intersection of the central flow line and the March 2014 terminus. Measurements are made at distances 1.25 and 3.5 km upstream from the terminus (positions A and B, respectively), and 2 km upstream from the top of the icefall (position C; **Figure 2A**). For analyses of interannual changes in driving stress and seasonal changes in velocity, we use a Lagrangian coordinate system with the origin placed at the intersection of the central flow line and the moving terminus. Distances 1.25 and 3.5 km upstream of the moving terminus position are labeled A$'$ and B$'$, respectively. By using a Lagrangian coordinate system, we avoid changes in the distance between reference points and the terminus (e.g., Howat et al., 2005).

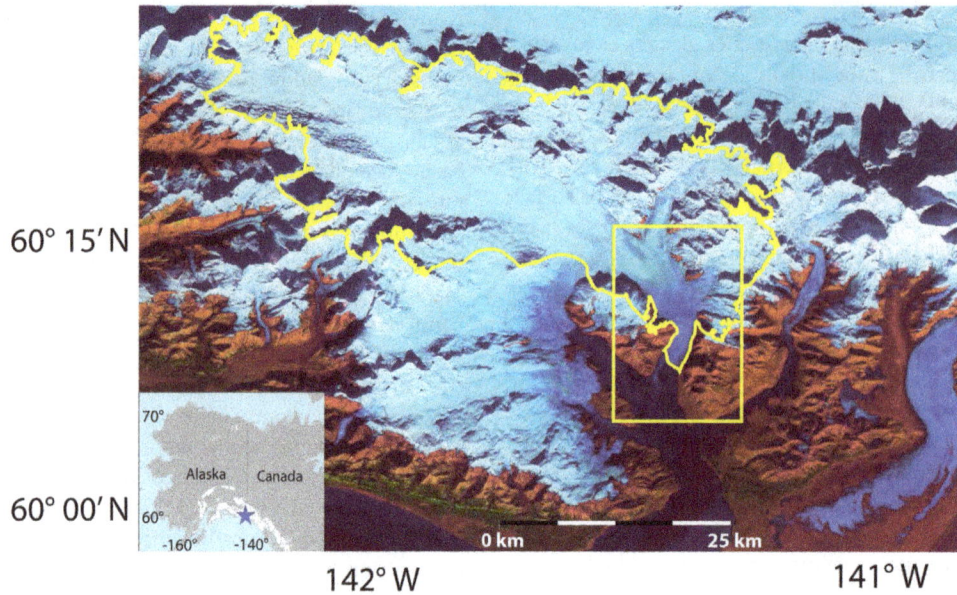

FIGURE 1 | Regional map of Icy Bay and surrounding glaciers. Yahtse Glacier is outlined in yellow, and the study area containing **Figures 2A**, **6** is outlined by a yellow box. Map image is an October 17, 2014 Landsat 8 composite of bands 7, 5, and 3.

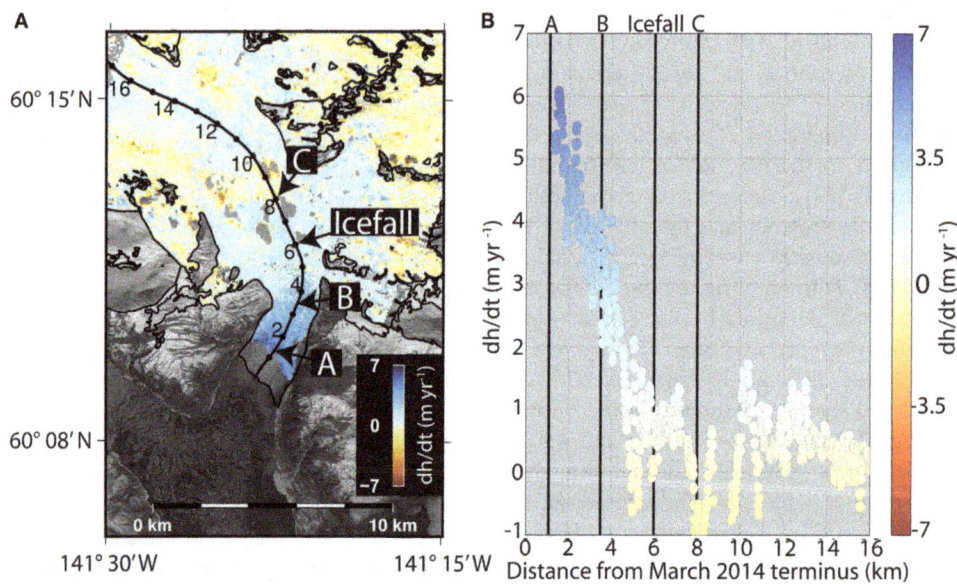

FIGURE 2 | $\frac{dh}{dt}$ is shown for the region of Yahtse Glacier upstream of its February 2000 terminus position **(A)** and in profile along the center streamline **(B)**. The locations of the icefall and Eulerian positions A, B, and C are marked by arrows, and the center streamline is shown by the black line with markers placed every 1 km **(A)**.

2.2. Ice Elevation Change Rates

Ice elevation change rates ($\frac{dh}{dt}$) were estimated using a weighted linear regression with a horizontally and vertically coregistered "stack" of eleven Advanced Spaceborne Thermal Emission Reflection radiometer (ASTER) Digital Elevation Models (DEMs) that span the years 2000–2011, two WorldView DEMs that span the years 2012–2014,

and the Shuttle Radar Topography Mission (SRTM) DEM collected during February 11-22, 2000. The acquisition dates and operational parameters of the images and satellites used are shown in **Figure 3** and **Table 1**, respectively. The methods used are fully described by Melkonian et al. (2013, 2016) and Willis et al. (2012a), and are briefly described here.

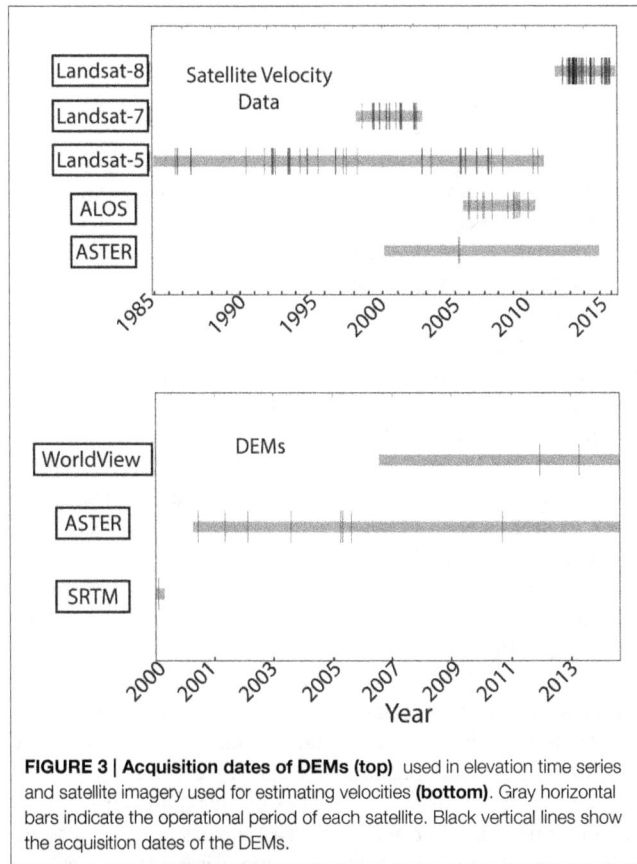

FIGURE 3 | Acquisition dates of DEMs (top) used in elevation time series and satellite imagery used for estimating velocities **(bottom)**. Gray horizontal bars indicate the operational period of each satellite. Black vertical lines show the acquisition dates of the DEMs.

TABLE 1 | Details of satellite imagery used in constructing the ice elevation time series.

Satellite	Resolution (m)	Count	Uncertainty Range (m)/Average (m)
SRTM	60	1	8
ASTER	15	11	9–25/15.2
WorldView	3	2	9–11/10

The SRTM DEM was used as a base image because it is a C-band radar product and is therefore not affected by cloud and snow coverage that can yield spurious elevations in the optical ASTER and WorldView DEMs. WorldView and ASTER DEMs were down-sampled to 60 m resolution and horizontally and vertically coregistered to off-ice elevations from the SRTM DEM using the Ames Stereo Pipeline toolkit (Broxton and Edwards, 2008; Shean et al., 2016). Off-ice pixels were identified using the glacier and rock outlines provided by the Randolph Glacier Inventory (version 4.0; Pfeffer et al., 2014). We assigned 1σ uncertainty to each DEM using the difference between the ASTER or WorldView DEM and the SRTM DEM off-ice elevations. Uncertainties of the ASTER DEMs range from 9 to 25 m and are 15.2 m on average, while Worldview DEMs uncertainties range from 9-11 m and have an average of 10 m (**Table 1**). We assigned an initial 5 m uncertainty to the SRTM DEM following Carabajal and Harding (2005) and Rodriguez

et al. (2006). Because the SRTM DEM is a C-band radar product, it is subject to snow and ice penetration that must be corrected before it is included in the DEM time series (e.g., Willis et al., 2012b; Melkonian et al., 2014; Berthier et al., 2016). Previous studies in the Juneau Icefield and Southern Patagonian Icefield found that the C-band SRTM DEM has a maximum penetration depth of 2-3 m in these regions compared to its X-band counterpart which, due to its smaller wavelength, is assumed to have a relatively small radar penetration depth (Willis et al., 2012b; Melkonian et al., 2014). Icy Bay, the Juneau Icefield, and the Southern Patagonian Icefield are similar in that they are regions of temperate glaciers with high mass-turnover rates. We therefore used 3 m as a conservative estimate of the uncertainty in the SRTM DEM's penetration depth, yielding a total uncertainty of 8 m (**Table 1**). Further work is needed to determine if this method under-estimates the penetration depth of the radar in Alaska, a situation that has been suggested to occur in the European Alps (Berthier et al., 2016).

We calculated $\frac{dh}{dt}$ values on a pixel-by-pixel basis by running a linear regression in which each elevation value is weighted by the inverse of the DEM uncertainty. We expect that the greatest elevation changes occur in the region of the advancing/retreating termini of Yahtse and Guyot Glaciers and used an iterative approach to identify an upper bound of +10 m year^{-1} and lower bound of -30 m year^{-1} for the $\frac{dh}{dt}$ in these regions. We filtered spurious elevations resulting from cloud coverage by excluding elevations from the final regression that deviate by more than $^{+10}_{-30}$ m year^{-1} from the first elevation in the time series. Because Yahtse is advancing into the ocean, regions between the most advanced and retracted terminus positions in the DEM time series will show a positive $\frac{dh}{dt}$ that does not necessarily represent ice thickening. We therefore limited $\frac{dh}{dt}$ results to regions upstream of the terminus as mapped in 2000 by the SRTM. We do not apply any correction for seasonal changes in elevation, as each DEM has an uncertainty ranging from 8 to 25 m (**Table 1**), and it is unclear that any seasonal change could be resolved (e.g., Wang and Kääb, 2015). Although the noise on individual ASTER DEMs is large, the time series approach using multiple dates spread over different seasons overcomes some of the issues of differencing only two DEMs, without applying seasonal corrections (Willis et al., 2012a; Wang and Kääb, 2015; Berthier et al., 2016). Finally, we limited $\frac{dh}{dt}$ results to pixels with a stack of at least seven DEMs to improve accuracy.

2.3. Bathymetry

Bathymetric soundings from 1981, shortly before the glacier reached its most retracted position (Post, 1983) were manually georeferenced and contoured using GDAL (http://www.gdal.org) and interpolated using Generic Mapping Tools (GMT; Wessel et al., 2013). However, the elevations of more recent glacier beds likely have changed since 1981 due to rapid erosion and sedimentation. Soundings made 1.5 km from the terminus in 2011 show a shallowing of 50 m compared to the 1981 soundings (Bartholomaus et al., 2013). We therefore assigned the glacier bed a crude time-dependent uncertainty estimate by assuming that the bathymetry has changed at a linear rate since the terminus

TABLE 2 | Operational parameters for satellite images used in velocity time series.

Satellite	Resolution (m)	Orbital repeat (days)	Reference window (m)	Search window (m)	Count
Landsat 5	30	16	480 × 480	960 × 960	27
Landsat 7	15	16	480 × 480	480 × 480	21
		32	480 × 480	690 × 690	6
Landsat 8	15	7	480 × 480	480 × 480	17
		9	480 × 480	480 × 480	14
		16	480 × 480	480 × 480	30
		23	480 × 480	480 × 480	15
		25	480 × 480	480 × 480	13
		32	480 × 480	690 × 690	6
ASTER	15	16	480 × 480	720 × 720	1
ALOS	8.3 × 3.3	46	498 × 396	830 × 660	13

began advancing in 1990. The uncertainty σ in meters as function of the year of interest t is

$$\sigma = 50 \frac{t - 1990}{2011 - 1990} \qquad (1)$$

This uncertainty reflects the ability of the advancing terminus to override or entrench itself in proglacial sediments (e.g., Kuriger et al., 2006; Motyka et al., 2006). Because deposition and erosion can both modify the elevation of the glacier bed, we do not make any assumptions as to whether the glacier bed has shallowed or deepened over time. We then used overridden seafloor topography for estimating the ice thickness and driving stress.

2.4. Driving Stress

Driving stress was calculated along the center streamline using a March 23, 2014 WorldView DEM, April 1, 2006 ASTER DEM, and the February 11-22, 2000 SRTM DEM with the gridded bathymetry. Topography was smoothed with a 400 m moving average, and surface slope was smoothed with a 1 km moving average to account for longitudinal coupling of resistive stresses (e.g., O'Neel et al., 2005). Ice thickness H was found as the difference between the ice surface topography and overridden bathymetry, and driving stress τ_D was calculated as

$$\tau_D = \rho_i g H \sin(\alpha) \qquad (2)$$

where ρ_i is the density of ice (917 kg m^{-3}), g is acceleration due to gravity (9.8 m s^{-2}), and α is the surface slope (e.g., Cuffey and Paterson, 2010).

2.5. Ice Velocities

We estimated horizontal ice velocities from 1985 to 2016 using pixel-tracking techniques with 163 ALOS, ASTER, and Landsat 5, 7, and 8 satellite image pairs. The acquisition dates of the imagery and the operational parameters for each satellite are summarized in **Figure 3** and **Table 2**. ASTER images were orthorectified to their own DEM by Land Processes Distributed Active Archive

Center (LPDAAC). Landsat 5, 7, and 8 imagery were provided orthorectified by the United States Geologic Survey (USGS). Raw ALOS SAR imagery were prepared for pixel-tracking using the Repeat Observation Interferometry PACkage (ROI_PAC: Rosen et al., 2004).

Once the images were orthorectified and prepared, we performed pixel-tracking through normalized cross-correlation using the "ampcor" function of the ROI_PAC software as implemented by Melkonian et al. (2014). Uncertainties for each velocity pair were calculated as the median velocity of the off-ice portions, which should be zero (e.g., Willis et al., 2012a), and velocity maps with uncertainties >0.5 m day^{-1} were excluded from the time series. The average uncertainty of the velocities used in the time series is 0.25 m day^{-1}. Full details on the methods used for filtering and other post processing are described by Willis et al. (2012a); Melkonian et al. (2013, 2016). We sampled velocities along a center streamline by taking the median within a 500 m section of the center streamline centered positions A, B, and C (**Figure 2A**).

Strain rates were calculated in the UTM coordinate system and then projected into the flow line coordinate system, in which the x-axis is aligned with ice flow and the y-axis is in the cross-flow direction. We use a sign convention in which longitudinal compression is negative. Due to the lower image quality of optical imagery predating the launch of the Landsat 8 satellite (e.g., Jeong and Howat, 2015) and the narrow shape of the terminus, we were only able to obtain useful strain rates in the region surrounding the center flow line. We assume that a 1D measurement of strain rates is a good approximation, as we are primarily interested in longitudinal strain rates, and the narrow shape of the terminus limits cross-flow variations.

2.6. Terminus Position

We tracked the terminus position at the centerline using the "Box Method" (e.g., Moon and Joughin, 2008; Mcnabb and Hock, 2014) with the Landsat images. In this method, we constructed a three-sided box formed by a fixed gate and sides that extend to a digitized terminus outline. The average length between the gate and the terminus was found by dividing the area of the box by the gate's width. Terminus shape was measured by outlining the terminus in the imagery manually. Because we are only interested in the qualitative changes in the terminus shape, we fit a smooth line to the terminus outline. Only cloud-free images are used in this study, and of those, only a sample are used to illustrate the seasonal changes in terminus shape.

3. RESULTS

3.1. Interannual Changes
3.1.1. Geometry and Driving Stress
Two patterns of $\frac{dh}{dt}$ occur on alternate sides of the icefall (**Figure 2**). Upstream of the icefall, $\frac{dh}{dt}$ is small and uniform along the centerline (± 1.5 m year^{-1}). Downstream of the icefall, $\frac{dh}{dt}$ increases at a linear rate and is greatest closest to the terminus (**Figure 2B**). Ice thickness did not change significantly between the years 2000 (277 ± 32 m) and 2014 (294 ± 66 m), however

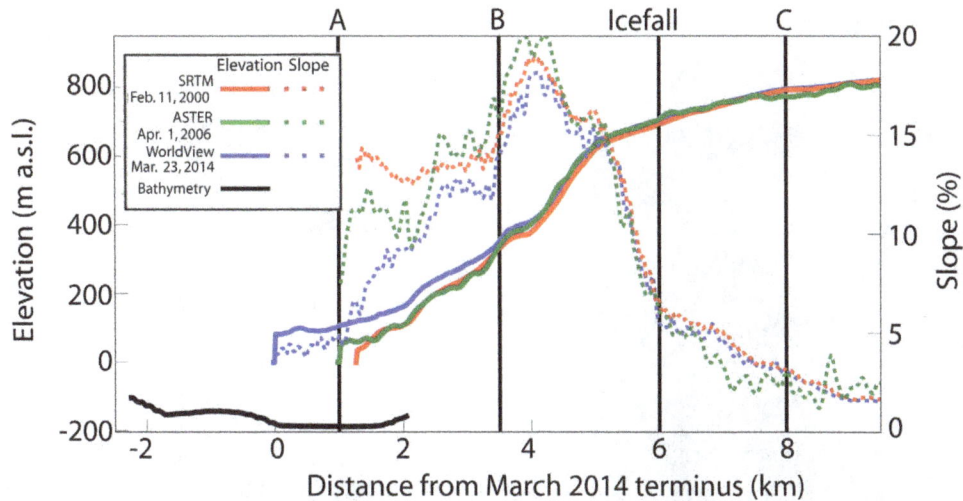

FIGURE 4 | Elevation and slope sampled along the center streamline for the February 11–22, 2000, SRTM, April 1, 2006 ASTER, and March 23, 2014 WorldView DEMs. Bathymetry (Post, 1983) is shown in black. Vertical black lines show the locations of the icefall and Eulerian positions A, B, and C.

surface slopes near the terminus decreased by ∼70% (**Figure 4**) and are correlated with a high positive $\frac{dh}{dt}$. Using Equation 2, we found that this decrease in slope drove a $60 \pm 20\%$ decrease in the driving stress along the first 500 m of the terminus between 2000 and 2014 (**Figure 5**). The terminus widened by ∼8% between 2000 and 2014, however due to uncertainties in the topography and bathymetry we did not find a significant change in the cross-sectional area and therefore did not observe a significant amount of lateral spreading that could contribute to the observed kinematic changes on the decadal scale.

3.1.2. Velocities

As ice flows across the icefall, it is accelerated from 2–5 m day^{-1} to 15–20 m day^{-1} (**Figure 6**), in agreement with the results of Burgess et al. (2013a) and Bartholomaus et al. (2013). Separating velocity results by season shows that while position B decelerated by ∼40% between 1996 and 2016 at a linear rate of −0.39 m day^{-1} year^{-1}, there was no significant decadal trend in velocities at position C during this time (**Figure 7B**; **Tables 3, 4**). Spring-time (March – May) velocity profiles from 2000 to 2016 (**Figure 7A**) show that the rate of deceleration was higher for regions closer to the terminus than at position B. Between 2000 and 2016, longitudinal strain rates averaged between 2.5 and 3.5 km in the Eulerian coordinate system transitioned from a tensile strain rate regime to one of compression (**Figure 8**).

3.2. Seasonal Variations

3.2.1. Velocities

Velocity profiles for November 2013 – November 2014 are shown in **Figure 9A**, and the timelines of velocities at positions A′, B′, and C for years 2013–2016 are shown in **Figure 9B**. Seasonal velocity variations at position C increased during the spring, peaked in early June, and decelerated to a minimum in September (**Figure 9B**). Downstream of the icefall, a different

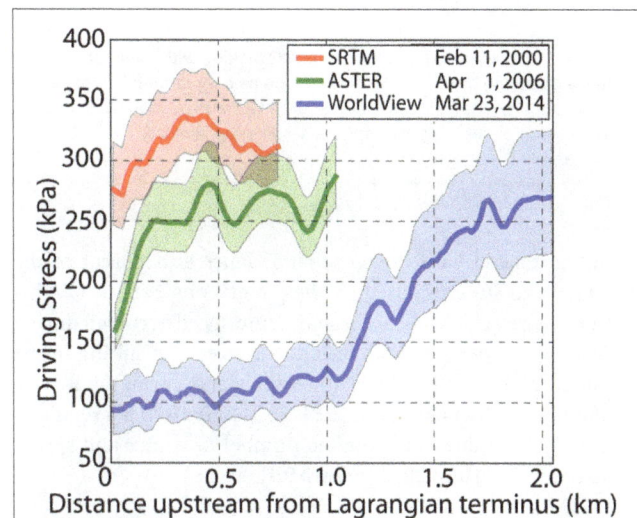

FIGURE 5 | Driving stress at the terminus in a Lagrangian coordinate system for February 11–22, 2000, April 1, 2006, and March 23, 2014 with uncertainty shown in the shaded regions. Driving stress along the first 750m upstream from the terminus decreased by 60 ± 20% between 2000 and 2014. Driving stress is calculated for a greater distance upstream of the terminus in 2014 than in previous years because in 2014 the terminus had overlapped a greater portion of the bathymetry.

pattern emerges. At A′ the timing of maximum velocity in 2013–2014 occurred when the terminus was in a retracted position in the winter (**Figures 9B**, **10B**). Deceleration at A′ between February and June was synchronous with a ∼200 m advance of the terminus. Further upstream at B′, velocities remained high during the winter and spring before they reached a maximum in early June and decreased throughout the summer (**Figure 9B**). The minima of all positions occurred in the fall. The 2013–2014 seasonal velocity pattern downstream of the icefall appears to

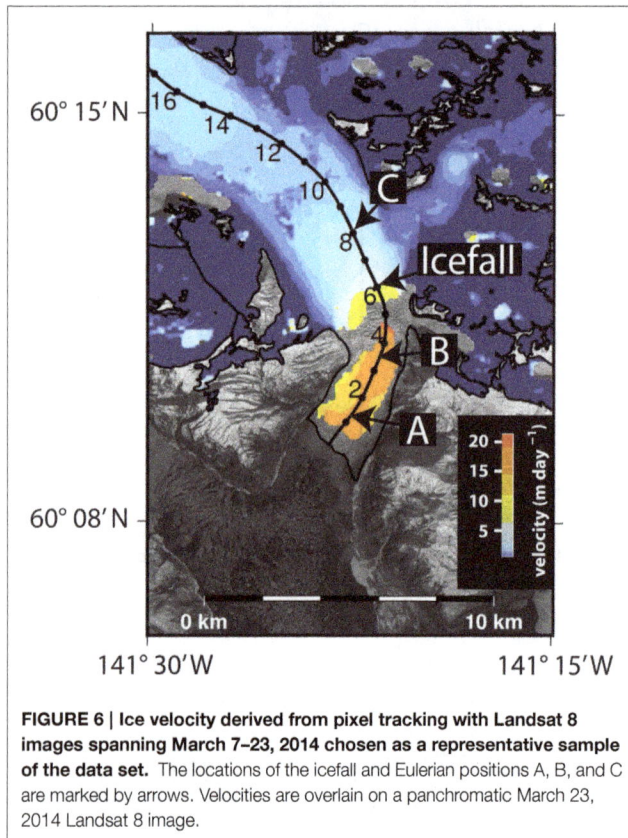

FIGURE 6 | Ice velocity derived from pixel tracking with Landsat 8 images spanning March 7–23, 2014 chosen as a representative sample of the data set. The locations of the icefall and Eulerian positions A, B, and C are marked by arrows. Velocities are overlain on a panchromatic March 23, 2014 Landsat 8 image.

be anticorrelated with the seasonal advance and retreat of the terminus and superimposed on the pattern observed at C. This effect is highest at A′, where the maximum speed occurred during the winter and decayed with distance upstream, resulting in the sustained high speeds during the winter and spring at B′. We cannot test whether or not the 2013–2014 A′ pattern is a regularly occurring event due to the limited number of winter and spring velocities in 2015 and 2016 (**Figure 9B**).

3.2.2. Terminus Position and Shape

Figure 10B shows the seasonal terminus position calculated using the 'Box Method' and represents the seasonal advance and retreat averaged over the entire width of the terminus. On average, the terminus retreated from the early-summer to late-fall and advanced from mid-winter to the early-summer. The shape of the terminus changed from smooth in the winter and spring to crenulated in June due to the development of a calving embayment on the western edge in 2014 and 2016 (**Figure 10A**). The embayment in each year grew during the summer and reached a maximum size in October (**Figure 10A**). A similar calving embayment and locus of turbid discharge appeared and developed during the melt season of 2000 and 2015 (not shown), and was observed previously by Bartholomaus et al. (2013) at Yahtse in September 2011 (see **Figure 1B** of the referenced paper). As the embayment grew throughout the summer, the remaining portions of the terminus remained at their advanced position. A comparison between the timeline

of average terminus positions (**Figure 10B**) and the terminus outlines (**Figure 10A**) shows that, although the width-averaged terminus position retreated during the summer, this was due to the growth of the calving embayment. The terminus did not retreat across its entire width until October – November (**Figure 10A**).

4. DISCUSSION

4.1. Seasonal Dynamics

Although C is positioned ∼8 km upstream of the March 2014 terminus and 2 km upstream of the icefall, the seasonality of velocities at C is similar to that reported near the termini of other Alaskan tidewater glaciers (e.g., Mcnabb et al., 2015; Stearns et al., 2015) — velocities are highest in late spring/early summer, at a minimum in the late fall, and intermediate during the winter. Spring acceleration is thought to be caused by surface meltwater that reaches an inefficient subglacial hydrologic system, leading to increased basal water pressure, increased separation at the ice-bed interface, and accelerated glacier sliding speeds (Iken and Bindschadler, 1986; Schoof, 2010). Over the course of the melt season, continued and increasing surface melt water routed to the glacier bed causes the subglacial hydrology to evolve toward increasingly efficient drainage, resulting in decreased basal water pressure and slower ice speeds (e.g., Bartholomaus et al., 2008). Intermediate velocity in winter may be the result of undrained basal water becoming trapped and pressurized as meltwater channels close in the fall (Schoof, 2010; Burgess et al., 2013b). The seasonal velocity pattern at C appears to respond to an increasingly efficient subglacial hydrologic system. This is consistent with (Bartholomaus et al., 2015a), who found that at Yahtse the lag time between melt input and glaciohydraulic tremor — a proxy for subglacial discharge — decreases over the course of the melt season. The coevolution of the decline in velocities at position C and the enlargement of the calving embayment between late May and October is evidence in support of an increasingly channelized subglacial hydrologic system that focuses subglacial discharge into a plume. The focused subglacial discharge at the glacier terminus locally increases submarine melt rates and iceberg calving (e.g., Sikonia and Post, 1979; Motyka et al., 2003; Ritchie et al., 2008; Bartholomaus et al., 2013). The appearance of a calving embayment in the western portion of the terminus and its growth throughout the melt season appears to be a regularly occurring event, as we observe similar sequences in the years 2000 and 2015 (not shown). A similar embayment and locus of turbidity were observed in September 2011 by Bartholomaus et al. (2013), **Figure 1B** of cited study. Future studies of embayment development at Yahtse Glacier, particularly its apparent geographic stability, can potentially reveal factors controlling the evolution of efficient subglacial conduits.

We do not have good constraints on the seasonal changes in ice thickness near Yahtse's terminus and therefore cannot explicitly calculate the seasonal changes in ice flux. However, at large-flux marine terminating glaciers in Greenland (e.g., Helheim; Bevan et al., 2015), near-terminus ice thickness was found to be thickest in the spring, following winter accumulation and advance, and lowest in the fall, following a period of ablation

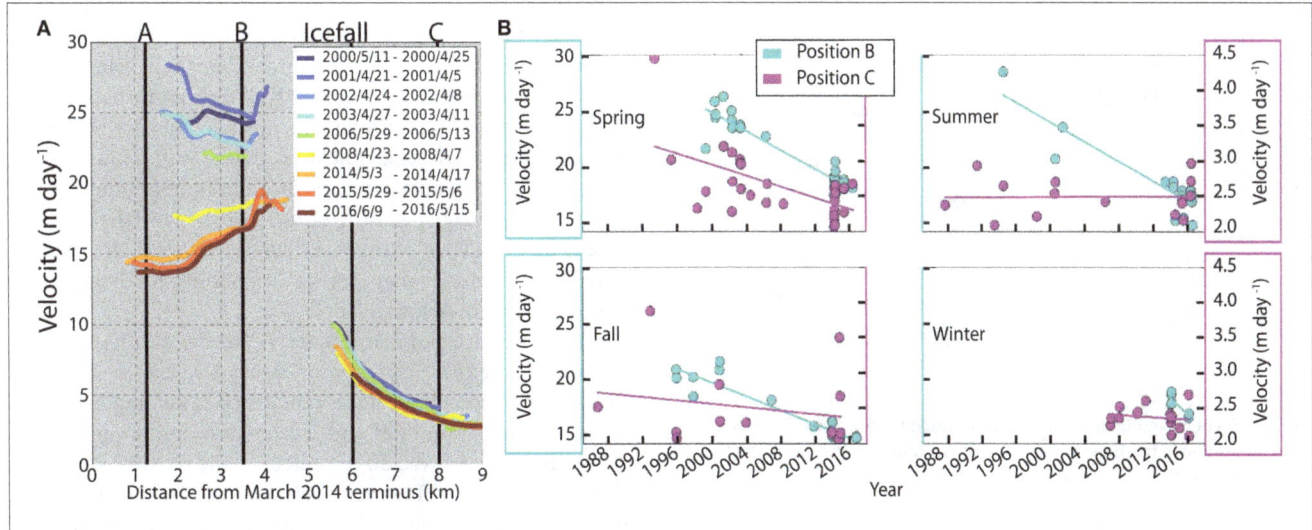

FIGURE 7 | Velocity profiles for select spring months (March-May) for years with velocity coverage downstream of the icefall (A), and timelines of velocities separated by season and sampled at Eulerian positions B and C **(B).** Left and right vertical axes of **(B)** are at different scales, in which right axes (magenta) correspond to velocities sampled at C and left axes (cyan) correspond to velocities sampled at B. Statistics of the regression are shown in **Tables 3, 4.**

TABLE 3 | Statistics for linear regression of velocities measured at position B.

Position B	acceleration (m day^{-1} year^{-1})	r^2	p-value	Count
Spring	−0.42 ± 0.16	0.87	2.3 × 10^{-13}	28
Summer	−0.45 ± 0.22	0.82	1.37 × 10^{-6}	16
Fall	−0.31 ± 0.12	0.88	6.9 × 10^{-10}	20
Winter	—	—	—	6

TABLE 4 | Statistics for linear regression of velocities measured at position C.

Position C	acceleration (m day^{-1} year^{-1})	r^2	p-value	N
Spring	−0.04 ± 0.05	0.37	2 × 10^{-4}	32
Summer	−0.00 ± 0.03	0.00	9.6 × 10^{-1}	15
Fall	−0.01 ± 0.06	0.05	4.5 × 10^{-1}	14
Winter	−0.01 ± 0.05	0.15	5.7 × 10^{-1}	15

and retreat. A similar pattern may occur at Yahtse. This would cause the seasonal variations in ice thickness to be in phase with the seasonal variations in velocity, resulting in high flux in the spring and low flux in the fall. We can therefore use the velocity near the terminus as a proxy for ice flux. The 2013–2014 seasonal velocity pattern at A′ (i.e. fastest in the winter, decelerating through the summer, and minimum in fall) is not a commonly observed pattern among Alaskan tidewater glaciers (e.g., Mcnabb et al., 2015), although previous studies have generally not resolved glacier velocities so close to the glacier front. At Yahtse, velocities at A′ accelerate following a full-width retreat of the terminus in October-November and decelerate from February through September as the terminus advances during the spring (**Figures 9, 10**). During the course of the melt season, Yahtse's terminus remains in an advanced position while the delivery of ice to the terminus decreases (**Figures 9, 10A**) and submarine melting increases (Bartholomaus et al., 2013). These two factors reach a critical point in October, leading to a full-width terminus retreat of ∼80 m (**Figure 9A**). Bartholomaus et al. (2015b) observed that Yahtse's calving flux peaks in the fall, coincident with the timing of the full-width terminus retreat observed here. Seasonal terminus retreat at Yahtse shows a dependence on tidewater glacier dynamics governing seasonal changes in ice flux (e.g., Stearns et al., 2015) as well as submarine melting and calving flux (e.g., Bartholomaus et al., 2015b). We interpret that the terminus retreats from the crest of its submarine moraine, which would result in a loss of backstress that could cause the winter acceleration. This is similar to the pattern of certain Greenland tidewater glaciers in response to the loss of backstress following the breakup of ice melange (Joughin et al., 2008; Moon et al., 2014). Interactions between the terminus and a submarine moraine have been observed in other years as well. In 2011 time-lapse photographs, Bartholomaus et al. (2012) observed a vertical component in Yahtse's seasonal advance, which they interpreted as the glacier moving up and over a submarine moraine.

Most Alaskan tidewater glaciers are in their retreat phase (Mcnabb and Hock, 2014) and would not override or press against a moraine crest, perhaps explaining why the pattern observed at A′ on Yahtse Glacier is largely absent from Alaskan tidewater glaciers. Interestingly, the seasonal pattern of speeds at the terminus of the advancing Hubbard Glacier is highest in mid-April and minimum in October-November (Stearns et al., 2015), which is closer to the C pattern of this study. However, this may be because speeds at Hubbard were sampled from a ∼ 5 x 5 km area which could average out this pattern.

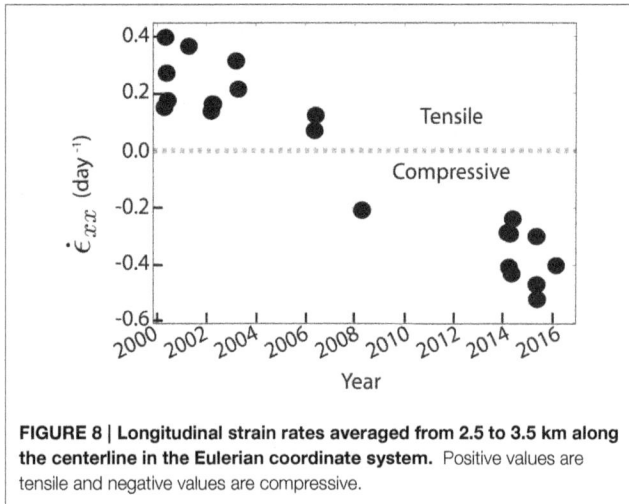

FIGURE 8 | Longitudinal strain rates averaged from 2.5 to 3.5 km along the centerline in the Eulerian coordinate system. Positive values are tensile and negative values are compressive.

Pfeffer (2007) proposed a criteria for determining whether or not a perturbation at the terminus would result in sustained retreat. If $\frac{\rho_i h}{\rho_w d} < 1.3$, where ρ_i is the density of ice, ρ_w is the density of sea water, h is the ice thickness, and d is the water depth, then a retreat from the moraine shoal will cause basal drag to decrease faster than the driving stress, leading to dynamic drawdown and rapid retreat. If this ratio is >1.3, then the driving stress will decrease faster than basal drag and the terminus will be stable against thinning or, in this case, a change in water depth associated with a seasonal separation from the moraine shoal. After applying the uncertainties from Equation 1 to the centerline bathymetry (**Figure 4**), we find a water depth of 180 ± 57 m. Bathymetric soundings from 2011 near Yahtse's terminus measured a water depth of 120 m (Bartholomaus et al., 2013). Because the water depth measured from the 2011 bathymetric soundings is near the bounds of our water depth estimate and has considerably smaller uncertainties, we use a water depth of 120 m and a centerline ice thickness of 200 ± 11 m (**Figure 4, Table 1**). Using an ice density of 917 kg m^{-3} and water density of 1029 kg m^{-3}, this criteria evaluates to 1.5 ± 0.1 and clearly, after years of observed advance, the terminus is stable against seasonal retreat despite its large 300 m magnitude.

4.2. Interannual Dynamics

If Yahtse's deceleration were due to interactions between the terminus and features in the seafloor topography (e.g., bumps, ridges, sills, etc) during the glacier's advance, we would expect to see a sudden deceleration in the velocity accompanied by a sharp decrease in terminus advance-rate as the glacier encounters topographic obstacles. Instead, Yahtse's advance is continuous and, while not constant, is well approximated by an average rate of 100 m year^{-1} (e.g., Mcnabb and Hock, 2014). Over the period of Yahtse's advance, we also find that the terminus decelerated at a linear rate of ~-0.4 m day^{-1} year^{-1} at position B (**Figure 7B**). The continuous rates of terminus advance and deceleration allow us to rule out interactions between the terminus and pinning points in the seafloor topography as a source of the observed interannual deceleration.

Interannual changes in velocity, thickness, and longitudinal strain rates that occur near the terminus attenuate with distance upstream and, in the case of interannual changes in velocity, may be absent above the icefall (**Figure 7B**). Stress perturbations that occur at the terminus may attenuate upstream due to the icefall's steep slopes and likely small thickness (Nye, 1960). Thinning that occurs in the accumulation area of Yahtse's large upper basin (Muskett et al., 2008; Larsen et al., 2015) may be the result of local mass balance changes, in synchrony with the climate. Therefore, the glacier may not be responding to changes that occur downstream of the icefall.

In the case of Columbia Glacier, O'Neel et al. (2005) found that, during retreat, its driving stress became increasingly supported by lateral drag at the expense of basal drag, requiring increased shear strain rates and acceleration during the glacier's multidecadal retreat. We do not calculate the resistive stresses at Yahtse in this study. However, if we assume that the changes in resistive stresses at Yahtse during its advance are the converse of the changes at Columbia Glacier during its retreat (i.e., the driving stress becomes increasingly supported by the basal drag relative to lateral drag), we can explain the kinematic changes we observe. As Yahtse advances, it likely terminates at the crest of a growing moraine (e.g., Powell, 1991; Motyka et al., 2006; Nick et al., 2007), resulting in decreased water depths, increased effective pressure, and increased basal drag (e.g., Pfeffer, 2007; Cuffey and Paterson, 2010). Increased basal drag at the terminus is often associated with decreased sliding rates (e.g., Pfeffer, 2007), and the combination of a large decrease in velocities at the terminus and small change in cross sectional area would drive down ice flux through the terminus, resulting in a decrease in frontal ablation. Because stress perturbations from the terminus are not transmitted upstream of the icefall, the upper basin would remain dynamically decoupled from changes at the terminus and the influx of ice from above the icefall would be nearly constant over time (e.g., **Figures 2B, 7B**). The steady influx of ice from above the icefall coupled with decreasing A′ velocities and frontal ablation is in agreement with the observed transition from tensile to compressive longitudinal strain rates (**Figure 8**). Increased surface elevation and decreased surface slopes observed between the Eulerian positions A and B (**Figures 2, 4**) are also consistent with dynamic thickening due to compressive longitudinal strain rates.

4.3. Sustained, Rapid Advance

The dynamic changes observed in this study are consistent with the fjord shallowing and development of a submarine terminal moraine. We suggest that moraine development serves as the foundation for understanding Yahtse's transition from retreat to its current advance phase. During the phase of rapid retreat, the terminus would not have sufficient time to build a stabilizing moraine (Powell, 1991). After ending its retreat at the base of the icefall, presumably because a steep bed slope placed the terminus close to the tidewater line, Yahtse remained in a stable phase from 1985 to 1990. In shallow water, this brief period of terminus stability presumably allowed for the development of a submarine moraine, which then enabled its advance phase (e.g., Powell,

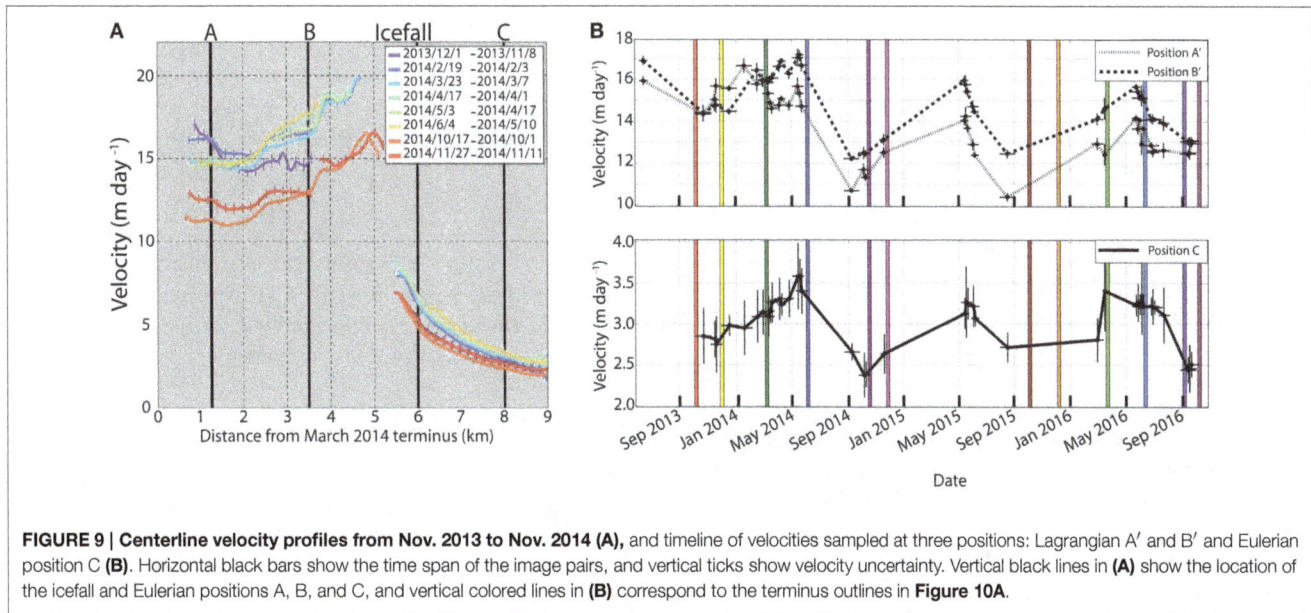

FIGURE 9 | **Centerline velocity profiles from Nov. 2013 to Nov. 2014 (A),** and timeline of velocities sampled at three positions: Lagrangian A′ and B′ and Eulerian position C **(B)**. Horizontal black bars show the time span of the image pairs, and vertical ticks show velocity uncertainty. Vertical black lines in **(A)** show the location of the icefall and Eulerian positions A, B, and C, and vertical colored lines in **(B)** correspond to the terminus outlines in **Figure 10A**.

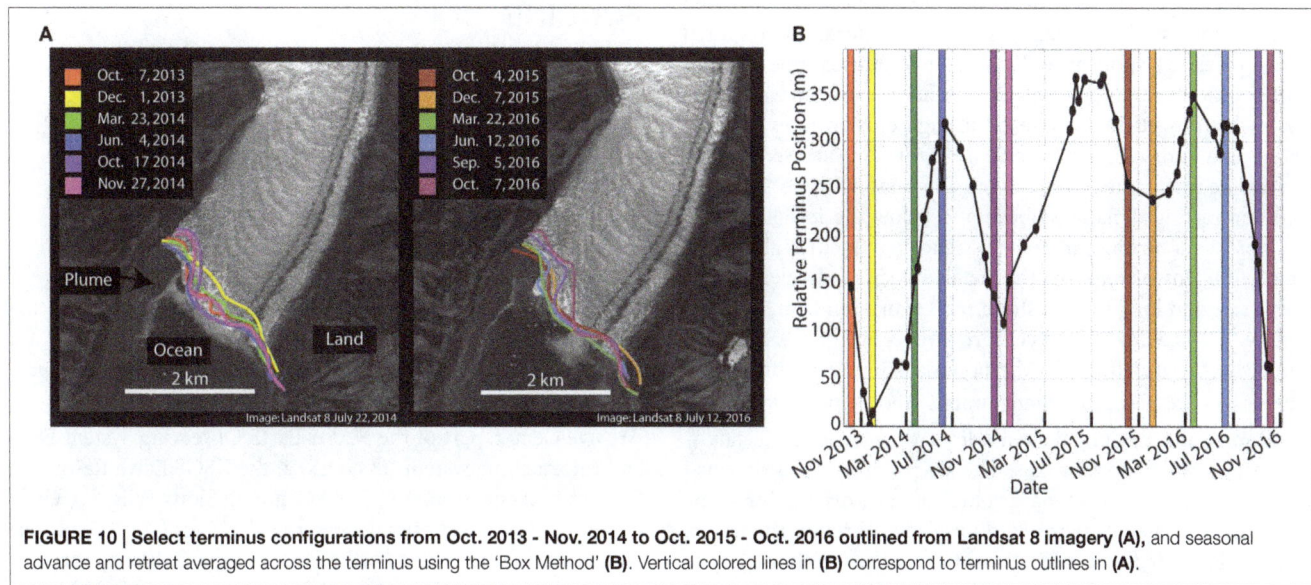

FIGURE 10 | **Select terminus configurations from Oct. 2013 - Nov. 2014 to Oct. 2015 - Oct. 2016 outlined from Landsat 8 imagery (A),** and seasonal advance and retreat averaged across the terminus using the 'Box Method' **(B)**. Vertical colored lines in **(B)** correspond to terminus outlines in **(A)**.

1991). Today, advance is ~3 times faster than the next fastest-advancing Alaskan tidewater glacier, Hubbard. Tidewater glacier advance has been suggested by previous studies to be facilitated by the progradation of the submarine terminal moraine (e.g., Motyka et al., 2006; Nick et al., 2007; Post et al., 2011; Goff et al., 2012) and we hypothesize that this is the case at Yahtse. Three factors support this interpretation:

1. Yahtse is the second largest tidewater glaciers in Alaska by area (e.g., Mcnabb et al., 2015) and terminates in a narrow ~2.5 km fjord. The region in between the icefall and terminus represents <2% of Yahtse's total surface area, and any eroded material produced up glacier is deposited in this focused region. By contrast, Hubbard's geometry is such that ice flows out of a ~2.5 km wide valley into a ~14 km wide terminal lobe

(e.g., Stearns et al., 2015), resulting in the lateral spreading of ice and sediments.

2. Large subglacial discharge due to Yahtse's large surface area and southern Alaska's high precipitation rates (Hill et al., 2015), coupled with fast ice flow near a narrow terminus would lead to rapid erosion and sedimentation, similar to observations made at the advancing Taku Glacier (Motyka et al., 2006).

3. Fast flow through a steep icefall is likely to produce rapid erosion of the rock beneath the icefall, leading to rapid sediment production which can be transported to the terminus.

Each of these three factors indicate that submarine moraine building and proglacial sedimentation may be unusually rapid at the terminus of Yahtse Glacier.

5. CONCLUSIONS

We have presented a record of kinematic changes on decadal and seasonal scales using satellite imagery from 1985 to 2016. We find that Yahtse's icefall and terminal moraine play significant roles in shaping the major kinematic changes at Yahtse on these timescales and also shape its advance.

Seasonal variations in velocity above the icefall show a dependence on subglacial hydrology. Velocities above the icefall are highest in early June and lowest in September, a pattern similar to many Alaskan tidewater glaciers and attributed to the development of an efficient/channelized drainage system during the melt season. The appearance and growth of a calving embayment and locus of turbidity at the terminus is synchronous with the summer deceleration of ice flow and is evidence for a plume of subglacial discharge focused by a channelized drainage system. Below the icefall, seasonal velocity variations in 2013–2014 show an additional, second-order dependence on the seasonal advance and retreat of the terminus. We suggest this is due to decreased backstress at the terminus during the fall as the terminus retreats into deeper water from its moraine shoal.

We suggest a number of geometric and dynamic factors could facilitate the rapid sedimentation at Yahtse's terminus required for the glacier's fast-advance. These include Yahtse's large area and funnel-like shape, high subglacial discharge rates, and rapid ice flow, which together likely result in high erosion rates across a broad area and sedimentation in a narrow and focused region. On the decadal time scale, as with the seasonal time scale, we find contrasting dynamics above and below the icefall. We do not observe a significant decadal trend in velocities above the icefall, but downstream of the icefall velocities decelerate at a linear rate and longitudinal strain rates transition from a tensile regime ($\sim +0.3$ day^{-1}) to a compressive regime (~ -0.4 day^{-1}). Similarly, the icefall marks the boundary between contrasting patterns in ice elevation change rates. Above the icefall, $\frac{dh}{dt}$ is small and uniform along the centerline, while below the icefall $\frac{dh}{dt}$ increases toward the terminus. The steep slope and likely small ice thickness at the icefall may prohibit stresses originating at the terminus from being transmitted up-glacier through the icefall. This could allow for continued advance at the terminus despite persistent thinning of the upper basin. As Yahtse advances, it likely terminates at the crest of an increasingly large moraine shoal, which would increase effective pressure at the terminus and drive down frontal ablation. Decreasing frontal ablation coupled with continued influx of ice is consistent with our observations of increasingly compressive longitudinal stain rates and dynamic thickening at the terminus.

6. EXTRA

6.0.1. Permission to Reuse and Copyright

Permission must be obtained for use of copyrighted material from other sources (including the web). Please note that it is compulsory to follow figure instructions.

AUTHOR CONTRIBUTIONS

MP initiated the project. WD and TB led the writing of the paper. WD led time series analyses, made all calculations, and made the figures. MW and WD collected and processed remote sensing data. MW and MP assisted in time series analyses, discussion, and writing.

FUNDING

This work was funded by NSF Grant EAR-0955547 and NASA grant NNX12AO31G from the Science Mission Directorate. DigitalGlobe imagery was provided via the Polar Geospatial Center under NSF PLR awards 1043681 and 1559691.

ACKNOWLEDGMENTS

We wish to thank Andrew Melkonian for his insight and support. We also thank the University of North Carolina at Chapel Hill Research Computing group for providing computational resources that have contributed to these research results. ASTER data were provided by the Land Processes Distributed Active Archive Center, part of the NASA Earth Observing System Data and Information System (EOSDIS) at the USGS Earth Resources Observation and Science (EROS) Center in Sioux Falls, SD, USA. Landsat data was downloaded via the USGS tool EarthExplorer. ALOS data was provided by the Japanese Space Agency through the Alaska Satellite Facility and NASA. The SRTM DEM was provided by NASA.

REFERENCES

Bartholomaus, T. C., Amundson, J. M., Walter, J. I., O'Neel, S., West, M. E., and Larsen, C. (2015a). Subglacial discharge at tidewater glaciers revealed by seismic tremor. *Geophys. Res. Lett.* 42, 6391–6398. doi: 10.1002/2015gl064590

Bartholomaus, T. C., Anderson, R. S., and Anderson, S. P. (2008). Response of glacier basal motion to transient water storage. *Nat. Geosci.* 1, 33–37. doi: 10.1038/ngeo.2007.52

Bartholomaus, T. C., Larsen, C. F., and O'Neel, S. (2013). Does calving matter? Evidence for significant submarine melt. *Earth Planet. Sci. Lett.* 380, 21–30. doi: 10.1016/j.epsl.2013.08.014

Bartholomaus, T. C., Larsen, C. F., O'Neel, S., and West, M. E. (2012). Calving seismicity from iceberg-sea surface interactions. *J. Geophys. Res. Earth Surf.* 117, 1–16. doi: 10.1029/2012JF002513

Bartholomaus, T. C., Larsen, C. F., West, M. E., O'Neel, S., Pettit, E. C., and Truffer, M. (2015b). Tidal and seasonal variations in calving flux observed with passive seismology. *J. Geophys. Res. Earth Surf.* 120, 2318–2337. doi: 10.1002/2015JF003641

Berthier, E., Cabot, V., Vincent, C., and Six, D. (2016). Decadal region-wide and glacier-wide mass balances derived from multi-temporal aSTER satellite digital elevation models. validation over the mont-blanc area. *Front. Earth Sci.* 4:63. doi: 10.3389/feart.2016.00063

Bevan, S. L., Luckman, A., Khan, S. A., and Murray, T. (2015). Seasonal dynamic thinning at Helheim Glacier. *Earth Planet. Sci. Lett.* 415, 47–53. doi: 10.1016/j.epsl.2015.01.031

Broxton, M. J. and Edwards, L. J. (2008). "The ames stereo pipeline: automated 3D surface reconstruction from orbital imagery," in *Lunar and Planetary Science Conference*, Vol. 39, 2419.

Burgess, E. W., Forster, R. R., and Larsen, C. F. (2013a). Flow velocities of Alaskan glaciers. *Nat. Commun.* 4:2146. doi: 10.1038/ncomms3146

Burgess, E. W., Larsen, C. F., and Forster, R. R. (2013b). Summer melt regulates winter glacier flow speeds throughout Alaska. *Geophys. Res. Lett.* 40, 6160–6164. doi: 10.1002/2013GL058228

Carabajal, C. C. and Harding, D. J. (2005). ICESat validation of SRTM C-band digital elevation models. *Geophys. Res. Lett.* 32, 1–5. doi: 10.1029/2005GL023957

Cuffey, K. M. and Paterson, W. S. B. (2010). *The Physics of Glaciers.* (Burlington, MA: Academic Press).

Goff, J. A., Lawson, D. E., Willems, B. A., Davis, M., and Gulick, S. P. S. (2012). Morainal bank progradation and sediment accumulation in disenchantment bay, alaska: response to advancing Hubbard Glacier. *J. Geophys. Res. Earth Surf.* 117, 1–15. doi: 10.1029/2011JF002312

Hill, D., Bruhis, N., Calos, S., Arendt, A., and Beamer, J. (2015). Spatial and temporal variability of freshwater discharge into the gulf of alaska. *J. Geophys. Res. Oceans* 120, 634–646. doi: 10.1002/2014JC010395

Howat, I. M., Joughin, I., Tulaczyk, S., and Gogineni, S. (2005). Rapid retreat and acceleration of Helheim Glacier, east Greenland. *Geophys. Res. Lett.* 32, 1–4. doi: 10.1029/2005GL024737

Iken, A. and Bindschadler, R. A. (1986). Combined measurements of subglacial water pressure and surface velocity of Findelengletscher, Switzerland: conclusions about drainage system and sliding mechanism. *J. Glaciol.* 32, 101–119. doi: 10.1017/S0022143000006936

Jeong, S. and Howat, I. M. (2015). Performance of Landsat 8 Operational Land Imager for mapping ice sheet velocity. *Remote Sens. Environ.* 170, 90–101. doi: 10.1016/j.rse.2015.08.023

Joughin, I., Howat, I. M., Fahnestock, M., Smith, B., Krabill, W., Alley, R. B., et al. (2008). Continued evolution of Jakobshavn Isbrae following its rapid speedup. *J. Geophys. Res. Earth Surf.* 113, 1–14. doi: 10.1029/2008JF001023

Kuriger, E. M., Truffer, M., Motyka, R. J., and Bucki, a. K. (2006). Episodic reactivation of large-scale push moraines in front of the advancing Taku Glacier, Alaska. *J. Geophys. Res. Earth Surf.* 111, 1–13. doi: 10.1029/2005JF000385

Larsen, C. F., Burgess, E., Arendt, A. A., O'Neel, S., Johnson, A. J., and Kienholz, C. (2015). Surface melt dominates Alaska glacier mass balance. *Geophys. Res. Lett.* 42, 5902–5908. doi: 10.1002/2015GL064349

Mcnabb, R. W. and Hock, R. (2014). Alaska tidewater glacier terminus positions, 1948-2012. *J. Geophys. Res. Earth Surf.* 119, 153–167. doi: 10.1002/2013jf002915

Mcnabb, R. W., Hock, R., and Huss, M. (2015). Variations in Alaska tidewater glacier frontal ablation, 1985-2013. *J. Geophys. Res. Earth Surf.* 120, 120–136. doi: 10.1002/2014JF003276

McNabb, R. W., Hock, R., O'Neel, S., Rasmussen, L. a., Ahn, Y., Braun, M., et al. (2012). Using surface velocities to calculate ice thickness and bed topography: a case study at Columbia Glacier, Alaska, USA. *J. Glaciol.* 58, 1151–1164. doi: 10.3189/2012JoG11J249

Meier, M. F. and Post, A. (1987). Fast tidewater glaciers. *J. Geophys. Res.* 92:9051. doi: 10.1029/JB092iB09p09051

Melkonian, A. K., Willis, M. J., Pritchard, M. E., Rivera, A., Bown, F., and Bernstein, S. A. (2013). Satellite-derived volume loss rates and glacier speeds for the Cordillera Darwin Icefield, Chile. *Cryosphere* 7, 823–839. doi: 10.5194/tc-7-823-2013

Melkonian, A. K., Willis, M. J., Pritchard, M. E., Rivera, A., Bown, F., and Bernstein, S. A. (2014). Satellite-derived volume loss rates and glacier speeds for the Juneau Icefield, Alaska. *J. Glaciol.* 60, 743–760. doi: 10.3189/2014JoG13J181

Melkonian, A. K., Willis, M. J., Pritchard, M. E., and Stewart, A. J. (2016). Recent changes in glacier velocities and thinning at Novaya Zemlya. *Remote Sens. Environ.* 174, 244–257. doi: 10.1016/j.rse.2015.11.001

Molnia, B. F. (2008). Glaciers of Alaska. *Satellite Image Atlas of the Glaciers of the World,* 554.

Moon, T. and Joughin, I. (2008). Changes in ice front position on Greenland's outlet glaciers from 1992 to 2007. *J. Geophys. Res.* 113:F02022. doi: 10.1029/2007JF000927

Moon, T., Joughin, I., Smith, B., Broeke, M. R., Berg, W. J., Noël, B., and Usher, M. (2014). Distinct patterns of seasonal Greenland glacier velocity. *Geophys. Res. Lett.* 41, 7209–7216. doi: 10.1002/2014GL061836

Motyka, R. J., Hunter, L., Echelmeyer, K. A., and Connor, C. (2003). Submarine melting at the terminus of a temperate tidewater glacier, LeConte Glacier, Alaska, USA. *Ann. Glaciol.* 36, 57–65. doi: 10.3189/172756403781816374

Motyka, R. J. and Truffer, M. (2007). Hubbard Glacier, Alaska: 2002 closure and outburst of Russell Fjord and postflood conditions at Gilbert Point. *J. Geophys. Res. Earth Surf.* 112, 1–15. doi: 10.1029/2006JF000475

Motyka, R. J., Truffer, M., Kuriger, E. M., and Bucki, A. K. (2006). Rapid erosion of soft sediments by tidewater glacier advance: Taku Glacier, Alaska, USA. *Geophys. Res. Lett.* 33, 1–5. doi: 10.1029/2006GL028467

Muskett, R. R., Lingle, C. S., Sauber, J. M., Rabus, B. T., and Tangborn, W. V. (2008). Acceleration of surface lowering on the tidewater glaciers of Icy Bay, Alaska, USA. from InSAR DEMs and ICESat altimetry. *Earth Planet. Sci. Lett.* 265, 345–359. doi: 10.1016/j.epsl.2007.10.012

Nick, F. M., van der Veen, C. J., and Oerlemans, J. (2007). Controls on advance of tidewater glaciers: Results from numerical modeling applied to Columbia Glacier. *J. Geophys. Res. Earth Surf.* 112, 1–11. doi: 10.1029/2006JF000551

Nye, J. (1960). "The response of glaciers and ice-sheets to seasonal and climatic changes," In *Proceedings of the Royal Society of London A: Mathematical, Physical and Engineering Sciences,* Vol. 256 (London: The Royal Society), 559–584.

O'Neel, S., Pfeffer, W. T., Krimmel, R., and Meier, M. (2005). Evolving force balance at Columbia Glacier, Alaska, during its rapid retreat. *J. Geophys. Res. Earth Surf.* 110, 1–18. doi: 10.1029/2005JF000292

Pachauri, R. K., Allen, M. R., Barros, V., Broome, J., Cramer, W., Christ, R., et al. (2014). *Climate Change 2014: Synthesis Report. Contribution of Working Groups I, II and III to the Fifth Assessment Report of the Intergovernmental Panel on Climate Change.* IPCC.

Pfeffer, W. T. (2007). A simple mechanism for irreversible tidewater glacier retreat. *J. Geophys. Res. Earth Surf.* 112, 1–12. doi: 10.1029/2006JF000590

Pfeffer, W. T., Arendt, A. A., Bliss, A., Bolch, T., Cogley, J. G., Gardner, A. S., et al. (2014). The randolph glacier inventory: a globally complete inventory of glaciers. *J. Glaciol.* 60, 537–552. doi: 10.3189/2014JoG13J176

Porter, S. C. (1989). Late Holocene fluctuations of the Fjord Glacier system in Icy Bay, Alaska, USA. *Arct. Alp. Res.* 21, 364–379. doi: 10.2307/1551646

Post, A. (1983). *Preliminary Bathymetry of Upper Icy Bay, Alaska.* Technical report. U.S. Geological Survey.

Post, A., O'Neel, S., Motyka, R. J., and Streveler, G. (2011). A complex relationship between calving glaciers and climate. *Eos Trans. Am. Geophys. Union* 92, 305–307. doi: 10.1029/2011EO370001

Powell, R. D. (1991). Grounding-line systems as second-order controls on fluctuations of tidewater termini of temperate glaciers. *Geol. Soc. Am. Spec. Pap.* 261, 75–94. doi: 10.1130/spe261-p75

Ritchie, J. B., Lingle, C. S., Motyka, R. J., and Truffer, M. (2008). Seasonal fluctuations in the advance of a tidewater glacier and potential causes: hubbard Glacier, Alaska, USA. *J. Glaciol.* 54, 401–411. doi: 10.3189/002214308785836977

Rodriguez, E., Morris, C. C., and Belz, J. J. (2006). A global assessment of the SRTM performance. *Photogramm. Eng. Remote Sens.* 72, 249–260. doi: 10.14358/PERS.72.3.249

Rosen, P. a., Hensley, S., Peltzer, G., and Simons, M. (2004). Updated repeat orbit interferometry package released. *Eos Trans. Am. Geophys. Union* 85:47. doi: 10.1029/2004EO050004

Schoof, C. (2010). Ice-sheet acceleration driven by melt supply variability. *Nature* 468, 803–806. doi: 10.1038/nature09618

Shean, D. E., Alexandrov, O., Moratto, Z. M., Smith, B. E., Joughin, I. R., Porter, C., and Morin, P. (2016). An automated, open-source pipeline for mass production of digital elevation models (DEMs) from very-high-resolution commercial stereo satellite imagery. *ISPRS J. Photogramm. Remote Sens.* 116, 101–117. doi: 10.1016/j.isprsjprs.2016.03.012

Sikonia, W. G. and Post, A. (1979). *Columbia Glacier, Alaska; Recent Ice Loss and Its Relationship to Seasonal Terminal Embayments, Thinning, and Glacier Flow.* Technical report. U.S. Geological Survey.

Stearns, L. A., Hamilton, G. S., Veen, C. J. V. D., Finnegan, D. C., Neel, S. O., Scheick, J. B., et al. (2015). Journal of Geophysical Research : earth surface glaciological and marine geological controls on terminus dynamics of Hubbard Glacier , southeast Alaska. *J. Geophys. Res. Earth Surf.* 120, 1065–1081. doi: 10.1002/2014JF003341

Wang, D. and Kääb, A. (2015). Modeling glacier elevation change from DEM time series. *Remote Sens.* 7, 10117–10142. doi: 10.3390/rs708 10117

Wessel, P., Smith, W. H. F., Scharroo, R., Luis, J., and Wobbe, F. (2013). Generic mapping tools: inproved version released. *EOS Trans. Am. Geophys. Union* 94, 409–410. doi: 10.1002/2013EO450001

Willis, M. J., Melkonian, A. K., Pritchard, M. E., and Ramage, J. M. (2012a). Ice loss rates at the Northern Patagonian Icefield derived using a decade of satellite remote sensing. *Remote Sens. Environ.* 117, 184–198. doi: 10.1016/j.rse.2011.09.017

Willis, M. J., Melkonian, A. K., Pritchard, M. E., and Rivera, A. (2012b). Ice loss from the Southern Patagonian Ice Field, South America, between 2000 and 2012. *Geophys. Res. Lett.* 39. doi: 10.1029/2012gl053136

Conflict of Interest Statement: The authors declare that the research was conducted in the absence of any commercial or financial relationships that could be construed as a potential conflict of interest.

The reviewer SD and handling Editor AH declare a shared institutional affiliation.

8

Circumpolar Mapping of Ground-Fast Lake Ice

Annett Bartsch[1,2]*, Georg Pointner[1,2], Marina O. Leibman[3,4], Yuri A. Dvornikov[3], Artem V. Khomutov[3,4] and Anna M. Trofaier[5]

[1] Zentralanstalt für Meteorologie und Geodynamik, Vienna, Austria, [2] Department of Geodesy and Geoinformation, Vienna University of Technology, Vienna, Austria, [3] Academic Department of Cryosophy, Earth Cryosphere Institute, Russian Academy of Sciences, Tyumen, Russia, [4] Academic Department of Cryosophy, Tyumen State University, Tyumen, Russia, [5] Svalbard Integrated Arctic Earth Observing System, University Centre in Svalbard, Longyearbyen, Norway

Edited by:
Marco Tedesco,
Lamont-Doherty Earth Observatory -
Columbia University, USA

Reviewed by:
Lukas Arenson,
BGC Engineering Inc., Canada
Klaus D. Joehnk,
Commonwealth Scientific and
Industrial Research Organisation
(CSIRO), Australia

*Correspondence:
Annett Bartsch
annett.bartsch@zamg.ac.at

Specialty section:
This article was submitted to
Cryospheric Sciences,
a section of the journal
Frontiers in Earth Science

Shallow lakes are common across the entire Arctic. They play an important role as methane sources and wildlife habitats, and they are also associated with thermokarst processes which are characteristic of permafrost environments. Many lakes freeze to the ground along their rims and often over the entire extent during winter time. Knowledge on the spatial patterns of ground-fast and floating ice is important as it relates to methane release, talik formation and hydrological processes, but no circumpolar account of this phenomenon is currently available. Previous studies have shown that ground-fast ice can easily be detected using C-band Synthetic Aperture Radar (SAR) backscatter intensity data acquired from satellites. A major challenge is that backscatter intensity varies across the satellite scenes due to incidence angle effects. Circumpolar application therefore requires the inclusion of incidence angle dependencies into the detection algorithm. An approach using ENVISAT ASAR Wide Swath data (approximately 120 m spatial resolution) has therefore been developed supported by bathymetric measurements for lakes in Siberia. This approach was then further applied across the entire Arctic for late winter 2008. Ground-fast ice fraction has been derived for (1) two million lake objects larger than 0.025 km^2 (post-processed GlobeLand30), (2) a 50 × 50 km grid and (3) within certain zones relevant for climate studies (permafrost type, last glacial maximum, Yedoma). Especially lakes smaller than approximately 0.1 km^2 may freeze completely to the ground. The proportion of ground-fast ice increases with increasing soil organic carbon content in the proximity of the lakes. This underlines the importance of such lakes for emission studies and the need to map the occurrence of ground-fast lake ice. Clusters of variable fractions of ground-fast ice occur especially in Yedoma regions of Eastern Siberia and Alaska. This reflects the nature of thaw lake dynamics. Analyses of lake depth measurements from several sites suggest that the used method yields the potential to utilize ground-fast lake ice information over larger areas with respect to landscape development, but results need to be treated with care, specifically for larger lakes and along river courses.

Keywords: thaw lakes, permafrost, synthetic aperture radar, Arctic, lake depth

INTRODUCTION

Shallow lakes are common features across the entire Arctic. They occur for example on the Alaskan North Slope (Jeffries et al., 1994), the Northwest Territories in Canada (Burn, 2002) and across Siberia (Lantuit, 2007; Morgenstern et al., 2013; Dvornikov et al., 2016). They are mostly associated with permafrost and especially thermokarst phenomena (thaw lakes) and develop at time scales of thousands of years growing in width and depth. Shallowness depends on age and ground ice conditions (West and Plug, 2008). Ground-fast ice is associated with shallow water (Grunblatt and Atwood, 2014). Lakes freeze to varying extent to the ground depending on their depth and temperature conditions during winter time. Ice thickness reaches its maximum in April with values of about 1–1.5 m on the Alaskan North Slope, where the mean annual air temperature (MAAT) was between $-10°$ and $-11°C$ in recent years (Surdu et al., 2015). Burn (2005) reported up to 2.1 m near Illisarvik, which has a MAAT of about $-11.2°C$ and less than 1 m near Inuvik with MAAT of about $-8.8°C$. As a result of recent climate change, changes in lake ice freezing depth have been observed on the Alaskan North Slope over the past decades, where a steady decline in ground-fast ice area is identified (Surdu et al., 2014). This leads to a decrease in ground-fast (or also called bottom-fast) lake ice area. Water depth is also one of the explaining factors for variation in methane fluxes across the Arctic (Wik et al., 2016). It is thus of interest to emission studies. As methane is released during winter from the non-frozen part of the lakes it is important to identify the extent of shallow water areas (Wik et al., 2011). Furthermore, lake ice properties impact hydrological processes. Spring melt-out dynamics differ between ground-fast ice and floating ice lakes (Arp et al., 2015). These differences in ice-out timing result in differences in evaporative losses across the landscape. The presence of unfrozen parts of the lake bottom (taliks) also determines the connection with the subpermafrost groundwater system (Burn, 2005).

Smith et al. (2007) have analyzed the density and fraction of lakes across the Arctic with respect to permafrost types, peatlands and the extent of the last glacial maximum based on the Global Lakes and Wetlands Database (GLWD) by Lehner and Döll (2004). It was found that lake abundance does not appear to decline in lock-step with permafrost fraction. The highest lake density was found in peatlands of the discontinuous and sporadic permafrost zone (Smith et al., 2007). The lake fraction is higher in permafrost peatland areas than in non-peatland areas. Lake occurrence as identified in the GLWD is in general higher in previously glaciated areas, giving an indication of lake history (Smith et al., 2007). An extension of such an analysis to depth properties may provide further insight into their development and relevance with respect to the discussion of Arctic lakes and wetlands under continued climate warming and permafrost thaw. The GLWD, however, only covers lakes larger than 10 ha. This leads to significant underestimation of lakes in the Arctic (Bartsch et al., 2008). Recently, Landsat based products (30 m) have been used instead (Paltan et al., 2015). Soil organic carbon content is also of relevance in this context. Methane emissions are high for thermokarst lakes in areas with carbon

rich permafrost (Wik et al., 2016). The Northern Circumpolar Soil Carbon Database (NCSCD, Hugelius et al., 2013) allows for a more detailed investigation of soil properties than a separation between peatland and non-peatland area. The database includes, among many other parameters, estimates of soil organic carbon within the top 3 m of the ground.

A baseline dataset of ground-fast ice for the entire Arctic is needed to identify target areas for the assessment of impacts of climate change. The suitability of satellite data for mapping ground-fast ice has been demonstrated by several studies. While initial studies on mapping ground-fast ice at high latitudes were undertaken with airborne measurements in the 1970s (Surdu et al., 2014), investigations based on satellite data were first enabled with the availability of C-Band ERS (European Remote Sensing satellite) data in the 1990s (e.g., Jeffries et al., 1994; Duguay et al., 2002). Studies have been published for the North Slope of Alaska (Wakabayashi et al., 1993; Jeffries et al., 1994; Arp et al., 2011, 2012; Engram et al., 2013; Surdu et al., 2014; Arp et al., 2015), Seward peninsula in Alaska (Engram et al., 2013), Manitoba (Duguay et al., 2002) and the MacKenzie Delta (Hirose et al., 2008; Yue et al., 2013) in Canada. Furthermore, the application of X- and L-band measurements for the detection of ground-fast ice has also been demonstrated over the North Slope (Engram et al., 2013; Jones et al., 2013).

The mapping principle is based on distinctive scattering behavior of microwave radiation. When ice is grounded, dielectric contrasts are reduced compared to the scattering at the ice water boundary (Duguay et al., 2002). Backscatter intensity is low due to absorption. This leads to a large difference between floating and ground-fast ice. The intensity of the return signal is higher for floating ice. Several studies have investigated the role of bubbles in the signal interaction. It has been postulated that backscatter of floating ice increases with the amount of bubbles in the ice. Scattering at especially vertically oriented, tubular bubbles and then off the ice/water interface, determine the magnitude of the returning signal (Jeffries et al., 2013). Tiny spherical bubbles can also increase backscatter (Duguay et al., 2002). At the same time a decrease in backscatter for increasing incidence angles is apparent for HH (horizontally sent and horizontally received) polarization at C-band. Duguay et al. (2002) suggest that backscatter differences between steep and shallow incidence angles increase as the amount of bubbles decreases. Variations of up to 7 dB for floating ice have been reported in C-band (HH) at comparably high incidence angles of $31°–37°$ (Walter et al., 2008). Duguay et al. (2002) find that backscatter intensity from floating ice measured at shallower incidence angles can be similar to that observed from ground-fast ice where only small amounts of bubbles are present. This may relate to a smooth ice water boundary taking into account findings of Atwood et al. (2015). Atwood et al. (2015) have used a coherent numerical model to test the impact of bubbles upon the lake ice backscatter. It was shown that the return signal represents roughness of the ice water boundary rather than bubbles within the ice.

Due to the relatively high difference in backscatter between ground-fast and floating ice, samples are commonly selected visually for classification purposes (e.g., Hirose et al., 2008) but bathymetric measurements are considered in support (e.g., Arp et al., 2015). Backscatter intensity in the surrounding of the lakes can be similar to lake ice. It is therefore common to mask them based on lake maps derived from summer acquisitions from optical or Synthetic Aperture Radar (SAR) data before classification (Grunblatt and Atwood, 2014). Landsat provides a relatively high spatial resolution for large area mapping and is suited to account for lakes across the entire Arctic (Paltan et al., 2015).

Backscatter intensity decreases considerably when the surface starts to melt (Mätzler et al., 1984). Mid winter melt events are common in many regions across the Arctic (Bartsch, 2010). Scenes which have been acquired under melting conditions need to therefore be excluded in order to avoid misclassification. Backscatter of floating ice at C-band may also vary across large lakes due to e.g., movements (fracturing at early ice growth stages) and low amount of bubbles (Jeffries et al., 2013). Such lakes need to therefore be excluded as well.

Applications of SAR data have so far been limited to specific regions due to data availability, local to regional interests in this phenomenon as well as technical challenges regarding acquisition geometry and processing demand. The usable acquisition period is rather short. It corresponds to late winter when maximum ice thickness is reached (Duguay et al., 2002). SAR data are not acquired in a consistent manner globally and across the Arctic (Bartsch et al., 2009). ENVISAT Advanced SAR (ASAR) data in Wide Swath (ASAR WS) mode were however acquired over large parts of the Arctic area as there was a high demand for these data by sea ice monitoring services. Scenes often extend to the south, also covering land areas (Bartsch et al., 2012). ASAR WS data therefore provide almost circumpolar coverage of the lake rich regions for some years. These scenes have been acquired in C-band (approximately 5.6 cm wavelength) and HH polarizations. Backscatter intensity varies strongly for floating ice in HH with respect to incidence angle (Atwood et al., 2015). This needs to be considered for large area operational classifications and has not been addressed in previous studies. The common approach is to only consider acquisitions with similar incidence angles over the study site (e.g., Hirose et al., 2008).

In order to get a circumpolar coverage, data from different incidence angles must be combined. A new approach, which takes incidence angle effects into account and is applicable to all acquisitions, therefore needs to be developed. The relationship between incidence angle and backscatter differs by surface type (Henderson and Lewis, 1998). Consequently, a location specific normalization is required, which can be applied when multiple acquisitions from different overlapping orbits, and thus points in time, are used (Wagner et al., 2008). This is, however, constrained by data availability. Only a few places such as the Yamal Peninsula are sufficiently covered by C-band data (Bartsch et al., 2012). An alternative is the normalization over the respective satellite image. Lakes may only cover fractions of such scenes (Bartsch et al., 2008). This means that most of the land surface types used in the normalization may have differing behavior toward incidence angles than lake ice. The scene based approach therefore leads to variations from scene to scene, which results in the need to classify each image separately, as for example carried out by Grunblatt and Atwood (2014) for the Alaskan

North Slope. Training data for supervised classification within each image would be required.

An approach that allows for circumpolar mapping by solving the issue of incidence angle variation between and across acquisitions is presented in this study. It is developed over the Yamal Peninsula, Siberia, strengthened by complementing it with bathymetric measurements and by taking into account that bubbling is common for lakes in this area (observation by authors). The method is eventually applied to the entire Arctic and results compared to lake depth data that have been taken from the literature. The purpose of this study is to provide an account for shallow lakes, establishing a baseline dataset for future climate change impact studies across the entire Arctic. This dataset enables the assessment of lake properties and associated characteristics such as methane emission and thermokarst processes. Some basic statistics are calculated, partially resembling the analyses of lake properties by Smith et al. (2007).

MATERIALS AND METHODS

Satellite Data

The Advanced Synthetic Aperture Radar (ASAR) data serve as source for the ground-fast ice classification. The ASAR instrument was mounted on the ENVISAT satellite platform and was operated by the European Space Agency (ESA) from 2002 until its breakdown in April 2012. The C-band sensor took measurements at a center frequency of 5.331 GHz and could be run in different acquisition modes with varying spatial and temporal resolutions and varying swath widths. In Wide Swath (WS) mode the spatial resolution is approximately 120 m (Closa et al., 2003) and the swath width is 400 km. Incidence angles ranged from 15° to 45°.

Auxiliary data are required to ensure that only frozen conditions are analyzed as the scattering mechanism changes when ice and snow start to melt. This information can also be obtained from C-band data, but requires sufficient temporal sampling (Park et al., 2011). This can only be achieved with coarser resolution scatterometer data. The Advanced Scatterometer (ASCAT) onboard the MetOp-A and MetOp-B satellite platforms is a scatterometer operating at a frequency of 5.255 GHz (C-band). ASCAT data are suitable for detecting freeze/thaw states globally (Naeimi et al., 2012) due to the operational nature of these missions and thus globally continuous records. Datasets of surface status are readily available for the Arctic. The ASCAT Surface State Flag (SSF) product by Paulik et al. (2014) provides freeze/thaw information for the Arctic at a temporal sampling of 1–2 days and 25 km resolution. It is used to constrain the analysis period to frozen surface conditions across the entire Arctic.

Bathymetric Measurements

Bathymetric measurement (surveyed in 2012, 2014, and 2015) are available for several lakes in central Yamal (Dvornikov, 2016; Dvornikov et al., 2016). Seven of these lakes are of sufficient size (with respect to ASAR WS spatial resolution) for comparison with the ASAR WS product. One of the lakes (labeled LK-003, location see **Figure 1**, area 1.08 km^2) is characterized by a very shallow shelf area wider than two pixels of ASAR WS. This lake has been surveyed during summer 2012 in order to produce a detailed map of the lake bottom. The lakes on central Yamal are formed within alluvial-lacustrine-marine plains and terraces (Leibman et al., 2015). Mean Annual Air Temperature (MAAT) at a meteorological station at 100 km distance, located on the coast, is about −8°C. Temporary measurements have indicated the representativeness of this station for the central Yamal study area (Leibman et al., 2015). Freezing index (−3,000 to −3,200 for 2012–2015, average −3,457 since 1966) and thawing index (+850 for 2012–2015, average +646 since 1966) values are similar to those at the Alaskan North Slope (Frauenfeld et al., 2007), resulting in similar freezing depth of the lakes in the order of 1.5 m. Thickness above 2 m was observed in the Canadian Arctic, but at the same time much lower MAAT (−11°C) (Burn, 2002) and lower freezing index (Frauenfeld et al., 2007) were detected. Here, measurements from lake LK-003 are used for assessment of the threshold function for ground-fast ice delineation.

The results were assessed using lake characteristics (lake depth and surface area) from several sites across the Arctic (see **Table 1**). Most measurements are available from the Alaskan North Slope (Hinkel, 2010, 2016). Thaw lakes have formed here in ice rich, glacio-marine silty sediments (Arp et al., 2011). This region has been intensively studied for ground-fast ice detection from SAR (Wakabayashi et al., 1993; Jeffries et al., 1994; Arp et al., 2011, 2012, 2015; Engram et al., 2013; Surdu et al., 2014). The lakes are located on the inner and outer coastal plain, representing different ages. A transect of lakes on Richards Island, which is located north of the Mackenzie Delta, has been surveyed by Burn (2002). They are underlain by glacial tills, eolian sands and glaciofluvial gravels. They can also be associated with different physiographic subdivisions. A few measurements are also available from different units (terraces) of the Lena Delta in Siberia (Lantuit, 2007; Morgenstern et al., 2013). The lakes on the central Yamal peninsula have a much larger maximum depth

than those in the other regions where records are available. They can be deeper than 15 m with a hole-like feature. The formation of deep holes is different to the common thermokarst processes and their transformation into lakes has recently been reported in this region (Kizyakov et al., 2015).

Land Cover

The 30 m Global Land Cover (GlobeLand30) dataset of the National Geomatics Center of China (NGCC) was obtained from   http://www.globallandcover.com/GLC30Download/. It constitutes the first global land cover dataset at a spatial resolution of 30 m. The dataset was derived by NGCC applying an advanced classification algorithm on Landsat Thematic Mapper (TM), Landsat Enhanced Thematic Mapper+ (ETM+)

TABLE 1 | Sources and characteristics of bathymetric (maximum lake depth) measurements.

References	Location	Number of lakes	Variation in maximum depth [m]
Dvornikov et al., 2016	Vaskiny Dachi, Central Yamal	7	6.3–16.9
Hinkel, 2010	Barrow, Outer Coastal Plain (OCP), North Slope, Alaska	5	1.5–2.5
Hinkel, 2016	Atkasuq and Reindeer Camp, Inner Coastal Plain (OCP), North Slope, Alaska	19	2.1–9.5
Lantuit, 2007	Arga Island, Second Terrace, Lena Delta	4	3.0–9.5
Morgenstern et al., 2013	Kurungnakh Island, Third Terrace, Lena Delta	3	3.6–4.2
Burn, 2002	Richards Island, Northwest Territories, Canada	12	2.2–13.1

Only lakes larger than 3 × 3 ASAR WS pixels and which are separate objects in the GlobLand30 derived lake mask are considered.

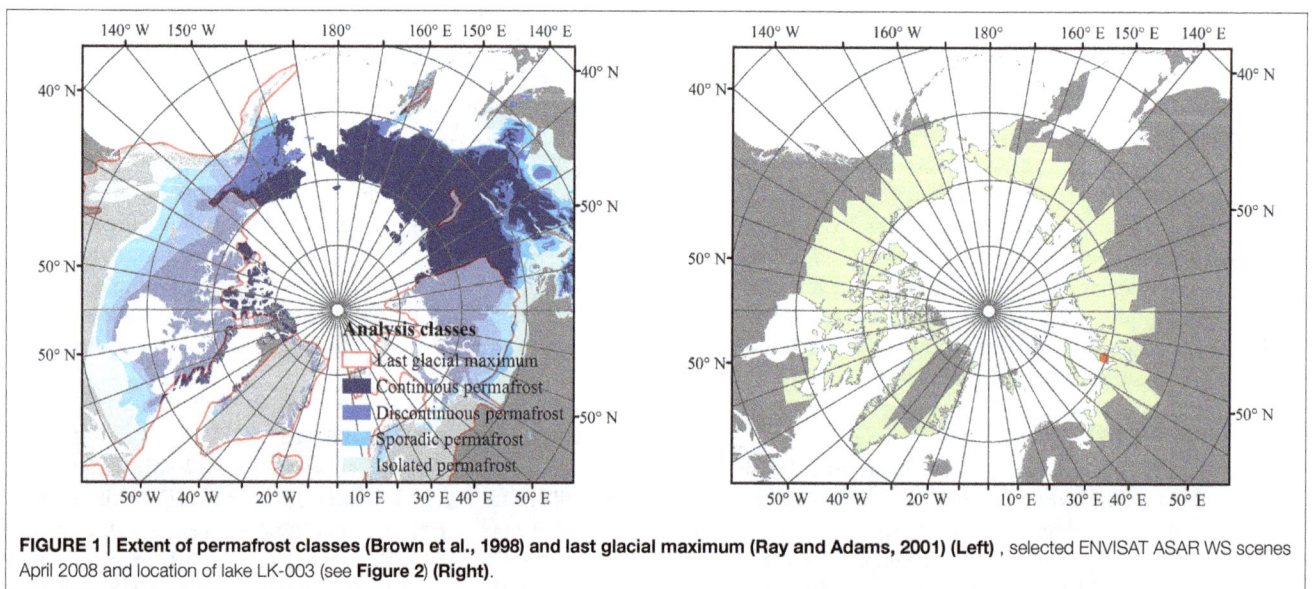

FIGURE 1 | Extent of permafrost classes (Brown et al., 1998) and last glacial maximum (Ray and Adams, 2001) (Left) , selected ENVISAT ASAR WS scenes April 2008 and location of lake LK-003 (see **Figure 2**) (Right).

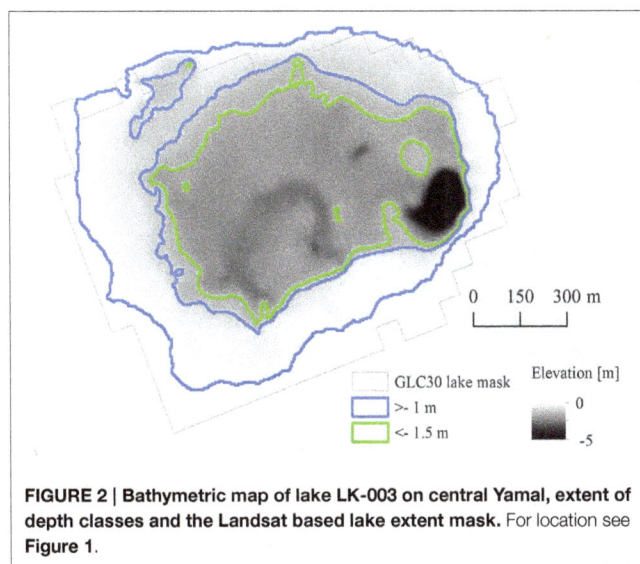

FIGURE 2 | Bathymetric map of lake LK-003 on central Yamal, extent of depth classes and the Landsat based lake extent mask. For location see **Figure 1**.

and Chinese Environmental Disaster Alleviation Satellite (HJ-1) multispectral images. The satellite data used for the land cover classification were mainly acquired over vegetation growing seasons from 2009 to 2011. According to the product description the overall positional accuracy is 75 m and the classification accuracy of the water body class is 92.09%. It is used for the masking of lakes in this study.

Boundaries of Selected Landscape Units

The Northern Circumpolar Soil Carbon Database (NCSCD) (Hugelius et al., 2013) is a polygon database containing soil information for all permafrost affected areas in the Northern Hemisphere. It was created by combining national and regional soil maps, and linking these soil maps to pedon data from field observations. It provides information on soil organic carbon content (SOCC) for different depth ranges. The SOCC values can range from zero to more than 100 kg m^{-3}. SOCC values are used to analyze ground-fast ice patterns with different SOCC classes.

The results are interpreted considering three different sets of landscape boundaries: The Permafrost extent classification by Brown et al. (1998), the extent of the last glacial maximum (LGM) by Ray and Adams (2001) as used in Smith et al. (2007) (**Figure 1**), as well as the borders between Yedoma and non-Yedoma regions (Strauss et al., 2016). Yedoma regions are a subcategory of the area outside of the LGM boundary. They were unglaciated during the last glaciation. Soil carbon could therefore accumulate together with ice lenses that form by segregation (Strauss et al., 2013). These ice lenses promote thermokarst processes, which are therefore common phenomena in these areas.

Methodology

The ASAR WS scenes were used to derive the floating and ground-fast ice fraction of lakes with almost circumpolar coverage for the month April, in which the peak of frozen lake ice is assumed (Surdu et al., 2014). In order to achieve the maximum possible extent for mapping the lakes, the total coverage and distribution of ASAR scenes in April of each year during ENVISAT's mission duration (2002–2012) was analyzed. It was established that April 2008 had the highest total number of scenes and also the largest total coverage of all years. This year was therefore chosen for this study.

In order to minimize errors in the classification due to the presence of wet snow on top of the frozen lakes, the ASAR scenes were compared to the ASCAT surface status data and only scenes with no presence of melting snow were used for further analysis. Consequently 51 ASAR scenes for April 2008 with the optimum extent (**Figure 1**) were selected as main input for the ground-fast lake ice classification.

Subsets (excluding oceans) were taken from these selected 51 ASAR scenes. The subsets have been radiometrically calibrated and orthorectified using the Next ESA SAR Toolbox (NEST). The images were resampled to the commonly used nominal resolution for this mode of 75 m, and the output backscatter coefficient σ^0 images were converted into units of decibels (dB). In addition, the projected local incidence angle was computed for each scene for further processing.

Samples were taken across an entire scene over the Yamal peninsula in western Siberia for ground-fast ice and floating ice classes, in order to determine the incidence angle dependent classification threshold. Ground-fast ice and floating ice was discriminated visually, which is a common procedure for this type of study (Atwood et al., 2015).

The bathymetric measurements of lake LK-003 was used to justify the manually selected samples. A Chartplotter Humminbird 788 cxi with internal GPS was used to collect both distances between sonar and lake bottom and position coordinates. The readings were taken along several profiles for the lake and then were filtered with 5 s and converted to ASCII format. In total, 1,652 points were recorded for the lake. A bathymetric map was produced by interpolation of the data (**Figure 2**). Two regions were separated from Lake LK-003. The first area represents depths smaller than 1 m, where it can be assumed that only ground-fast lake ice is present in April. The second area represents depths larger than 1.5 m, where mostly floating ice is assumed to be present (**Figure 2**). This leaves a buffer area which is required due to the unknown exact ice depth as well as potential positional inaccuracies in the satellite data (see also discussion on this approach in Grunblatt and Atwood, 2014). From these regions, samples were taken from additional 27 ASAR WS scenes acquired from different but overlapping orbits in April 2008 with their respective incidence angles. Only a limited range of incidence angles is available for a single scene over the lake area, but the 27 images represent the variations across almost the entire swath (ASAR WS incidence angle range). These samples were used to obtain a separate empirical threshold function in order to assess the results from the samples taken manually.

The dependency of σ^0 on the local projected incidence angle was analyzed for each of the sample classes. A third degree polynomial function was fit to both classes and a common threshold function was determined by averaging the two sample functions to distinguish between ground-fast and floating ice at

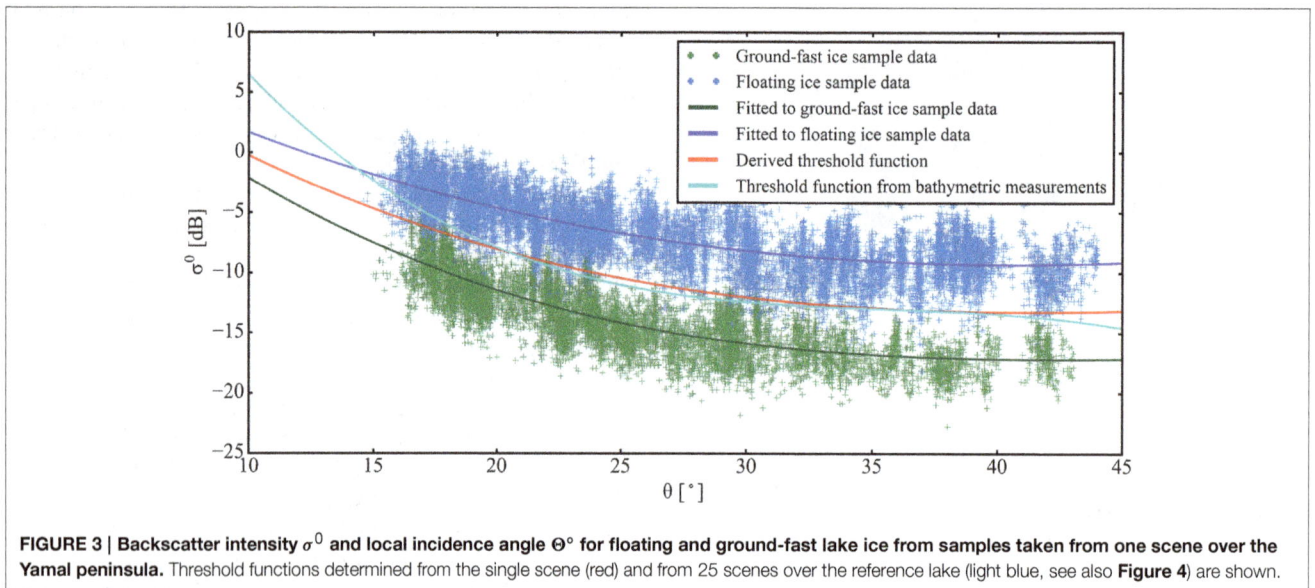

FIGURE 3 | Backscatter intensity σ^0 and local incidence angle $\Theta°$ for floating and ground-fast lake ice from samples taken from one scene over the Yamal peninsula. Threshold functions determined from the single scene (red) and from 25 scenes over the reference lake (light blue, see also **Figure 4**) are shown.

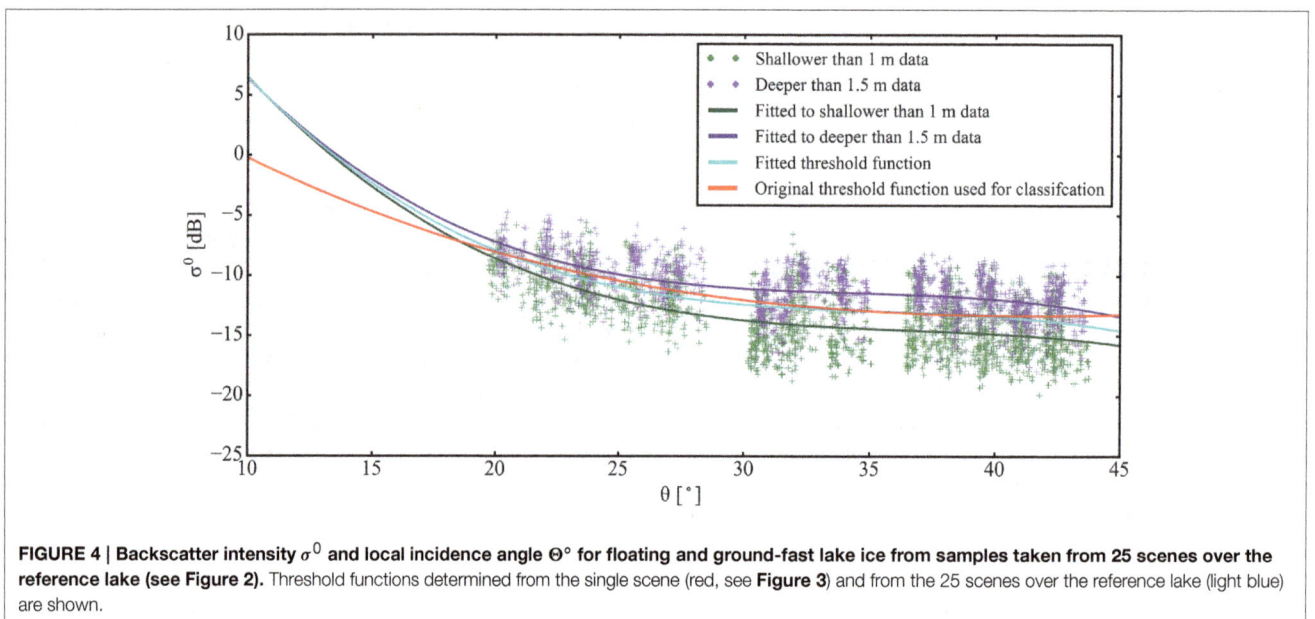

FIGURE 4 | Backscatter intensity σ^0 and local incidence angle $\Theta°$ for floating and ground-fast lake ice from samples taken from 25 scenes over the reference lake (see Figure 2). Threshold functions determined from the single scene (red, see **Figure 3**) and from the 25 scenes over the reference lake (light blue) are shown.

different incidence angles (**Figures 3, 4**). This threshold function was then used to classify each pixel as floating or ground-fast ice depending on σ^0 and the projected local incidence angle values. A mosaic was produced from the classified scenes, where the topmost scene is the latest (maximum freezing stage).

The GlobLand30 land cover map was post-processed in order to obtain a lake mask. Reclassification was performed, so that only the water body class was extracted from the GlobeLand30 dataset and converted into vector format. Large rivers and bays were removed manually from the dataset. Large lakes like the Great Slave Lake, Great Bear Lake and lakes located in the proximity of the Putorana Plateau were removed from the dataset as well. Those are reservoirs or glacially formed, comparably deep lakes, which show partially lower backscatter

intensity for floating ice than thaw lakes. An example is shown in **Figure 5**.

In order to consider the positional accuracy of the land cover dataset and resolution of ASAR WS data, lakes with an area smaller than 25,000 m^2 were also removed from the lake dataset. The final vector lake polygon dataset represents the lake mask containing all lake features for which the coverage of floating and ground-fast lake ice was classified using the empirically derived threshold function. It contains about two Mio objects covering in total approximately 0.67 Mio km^2. Zonal statistics were calculated to obtain the percentage of ground-fast and floating lake ice for each of the lake polygons. A lake is considered as one that partially freezes to the ground when more than one pixel has been mapped as ground-fast, while it

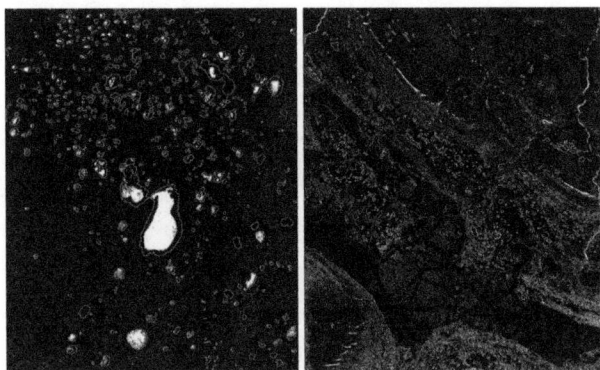

FIGURE 5 | Examples of backscatter intensity images overlain by the applied lake mask (yellow) and excluded lake (red). Left: lakes SW of the Lena Delta, Siberia. Dark rims correspond to ground-fast ice, bright centres to floating ice. **Right:** Chantaika-reservoir west of the Putorana-Plateau. A pattern of fractioning (bright areas) as well as low backscatter return due to low bubble amount is visible.

is determined as completely frozen when 95% is ground-fast. This accounts for spatial inaccuracies of the lake mask (compare **Figure 2**). Especially the inclusion of land area in the analyses may increase the floating ice proportion due to the comparably high backscatter of the surroundings.

The results were assessed using lake depth measurements available from the studies listed in **Table 1**. The fraction of ground-fast ice was then compared to maximum lake depth as well as lake area as available from the GlobLand30 mask. Only lakes which are larger than nine ASAR WS pixels and are distinct objects in the mask were considered.

Statistics were calculated in addition on a raster of 50 × 50 km, in order to visualize the circumpolar classification results and view the total distributions of the two lake ice types over different regions. This grid size also corresponds to a common cell size for regional climate models.

In order to analyse the results for the whole circumpolar region, zonal statistics were calculated for SOC content classes with respect to the different landscape units (Yedoma, LGM, permafrost extent). SOC content values for 0–300 cm depth were extracted from the NCSCD (Hugelius et al., 2013) and assigned to each lake. Different properties, such as number of lakes or average fraction of ground-fast ice area per lake were assessed for these classes. Furthermore, frequency-size distributions of lakes with different fractions of ground-fast ice were analyzed.

RESULTS

The first part of the results section deals with the new retrieval method, which is supported by the analyses of two different ways for sample collection. The second part describes the comparison to *in situ* measurements in order to evaluate this approach (Yamal bathymetric records) and to provide a basis for evaluating the applicability of the results (comparison to lake depth across several sites in the Arctic). The derived circumpolar dataset of

ground-fast lake ice fraction is presented for a defined grid, for single lake objects and within specific landscape units.

Retrieval Function

The obtained threshold functions from the visual sample, which was collected across a scene (**Figure 3**), and the bathymetric classification using multiple acquisitions (**Figure 4**) show similar values for incidence angles between 20° and 40°. There are deviations below and above these values due to limited availability of samples from overlapping orbits for lake LK-003. There is little variation of backscatter with incidence angle for values larger than 30°. The backscatter difference between floating and ground-fast lake ice is smaller for low incidence angles. The difference between ground-fast and floating ice is smaller for the LK-003 sample. Values range from 2 to 3 dB compared to 7 to 8 dB from the visual sample. The obtained threshold functions are, however, similar to each other.

Comparison to *in situ* Records

A certain ground-fast fraction can be recorded for all investigated sites where *in situ* measurements are available (**Figure 6**). A frozen fraction of 100% is obtained in two cases, both lakes are very shallow and are located on the Alaskan North slope where complete freeze up is common (e.g., Surdu et al., 2014). Lakes with a maximum depth of more than 10 m have a considerable ground-fast ice fraction with values larger than 20%. Lakes shallower than 4 m have a fraction from 5 to 100%. Different regions have distinct characteristics. Arga Island (Lena Delta second terrace) and Reindeer Camp (North Slope Inner Coastal Plain) have similar ground-fast ice fractions, which are low (mostly less than 20%) compared to the other regions. Barrow and Atkasuq, although on different plains, show similar patterns. There is a tendency that ground-fast ice fraction increases with increasing depth in areas with deep lakes (more than 4 m) such as Richards Island and specifically central Yamal. The samples from the region with the aligned lakes on Richards Island have similar depth but show large differences in ground-fast ice fraction. The "outlier," the lake with the very high value of more than 90%, is actually located at the border to a different physiographic subdivision, the "Tununuk low hills." There seems to be no general relationship between lake size and ground-fast ice fraction.

LK-003 on Yamal has about 50% ground-fast ice fraction, 45% of the lake is shallower than 1 m. In general, lakes on Yamal are found to have a ground-fast ice fraction higher than 20% (**Figure 7**). For three of the seven lakes (including LK-003), the derived fraction agrees (±5%) with the corresponding area where water depth is shallower than 1 m. For the remaining lakes, there is moderately better agreement with the −1.5 m depth area.

Circumpolar Results

Figure 8 shows the classified total lake area, the distribution of ground-fast and floating lake ice for the 50 km grid. The majority of the lake area lies within Nunavut, the North-West-Territories in Canada, the Alaskan North Slope and the Kolyma region in Siberia. Since total lake coverage in the above mentioned regions is high, the total area of ground-fast and floating lake ice is also

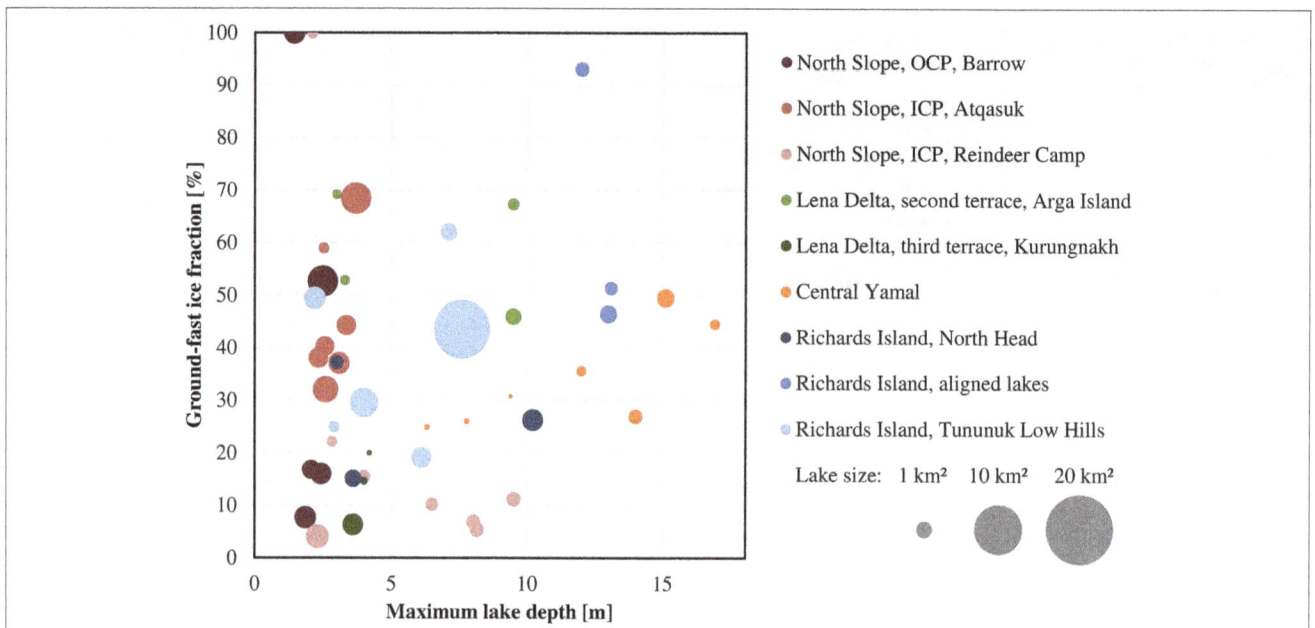

FIGURE 6 | ASAR WS derived ground-fast ice fraction in % vs. maximum lake depth (sources: Burn, 2002; Lantuit, 2007; Hinkel, 2010, 2016; Morgenstern et al., 2013; Dvornikov et al., 2016). Circle size represents lake area of the lake mask from GlobLand30. Maximum is 13.84 km², a lake on Richards Island which represents at least two merged lakes.

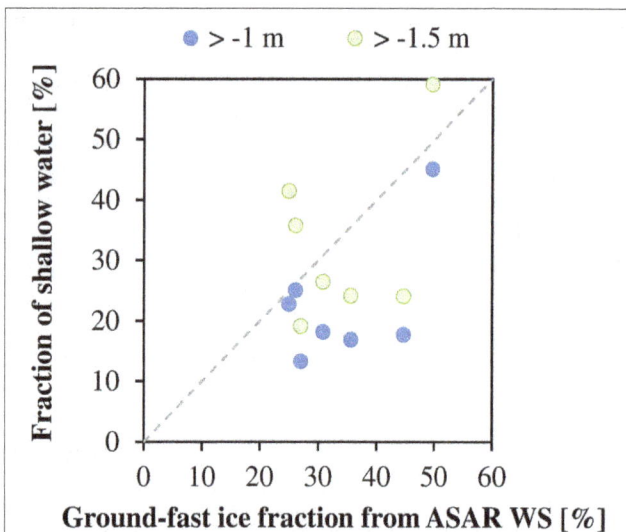

FIGURE 7 | ASAR WS derived ground-fast ice fraction vs. depth class fractions from bathymetric data (less than 1 m depth in blue, less than 1.5 m depth in green) in central Yamal. The gray dashed line represents a 1:1 relationship.

considerably higher than in other areas. The fraction of ground-fast lake ice (the percentage of ground-fast lake ice area compared to the overall lake area within the 50 km cell) for each raster cell differs from the total ground-fast area pattern (**Figure 8**).

The developed approach provides a spatially consistent map across the Arctic (**Figure 9**). Most lake rich areas are characterized by a mixture of lakes with different percentages of

ground-fast lake ice, except for locations that are known to have abundant shallow lakes such as the Lena Delta's second terrace in Siberia (Arga island, late Pleistocene to to Early Holocene age, Schwamborn et al., 2002) or areas with (deeper) lakes of glacial origin such as parts of Canada. The Yedoma region in Eastern Siberia (**Figure 9A**) comprises a high number of lakes with low ground-fast ice fraction, but this area also includes lakes that have a ground-fast ice fraction that is up to 50%. These lakes are found to follow a random-location pattern. Very few lakes have a higher fraction. The region near Barrow on the Alaskan North Slope (**Figure 9B**) shows distinct regional patterns of ground-fast ice fraction. This has also been described in the literature by Arp et al. (2011). It is common that lakes freeze to the ground almost completely in this area. Higher ground-fast ice fractions are constrained to comparably smaller lakes on central Yamal (**Figure 9C**) but this area also contains regions which demonstrate a mixed spatial pattern similar to the Alaskan site. A common feature is the proximity to the coast.

In general, smaller lakes tend to have high ground-fast ice fraction, but there are differences regarding the actual magnitude and number of lakes affected. More than 10% of lakes of sizes between 0.025 and 0.125 km² freeze almost completely to the ground, where at least 95% of the ice cover is ground-fast (**Figure 10** top). The proportion is smaller for larger lakes, with decreasing values for increasing lake size. The total number of lakes with such an extent of ground-fast ice is however very small for larger lakes (**Figure 10** middle). About 250,000 lakes (about 12.5% of all considered lakes) fall into the category 0.025–0.125 km², which corresponds to about three to thirteen ASAR WS pixels. Less than 20% of that amount accounts for sizes between 0.075 and 0.125 km². This difference can not be

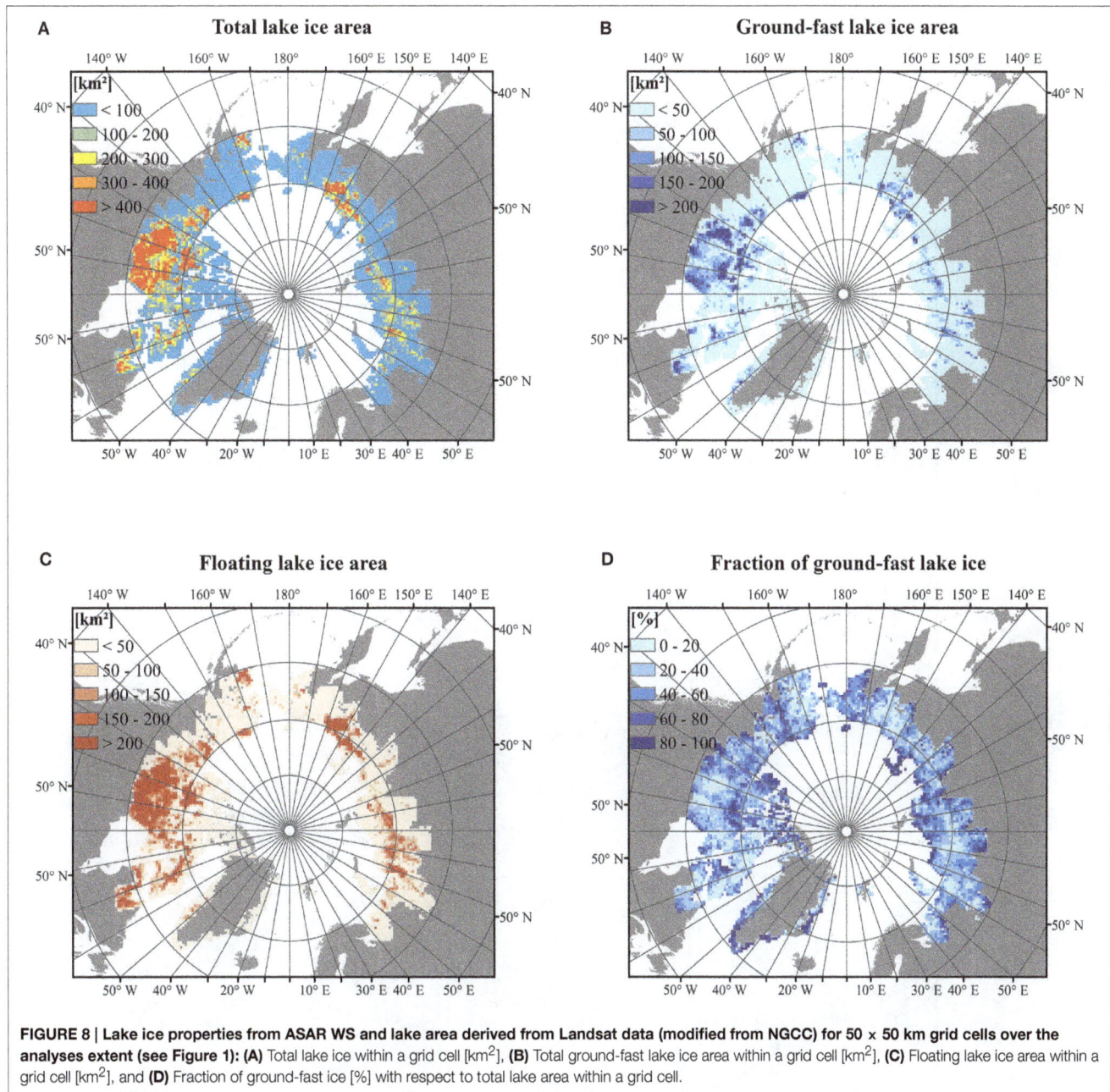

FIGURE 8 | Lake ice properties from ASAR WS and lake area derived from Landsat data (modified from NGCC) for 50 × 50 km grid cells over the analyses extent (see Figure 1): (A) Total lake ice within a grid cell [km²], **(B)** Total ground-fast lake ice area within a grid cell [km²], **(C)** Floating lake ice area within a grid cell [km²], and **(D)** Fraction of ground-fast ice [%] with respect to total lake area within a grid cell.

observed for all lakes which only partially freeze to the ground (**Figure 10** bottom). Here, the number of lakes is similar in both size categories. About 800,000 lakes (about one quarter of all analyzed objects) with sizes between 0.025 and 0.075 km² have at least one pixel identified as ground fast ice. At least 70% of analyzed lakes froze partially to the ground, while at least 20% froze up almost completely in 2008. The distribution function between lake size and number of lakes differs between the almost completely and the partially ground-fast lakes. The first shows an exponential relationship with most of the lakes within the smallest category (**Figure 10** middle). This is less pronounced in the second case (**Figure 10** bottom). **Figure 11** provides some

insight into variations of ground-fast ice for lakes of different sizes. The median of ground-fast ice fraction is 40% for the smallest area class and more than 25% for the large lake class (more than 0.475 km²). This confirms that mostly small lakes freeze-up almost completely, but it also shows that there is more variation in ground-fast ice fraction for smaller lakes. In general, ground-fast ice fraction is more often close to zero.

The number of lakes, the total ground-fast area and the fraction with respect to the total lake area are shown for ranges of soil organic carbon classes in **Figure 12**. It accounts for the entire analyzed area (light blue line) as well as for different zones (permafrost types, last glacial maximum) as also discussed in

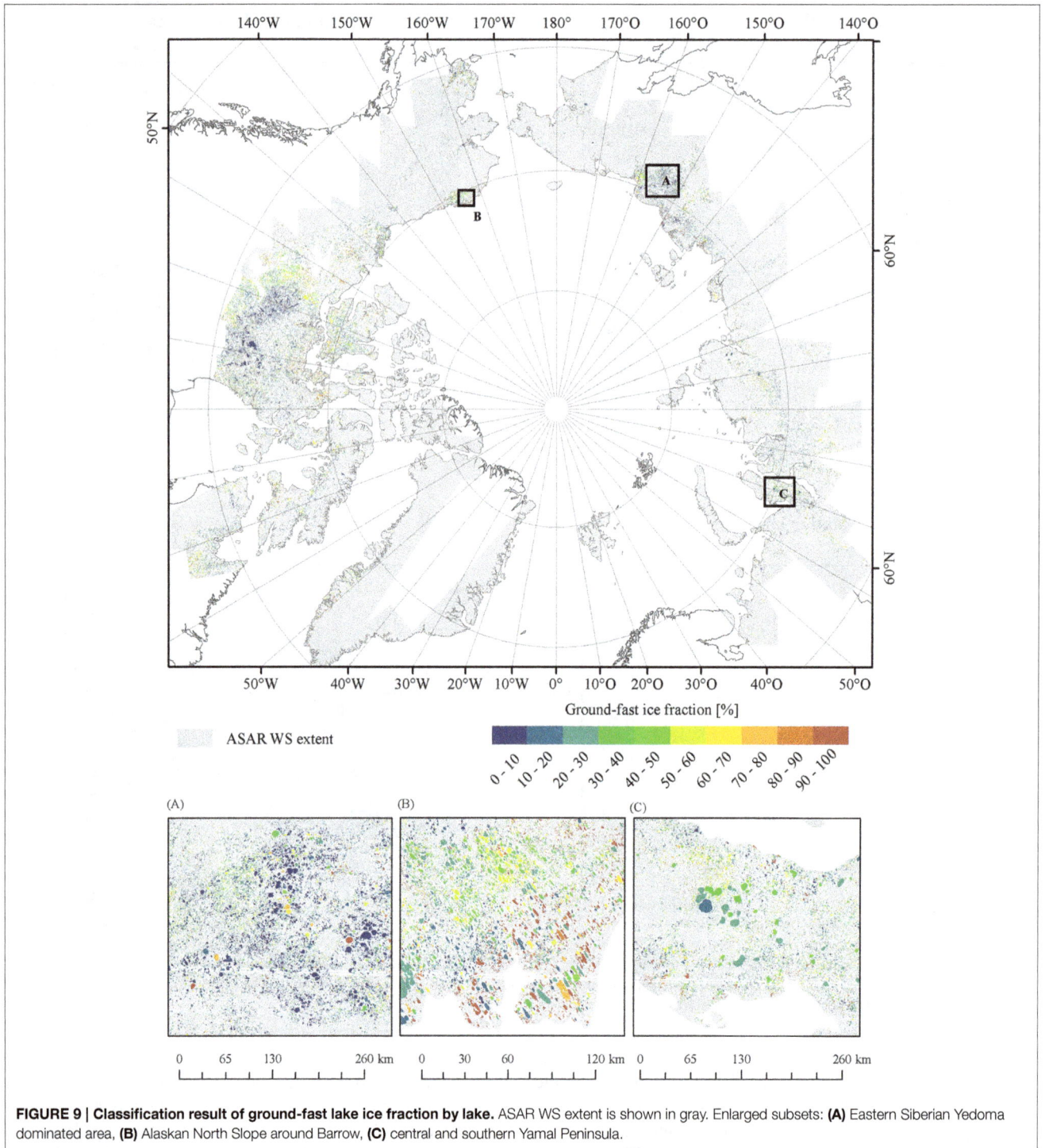

FIGURE 9 | Classification result of ground-fast lake ice fraction by lake. ASAR WS extent is shown in gray. Enlarged subsets: **(A)** Eastern Siberian Yedoma dominated area, **(B)** Alaskan North Slope around Barrow, **(C)** central and southern Yamal Peninsula.

Smith et al. (2007). Yedoma areas are further delineated due to their specific history regarding ground ice and soil development. Most lakes of the continuous permafrost zone can be found in areas with total SOC of 80–90 kg m^{-3}. The bin with the lowest number of lakes is 60–70 kg m^{-3}. Values for the extent of the last glacial maximum are similar to continuous permafrost except for this specific SOC content class. The same applies to the total area of ground-fast ice. The ratio of ground-fast ice to total lake area increases with SOC for continuous permafrost only. Areas with up to 70 kg m^{-3} are characterized by fractions of 20–30% and up to almost 40% for higher values. The Yedoma area, which mostly overlaps with high SOC content, does not differ from non-Yedoma area with respect to frozen ground fraction.

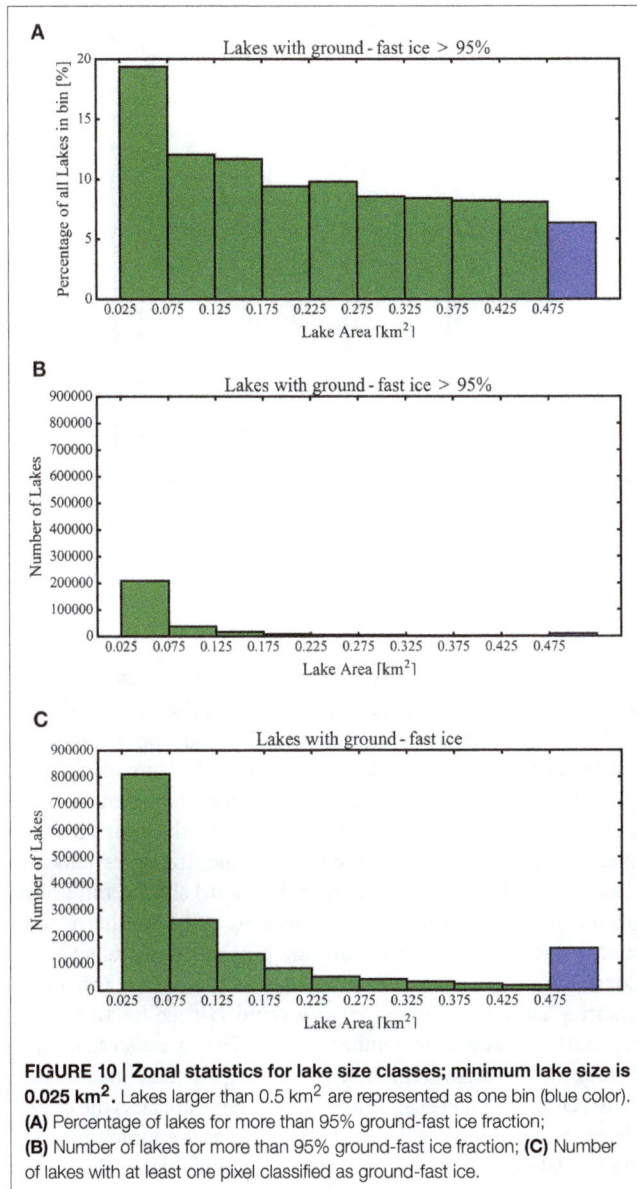

FIGURE 10 | Zonal statistics for lake size classes; minimum lake size is 0.025 km^2. Lakes larger than 0.5 km^2 are represented as one bin (blue color). **(A)** Percentage of lakes for more than 95% ground-fast ice fraction; **(B)** Number of lakes for more than 95% ground-fast ice fraction; **(C)** Number of lakes with at least one pixel classified as ground-fast ice.

DISCUSSION

Methodological Constraints

The developed method enables a first circumpolar account of ground-fast lake ice. Some issues related to the determined function and used auxiliary data do however remain unsolved. Due to a lack of relevant data, the presented approach for mapping of ground-fast ice does not consider potential variations in incidence angle behavior for floating ice caused by the amount of bubbles as described by Duguay et al. (2002). The difference between the two states (ground-fast/floating) is in the order of 7–8 dB for the visual sample. This is similar to values reported from Manitoba by Duguay et al. (2002). The lower values of the LK-003 sample may point to less bubbles in the ice taking into account findings from Walter et al. (2008). Ground-fast ice values are however also closer to the threshold function. The differences

between the two sample sets may arise from the fact that the visual sample contains well distinguishable regions, whereas the LK-003 sample is obtained from actual bathymetric data.

No specific effects on the ground-fast ice backscatter are reported in the literature. The obtained threshold function is therefore expected to be applicable to areas with higher bubble content. More critical are lakes without bubbles, as the separability from floating ice is lower at shallow incidence angles (Duguay et al., 2002). The percentage of all lakes that freeze up to at least 95% of their respective lake area (20%), as established here, may actually be too high due to this low incidence angle effect.

The determined relationships between incidence angle and backscatter are close to the same for ground-fast and floating ice for Yamal. Backscatter decreases with similar magnitude for incidence angles of 30° and larger (**Figure 3**). The separability, however, decreases with decreasing incidence angle for both cases (bathymetric and visual sample). This effect is more pronounced for the LK-003 sample (**Figure 4**), which may point to comparably low amount of bubbles according to Duguay et al. (2002).

The algorithm is known to not be applicable over larger and deeper lakes where ice properties differ from smaller lakes and floating ice can have comparably low backscatter values. Deeper lakes are in general associated with fewer trapped bubbles in the ice, which leads to rather low backscatter values (Jeffries et al., 2013). Large lakes were therefore manually set to 0% ground-fast ice fraction. The possibility that there are additional lakes with similar properties that are still contained within the datasets remains. This may in particular be the case in the region around the Great Slave Lake in Northern Canada. Furthermore, some of the excluded lakes may actually have ground-fast ice along the margins, which is not accounted for within the dataset. The majority of analyzed lakes is, however, smaller and can be expected to be correctly identified. This is supported by comparable results from the Alaskan North Slope (**Figure 6**). Jeffries et al. (2013) demonstrate the low backscatter effect for Teshekpuk lake on the North Slope with an acquisition from early winter when ice is forming. Floating ice cannot be detected at this stage due to the low amount of bubbles. Teshekpuk lake has a size of more than 800 km^2 and a maximum depth of 6.6 m. Our results show that the effect is reduced in late winter. A ground-fast ice fraction of 25% is obtained with the presented method, which is a likely value for a lake of this depth (compare to **Figure 6**).

The accuracy of the classification also depends on the lake mask. Small river courses are still partially contained in the datasets since such features have been removed manually. This may lead to the inclusion of false small lakes in some areas. Location accuracy is an issue as well. The used mask does for example not fully agree with the extent of the lakes as can be seen for lake LK-003 in **Figure 2**. This may also explain the comparably large amount of lakes with low ground-fast ice fraction in the lowest lake size class (**Figure 11**). This could be an effect of a mismatch between the lake mask and the actual lake extent as well as mixed pixel effects. The small lakes consist of a few pixels only, and therefore the proportion of mixed

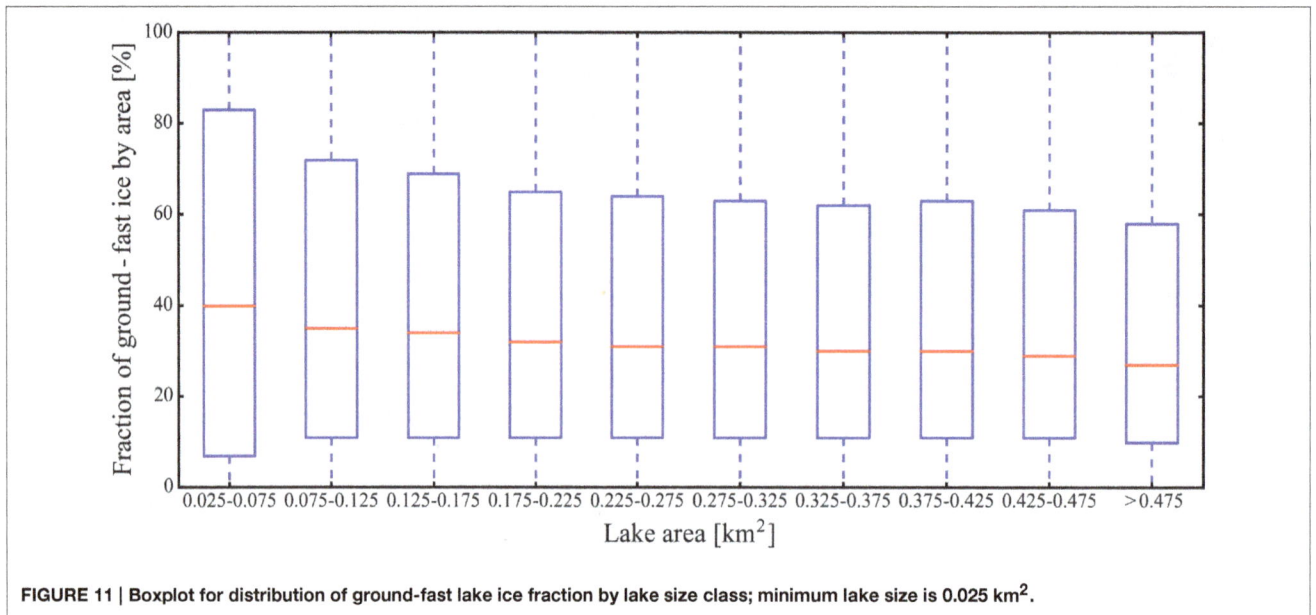

FIGURE 11 | Boxplot for distribution of ground-fast lake ice fraction by lake size class; minimum lake size is 0.025 km².

pixels is comparably large. The backscatter of the surrounding area is usually more similar to floating ice, which leads to an underestimation of ground-fast ice for the lake object.

While the spatial extent of the derived map covers the largest possible land area for 2008, and the territory north of the treeline is well represented, it does however not correspond to specific latitudinal or landscape unit boundaries. Data availability from ENVISAT ASAR WS for April 2008 is comparably good but not all lake rich permafrost regions could be covered. Apart from the above mentioned limitations of the algorithm, these data gaps further restrict a circum-Arctic analysis, and hence not all ground-fast lake ice has been captured by this study. The results can therefore not completely complement accounts such as by Smith et al. (2007). The combination of different years may allow for an estimate of where ground-fast ice occurs, but ground-fast ice fraction can vary from year to year (Surdu et al., 2014). A better coverage could potentially be achieved by consolidating this study with additional satellite data, e.g. Radarsat data. These data are also acquired in the C-band but were not available for the present study. Furthermore, the Sentinel-1 satellites could contribute to future updates, starting from April 2017. With the activation of the second sensor (Sentinel-1B launched in 2016), spatial coverage would be sufficient. The availability of HH polarization data will however be lower than for ASAR WS (Potin et al., 2014). The remaining areas are covered by acquisitions in VV (vertically sent and vertically received), which may require adjusting the parametrization.

Bathymetry vs. Ground-Fast Ice Fraction

The comparison with bathymetric data from Yamal suggests that freezing depth is between 1 to 1.5 m (**Figure 7**). This is in the same order of magnitude as lake ice thickness measurements from the Alaskan North Slope (Arp et al., 2012; Grunblatt and Atwood, 2014, e.g.,) and implies similar climatic conditions. There are variations between the lakes, which may result from the

comparably low separability as discussed above for lake LK-003 as well as the resolution and accuracy of the Landsat lake mask.

The results suggest that lakes at Atkasuq and Barrow are similar, although they are located on coastal plains of different age. The lakes on the Alaskan North Slope represent a more than 100 km long transect perpendicular to the coastline, and thus multiple stages of Cenozoic marine transgressions are present here. Deeper lakes (deeper than 6 m) show a rather low ground-fast ice fraction. They are located at the southern, most inland part of the transect. They are part of the so called Inner Coastal Plain (ICP), but Atkasuq is also located on this plain. Atkasuq lakes have much higher ground-fast ice fractions than the southern lakes near Reindeer Camp. They are also shallower, similar to the Barrow region, which belongs to the Outer Coastal Plain (OCP) (Hinkel et al., 2005). The Barrow area has the largest diversity of ground-fast ice fraction, while at the same time only shallow lakes occur here. This result might be biased due to the limited sample from this region, but lakes are in general very shallow in this area (Arp et al., 2011). The mapped range of ground-fast ice fraction on the OCP agrees with results from previous studies (Arp et al., 2012).

Deeper lakes with a frozen fraction of more than 20% are found on Yamal as well as Richards Island. The lakes on Yamal have usually a deeper (but small) circular part and shallow more or less extensive shelves. A good example of an extensive shelf is lake LK-003 (**Figure 2**). It has a maximum depth of 15 m and a ground-fast fraction of 50%. A common feature of central Yamal and Richards Island are distinct ridges (Burn, 2002; Leibman et al., 2015). Thaw slumps and erosion along lake shores are a common feature at the Yamal site. Material is deposited in the lakes forming prominent shelves. Erosion rates of lakes in permafrost areas are in general positively correlated with lake depth (Arp et al., 2011). The results from ASAR WS confirm the presence of distinct littoral terraces for the deeper lakes. The ground-fast ice fractions can be expected to represent the

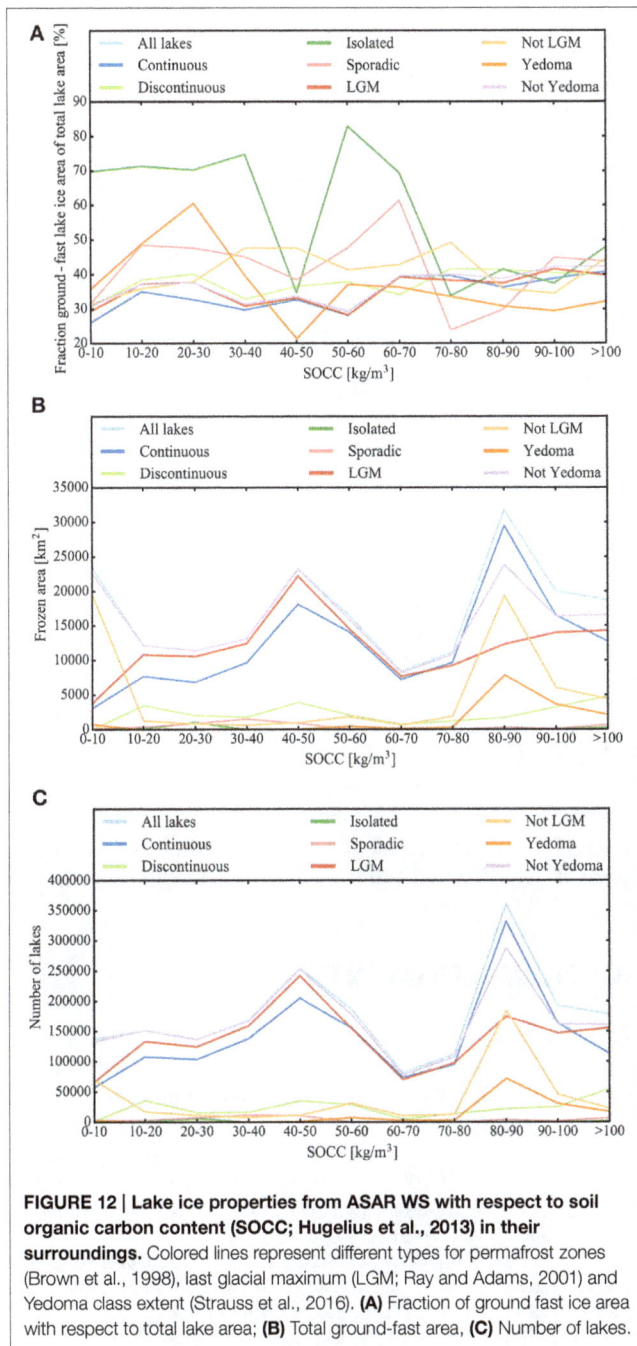

FIGURE 12 | Lake ice properties from ASAR WS with respect to soil organic carbon content (SOCC; Hugelius et al., 2013) in their surroundings. Colored lines represent different types for permafrost zones (Brown et al., 1998), last glacial maximum (LGM; Ray and Adams, 2001) and Yedoma class extent (Strauss et al., 2016). **(A)** Fraction of ground fast ice area with respect to total lake area; **(B)** Total ground-fast area, **(C)** Number of lakes.

littoral terraces in these regions. The fraction could therefore be used to provide an estimate of the average terrace width. This width in relation to the radius of the central pool can be subsequently used to determine talik formation (Burn, 2002). A certain unfrozen lake area is required for penetration of the talik through the permafrost beneath a lake. The actual relationship depends on temperature and the width of the unfrozen central pool. Investigations from Richards Island (Canada), for example, suggest a maximum talik of the order of 150 m for a terrace width of the same size (Burn, 2002). This terrace width is also of the same order of magnitude as the ASAR WS spatial resolution. The

talik depth is expected to be higher for smaller terraces. Higher spatial resolution SAR missions such as TerraSAR-X could also be used (Jones et al., 2013; Antonova et al., 2016) to identify these smaller littoral terraces, as well as to identify any subtle changes locally. However, the application of such data over larger regions across the Arctic is not possible due to their limited spatial coverage.

Regional Patterns

The relatively high fraction of ground-fast ice in high SOC content regions points to a larger proportion of shallow lakes. These are also lake-rich regions in which a high density of small lakes are found. This underlines the importance of these lakes for methane flux studies. While winter fluxes might be reduced in such areas due to freeze up, shallow lakes show higher methane fluxes during the ice free season (Wik et al., 2016). However, the present satellite data analysis cannot account for the depth of any unfrozen lake areas. This is made evident by the bathymetric data.

Spatial clusters of different ground-fast ice fraction classes are specifically found in the Yedoma regions of Eastern Siberia and Alaska (**Figure 9**). The co-occurrence of lakes of different depths is indicative of the nature of thaw lake development, where it is common for lakes at different stages in their development cycle to exist next to each other (West and Plug, 2008). Deeper, older lakes can be found next to shallower and probably younger lakes, or lakes that have developed over different ground material.

The higher fraction of ground-fast lake ice in the 50 km grid in coastal areas could be partially explained by incomplete removal of ocean water surfaces during the lake mask preparation. Some areas located close to coastlines are known to have an abundance of shallow lakes. This abundance, as found on the North Slope and on parts of the Lena Delta, point to comparably young lakes of Pleistocene age (Black, 1969; Schwamborn et al., 2002).

Shallow lakes, which freeze to the ground, cannot be exclusively associated with thaw lake development. Many lakes, such as those on the lower terraces at Yamal, are part of floodplains. Higher ground-fast ice fraction has been identified for water bodies on floodplains of large rivers like the Ob on Yamal. These are located in non-continuous permafrost regions. Areas detected as ground-fast ice outside the continuous zone are therefore not expected to be related to thaw lakes. Due to the lack of ASAR WS data in southern areas, the number of samples within the non-continuous permafrost zones is in general much smaller than those within the continuous permafrost zones. The obtained results for these areas (zonal statistics presented in **Figure 12**) may thus not be completely representative.

Ground-fast ice fraction information may to some extent support the identification of landscape units, for example areas of adjacent lakes with similar patterns (terraces) or areas with mixed ground-fast fractions that indicate different lake development stages (such as parts of the Yedoma areas). The available validation datasets mostly represent marine and alluvial terraces. Especially samples from more continental lake-rich areas, such as the East Siberian Yedoma region, would be required for a more complete picture. An additional constraint for the interpretation of the results arises from larger lakes, which are not considered. Furthermore, ponds, which are also very common in permafrost regions, are also excluded from the analyses due to the spatial resolution limitation of the satellite datasets. Varying lake size distribution functions have also been reported for thaw ponds in the Arctic (Muster et al., 2013) and are associated with specific landscape types.

CONCLUSIONS

Space-borne synthetic aperture radar (SAR) data can be used to determine the fraction of circumpolar ground-fast ice and to create a spatially consistent mosaic, if incidence angle dependencies of the radar data are taken into consideration. Because radar backscatter values vary with incidence angle, and inconsistencies may arise when delineating ice-covered lakes from their surroundings, it is important to only consider the backscatter intensity of the lakes, masking the signal that arises from the surrounding land areas as well as the ocean in coastal regions, where masking inaccuracies need to be addressed separately.

Using this approach, it was established that in general small lakes (approximately 0.1 km^2) freeze almost entirely to the ground, while approximately 20% of all the lakes considered in this study (this excludes large lakes such as Great Slave Lake and Great Bear Lake) have ice covers that are ground-fast for at least 95% of their surface extent. No clear relationship between maximum lake depth (from bathymetric data and literature) and ground-fast ice fraction (from SAR data) was found. This result may however not hold locally, where higher spatial resolution data and more extensive bathymetric measurements would provide a more detailed picture.

To improve the understanding of lake formation, lake size distributions have already been analyzed with respect to glaciation history, permafrost zones and peatland areas. The present study's results further demonstrate that ground-ice fraction is related to lake size, where spatial patterns of lakes with ground-fast ice are also linked to certain landscape units. These findings are also supported by lake depth data from several sites across the Arctic. Useful information on the extent of the littoral terraces of lakes can also be determined using the presented method to establish ground-fast ice cover. Furthermore, preliminary investigation suggests that the proportion of ground-fast ice also increases with SOC content within the ground surrounding these lakes. It is anticipated that SAR derived ground-fast ice fractions, combined with auxiliary data on lake depth, will be beneficial to studies on methane upscaling from lakes, where winter and summer methane fluxes are driven by changes in frozen ground conditions.

AUTHOR CONTRIBUTIONS

AB has developed the concept for the presented study and wrote the majority of the manuscript. GP conducted all data analyses, contributed to the literature survey, compilation of the figures and writing of the methods part of the manuscript. YD and AK collected and processed the bathymetric data on the Yamal peninsula. ML is the supervising scientist for the Yamal surveys and contributed to the writing of related texts. AT has contributed to the writing of the manuscript, specifically the conclusions.

FUNDING

This work was supported by the Austrian Science Fund (FWF) under Grant [I 1401] and the Russian Foundation for Basic Research Grant 13-05-91001-ANF-a (Joint Russian–Austrian project COLD–Yamal).

ACKNOWLEDGMENTS

The 30 m land cover data set is provided by the National Geomatics Center of China. (DOI:10.11769/GlobeLand30.2000.db; DOI:10.11769/GlobeLand30.2010.db).

REFERENCES

Antonova, S., Duguay, C. R., Kääb, A., Heim, B., Langer, M., Westermann, S., et al. (2016). Monitoring bedfast ice and ice phenology in lakes of the Lena river delta using TerraSAR-X backscatter and coherence time series. *Remote Sens.* 8:903. doi: 10.3390/rs8110903

Arp, C. D., Jones, B. M., Liljedahl, A. K., Hinkel, K. M., and Welker, J. A. (2015). Depth, ice thickness, and ice-out timing cause divergent hydrologic responses among Arctic lakes. *Water Resour. Res.* 51, 9379–9401. doi: 10.1002/2015WR017362

Arp, C. D., Jones, B. M., Lu, Z., and Whitman, M. S. (2012). Shifting balance of thermokarst lake ice regimes across the Arctic Coastal Plain of northern Alaska. *Geophys. Res. Lett.* 39:L16503. doi: 10.1029/2012GL052518

Arp, C. D., Jones, B. M., Urban, F. E., and Grosse, G. (2011). Hydrogeomorphic processes of thermokarst lakes with grounded-ice and floating-ice regimes on the Arctic coastal plain, Alaska. *Hydrol. Processes* 25, 2422–2438. doi: 10.1002/hyp.8019

Atwood, D. K., Gunn, G. E., Roussi, C., Wu, J., Duguay, C., and Sarabandi, K. (2015). Microwave Backscatter From Arctic Lake Ice and Polarimetric Implications. *Geosci. Remote Sens. IEEE Trans.* 53, 5972–5982. doi: 10.1109/TGRS.2015.2429917

Bartsch, A. (2010). Ten Years of SeaWinds on QuikSCAT for Snow Applications. *Remote Sens.* 2, 1142–1156. doi: 10.3390/rs2041142

Bartsch, A., Pathe, C., Wagner, W., and Scipal, K. (2008). Detection of Permanent Open Water Surfaces in Central Siberia with ENVISAT ASAR Wide Swath Data with Special Emphasis on the Estimation of Methane Fluxes from Tundra Wetlands. *Hydrol. Res.* 39, 89–100. doi: 10.2166/nh.2008.041

Bartsch, A., Trofaier, A., Hayman, G., Sabel, D., Schlaffer, S., Clark, D., and Blyth, E. (2012). Detection of Open Water Dynamics with ENVISAT ASAR in Support of Land Surface Modelling at High Latitudes. *Biogeosciences* 9, 703–714. doi: 10.5194/bg-9-703-2012

Bartsch, A., Wagner, W., Scipal, K., Pathe, C., Sabel, D., and Wolski, P. (2009). Global Monitoring of Wetlands - the Value of ENVISAT ASAR Global Mode. *J. Environ. Manag.* 90, 2226–2233. doi: 10.1016/j.jenvman.2007.06.023

Black, R. (1969). Geology, Especially Geomorphology, of Northern Alaska. *Arctic* 22, 283–299. doi: 10.14430/arctic3220

Brown, J., Ferrians, O. J. Jr., Heginbottom, J. A., and Melnikov, E. S. (1998). *Circum-Arctic Map of Permafrost and Ground-Ice Conditions.* Boulder, CO: National Snow and Ice Data Center/World Data Center for Glaciology. Digital Media.

Burn, C. R. (2002). Tundra lakes and permafrost, Richards Island, western Arctic coast, Canada. *Can. J. Earth Sci.* 39, 1281–1298. doi: 10.1139/e02-035

Burn, C. R. (2005). Lake-bottom thermal regimes, Western Arctic Coast, Canada. *Permafrost Periglac. Process* 16, 355–367. doi: 10.1002/ppp.542

Closa, J., Rosich, B., and Monti-Guarnieri, A. (2003). "The ASAR Wide Swath Mode products," in *Geoscience and Remote Sensing Symposium, 2003. IGARSS '03. Proceedings. 2003 IEEE International* (Toulouse), vol. 2, 1118–1120. doi: 10.1109/IGARSS.2003.1294030

Duguay, C. R., Pultz, T. J., Lafleur, P. M., and Drai, D. (2002). RADARSAT backscatter characteristics of ice growing on shallow sub-Arctic lakes, Churchill, Manitoba, Canada. *Hydrol. Process.* 16, 1631–1644. doi: 10.1002/hyp.1026

Dvornikov, Y. (2016). *The Processes of Thermodenudation in Cryolithozone and the Dissolved Organic Matter as their Indication [In Russian]*. PhD thesis, Earth Cryosphere Institute SB RAS.

Dvornikov, Y., Leibmann, M., Heim, B., Bartsch, A., Haas, A., Khomutov, A., et al. (2016). Geodatabase and WebGIS project for long-term permafrost monitoring at the Vaskiny Dachi research station, Yamal, Russia. *Polarforschung* 85, 107–115. doi: 10.2312/polarforschung.85.2.107

Engram, M., Anthony, K. W., Meyer, F. J., and Grosse, G. (2013). Characterization of L-band synthetic aperture radar (SAR) backscatter from floating and grounded thermokarst lake ice in Arctic Alaska. *Cryosphere* 7, 1741–1752. doi: 10.5194/tc-7-1741-2013

Frauenfeld, O. W., Zhang, T., and McCreight, J. L. (2007). Northern hemisphere freezing/thawing index variations over the twentieth century. *Int. J. Climatol.* 27, 47–63. doi: 10.1002/joc.1372

Grunblatt, J., and Atwood, D. (2014). Mapping lakes for winter liquid water availability using SAR on the North Slope of Alaska. *Int. J. Appl. Earth Observ. Geoinformat.* 27, 63–69. Special Issue on Polar Remote Sensing 2013. doi: 10.1016/j.jag.2013.05.006

Henderson, F. M., and Lewis, A. J., (eds.). (1998). *Principles and Applications of Imaging Radar*, Vol. 2 *of Manual of Remote Sensing, 3 Edn*. Hoboken, N. J: John Wiley.

Hinkel, K. (2010). *Water Temperature and Bathymetric Depth Soundings for Lakes near Barrow, Version 1.0*. Boulder: UCAR/NCAR - Earth Observing Laboratory. (Accessed 19, Aug 2016). doi: 10.5065/D6WW7FS9

Hinkel, K. (2016). *Shapefiles - Lake Bathymetry, Watershed Delineation, Lake surface Area, Site Locations (CALON)*. University of Concinnati; NSF Arctic Data Center. doi: 10.5065/D60P0X57

Hinkel, K. M., Frohn, R. C., Nelson, F. E., Eisner, W. R., and Beck, R. A. (2005). Morphometric and spatial analysis of thaw lakes and drained thaw lake basins in the Western Arctic Coastal Plain, Alaska. *Permafrost Periglac. Process.* 16, 327–341. doi: 10.1002/ppp.532

Hirose, T., Kapfer, M., Bennett, J., Cott, P., Manson, G., and Solomon, S. (2008). Bottomfast ice mapping and the measurement of ice thickness on tundra lakes using C-Band Synthetic Aperture Radar Remote Sensing. *J. Am. Water Res. Assoc.* 44, 285–292. doi: 10.1111/j.1752-1688.2007.00161.x

Hugelius, G., Tarnocai, C., Broll, G., Canadell, J. G., Kuhry, P., and Swanson, D. K. (2013). The Northern Circumpolar Soil Carbon Database: spatially distributed datasets of soil coverage and soil carbon storage in the northern permafrost regions. *Earth Sys. Sci. Data* 5, 3–13. doi: 10.5194/essd-5-3-2013

Jeffries, M. O., Morris, K., and Kozlenko, N. (2013). "Ice characteristics and processes, and remote sensing of frozen rivers and lakes," in *Remote Sensing in Northern Hydrology: Measuring Environmental Change, Number 163 in Geophysical Monograph*, eds C. R. Duguay and A. Pietroniero (Washington, DC: American Geophysical Union), 63–90.

Jeffries, M. O., Morris, K., Weeks, W. F., and Wakabayashi, H. (1994). Structural and stratigraphie features and ERS 1 synthetic aperture radar backscatter characteristics of ice growing on shallow lakes in NW Alaska, winter 1991–1992. *J. Geophys. Res. Oceans* 99, 22459–22471. doi: 10.1029/94JC01479

Jones, B. M., Gusmeroli, A., Arp, C. D., Strozzi, T., Grosse, G., Gaglioti, B. V., et al. (2013). Classification of freshwater ice conditions on the Alaskan Arctic Coastal Plain using ground penetrating radar and TerraSAR-X satellite data. *Int. J. Remote Sens.* 34, 8267–8279. doi: 10.1080/2150704X.2013.834392

Kizyakov, A. I., Sonyushkin, A. V., Leibman, M. O., Zimin, M. V., and Khomutov, A. V. (2015). Geomorphological conditions of the gas-emission crater and its dynamics in Central Yamal. *Earth's Cryosphere* XIX, 15–25. Available online at: http://www.izdatgeo.ru/pdf/earth_cryo/2015-2/13_eng.pdf

Lantuit, H. (2007). *Geometrical Features of Lakes in the Northern Part of Arga Island*. Bremerhaven. doi: 10.1594/PANGAEA.611521

Lehner, B., and Döll, P. (2004). Development and Validation of a Global Database of Lakes, Reservoirs and Wetlands. *J. Hydrol.* 296, 1–22. doi: 10.1016/j.jhydrol.2004.03.028

Leibman, M., Khomutov, A., Gubarkov, A., Mullanurov, D., and Dvornikov, Y. (2015). The research station Vaskiny Dachi, Central Yamal, West Siberia, Russia - A review of 25 years of permafrost studies. *Fennia* 193, 3–30. doi: 10.11143/45201

Mätzler, C., Aebischer, H., and Schanda, E. (1984). Microwave dielectric properties of surface snow. *IEEE J. Ocean. Eng.* OE-9, 366–371. doi: 10.1109/JOE.1984.1145644

Morgenstern, A., Fedorova, I., Roessler, S., and Ivlev, P. (2013): Lake bathymetry, Kurungnakh Island, Lena Delta. doi: 10.1594/PANGAEA.848485. in supplement to: Morgenstern, A., Ulrich, M., Günther, F., Roessler, S., Fedorova, I., Rudaya, N. A., Wetterich, S., Boike, J., et al. (2013): Evolution of thermokarst in East Siberian ice-rich permafrost: a case study. *Geomorphology* 201, 363–379. doi: 10.1016/j.geomorph.2013.07.011

Muster, S., Heim, B., Abnizova, A., and Boike, J. (2013). Water body distributions across scales: a remote sensing based comparison of three arctic tundra wetlands. *Remote Sens.* 5, 1498–1523. doi: 10.3390/rs5041498

Naeimi, V., Paulik, C., Bartsch, A., Wagner, W., Kidd, R., Park, S. E., et al. (2012). ASCAT Surface State Flag (SSF): extracting information on surface freeze/thaw conditions from backscatter data using an Empirical Threshold-Analysis Algorithm. *IEEE Trans. Geosci. Remote Sens.* 50, 2566–2582. doi: 10.1109/TGRS.2011.2177667

Paltan, H., Dash, J., and Edwards, M. (2015). A refined mapping of Arctic lakes using Landsat imagery. *Int. J. Remote Sens.* 36, 5970–5982. doi: 10.1080/01431161.2015.1110263

Park, S.-E., Bartsch, A., Sabel, D., Wagner, W., Naeimi, V., and Yamaguchi, Y. (2011). Monitoring Freeze/Thaw cycles using ENVISAT ASAR global mode. *Remote Sens. Environ.* 115, 3457–3467. doi: 10.1016/j.rse.2011.08.009

Paulik, C., Melzer, T., Hahn, S., Bartsch, A., Heim, B., Elger, K., et al. (2014). *Circumpolar Surface Soil Moisture and Freeze/Thaw Surface Status Remote Sensing Products (version 4) with Links to Geotiff Images and NetCDF Files (2007-01 to 2013-12)*. Bremerhaven: Department of Geodesy and Geoinformatics, TU Vienna. doi: 10.1594/PANGAEA.832153

Potin, P., Rosich, B., Roeder, J., and Bargellini, P. (2014). "Sentinel-1 Mission operations concept," in *2014 IEEE Geoscience and Remote Sensing Symposium* (Quebec City, QC), 1465–1468. doi: 10.1109/IGARSS.2014.6946713

Ray, N., and Adams, J. (2001). A GIS-based vegetation map of the world at the Last Glacial Maximum (25,000-15,000 BP). *Internet Archaeol* 11. doi: 10.11141/ia.11.2

Schwamborn, G., Rachold, V., and Grigoriev, M. N. (2002). Late Quaternary sedimentation history of the Lena Delta. *Quat. Int.* 89, 119–134. doi: 10.1016/S1040-6182(01)00084-2

Smith, L. C., Sheng, Y., and MacDonald, G. M. (2007). A First Pan-Arctic Assessment of the Influence of Glaciation, Permafrost, Topography and Peatlands on Northern Hemisphere Lake Distribution. *Permafrost Periglacial Process.* 18, 201–208. doi: 10.1002/ppp.581

Strauss, J., Laboor, S., Fedorov, A. N., Fortier, D., Froese, D., Fuchs, M., et al. (2016): Database of Ice-Rich Yedoma Permafrost (IRYP), link to FileGDB. doi: 10.1594/PANGAEA.861731. in: Strauss, J., Laboor, S., Fedorov, A. N., Fortier, D., Froese, D., Fuchs, M., et al. (2016): Database of Ice-Rich Yedoma Permafrost (IRYP). doi: 10.1594/PANGAEA.861733

Strauss, J., Schirrmeister, L., Grosse, G., Wetterich, S., Ulrich, M., Herzschuh, U., et al. (2013). The deep permafrost carbon pool of the Yedoma region in Siberia and Alaska. *Geophys. Res. Lett.* 40, 6165–6170. doi: 10.1002/2013GL058088

Surdu, C. M., Duguay, C. R., Brown, L. C., and Fernández Prieto, D. (2014). Response of ice cover on shallow lakes of the North Slope of Alaska to contemporary climate conditions (1950-2011): radar remote-sensing and numerical modeling data analysis. *Cryosphere* 8, 167–180. doi: 10.5194/tc-8-167-2014

Surdu, C. M., Duguay, C. R., Pour, H. K., and Brown, L. C. (2015). Ice Freeze-up and Break-up Detection of Shallow Lakes in Northern Alaska with Spaceborne SAR. *Remote Sens.* 7, 6133–6159. doi: 10.3390/rs70506133

Wagner, W., Pathe, C., Doubkova, M., Sabel, D., Bartsch, A., Hasenauer, S., et al. (2008). Temporal stability of soil moisture and radar backscatter observed by the Advanced Synthetic Aperture Radar (ASAR). *Sensors* 8, 1174–1197. doi: 10.3390/s80201174

Wakabayashi, H., Weeks, W., and Jeffries, M. (1993). "A C-band backscatter model for lake ice in Alaska," in *Geoscience and Remote Sensing Symposium, 1993. IGARSS '93. Better Understanding of Earth Environment., International*, Vol.3 (Tokyo), 1264–1266. doi: 10.1109/IGARSS.1993.322103

Walter, K. M., Engram, M., Duguay, C. R., Jeffries, M. O., and Chapin, F. (2008). The potential use of synthetic aperture radar for estimating Methane Ebullition from Arctic Lakes. *J. Am. Water Res. Assoc.* 44, 305–315. doi: 10.1111/j.1752-1688.2007.00163.x

West, J. J., and Plug, L. J. (2008). Time-dependent morphology of thaw lakes and taliks in deep and shallow ground ice. *J. Geophys. Res.* 113:F01009. doi: 10.1029/2006JF000696

Wik, M., Crill, P., Bastviken, D., Danielsson, A., and Norbäck, E. (2011). Bubbles trapped in arctic lake ice: Potential implications for methane emissions. *Geophys. Res. Biogeosci.* 116:G03044. doi: 10.1029/2011JG001761

Wik, M., Varner, R. K., Anthony, K. W., MacIntyre, S., and Bastviken, D. (2016). Climate-sensitive northern lakes and ponds are critical components of methane release. *Nat. Geosci.* 9, 99–105. doi: 10.1038/ngeo2578

Yue, B., Chamberland, J., and Mulvie, J. (2013). Bottom-fast ice delineation with PolSAR and InSAR techniques in the Mackenzie Delta region, Northwest Territories, Canada. *Can. J. Remote Sens.* 39, 341–353. doi: 10.5589/m13-042

Conflict of Interest Statement: The authors declare that the research was conducted in the absence of any commercial or financial relationships that could be construed as a potential conflict of interest.

9

Inference of Soil Hydrologic Parameters from Electronic Soil Moisture Records

David G. Chandler[1]*, Mark S. Seyfried[2], James P. McNamara[3] and Kyotaek Hwang[1]

[1] *Civil and Environmental Engineering, Syracuse University, Syracuse, NY, USA,* [2] *USDA Agricultural Research Service, Northwest Watershed Research Center, Boise, ID, USA,* [3] *Department of Geosciences, Boise State University, Boise, ID, USA*

Key Points

- Soil hydrologic parameters including field saturation, field capacity, initiation of plant water stress and plant extraction limits can be reliably determined from electronic soil moisture sensor records.
- Soil profile wetting and drying occurs along a regular continuum of soil moisture following the advance of the wetting from to the effective base of the soil profile.
- Frozen soil conditions and interactions between energy and water limited water balances complicate interpretations of fluxes in the soil-plant-atmosphere continuum.

Edited by:
Theresa Blume,
Helmholtz-Zentrum
Potsdam—Deutsches
Geoforschungszentrum (HZ),
Germany

Reviewed by:
Guy Jean-Pierre Schumann,
Remote Sensing Solutions, USA
Ryan D. Stewart,
Virginia Tech, USA

*Correspondence:
David G. Chandler
dgchandl@syr.edu

Specialty section:
This article was submitted to
Hydrosphere,
a section of the journal
Frontiers in Earth Science

Soil moisture is an important control on hydrologic function, as it governs vertical fluxes from and to the atmosphere, groundwater recharge, and lateral fluxes through the soil. Historically, the traditional model parameters of saturation, field capacity, and permanent wilting point have been determined by laboratory methods. This approach is challenged by issues of scale, boundary conditions, and soil disturbance. We develop and compare four methods to determine values of field saturation, field capacity, plant extraction limit (PEL), and initiation of plant water stress from long term *in-situ* monitoring records of TDR-measured volumetric water content (Θ). The monitoring sites represent a range of soil textures, soil depths, effective precipitation and plant cover types in a semi-arid climate. The Θ records exhibit attractors (high frequency values) that correspond to field capacity and the PEL at both annual and longer time scales, but the field saturation values vary by year depending on seasonal wetness in the semi-arid setting. The analysis for five sites in two watersheds is supported by comparison to values determined by a common pedotransfer function and measured soil characteristic curves. Frozen soil is identified as a complicating factor for the analysis and users are cautioned to filter data by temperature, especially for near surface soils.

Keywords: saturation, field capacity, plant water stress, permanent wilting point, hysteresis

INTRODUCTION

Soil moisture is an important control on hydrologic function, as it governs vertical fluxes from and to the atmosphere, groundwater recharge, and lateral fluxes through the soil (Loik et al., 2004; Grayson et al., 2006; Vereecken et al., 2008). Soil moisture is also an excellent metric of hydrologic model performance, as it integrates temporal variation in precipitation and evaporation and is responsive to topography and soil physical properties governing fluxes of water (Wood et al., 1992; Cuenca et al., 1996; Rodriguez-Iturbe et al., 1999, 2001; Laio et al., 2001; Western et al., 2004; Seneviratne et al., 2010; Bandara et al., 2013; Or et al., 2015; Paniconi and Putti, 2015). As such, the definition and quantification of physical properties that govern the ability of a soil to retain water against gravity drainage, evaporation, and transpiration have long been of interest to agronomists and physical scientists (Briggs and McLane, 1907; Buckingham, 1907). However, measurement of relevant soil physical properties has historically been conducted in laboratories, which may not represent the behavior of water in field conditions. Techniques to measure soil moisture in the field have improved dramatically in recent decades so that near-continuous measurements are now routine. In this paper we demonstrate how multiple years of near-continuous measurements of soil water content can improve estimates of hydrologically relevant soil properties.

The physical properties of soils determine several important thresholds in Θ, defined as the volumetric fraction of water in a unit volume of soil ($m^3\ m^{-3}$): Soil saturation is conventionally defined as the condition when the entire pore volume is filled with water (Θ_s), which typically occurs only during subirrigation. The Θ at which gravity drainage effectively ceases has been termed field capacity (FC; Veihmeyer and Hendrickson, 1949), and the Θ at which transpiration ceases has been termed permanent wilting point (PWP; Briggs and Shantz, 1912). The difference in the value of FC and PWP is often used to estimate plant available water (Hansen et al., 1980) and the difference in the value of Θ_s and PWP is similarly used to estimate root zone storage (Seneviratne et al., 2010). Unfortunately, most measurements of these important field-scale properties are performed in laboratories on small samples.

The prevailing concepts of FC and PWP are a legacy of efforts to relate soil hydraulic and hydrologic parameters to standardized measurements of soil physical properties. Briggs and McLane (1907), introduced a "moisture equivalent" as a water to soil mass fraction for samples that were air dried, screened through a 2 mm sieve, packed in cylinders, saturated, and spun to achieve a force 1,000 times the force of gravity. These results were later expanded to analysis of PWP of wheat seedlings in a controlled study over a range of soil textures (Briggs and Shantz, 1912). Meinzer (1923) later defined "specific retention" as the volumetric fraction of water retained after long term drainage from saturation, which has been determined by measurement in small cells of disturbed soils (Hazen, 1892), and *in situ* (Ellis and Lee, 1919). Piper (1933) aptly concluded that experimental scale and depth of the capillary fringe exerted important controls on the results. Subsequently, Richards and Weaver (1944) built on the work of Buckingham (1907) to systematize measurement of matric potential (Ψ)-Θ relationships. Using the pressure plate technique, they equated the Θ at -33 kPa to the "moisture equivalent" for 71 near surface samples of irrigated soils (Richards and Weaver, 1944). The use of -10 kPa to determine Θ at field capacity was reserved for coarse and volcanic soils. These advancements led to the development of capillary permeability relations for isotropic porous media (Burdine and Redford, 1952; Brooks and Corey,

1964) and continuous Ψ-Θ conductivity models (Mualem, 1976) that remain in use.

The validity of these historic terms as physical constants has been widely criticized (Miller and McMurdie, 1953) resulting in refinements and additional terms for use in fields beyond agronomy. For example, it is understood that saturation may never be achieved in well drained soils under field conditions (Wang et al., 1998). Field saturation, Θ_{fs}, accounts for gas filled soil pore space that may be occluded within the pore matrix at approximately the bubbling pressure (Reynolds et al., 2002) and can occur due to air encapsulation (Fayer and Hillel, 1986), fluctuating groundwater levels (Marinas et al., 2013), and biogenic gases (Morse et al., 2015).

Field capacity has been criticized because the end point for gravity drainage is unclear and possibly method dependent due to differences in boundary conditions and scale (Colman, 1947; Hillel, 1980; Cassel and Nielsen, 1986; Kirkham, 2005). As a result, Ψ-values from -33 to -5 kPa have been proposed to define FC (Richards and Weaver, 1944; Salter and Haworth, 1961; Linsley and Franzini, 1972; Romano and Santini, 2002; Kirkham, 2005; Nemes et al., 2011). More recently, Assouline and Or (2014) proposed a method to relate the slope of the soil characteristic curve (n), the air entry value (Ψ_{ae}) and the residual water of a soil (Θ_r) (van Genuchten, 1980) to a soil specific FC, based on the balance between capillary and gravitational forces. This dynamical approach seeks to transcend the use of static arbitrary values.

PWP has been criticized because the actual wilting point is species and soil texture dependent and, in the case of many native plants adapted to dry conditions, transpiration can cease with no wilting at Ψ in the range of -3 to -5 MPa for both tree and grass species (Scholes and Walker, 1993; Damesin and Rambal, 1995; Sperry et al., 2002; Rambal et al., 2003). To address these issues, Seyfried et al. (2009) used the term "plant extraction limit" (PEL) in place of PWP to describe the Θ below which plants cannot extract water, and evaporation is the primary means of soil water loss. Therefore, Θ at PEL is likely to be equal to or greater than Θ at the limit of capillary continuity (Lehmann et al., 2008; Assouline and Or, 2014), since evaporation can continue following the cessation of plant water use.

Furthermore, between FC and PEL, there is range of declining Ψ over which plant stress increases and evapotranspiration decreases (Rodriguez-Iturbe et al., 1999). This range of Ψ and the related range of Θ is associated with a shift from energy-limited to soil moisture potential-limited evapotranspiration (Budyko, 1956; Koster et al., 2009). The threshold in water content between these states in terms of Θ has been defined as s* (Laio et al., 2001; Rodriguez-Iturbe et al., 2001), Θ_{crit} (Koster et al., 2004; Seneviratne et al., 2010), or Θ_{ws} (Smith et al., 2011) and conceptualized as an inflection point in evaporation-soil water content relation models. Similarly, Buckingham (1907) proposed that evaporation from soil is initially limited by moisture content, and secondly by vapor diffusion. Various formulations (Idso et al., 1974; Brutsaert and Chen, 1995) and methods have since been used to determine energy limited (Stage I) and supply/transport limited (Stage II) evaporation from soils (Salvucci, 1997). Stage III is vapor diffusion transport (Metzger and Tsotsas, 2005; Lehmann et al., 2008). During the second and third stages, the vaporization plane (e.g., drying front) will move downward from the soil surface as evaporation proceeds (Shokri et al., 2008; Shokri and Or, 2011). Therefore, the annual endpoint of soil drying may range from a value less than FC, to near the hygroscopic point, depending on the characteristics of local climate, soil, and plants (Laio et al., 2001).

The current capacity to resolve soil hydrologic properties and Θ at scales sufficient to parameterize and validate distributed hydrologic models remains challenged by the relatively limited data available to define spatial heterogeneity and effective scales of measurement (Njoku et al., 2003; Reichle et al., 2007; Huang et al., 2016). Since transitions in fluxes in agronomic and ecohydrologic models are often indicated by FC, Θ_{ws}, and PEL either indirectly (Liang et al., 1996) or directly (e.g., Wigmosta et al., 1994; Boote et al., 2008; Seneviratne et al., 2010; Paniconi and Putti, 2015) and are bounded by Θ_s (or Θ_{fs}) and either PEL or some lower limit to Θ, we consider these important variables for hydrologic modeling. Recent approaches to determine soil moisture parameters include inverse modeling, remote sensing and other synthetic approaches (Santanello et al., 2007; Montzka et al., 2011; Bandara et al., 2013). However, soil hydrologic properties would ideally be determined by field measurement (Romano and Santini, 2002). Hillel (1980) recommended that properties such as FC must be measured repeatedly in the field, after the entire depth of the soil profile is wetted and that laboratory methods based on Ψ equivalent to -1.5 MPa, -33 kPa, or -10 kPa matric potential are arbitrary static measurements that are not representative of systems with external controls on the soil boundary conditions, such as poor drainage or evaporation.

Given the increasing availability of electronic measurements of soil water content and considerable theoretical and conceptual controversy regarding soil water content constants, the primary objective of this paper is to determine if temporal patterns of measured soil water content values are consistent with the concepts of Θ_s, FC, and PEL. The proposed methods presented here are intended to determine consistent and physically/ecologically relevant estimates of these soil hydrologic parameters directly from long term field measurements of Θ. We pose the hypothesis: Data attractor values and inflections in Θ time series records represent changes in dQ/dt that are consistent with the concepts of FC, Θ_{ws}, and PEL. This is based on prior observations that FC and PEL exist as quasi-stable states (Miller and Klute, 1967; Cassel and Nielsen, 1986; Kirkham, 2005) following periods of drainage and evaporation. These states are represented by greater data frequency (data attractors) at intermediate and low values of Θ. That is, following input events, Θ rapidly trends toward FC where it is relatively stable due to capillary forces until decreases by evapotranspiration or increased by input flux greater than the associated unsaturated hydraulic conductivity [K(Θ)]. Then, as soils dry they stabilize near PEL. Thus, FC may be better represented in the absence of strong ET, such as during winter in mid-latitude regions, and PEL is best determined when energy does not limit transpiration. To test the hypothesis, we first compare soil hydrologic property values determined by four methods of analysis to and then assess the accuracy of the estimated values to values determined from a range physical methods at traditional soil water potentials. Finally, we evaluate the applicability of the tested approaches to other environments.

Setting

The analysis exploits the wide range and regular seasonal changes in soil moisture at sites near Boise ID, USA. In semiarid Mediterranean climates, seasonal patterns in soil moisture storage and streamflow follow seasonal changes in precipitation, temperature, and solar radiation (McNamara et al., 2005: Flerchinger et al., 2010; Seyfried et al., 2011). In this system, the water year begins with dry soil following the annual summer drought. As autumn rains arrive, the soil wets downward from the soil surface and soil profile water storage increases over a long wet period due to winter rains or snow melt. In spring, warming temperatures drive snowmelt and the annual hydrograph. Ultimately, precipitation decreases and plant water use depletes soil moisture as summer proceeds. In many years, this hydrologic setting imposes different soil surface boundary conditions in each season; intermittent surface wetting of the dry soil in fall, low flux snow melt in winter, high flux snow melt in spring and progressive soil drying over the summer (**Figure 1**).

Sites

Data were collected from five sites at two watersheds in southwest Idaho, Reynolds Creek Experimental Watershed (RCEW) and Critical Zone Observatory operated by the USDA Northwest Watershed Research Center and Dry Creek Experimental Watershed (DCEW) operated by Boise State University. Data for RCEW are available at http://www.ars.usda.gov/Research/docs.htm?docid=8281 and from Seyfried et al. (2001) and Chauvin et al. (2011). Data for DCEW are available at http://earth.boisestate.edu/drycreek/.

The Upper Sheep Creek watershed is located within the Reynolds Creek Experimental Watershed, in the Owyhee Mountains of southwestern Idaho. The site ranges in elevation from 1,840 to 2,036 mamsl and is underlain by basalt bedrock. Mean annual precipitation is 426 mm, about 60% of which is snow. The snow is redistributed by wind, resulting in snow accumulations of 3–4 m in drifts and <15 cm in scour zones. Deposition patterns are persistent and result in distinctive snow-soil-vegetation complexes (Flerchinger et al., 1998; Prasad et al., 2001; Chauvin et al., 2011). The low sagebrush complex (Low Sage) is found on shallow (<50 cm deep) soils on mostly west-facing slopes and ridges. Soil textures are loam in the upper 10–20 cm and clay loam and gravel in the subsoil below the argillic horizon. The mountain sagebrush complex (High Sage) is on deeper soils, about 1 m thick, and mostly on north-facing slopes. The aspen complex (Aspen) is on deep (>2 m) soils and are closely associated with snow drifts. Soils in the High Sage and Aspen complexes were formed in aeolian deposits and are highly uniform with depth, mostly categorized as silt loam. Upper Sheep Creek was burned by a prescribed fire in summer 2007. The Big Sage area burned completely, and was replaced with grasses

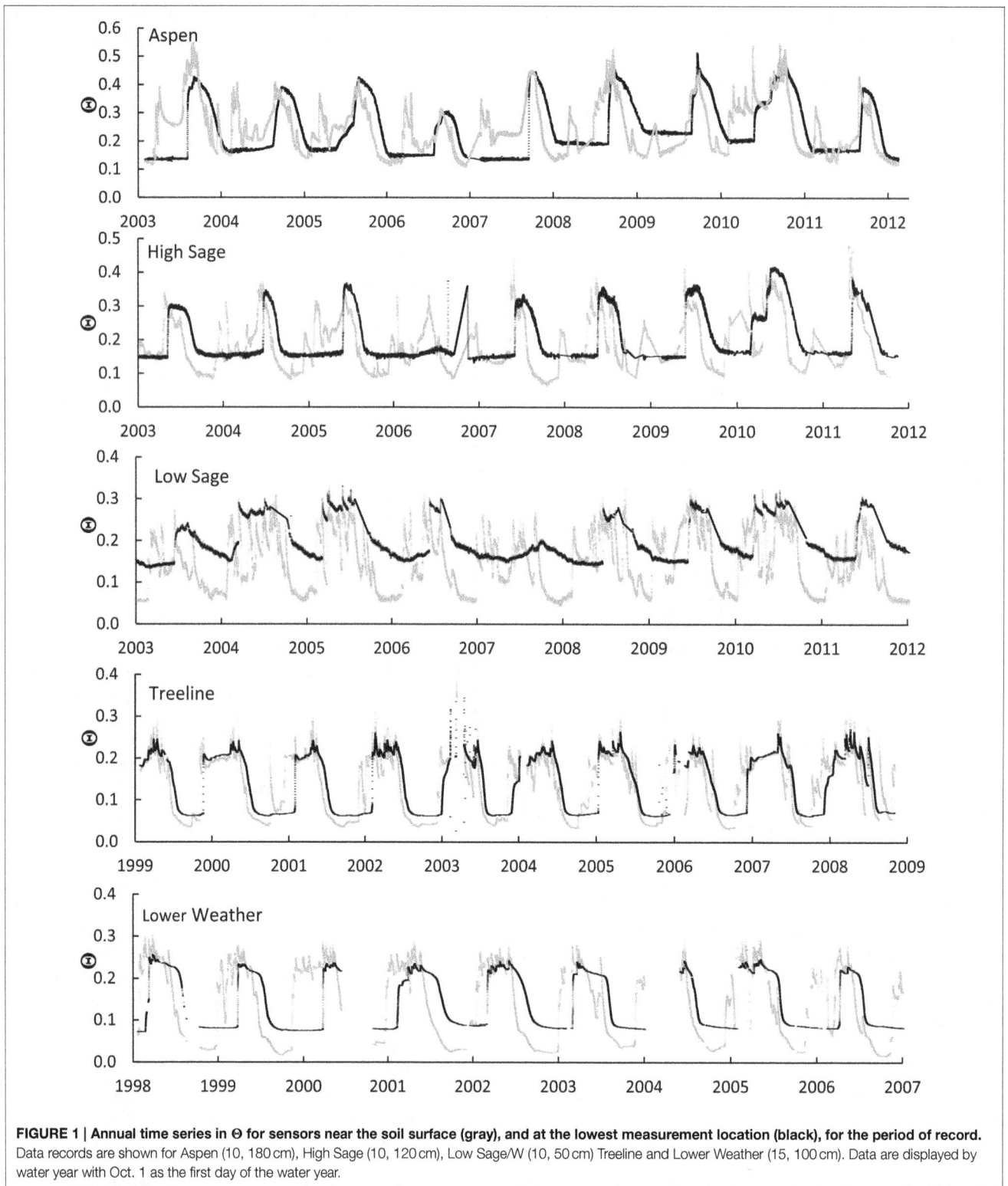

FIGURE 1 | Annual time series in Θ for sensors near the soil surface (gray), and at the lowest measurement location (black), for the period of record. Data records are shown for Aspen (10, 180 cm), High Sage (10, 120 cm), Low Sage/W (10, 50 cm) Treeline and Lower Weather (15, 100 cm). Data are displayed by water year with Oct. 1 as the first day of the water year.

and forbs by 2008. The Aspen was cut down and recovered, in a hydrologic sense, by 2009.

The Treeline and Lower Weather sites are located in the Dry Creek Experimental Watershed (DCEW) in the foothills of the Rocky Mountains of southwestern Idaho. The elevation of these sites is 1,620 mamsl at Treeline and 1,150 mamsl at Lower Weather. Both sites are underlain by weathered granite from the Idaho Batholith, although the Lower Weather site is below the shoreline of prehistoric Lake Idaho. Mean annual precipitation is 520 mm at Treeline and 300 mm at Lower Weather. Over the period of measurement, snow accounted for ~50% of precipitation at Treeline (Kormos et al., 2014) but Lower Weather receives mostly rain. Similar to Upper Sheep Creek, geologic processes including wind scour have formed deeper soils on north facing slopes at Treeline and snow accumulation up to 2 m currently enhances springtime soil water where the sensors for this study are located. The vegetation at the site is typical of a transition between lower elevation grass-lands and higher elevation forests; mountain big sagebrush and ceanothus shrubs, prunus subspecies, forbs, and grasses. The Lower Weather site is not subject to drifting due to the limited snowfall at the lower elevation. Soil texture at DCEW is sandy loam and ranges in depth up to 1.3 m (Gribb et al., 2009; Tesfa et al., 2009).

METHODS

Data Collection and Conditioning

Soil water content data were collected from representative sites of each of the five soil/vegetation complexes. These sites represent various controls of soil texture, soil depth, vegetation, and annual precipitation (Flerchinger et al., 2010) on Θ (**Figure 1**). In each instance, a pit was hand-excavated to bedrock or saprolite and two parallel profiles of soil moisture instruments, separated by 2 m, were installed horizontally into the pit face. A summary of sensor depths, soil texture and mean annual precipitation or estimated annual soil water input is provided in **Table 1**. At Upper Sheep Creek, hourly data were collected with time domain reflectometers (TDR 100, Campbell Sci., Logan UT, USA), with 3-prong, 30 cm long rods. Data are reported on water year basis. At Dry Creek, hourly data were collected with frequency domain reflectometers (CS615, Campbell Sci., Logan UT, USA) calibrated with less frequent samples from co-located TDR waveguides (Chandler et al., 2004). Data were prepared for analysis by removing most electronic noise, out of range data or data from failed sensors (Nayak et al., 2008; Kormos et al., 2014) and are reported by calendar year. Data cleaning did not include removal of records during periods of frozen soil, which were most prominent within 15 cm of the soil surface and affected a limited number of sensors. Frequency distribution of Θ from single sensors was done by constructing histograms from the conditioned data. The frequency analysis of temporally coincident data from paired sensors was done by binning data in two dimensional frequency arrays with increments of 0.01 Θ for $0.0 < \Theta < 0.65$.

Limited laboratory and field analyses are used to relate the results of the developed methods to traditional capillary pressure-saturation techniques. Measurements from a 66 kPa field tensiometer co-located with the 60 cm TDR sensor at Aspen site represent the range from field-saturated conditions just after snowmelt on May 20, 2004 to the time of removal on August 8, 2004. Laboratory measurements of Ψ over the ranges of 1.4–430 kPa (HYPROP, UMS GMBH, Munich Germany), and Ψ and Θ (corrected for bulk density) 0.2–115 MPa (dewpoint psychometer, Decagon Devices, Pullman WA, USA) and Θ were made to construct characteristic curves for soil samples

TABLE 1 | Sensor location, mean annual soil water input (SWI), and soil texture.

Site (replicate)	SWI (mm)	Texture	Sensor depths (cm)
Aspen (East)	1,000	Silt loam	10, 30, 60, 90, 120, 150, 165, 180
Aspen (West)		Silt loam	10, 30, 60, 90, 120, 150, 165, 180
High sage (East)	530	Silt loam	10, 30, 60, 90, 105, and 120
High sage (West)		Silt loam	10, 30, 60, 90, 105, and 120
Low sage (East)	410	Loam	10, 30, 40, 50
Low sage (West)		Clay loam, gravel	10, 30, 40, 50
Treeline (Pit 3)	570	Sandy loam, gravel	5, 15, 30, 60, 100
Treeline (Pit 4)		Sandy loam, gravel	5, 15, 30, 45, 65
Lower weather (Pit 1)	390	Coarse sandy loam	5, 15, 30, 50, 100
Lower weather (Pit 2)		Coarse sandy loam	5, 15, 30, 60, 100

collected at depths of 0–3 and 41–43 cm. At Treeline, soil water potential was previously measured in a field experiment to determine FC by surface irrigation from dry initial conditions. The experiment was interrupted by rain during the drainage period and terminated when the tensiometer lost capillary connection to the soil at ~33 kPa. Soil water potential was monitored by automated tensiometers (Gribb et al., 2009).

Values of Θ measured during the natural cycle of wetting and drying provide the basic data used in this analysis. In all cases the soils are well drained, therefore there is no attractor at or near soil saturation. We define attractors as values of Θ that occur at high frequency and develop four analyses that are based on observations made in a climate with distinct annual wet and dry phases. Field Saturation is the maximum observed value of Θ. Since complete soil saturation is infrequent in well drained field soils, the maximum annual value of Θ_{fs} is likely to range between FC and Θ_s. Field Capacity is a moisture state that prevails under wet, low flux conditions. In well-drained conditions, saturated soils drain quickly to near FC, then more slowly. Where soil water potential (Ψ) is seasonally affected by plant water uptake, FC may vary slightly with diel changes in evapotranspiration. Conversely, dry soils wet to a Θ somewhat greater than FC before drainage begins. These two processes enhance the local data frequency near FC and result in a data attractor that we define as the "wet attractor." **Figure 2A** shows a wet attractor near Θ of 0.20 throughout the winter months when drainage is minimal and transpiration is insignificant. We propose that this attractor is approximately equivalent to FC. Several other descriptive and mechanistic definitions of FC have been developed and are presented by Assouline and Or (2014), who suggest the stability of soil water retention is a function of disruption of liquid phase continuity. Plant Extraction Limit is the Θ at which existing vegetation does not extract soil water by transpiration. Conceptually similar to the PWP, PEL allows for variations among soil-plant combinations and acknowledges that the vegetation doesn't necessarily wilt when transpiration ceases (or becomes very slow). PEL is expected to be an attractor in environments with extended dry periods and perennial vegetation. Under these conditions, Θ declines during the dry

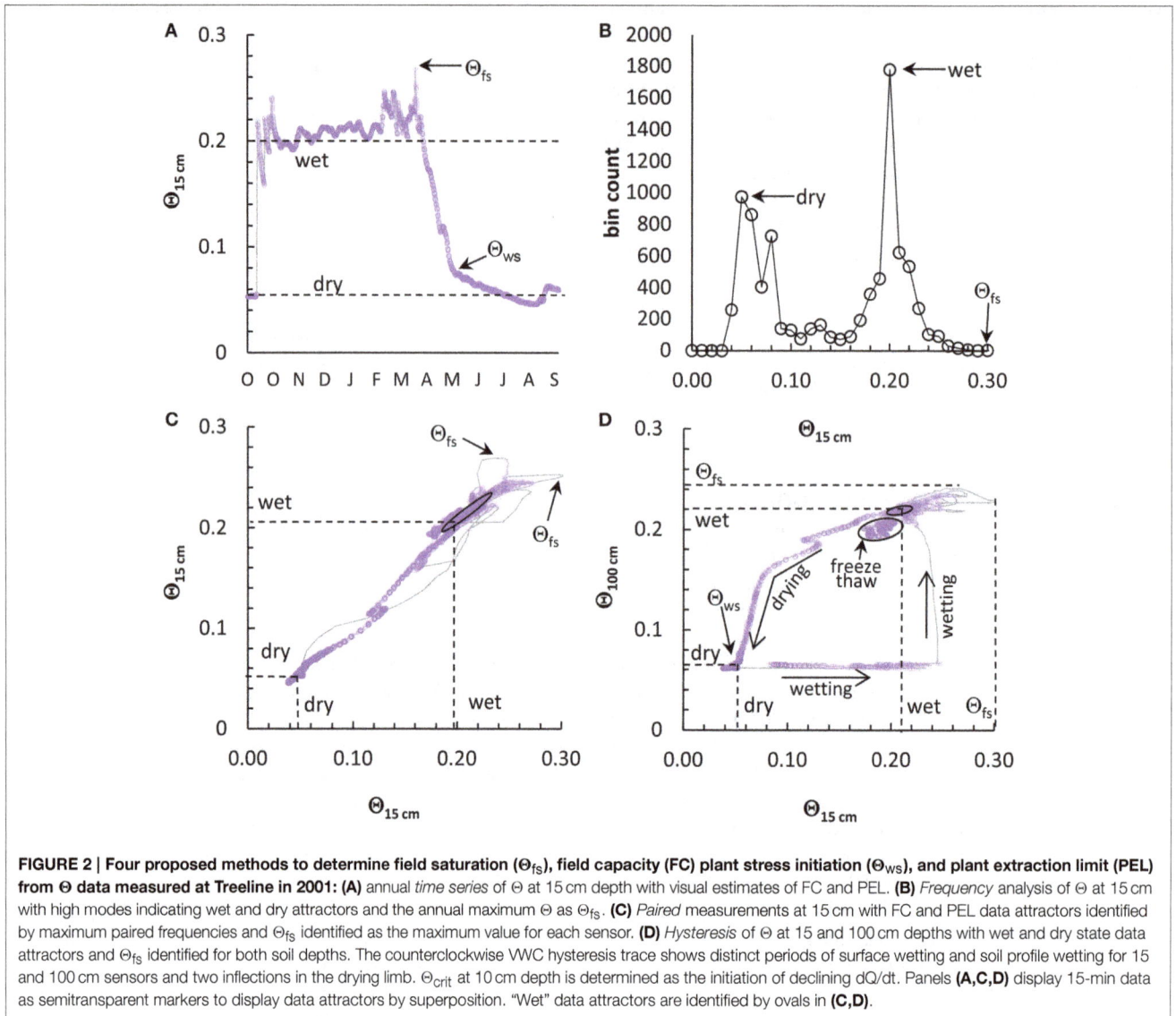

FIGURE 2 | Four proposed methods to determine field saturation (Θ_{fs}), field capacity (FC) plant stress initiation (Θ_{ws}), and plant extraction limit (PEL) from Θ data measured at Treeline in 2001: **(A)** annual *time series* of Θ at 15 cm depth with visual estimates of FC and PEL. **(B)** *Frequency* analysis of Θ at 15 cm with high modes indicating wet and dry attractors and the annual maximum Θ as Θ_{fs}. **(C)** *Paired* measurements at 15 cm with FC and PEL data attractors identified by maximum paired frequencies and Θ_{fs} identified as the maximum value for each sensor. **(D)** *Hysteresis* of Θ at 15 and 100 cm depths with wet and dry state data attractors and Θ_{fs} identified for both soil depths. The counterclockwise VWC hysteresis trace shows distinct periods of surface wetting and soil profile wetting for 15 and 100 cm sensors and two inflections in the drying limb. Θ_{crit} at 10 cm depth is determined as the initiation of declining dQ/dt. Panels **(A,C,D)** display 15-min data as semitransparent markers to display data attractors by superposition. "Wet" data attractors are identified by ovals in **(C,D)**.

season until PEL is reached, then remains unchanged until replenished with new rainfall or snowmelt except near the soil surface, which may be subject to direct evaporation and may proceed to values less than the PEL. These processes result in a "dry attractor" in the data. Plant Water Stress Initiation (Θ_{ws}) is not an attractor, but is a decrease in plant water use as governed by complex physiological traits that control stomatal aperture (Osakabe et al., 2014), and is an inflection in soil moisture records (Rodriguez-Iturbe et al., 1999; Smith et al., 2011) determined as $d^2Q/dt2 > 0$.

Analysis Methods

Four methods of analysis to determine the parameters PEL, FC, and Θ_{fs} are presented. Throughout the remainder of the paper, these analyses will be referred to as *time series, frequency, paired,* and *hysteresis,* respectively.

Visual inspection of Θ records in *time series* is a simple approach to identify approximate values for each parameter for

individual soil depths. **Figure 2A** shows this approach for one year of data from a 15 cm sensor at Treeline. At this site the minimal overlap between periods of slow drainage in winter and consistently low Θ in spring and summer. This allows subjective approximation of the wet and dry attractors, and clearly shows the maximum annual Θ, and an inflection in $\Theta(t)$ indicating a reduction in plant water use.

Frequency analysis of Θ distributions for single sensors show modal peaks in Θ at dry and wet soil moisture attractors for water year (**Figure 2B**) or period of record data. The maximum value of Θ is assumed to be equivalent to Θ_{fs}.

Paired sensors at similar depth, are often used to determine an average value of Θ, increase spatial representation of the measurement, or provide redundant data in case of instrument failure. These synchronous data can be plotted as x-y coordinates to determine high frequency attractors and extreme values. **Figure 2C** shows example data for a water year from paired sensors at 15 cm depth in soil pits separated by 2 m. Ideally, data pairs of Θ from paired sensors at similar depth will fall on a 1:1 line, bounded by the minimum and maximum Θ at the sensor depth. However, differences in the timing of wetting and drying result in divergence and spreading around a 1:1 line. In the example for one water year of data, Θ attractors emerge at dry Θ ($\approx$0.05) and wet Θ (0.19–0.24) attractor values, respectively and the greatest measured values of Θ (0.30, 0.27) for Θ_{fs} on the respective axes of each sensor. Although, similar to the frequency analysis, the paired sensor analysis leverages twice as many sample locations per data point. The use of data pairs decreases the influence of erroneous or extreme values on the frequency of values near the data attractors on the central trend. Additionally, the resulting variance around the trend may provide the user greater guidance on the range of local variability in soil hydrologic properties than would a single sample location.

Hysteresis is often observed in the temporal behavior of Θ over seasonal wetting and drying periods for sensors at two depths in a vertical profile. The lag in both wetting and drying periods between the two depths enhances the relative frequency of wet and dry attractors, near the intersection of the wetting and drying limbs of the hysteresis loop. For example, **Figure 2D** shows hysteresis between Θ at 15 and 100 cm at Treeline. For the climate of this study, the water year begins near the end of the dry season, with Θ at near minimum values for both the surface and bottom of the soil profile. Accordingly, the Θ trace for a water year begins near the low Θ attractor. As autumn precipitation increases Θ at 15 cm depth, the Θ at 100 cm depth remains nearly constant until the wetting front arrives. For a water year at the example site, the horizontal wetting trace provides strong visual support for the interpretation of low Θ attractor from the small range in the ordinate. Along the right side of **Figure 2D**, wetting at the deep sensor is shown as the vertical trace up to the wet attractor, found at the intersection of the wetting and drying traces and identified by the black oval. As above, the maximum values on either axis represent Θ_{fs}. Finally, the drying limb terminates at the dry Θ attractor and provides visual support for identification of the dry Θ attractor for depth of the shallow sensor.

RESULTS

We found wet and dry attractors within the expected ranges of FC and PEL by each of the presented methods, equated Θ_{fs} to the maximum value of Θ and interpreted the inflection during the drying phase, $d^2Q/dt2$ as Θ_{ws}. Results from the *time series, frequency, paired sensors,* and *hysteresis* approaches are in general agreement, although each has a different bias. Below, we first present detailed graphical results from the *paired* and *hysteresis* analyses to demonstrate the attractors for the Aspen, Treeline, and Low Sage sites, which span the range of soil texture and climate among the study sites. Then we compare values determined by these methods at three depths for each site with results from the *time series* and *frequency* analyses. Example data from soil physics experiments performed in the laboratory and from *in situ* measurements at Aspen and Treeline sites are then provided to relate the extreme values and data attractors to PEL, FC, and Θ_{fs}.

Paired Sensors

Figure 3 shows two visualizations of *paired* sensor analyses over the study period for Aspen (90 cm), Low Sage (10 cm), and Treeline (15 cm). Here, and below, we present data that demonstrate analyses over the broadest range in soil texture, soil depth, and vegetation class. For the Aspen the top panel shows semi-transparent data points with makers to represent the data attractors and the bottom panel shows color maps of log-transformed data frequency values. Many of the paired sensors at a depth, such as Aspen, 90 cm show a strong linear correlation between sensors (central trend) with occasional hysteresis traces outside of the central trend. Shallow paired sensors such as at Low Sage and Treeline (**Figure 3**) show much greater variability in Θ around the central trend. These features result from the temporally local hysteresis between sensors during input events, repeated over the annual period of changes in Θ.

Comparison of plots of raw data in the upper panel of **Figure 3** with the log transformed frequency colormaps in the lower panel demonstrates the greater visual support of the transformed data for identification of attractor values. Whereas, the point values can be selected directly from the frequency value matrix, as was done to identify attractors for the upper panel, the colormaps simplify the interpretation of the frequency value matrix. In **Figure 3** (lower panel) dry attractors for the period of record are clearly shown as dark red points in the colormaps, often near the minimum Θ, and are generally surrounded by a set of high frequency values which may include two or three modes. Similarly, the period of record wet attractor value is often diffuse, with weak frequency modes for the Aspen (silt loam) and Low Sage (clay loam) sites, but with a more distinct attractor for the Treeline (sandy) site. Variability between sensors extends to Θ_{fs}, which varies more by depth and year than the other paired analyses because it is an extreme value. The minimum and maximum of annual estimates (**Table 2**) bound the period of record estimates, and provide an approximation of the range in variability of the determined values. For the Treeline site, the range in annual estimates for individual sensors within the paired sensor configuration was zero to 0.02 Θ for both the attractors and Θ_{fs}, likely due to the high sand content at this site. At Low Sage and Aspen, the range increased progressively to 0.01–0.05 Θ and 0–0.11 Θ, respectively, primarily due to annual variability in Θ at depth. Spatial differences in $\Theta(t)$ resulted in variance from unity slope (1.00) for many sites and depths, but were similar within soil texture class: For the Aspen and High Sage sites the slope of the relation between paired sensors increased from a slope of 1.00–1.03 near the soil surface to as much as 1.33 at depth. At Low sage the slope of the relationship ranged from 1.10 to 1.19 from 10 to 40 cm depth, and 1.36 at 50 cm, in the C horizon, which is very stony. The Treeline and Lower Weather sites have low bulk density surface soil, with frequent macropores the slope of the trend at 5 cm depth (1.17–1.29) is much greater than for the deeper, stony soils (1.12).

Hysteresis plots of Θ for period of study data at Aspen, Low Sage, and Treeline show more complex wet and dry data attractors than are shown for an annual cycle in **Figure 2D**. **Figure 4** presents visualizations of hysteresis

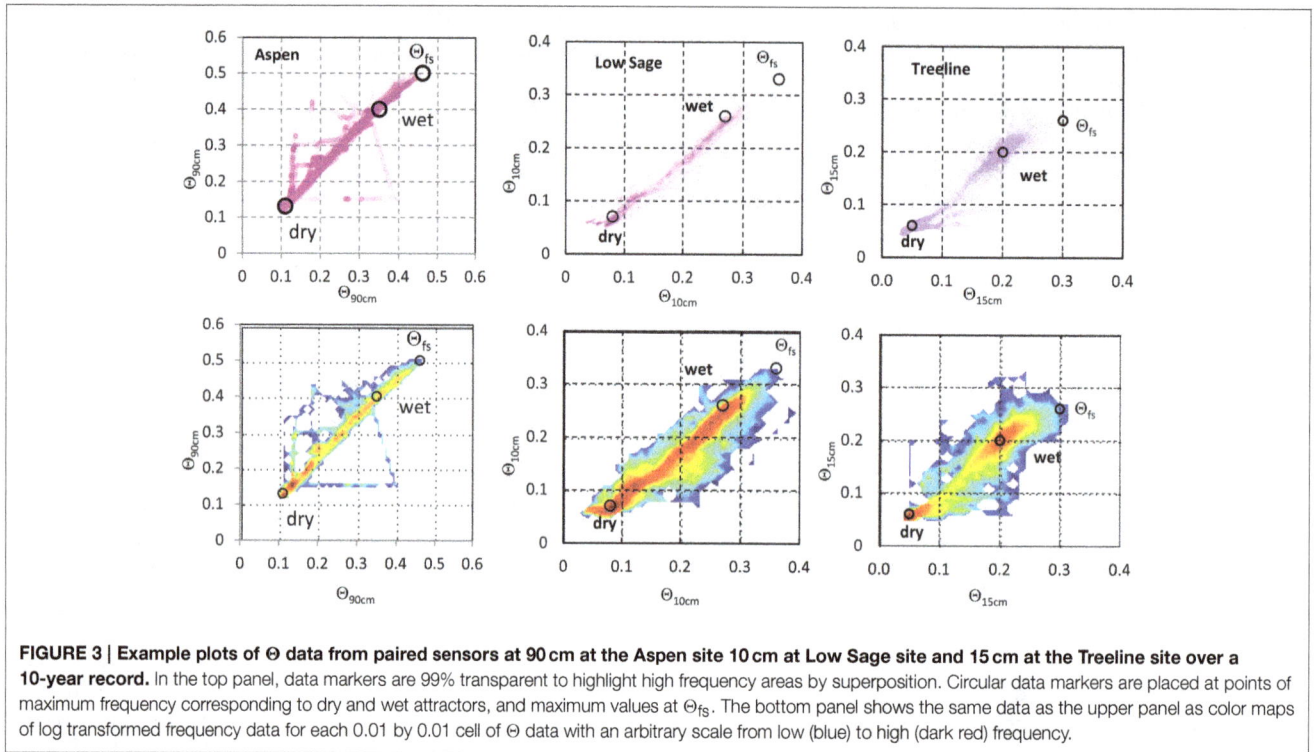

FIGURE 3 | Example plots of Θ data from paired sensors at 90 cm at the Aspen site 10 cm at Low Sage site and 15 cm at the Treeline site over a 10-year record. In the top panel, data markers are 99% transparent to highlight high frequency areas by superposition. Circular data markers are placed at points of maximum frequency corresponding to dry and wet attractors, and maximum values at Θ_{fs}. The bottom panel shows the same data as the upper panel as color maps of log transformed frequency data for each 0.01 by 0.01 cell of Θ data with an arbitrary scale from low (blue) to high (dark red) frequency.

TABLE 2 | Period of study maximum and minimum annual values of wet and dry data attractors (Θ) and Θ_{fs} for sensor *pairs* at a range of depths (cm) at selected sites.

Site	Depth cm	Wet min	Wet max	Dry min	Dry max	Θ_{fs} min	Θ_{fs} max
Aspen	10	0.35	0.38	0.15	0.15	0.50	0.50
Aspen	90	0.40	0.41	0.12	0.15	0.46	0.51
Aspen	180	0.37	0.45	0.14	0.15	0.41	0.52
Low sage	10	0.28	0.30	0.06	0.07	0.33	0.35
Low sage	30	0.30	0.32	0.12	0.17	0.36	0.37
Low sage	50	0.28	0.32	0.17	0.19	0.34	0.36
Treeline	15	0.17	0.17	0.02	0.03	0.22	0.24
Treeline	30	0.22	0.22	0.03	0.03	0.31	0.31
Treeline	65	0.22	0.22	0.07	0.07	0.26	0.26

between shallow and mid-profile depths, mid-profile and deep sensors, and log transformed colormaps of selected plots.

Despite differences in soil depth, texture, annual water input from rain and snow, and vegetation cover types, the *hysteresis* plots for most sites are similar in form, with a few important differences. All *hysteresis* plots with a near surface measurement (e.g., **Figures 2D, 4A–C**) show a marked inflection in the drying limb, whereas hysteresis plots for lower depths (e.g. **Figures 4D,E**) are more triangular, with the constant slope of the drying limb extending to near the dry attractor, indicating more uniform drying across depth. For some sites (e.g., **Figure 4F**) the hysteresis plot can vary annually between these forms, depending

on the timing of precipitation. As a result, the range of the wet attractor over the depth of the soil profile is from 0.13 to 0.08 Θ at Aspen and less (0.02–0.07 Θ) at drier sites such as Low Sage and Treeline (**Table 3**). Similarly, the distribution of Θ frequency values within the hysteresis loop is quite different among Aspen, Low Sage and Treeline (**Figures 4G–I**). The differences in the distribution Θ frequency is directly related to differences in depth and timing of precipitation and snowmelt and soil drainage rates for the different soil textures.

Figure 5 compares the range of estimated values for Θ_{fs}(**Figure 5A**), and the wet (**Figure 5B**) and dry (**Figure 5C**) attractors for selected soil depths at the study sites. Data are presented as maxima and minima of all estimates for time series, frequency, paired, and hysteresis analyses. Representative estimated values for each parameter were determined by comparison of the average of annual estimated values to period of record estimated values. Θ_{fs} values are equated to the maximum Θ found by long term (decadal) *frequency* analysis (**Figure 5A**) and exceed average values from the *paired* and *hysteresis* analyses by 0.02–0.16 Θ. For the wet and dry attractors, value estimates from the *paired* and *hysteresis* analyses are often similar, despite differences in range, both of which are narrower than the value ranges for *time series* and *frequency* analyses (**Figures 5B,C**). The representative estimated value was determined as the most common value among period of record estimate, the average value of annual estimates of *paired* analyses, and the average value of six period of record *hysteresis* estimates. This approach generally resulted in a set of values which differed by <0.02 Θ. Context for the results in **Figure 5** is provided by Θ estimates at 0 kPa, 33 kPa, and 1.5 MPa for the various soil textures by

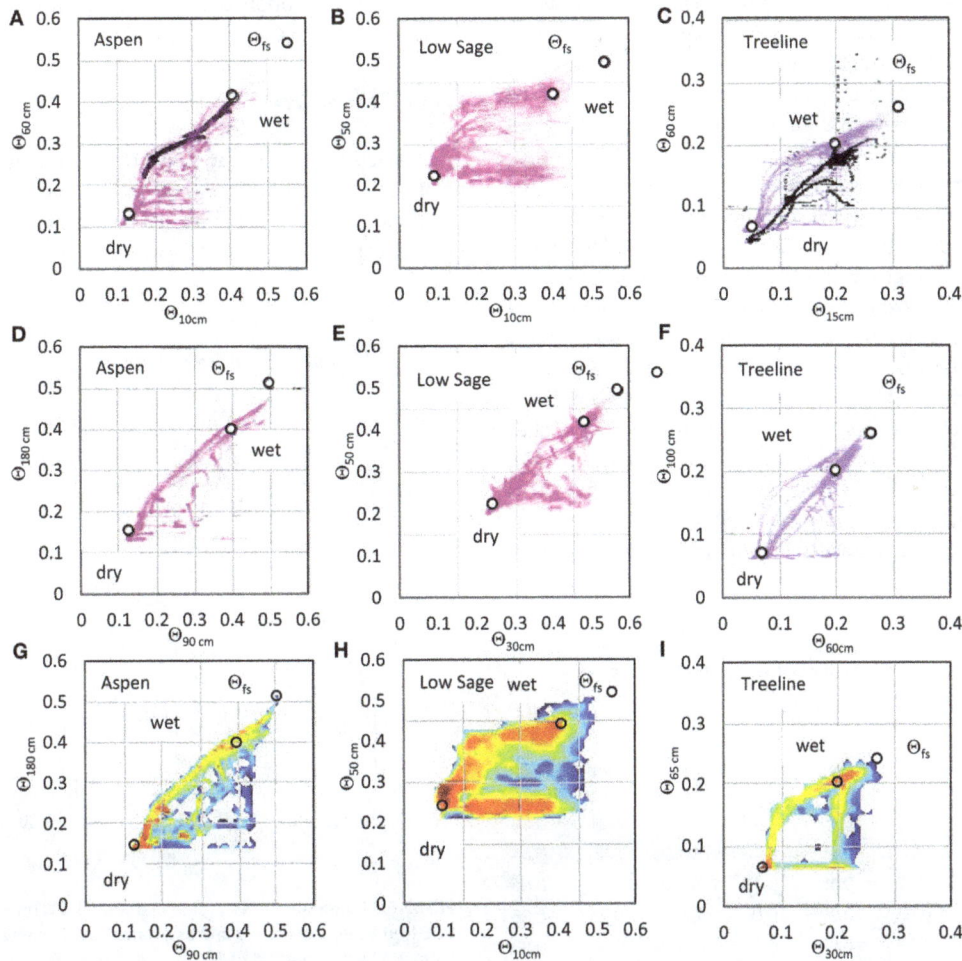

FIGURE 4 | Example plots of Θ data from sensors at different depths in one soil pit each at Aspen and Low Sage and for two pits at Treeline, using a range of sensor depth combinations. Data are shown with the deeper sensor on the y-axis to maintain a counterclockwise hysteresis for all plots. Data markers are semi-transparent to highlight high frequency areas by superposition. Circular data markers are placed at points of maximum frequency corresponding to PEL and FC and maximum values at Θfs. As in **Figure 3**, the highest frequency Θ data values are displayed as circular markers for the dry and wet attractors and Θfs. Frames **(B,D)** are shown as colormaps in frames **(G,H)**. Plot **(I)** at Treeline pit 4 is displayed for the 30 cm depth analysis due to a failed sensor in pit 3 to compare with pit 3 data in **(C,F)**. Additional data overlays of field tensiometer data **(A,C)** support later discussion.

a commonly used pedotransfer function (Saxton and Rawls, 2006).

The inherent uncertainty in the determining representative attractor values is complicated by physical heterogeneity in soil properties and the somewhat subjective assignment of a single value from within a data attractor that may be weak, have multiple maxima (**Figure 4G**) or a wide value range (**Figure 4H**). **Figure 6** provides a summary of data attractor results from **Figure 5** with the addition of estimated point values of Θ_{ws}, determined as single values from *time series* analysis. The Θ_{ws} values ranged from 0.07 to 0.37 among sites, with the values from Treeline and Lower Weather ranging from 0.06 to 0.10, slightly less than or equal to the values determined by Smith et al. (2011) for sites in Dry Creek watershed. The Θ_{ws} values ranged from a seven to 38% of the difference between PEL and FC.

Relation of Inferred Values to Physical Measurements

Table 4 compares estimated values of wet and dry attractors to data from soil water retention curves, field tensiometry and an intensive soil irrigation, and drainage experiment. We found that the example dry and wet attractors for Aspen and Treeline sites are equivalent to FC and PEL, respectively. **Figure 7** shows the soil water retention curves for Aspen for samples from 1–3 cm (blue) and 41–43 cm (brown) and field tensiometer data at 60 cm (black). The wet attractor Θ values for both depths at Aspen (0.41, 0.41) nearly matched the Θ value (0.42, 0.42) at 10 kPa, commonly used as an approximation of FC. Similarly, the dry attractor Θ values (0.13, 0.13) for both depths at Aspen nearly matched the 1.5 MPa Θ values (0.12, 0.12). For Treeline at 60 cm the hysteresis method wet attractor Θ (0.21) compares well with the measured Θ (0.19) corresponding to a tensiometer value of 10 kPa. Although, not exhaustive, this evidence strongly supports the hypothesis that data attractors represent FC and PEL, and that other important soil hydrologic parameters can be determined from Θ records for sites with different soils, wetness and plant cover.

Additional inspection of **Figure 7** and results from the frequency analysis of Θ are presented here to clarify the interpretation of Θ_s, and Θ_{fs}. For the Aspen the laboratory saturated (0 kPa) Θ ranges from Θ_s of 0.64 at 1–3 cm to 0.50 at 41–43 cm depth. The *frequency* method estimated value of Θ_{fs} at 10 cm depth (0.55) falls between these values and is substantially less than Θ_s for the soil surface measurement. This difference is likely due to either a decrease in soil porosity between 2 and 10 cm depth or a difference between Θ at Θ_s and Θ_{fs}. At greater depth the difference between the *frequency* method value for Θ_{fs} at 60 cm (0.51) and the value at for 0 kPa at 41–43 cm (0.50) is negligible, indicating Θ_{fs} is equivalent to Θ_s at these depths. For sandy soils at Treeline, constant irrigation achieved Θ_s at 60 cm

($\Theta = 0.39$) and upon drainage, the air entry value (0.36) was similar to Θ_{fs} (0.34) as determined by the *frequency* method.

DISCUSSION

The objectives of this paper are to evaluate four approaches to determine if temporal patterns of measured soil water values are consistent with the concepts of Θ_s, FC, PEL, and Θ_{ws}. The method exploits the increasing availability of electronic soil water

TABLE 3 | Period of study maximum and minimum annual values of wet and dry data attractors (Θ) and Θ_{fs} for *hysteresis* analysis at a range of depths (cm) at selected sites.

Site	Depth	Wet min	Wet max	Dry min	Dry max	Θ_{fs} min	Θ_{fs} max
Aspen	10	0.32	0.45	0.12	0.15	0.51	0.55
Aspen	90	0.37	0.45	0.12	0.15	0.45	0.54
Aspen	180	0.37	0.45	0.15	0.15	0.42	0.52
Low sage	10	0.25	0.27	0.05	0.08	0.34	0.35
Low sage	30	0.25	0.32	0.07	0.17	0.32	0.38
Low sage	50	0.27	0.32	0.08	0.19	0.34	0.37
Treeline	15	0.15	0.17	0.03	0.03	0.22	0.24
Treeline	30	0.15	0.20	0.03	0.08	0.25	0.28
Treeline	65	0.20	0.22	0.06	0.07	0.22	0.26

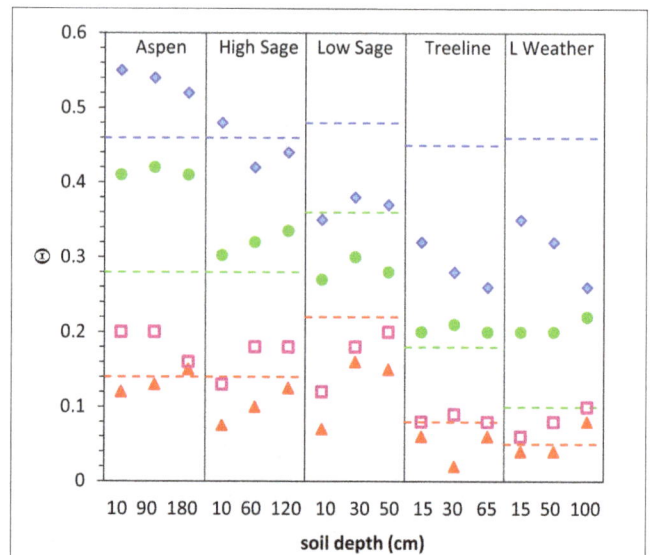

FIGURE 6 | Summary of Θ_{fs} (blue diamonds), FC (green circles), and PEL (red triangles) values and estimated point values of Θ_{ws}, (violet squares) by the method of Smith et al. (2011). Results are shown for representative depths at each site and compared to Θ values estimated by a pedotransfer function based on soil texture for 0 kPa (blue dashed line), 33 kPa (green dashed line), and 1500 kPa (red dashed line) by Saxton and Rawls (2006).

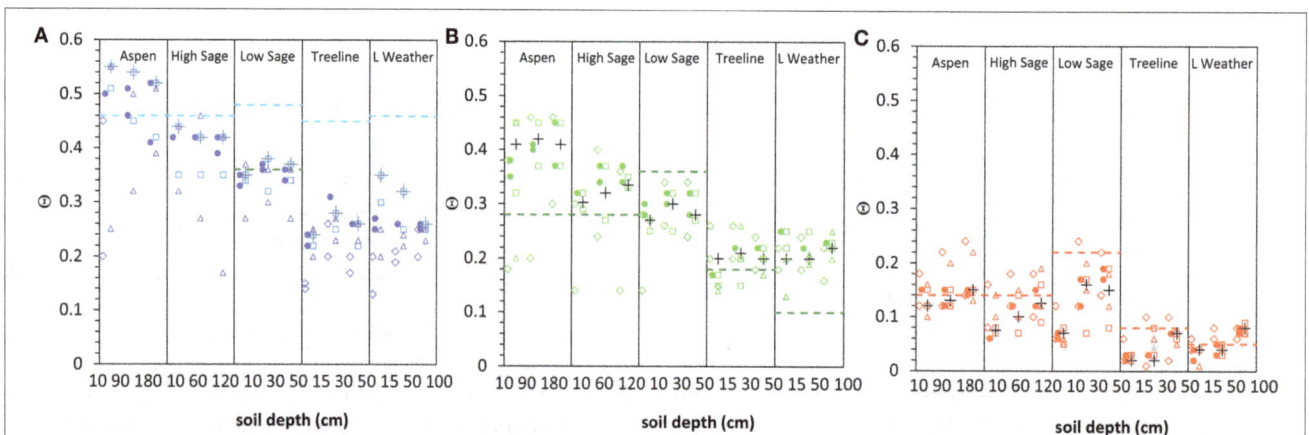

FIGURE 5 | Range of values for Θ_{fs} (A), wet attractor **(B)**, and dry attractor **(C)** for three soil depths at the study sites. Data are presented as maxima and minima of all estimates for time series (triangle), frequency (diamond), paired (circle) and hysteresis (square) analyses. Representative estimated values (cross) are shown as point values and Θ estimates at 0 kPa, 33 kPa and 1.5 MPa (dashed lines) for the various soil textures by a commonly used pedotransfer function (Saxton and Rawls, 2006) are provided for context.

TABLE 4 | Comparison of estimated values of soil hydrologic parameters Θ_{fs}, FC, and PEL from Θ records by frequency, paired sensors, and hysteresis methods to Θ-values corresponding to common soil water potentials used in laboratory experiments.

Site	Depth (cm)	Value	Frequency (Θ)	Pair (Θ)	Hysteresis (Θ)	0 (kPa)	10 (kPa)	33 (kPa)	1.5 (MPa)
Aspen	3–10	Θ_{fs}	**0.55**	0.50	0.53	0.64			
Aspen	40–60	Θ_{fs}	**0.51**	0.49	0.50	**0.50**			
Treeline	60	Θ_{fs}	**0.34**	0.26	0.25	0.36–0.39			
Aspen	3–10	FC	**0.28**	0.37	0.41		0.42	0.26	
Aspen	40–60	FC	0.32	0.40	0.41		0.49	0.34	
Treeline	60	FC	0.21	0.21	0.21		**0.19**	0.13	
Aspen	3–10	PEL	0.15	0.15	**0.13**			0.26	**0.12**
Aspen	40–60	PEL	0.15	0.13	**0.13**			0.34	**0.12**
Treeline	60	PEL	0.07	0.07	**0.07**			0.13	n/a

Bold type is used to indicate the closest match in between field determined and laboratory determined values of Θ_{fs}, FC, and PEL.

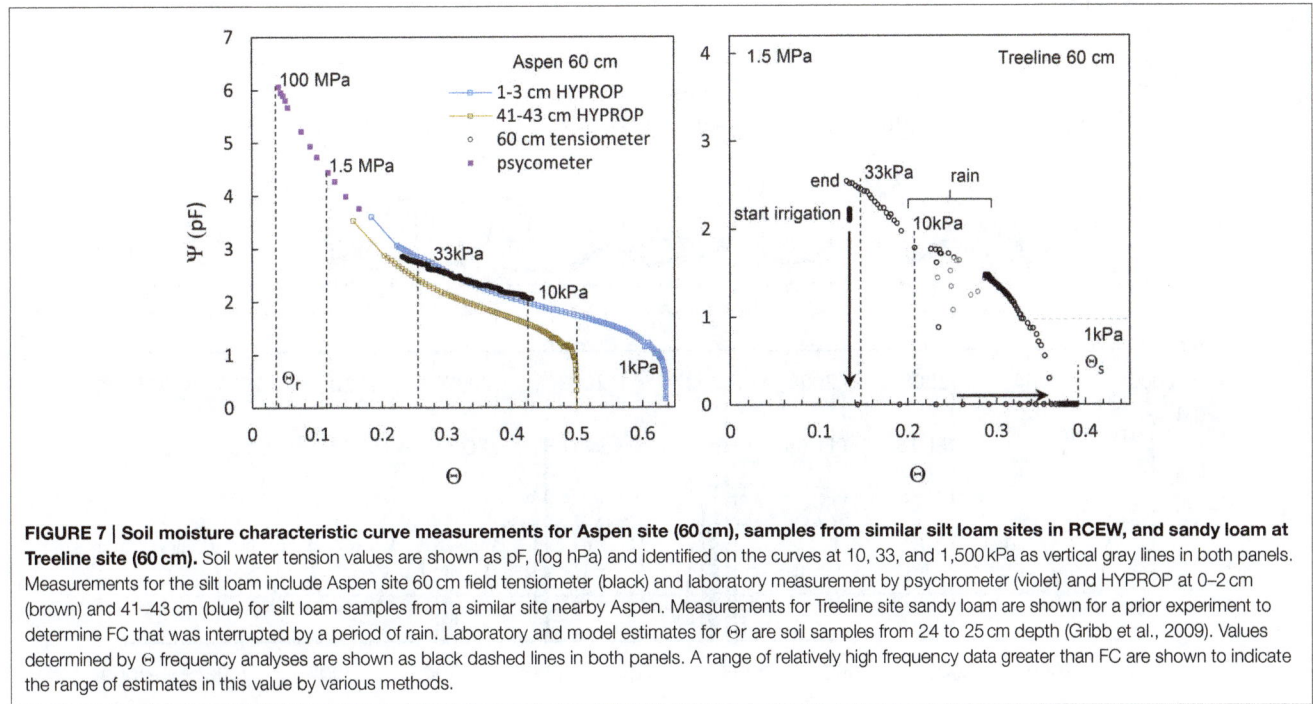

FIGURE 7 | Soil moisture characteristic curve measurements for Aspen site (60 cm), samples from similar silt loam sites in RCEW, and sandy loam at Treeline site (60 cm). Soil water tension values are shown as pF, (log hPa) and identified on the curves at 10, 33, and 1,500 kPa as vertical gray lines in both panels. Measurements for the silt loam include Aspen site 60 cm field tensiometer (black) and laboratory measurement by psychrometer (violet) and HYPROP at 0–2 cm (brown) and 41–43 cm (blue) for silt loam samples from a similar site nearby Aspen. Measurements for Treeline site sandy loam are shown for a prior experiment to determine FC that was interrupted by a period of rain. Laboratory and model estimates for Θr are soil samples from 24 to 25 cm depth (Gribb et al., 2009). Values determined by Θ frequency analyses are shown as black dashed lines in both panels. A range of relatively high frequency data greater than FC are shown to indicate the range of estimates in this value by various methods.

data to address the considerable controversy, both theoretical and conceptual, regarding soil water content constants. We found good agreement between soil hydrologic parameter values determined from attractors found in *in situ* Θ sensor records and from conventional laboratory techniques, for a limited number of samples representing sandy loam and silt loam soils (**Figure 7**). These results support the hypothesis that data attractors represent FC and PEL, and that other important soil hydrologic parameters can be determined from Θ records for sites with different soils, wetness, and plant cover. We demonstrate the results of the analyses by overlaying the representative parameter values on the time series data from which the values were derived (**Figure 8**) to support discussion of the determination of values for various soil hydrologic properties, applicability of this approach to other sites and caveats for use where frozen soils occur. In the following sections we comment

on the value of the approaches for estimating each property (Θ_s, FC, PEL, and Θ_{ws}), followed by a discussion of uncertainty and errors.

Saturation (Θ_s)

Saturation was uncommon in our data. All of the study sites are well-drained and did not show evidence of a water table. Therefore, it is not surprising that we found no attractor around saturation. However, near surface soil pipes and macropores may intermittently fill and appear as data outliers. This may explain why the maximum measured values of Θ can be quite different from the pedotransfer function estimates of saturation, especially near the soil surface. Results from our analyses for Aspen and Treeline sites show maximum Θ (Θ_{fs}) values of 0.55, 0.51, and 0.34 (**Table 4**) from controlled experiments and values of 0.64, 0.50, and 0.39 from field measurements. The difference in peak

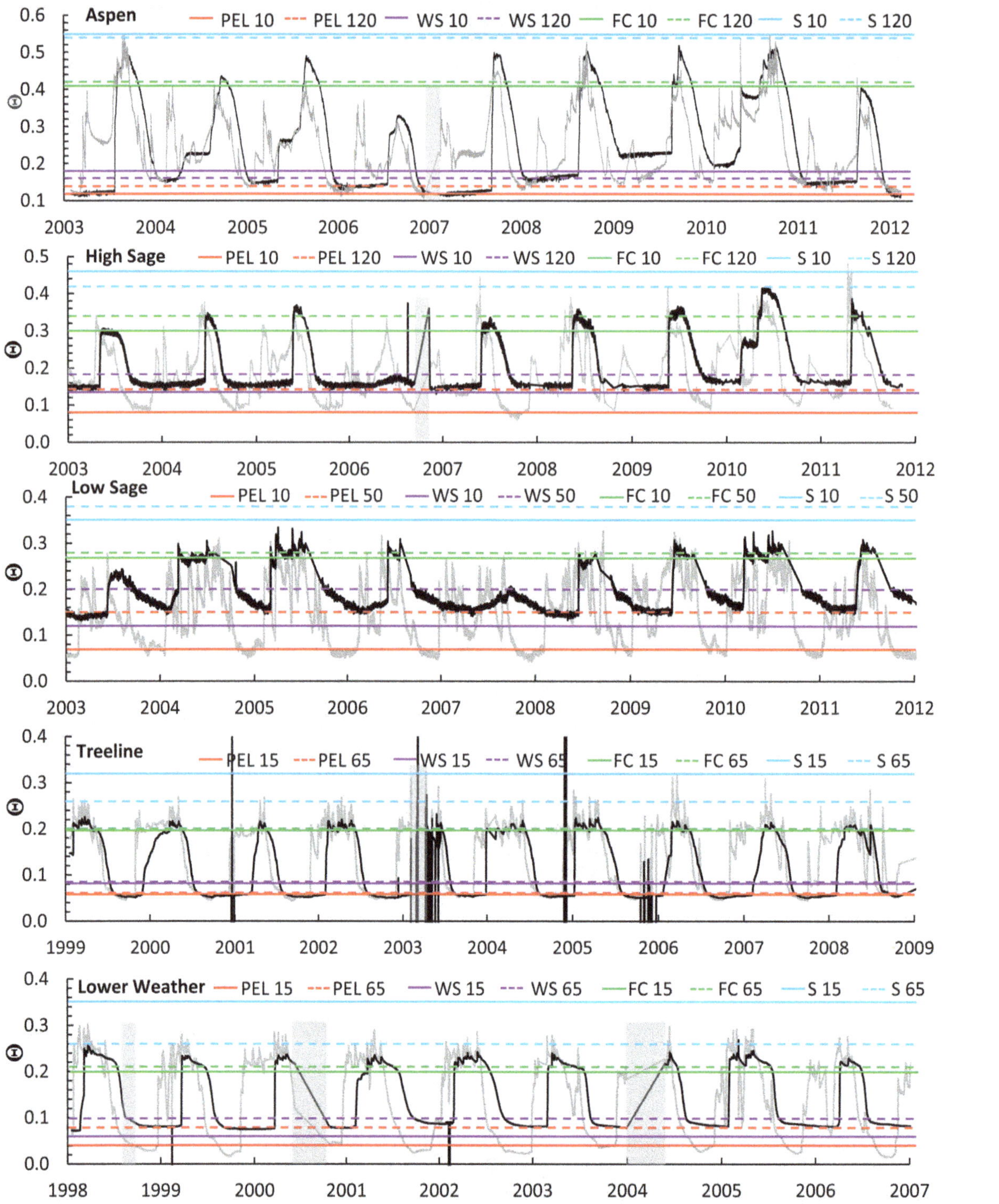

FIGURE 8 | Time series data and parameter values determined by frequency analysis for Θ_{fs} (blue lines), hysteresis analysis for FC (green lines), and PEL (red lines), and time series analysis for Θ_{ws} (violet lines). Depth of measurements are indicated in each panel for the shallow (gray) and deep (black) sensor locations.

saturation for shallow soil at Aspen may arise from differences in porosity between the surface 3 cm and at 10 cm depth, or the greater capacity for occluded soil gas below the very dense mat of fine roots within above the mineral soil. We note that

complete saturation of soils requires extensive irrigation, either in the field (Ellis and Lee, 1919; Hillel et al., 1972) or laboratory (Hazen, 1892; Watson, 1966) to ensure near complete removal of gas from the soil pores. Analysis of both laboratory and field measurements should be repeated for multiple depths in the soil profile.

At most sites, field saturation, Θ_{fs}, is a transient state achieved during infrequent periods of high flux and can range from FC to Θ_s in annual records (**Figure 8**). Assuming a consistent pore size distribution over time, it is logical to select the greatest recorded *in situ* value as Θ_{fs} either as a tail of the frequency analysis over the period of record. This selection requires judgment to determine if the extreme value is a valid measurement point or an artifact of measurement. This is likely an appropriate proxy value (**Figure 5A**) for most modeling purposes as long as the site remains freely draining. Θ_{fs} decreases with depth for all sites except Low Sage (which had a strong textural contrast), presumably in response to decreased porosity, decreased organic matter increased occluded volume by gravel or attenuation of flux by storage. These effects are lumped by the approach presented here. Due to the dependence of Θ on flux for well-drained sites, determination of Θ_{fs} is likely improved with longer data series and shorter time step data, especially for semi-arid or drier sites.

Field Capacity (FC)

The most consistent representation of FC often corresponded with the intersection of the wetting and drying limbs of the hysteresis loops (**Figure 2D**) and near the low edge of the wet data attractor in the paired analysis, as shown for Aspen and Treeline (**Figure 3**). These constraints on selection of FC recognize that this attractor requires that soil wetting is followed by a period of several days with no evaporation. Therefore, FC is most evident after cool season precipitation events, which are dominant in the study environment. During snowmelt, radiation and temperature patterns drive diel fluctuations of melt water flux to the soil that can to lead to positive bias in estimation of FC.

We found very similar values for FC at all depths within Aspen, Treeline, and Lower Weather but variability in FC with depth at High Sage and Low Sage (**Figure 6**). Although, texture is a first-order control on FC, the values from our analyses were often quite different from the pedotransfer function predictions, and even the same texture: FC at High Sage ranges from 74 to 82% of FC at Aspen for the surface and deep sensors, respectively, which is remarkable, given the close proximity and nearly identical soils at these sites. The difference in attractor values between Aspen and High Sage is due to the different soil water input environments. Aspen site is located beneath a seasonal snow drift that is typically 3 m deep. This results in extended periods during which daily water inputs from snowmelt maintain soil water contents at values greater than FC. Winter snow cover at High Sage site is typically 50 cm. It therefore melts much earlier and is subject to periodic cool season rainfall events that allow for drainage to FC, thus shifting the attractor downward to a value more consistent with the concept of FC. The pedotransfer function predictions appear reasonable (with hindsight) for sites other than Aspen (**Figure 6**), but appear to correspond better to Θ_{fs} values presented for Low Sage and Treeline. These differences in representation of the mobile water fraction in soil are likely attributable to soil plant interactions such as litter

cover and secondary soil structure and are important for careful parameterization for ecohydrological models.

Plant Extraction Limit (PEL)

PEL, like Θ_{ws} is an ecohydrologic parameter that represents an interaction among plant phenology, the surface energy balance, and soil water potential. For this study, the land surface cover is dominated by perennial vegetation with varying extent of litter and bare soil across sites, soil water input is primarily derived from spring snowmelt resulting in an extended dry growing season, and peak PET is in July (Seyfried et al., 2011). These conditions, in conjunction with the low mean annual precipitation outside of snow drifts all facilitate very dry soil states required for development of an attractor at PEL. The dry attractor values in this study were generally slightly less than those predicted by the pedotransfer function, and converse from Θ_{fs}, tended to increase with depth.

Our initial premise was that PEL at any depth is represented by a constant value of Θ for an extended period during the growing season, as shown for Treeline and Low Sage (**Figure 8**). For other sites, PEL may not be consistently represented across years for either the shallow or deep sensors for two reasons: First, Aspen is often energy limited and retains water at depth in most years, only achieving PEL in 2003, 2007, and 2012 (**Figure 8**). Similarly, other sites with insufficient energy to evaporate the mean annual precipitation are unlikely to reach PEL at depth. Second, Θ less than PEL commonly occurs in near surface soils. Records from Treeline and Lower Weather show Θ as low as 0.01 for sensors at 5 cm depth, but a PEL of 0.06 to 0.08 at depth (**Figure 8**). Evaporative drying near the soil surface can decrease Θ below PEL. In this case, when the evaporative potential at the leaf surface is balanced or exceeded by soil water tension, plant water use is negligible and vapor diffusion governs evaporation (Buckingham, 1907; Salvucci, 1997). This transition from viscous capillary (S1) to vapor diffusion (S2) control on evaporation (Brutsaert and Chen, 1995; Lehmann et al., 2008) is apparent in data, and complicates the interpretation of the dry attractor and PEL.

Thus, interpretation of the range of the dry attractor requires consideration, as with the wet attractor. Lower Weather receives the greatest energy and least precipitation among the study sites, and has coarse textured soils with a nearly bare surface, making it a good case study for dry soils. Initial estimates of the dry attractor Θ range from 0.04 at 15 cm to 0.08 at 100 cm (**Figure 8**). The much lower values of the dry attractor for 5 cm sensors at $0.00 < \Theta < 0.02$ (**Figure 8**) likely represents the limit of evaporation via vapor diffusion near the surface, for the available seasonal energy. We found a gradient $0.00 < \Theta < 0.08$ develops from 5 to 100 cm depth in response to the vapor diffusion gradient above PEL at 100 cm depth. These observations lead to the conclusion that near the soil surface, the shift from evapotranspiration to evaporation below PEL is difficult to distinguish using a data attractor.

Plant Water Stress (Θ_{ws})

Important features of soil drying below FC include the initiation of plant water stress, Θ_{ws}, the endpoint of plant water use, PEL, and the transition to evaporative drying by vapor diffusion. We found that the semi-arid climate of the study sites provided ideal conditions for soil drying through spring and summer, as shown by a nearly constant drying trend over approximately the middle half of the range in Θ for all sites (except Low Sage). Yet, a notable difference is evident between drying patterns for shallow and deep sensor locations: Soils at depths of 15 cm or less dried first and fastest at near steady state from below FC to near PEL (**Figure 8**). For soils below 15 cm the drying function $\Theta(t)$ was quasi-sinusoidal between FC and PEL, with a shift to faster drying when the shallow sensor reached Θ_{ws}. We attribute this initial increase in the rate of deep soil drying to consistent (energy-limited) plant water use from a smaller storage volume following surface drying below Θ_{ws}. For all sites, Θ_{ws} occurred annually for the shallow sensor depths, but was less consistent for the deep sensors. For example, Θ records from the deep sensors Aspen and High Sage regularly show linear declines until approximately Oct 1, when the water year ends with the onset of freezing nighttime air temperatures and abrupt termination of plant water use at these high elevation sites, e.g., WY 2008–2010 (**Figure 8**). Thus, as Θ_s may not be consistently achieved due to surface infiltration rates that seldom exceed soil profile drainage flux, Θ_{ws} may not be consistently achieved due to insufficient seasonal energy. Yet, the occasional dry years (e.g., WY 2006) at Aspen and High Sage are clearly water limited and are sufficient to estimate Θ_{ws} for the high elevation sites. Low Sage and the lower, elevation Treeline and Lower Weather sites are annually water limited and show consistent inflections indicating Θ_{ws} during July and August with values ranging from 0.20 for the deep clay soils at Low Sage to 0.07 for deep soils at Treeline (**Figures 6, 8**).

Uncertainty and Errors

One approach to the uncertainty in evaluating contemporaneous soil moisture states is to visualize $\Theta(z)$ at a daily to weekly time step. **Figure 9** shows $\Theta(z)$ for wetting and drying periods over one water year at Aspen and Treeline. Although, this approach is somewhat difficult to implement over multiple years, it provides insight on the gradation of soil hydrologic properties with depth and the different dynamics of wetting and drying across sites. This type of diagram is common in texts, but typically focuses attention on the dynamics of wetting and drying, rather than the predominant soil moisture states. For both sites, a vertical profile at FC emerges after the wetting front arrives at the base of the soil. Yet $\Theta(z)$ may not reach Θ_{fs}, as determined from the long term record for any depth in a given year. During soil drying, similar increasing gradients in $\Theta(z)$ emerge slightly below FC at both sites in response to plant water use, and indicate a rapid decline in drying associated with Θ_{ws}, near PEL. The much sharper inflection in $\Theta(z)$ near the soil surface at Treeline than Aspen reflects the difference in surface evaporation between sites. Finally, the much broader band of overlap among $\Theta(z)$ profiles at Treeline reflects the longer period and greater depth of plant water stress at that site.

Each approach to determining soil hydrologic properties has a different bias, but all are in general agreement. The applicability of measured data for this model parameterization depends on the nature of soil water dynamics under consideration. Selection of

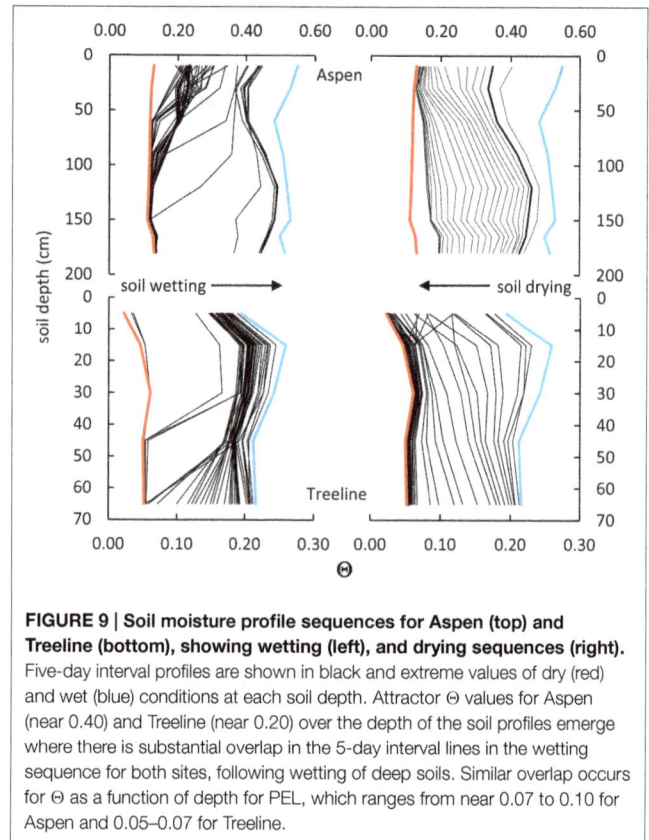

FIGURE 9 | Soil moisture profile sequences for Aspen (top) and Treeline (bottom), showing wetting (left), and drying sequences (right). Five-day interval profiles are shown in black and extreme values of dry (red) and wet (blue) conditions at each soil depth. Attractor Θ values for Aspen (near 0.40) and Treeline (near 0.20) over the depth of the soil profiles emerge where there is substantial overlap in the 5-day interval lines in the wetting sequence for both sites, following wetting of deep soils. Similar overlap occurs for Θ as a function of depth for PEL, which ranges from near 0.07 to 0.10 for Aspen and 0.05–0.07 for Treeline.

specific values for the wet and dry attractors by *time series* analysis is more subjective than by *frequency* analysis, as it depends on selection of a single value from a range of Θ near the extreme values. The two depth *hysteresis* method provides estimates that closely match the Θ at standard soil water potential values commonly used to define FC and PEL. The improvement over the other methods for determining these wet and dry attractors is related to the reduced overlap between valid data, and the invalid data and data errors which can skew the frequency distribution of the attractor. In particular, the juxtaposition of Θ for a sensor depth (a) with stable Θ to a sensor at a different depth (b) with either increasing or decreasing Θ develops a linear trace toward a point of inflection where the rate of change in sensor a increases and sensor b decreases. The relatively slow rate of change near the inflection enhances the frequency of (a,b) data at the attractor. Error values tend to fall at the margins of the hysteresis loop which reduces error distortion of the main attractors (e.g., **Figure 4**). Unlike the hysteresis analysis, the *paired sensor* approach can also be biased by any temporal lag in wetting front depth between the sensors.

Soil depth and cover effects are clearly represented in the presented analyses. We found the sites with very well drained soils (Treeline, Lower Weather) showed less uncertainty in Θ_{fs} and FC near the bottom of the soil profile (**Figure 5**), likely as a result of faster transit times and less retention. Similarly, the control of surface cover on evaporation is clearly represented in general across sites. PEL is not strictly a soil property, but also depends on vegetation where roots are active. The soil at High Sage has the same texture as Aspen, but the surface cover is dense shrub canopy 0.5–1.5 m over with 2–5 cm of litter and bare interspace and associated decreasing values of PEL at 60 cm (0.10) and 10 cm (0.08). This cover appears to effectively limit surface evaporation. Low Sage has a soil textural gradient from silt loam to clay loam and a bare soil surface and results in the strongest gradient in PEL (0.07–0.14) over the least depth of soil, 50 cm. Treeline and Lower Weather are both sandy sites with shrub cover and low bulk density surface, which supports extensive drying. Difference in PEL following vegetation removal by prescribed fire in 2007 are clear at the Aspen and High Sage sites.

There are two primary effects. First, electronic soil water sensors are based on relating measured permittivity to liquid water content, which is generally a robust relationship because the permittivity of liquid water (80) is so much greater than soil (5) or air (1). However, the permittivity of ice (3.1) is very similar to that of air, so frozen soil appears to have less water content and effectively approximates the liquid water content of the soil (Seyfried and Murdock, 1996). This effect is apparent when dramatic "drying" events during the winter appear as noise and stray from the FC attractor. Freezing effects are apparent in all four panels in **Figure 2** and can be clearly seen by comparison of the soil moisture trends at 15 cm and 30 cm in **Figure 9**. Second, freezing effectively reduces the soil water potential. When the near-surface soil freezes, the normal downward hydraulic gradient is reversed and water moves upward toward the freezing front. This can cause either an increase or decrease in the measured soil water content, depending on the proximity of the sensor to the freezing front (Hillel, 1980). These are most pronounced near the soil surface, which commonly freezes in this environment. In this study the freezing front rarely exceeded 30 cm. There is evidence of occasional upward gradient effects apparent deeper in the soil profile. These artifacts can introduce erroneous data into the analysis. This is more problematic for the *frequency* and *paired* analyses, in which the surface sensors are likely to be similarly affected, but less important for the *hysteresis* analysis. We suggest that soil temperature records, which are commonly collected along with TDR or other soil moisture measurements, could be used to censor data during periods of soil frost prior to conducting the presented analyses.

CONCLUSIONS

We demonstrate a novel approach to extract important soil hydrologic parameters directly from field data. The approach takes advantage of the increasing availability of continuously monitored Θ. Because it is based on *in-situ* data, results are directly related to the local climate, soils and vegetation and does not rely on assumed pedotransfer functions or proxy measurements such as Ψ. The approach builds primarily on concepts of Θ_s, FC, and PEL inherited from irrigated agriculture, but regards these quantities in terms of the amount of water retained in the soil, rather than in terms of soil water potentials. This approach is consistent with mass-based modeling schemes

and provides detail for estimation of other additional transitional and extreme states Θ_{fs}, Θ_{ws}. However, the applicability of measured data for this model parameterization depends on the nature of soil water dynamics under consideration. The implicit assumptions are that drainage from Θ_s or Θ_{fs} to FC is relatively fast in the absence of soil water input, and below that threshold 1D unsaturated flow responds primarily to plant water uptake, which decreases progressively below Θ_{ws} and ceases at PEL. The endpoint of drying in any annual cycle may range from a value greater than PEL to Θ_r, depending on plant type and soil texture and depth. These assumptions are supported by observations and data from the sites in this study, which indicate a downward wetting front progression for every major rainfall or snow melt and the regular and extensive periods of soil wetting and drying. Application of the developed approach in more humid climates, or in the presence of a persistent phreatic surface is expected to improve estimation of Θ_s, but may provide little guidance on PEL or Θ_{ws}. Preferential flow is not specifically addressed, but the non-unity slopes in the paired analysis indicate that variable rates of wetting front advance can complicate interpretation of this approach. Similarly, vertical bypass flow in individual sensor profiles may complicate interpretation of FC and Θ_{ws}, but not PEL.

We found that the frequency of measured values tended to be greatest around observable "attractors" that correspond to the specific hydrologic parameters of interest. Of the four approaches analyzed, we found the hysteresis analysis approach is the most robust predictor of data attractors and frequency analysis is the simplest approach to determine the extreme values. Nevertheless, we consider the construction of a two dimensional Θ frequency matrix as the most effective and practical single approach to parameter value identification. This approach enables analysis from a single profile of sensors and provides the greatest visual support to estimate parameter values. The data and analyses presented here clearly demonstrate how near surface Θ often differs from the deeper soil at many sites, due to differences in macroporosity, soil freezing, hysteresis in wetting. These controls on soil profile storage may confound attempts to measure soil profile storage by remote sensing methods, and opportunities to overcome some of these obstacles will be addressed in a successive paper.

AUTHOR CONTRIBUTIONS

DC: Conceived research concept and analyses, conducted fieldwork, was primary author. MS: Contributed to research concept, conducted fieldwork and laboratory experiments, was second author. JM: Contributed to research concept, conducted fieldwork, contributing author. KH: Conceived research concept and analyses.

ACKNOWLEDGMENTS

The authors would like to thank two reviewers who provided extensive and valuable comments to the manuscript and finding from USDA NRI Grant 2001-35102-11031, USDA SGP award 2005-34552-15828.

REFERENCES

Assouline, S., and Or, D. (2014). The concept of field capacity revisited: defining intrinsic static and dynamic criteria for soil internal drainage dynamics. *Water Resour. Res.* 50, 4787–4802. doi: 10.1002/2014WR015475

Bandara, R., Walker, J. P., and Rüdiger, C. (2013). Towards soil property retrieval from space: a one-dimensional twin-experiment. *J. Hydrol.* 497, 198–207. doi: 10.1016/j.jhydrol.2013.06.004

Boote, K. J., Sau, F., Hoogenboom, G., and Jones, J. W. (2008). "Experience with water balance, evapotranspiration, and predictions of water stress effects in the CROPGRO model," in *Response of Crops to Limited Water: Understanding and Modeling Water Stress Effect on Plant Growth Processes*, eds R. Ahuja, V. R. Reddy, S. A. Saseendran, and Q. Yu (Madison, WI: American Society of Agronomy), 59–103.

Briggs, L. J., and McLane, J. W. (1907). The moisture equivalent of soils. *U.S. Dept. Agr. Bur. Soils Bul.* 46, 1–23.

Briggs, L. J., and Shantz, H. L. (1912). The wilting coefficient for different plants and its indirect determination. *Bot. Gaz.* 53, 20–37. doi: 10.5962/bhl.title.64958

Brooks, R. H., and Corey, A. T. (1964). Hydraulic properties of porous media and their relation to drainage design. *Trans. ASAE* 7, 26–28. doi: 10.13031/2013.40684

Brutsaert, W., and Chen, D. (1995). Desorption and the two stages of drying of natural tallgrass prairie. *Water Resour. Res.* 315, 1305–1313. doi: 10.1029/95WR00323

Buckingham, E. (1907). *Studies on the Movement of Soil Moisture, Bureau of Soils, Bulletin 38*. Washington, DC: U.S. Department of Agriculture.

Budyko, M. I. (1956). *Heat Balance of the Earth's Surface*. Leningrad: Gidrometeoizdat.

Burdine, J. A., and Redford, E. S. (1952). Selected articles and documents on american government and politics. *Am. Polit. Sci. Rev.* 46, 254–257. doi: 10.1017/S0003055400069239

Cassel, D. K., and Nielsen, D. R. (1986). "Field capacity and available water capacity," in *Methods of Soil Analysis: Part 1—Physical and Mineralogical Methods*, ed A. Klute (Madison, WI: American Society of Agronomy and Soil Science Society of America), 901–926.

Chandler, D. G., Seyfried, M., Murdock, M., and McNamara, J. P. (2004). Field calibration of water content reflectometers. *Soil Sci. Soc. Am. J.* 68, 1501–1507. doi: 10.2136/sssaj2004.1501

Chauvin, G. M., Flerchinger, G. N., Link, T. E., Marks, D., Winstral, A. H., and Seyfried, M. S. (2011). Long-term water balance and conceptual model of a semi-arid mountainous catchment. *J. Hydrol.* 400, 133–143. doi: 10.1016/j.jhydrol.2011.01.031

Colman, E. A. (1947). A laboratory procedure for determining the field capacity of soils. *Soil Sci.* 63, 277–283. doi: 10.1097/00010694-194704000-00003

Cuenca, R. H., Ek, M., and Mahrt, L. (1996). Impact of soil water property parameterization on atmospheric boundary layer simulation. *J. Geophys. Res. Atmos.* 101, 7269–7277. doi: 10.1029/95JD02413

Damesin, C., and Rambal, S. (1995). Field study of leaf photosynthetic performance by a Mediterranean deciduous oak tree (*Quercus pubescens*) during a severe summer drought. *New Phytol.* 131, 159–167. doi: 10.1111/j.1469-8137.1995.tb05717.x

Ellis, A. J., and Lee, E. H. (1919). *Geology and Ground Waters of the Western Part of San Diego County, California*. Water Supply Paper, U.S. Geology Survey, 23–24.

Fayer, M. J., and Hillel, D. (1986). Air encapsulation: I. Measurement in a field soil. *Soil Sci. Soc. Am. J.* 50, 568–572 doi: 10.2136/sssaj1986.03615995005000030005x

Flerchinger, G. N., Cooley, K. R., Hanson, C. L., and Seyfried, M. S. (1998). A uniform versus an aggregated water balance of a semi-arid watershed. *Hydrol. Process.* 12, 331–342. doi: 10.1002/(SICI)1099-1085(199802)12:2<331::AID-HYP580>3.0.CO;2-E

Flerchinger, G. N., Marks, D., Reba, M. L., Yu, Q., and Seyfried, M. S. (2010). Surface fluxes and water balance of spatially varying vegetation within a small mountainous headwater catchment. *Hydrol. Earth Syst. Sci.* 14, 965–978. doi: 10.5194/hess-14-965-2010

Grayson, R. B., Western, A. W., Walker, J. P., Kandel, D. G., Costelloe, J. F., and Wilson, D. J. (2006). "Controls on patterns of soil moisture in arid and semi-arid systems," in *Dryland Ecohydrology*, eds P. D'Odorico and A. Porporato (Dordrecht: Springer), 109–127.

Gribb, M. M., Forkutsa, I., Hansen, A., Chandler, D. G., and McNamara, J. P. (2009). The effect of various soil hydraulic property estimates on soil moisture simulations. *Vadose Zone J.* 8, 321–331. doi: 10.2136/vzj2008.0088

Hansen, V. E., Israelsen, O. W., and Stringham, G. E. (1980). *Irrigation Principles and Practices, 4th Edn.* New York, NY: Wiley.

Hazen, A. (1892). *Experiments upon the Purification of Sewage and Water at the Lawrence Experiment Station*. Massachusetts State Board Health, Twenty-third Annual Report for 1891, Wright & Potter, Printing Co.

Hillel, D. (1980). *Applications of Soil Physics*. New York, NY: Academic Press.

Hillel, D., Krentos, V. D., and Stylianou, Y. (1972). Procedure and test of an internal drainage method for measuring soil hydraulic characteristics *in situ*. *Soil Sci.* 114, 395–400. doi: 10.1097/00010694-197211000-00011

Huang, X., Shi, Z. H., Zhu, H. D., Zhang, H. Y., Ai, L., and Yin, W. (2016). Soil moisture dynamics within soil profiles and associated environmental controls. *Catena* 136, 189–196. doi: 10.1016/j.catena.2015.01.014

Idso, S. B., Reginato, R. J., Jackson, R. D., Kimball, B. A., Nakayama, F. S., et al. (1974). The three stages of drying of a field soil. *Soc. Am. Proc.* 38, 831–836.

Kirkham, M. B. (2005). *Principles of Soil and Plant Water Relations*. San Diego, CA: Academic Press.

Kormos, P. R., Marks, D., Williams, C. J., Marshall, H. P., Aishlin, P., and Chandler, D. G. (2014). Soil, snow, weather, and sub-surface storage data from a mountain catchment in the rain–snow transition zone. *Earth Syst. Sci. Data* 6, 165–173. doi: 10.5194/essd-6-165-2014

Koster, R. D., Dirmeyer, P. A., Guo, Z. C., Bonan, G., Chan, E., Cox, P., et al. (2004). Regions of strong coupling between soil moisture and precipitation. *Science* 305, 1138–1140. doi: 10.1126/science.1100217

Koster, R. D., Schubert, S. D., and Suarez, M. J. (2009). Analyzing the concurrence of meteorological droughts and warm periods, with implications for the determination of evaporative regime. *J. Clim.* 22, 3331–3341. doi: 10.1175/2008JCLI2718.1

Laio, F., Porporato, A., Ridolfi, L., and Rodriguez-Iturbe, I. (2001). Plants in water-controlled ecosystems: active role in hydrologic processes and response to water stress: II. Probabilistic soil moisture dynamics. *Adv. Water Resour.* 24, 707–723. doi: 10.1016/S0309-1708(01)00005-7

Lehmann, P., Assouline, S., and Or, D. (2008). Characteristics lengths affecting evaporative drying from porous media. *Phys. Rev. E* 77:056309. doi: 10.1103/PhysRevE.77.056309

Liang, X., Wood, E. F., and Lettenmaier, D. P. (1996). Surface soil moisture parameterization of the VIC-2L model: evaluation and modification. *Global Planet. Change* 13, 195–206. doi: 10.1016/0921-8181(95)00046-1

Linsley, R. K., and Franzini, J. B. (1972). *Water Resources Engineering*. New York, NY: McGraw-Hill.

Loik, M. E., Breshears, D. D., Laurenroth, W. K., and Belnap, J. (2004). A multi-scale perspective of water pulses in dryland ecosystems: climatology and ecohydrology of the western USA. *Oecologia* 141, 269–281. doi: 10.1007/s00442-004-1570-y

Marinas, M., Roy, J. W., and Smith, J. E. (2013). Changes in entrapped gas content and hydraulic conductivity with pressure. *Ground Water* 51, 41–50. doi: 10.1111/j.1745-6584.2012.00915.x

McNamara, J. P., Chandler, D., Seyfried, M., and Achet, S. (2005). Soil moisture states, lateral flow, and streamflow generation in a semi-arid, snowmelt-driven catchment. *Hydrol. Process.* 19, 4023–4038. doi: 10.1002/hyp.5869

Meinzer, O. E. (1923). *Outline of Ground-Water Hydrology, with Definitions*. United States Geological Survey, Water Supply Paper 494, p. 29.

Metzger, T., and Tsotsas, E. (2005). Influence of pore size distribution on drying kinetics: a simple capillary model. *Drying Technol.* 23, 1797–1809. doi: 10.1080/07373930500209830

Miller, R. D., and McMurdie, J. L. (1953). Field capacity in laboratory columns. *Soil Sci. Soc. Am. J.* 17, 191–195. doi: 10.2136/sssaj1953.03615995001700030003x

Miller, E. E., and Klute, A. (1967). "The dynamics of soil water: part I—mechanical forces," in *Irrigation of Agricultural Lands*, eds R. M. Hagan, H. R. Haise, and T. W. Edminster (Madison, WI: American Society of Agronomy), 209–244.

Montzka, C., Moradkhani, H., Weihermüller, L., Franssen, H.-J. H., Canty, M., and Vereecken, H. (2011). Hydraulic parameter estimation by remotely-sensed top soil moisture observations with the particle filter. *J. Hydrol.* 399, 410–421. doi: 10.1016/j.jhydrol.2011.01.020

Morse, J. L., Duran, J., Beall, F., Enanga, E. M., Creed, I. F., Fernandez, I., et al. (2015). Soil denitrification fluxes from three northeastern North

American forests across a range of nitrogen deposition. *Oecologia* 177, 17–27. doi: 10.1007/s00442-014-3117-1

Mualem, Y. (1976). *A Catalogue of the Hydraulic Properties of Unsaturated Soils*. Haifa: Technion, Israel Institute of Technology, Technion Research & Development Foundation.

Nayak, A., Chandler, D. G., Marks, D., McNamara, J. P., and Seyfried, M. (2008). Correction of electronic record for weighing bucket precipitation gauge measurements. *Water Resour. Res.* 44:W00D11. doi: 10.1029/2008WR006875

Nemes, A., Pachepsky, Y. A., and Timlin, D. J. (2011). Towards improving global estimates of field soil water capacity. *Soil Sci. Soc. Am. J.* 75, 807–812. doi: 10.2136/sssaj2010.0251

Njoku, E. G., Jackson, T. J., Lakshmi, V., Chan, T. K., and Nghiem, S. V. (2003). Soil moisture retrieval from AMSR-E. *IEEE Trans. Geosci. Remote Sens.* 41, 215–229. doi: 10.1109/TGRS.2002.808243

Or, D., Lehmann, P., and Assouline, S. (2015). Natural length scales define the range of applicability of the Richards equation for capillary flows. *Water Resour. Res.* 51, 7130–7144. doi: 10.1002/2015WR017034

Osakabe, Y., Osakabe, K., Shinozaki, K., and Tran, L.-S. P. (2014). Response of plants to water stress. *Front. Plant Sci.* 5:86. doi: 10.3389/fpls.2014.00086

Paniconi, C., and Putti, M. (2015). Physically based modeling in catchment hydrology at 50: survey and outlook. *Water Resour. Res.* 51, 7090–7129. doi: 10.1002/2015WR017780

Piper, J. R. (1933). Notes on the relation between the moisture-equivalent and the specific retention of water-bearing materials. *EOS Trans. AGU* 14, 481–487. doi: 10.1029/tr014i001p00481

Prasad, R., Tarboton, D. G., Liston, G. E., Luce, C. H., and Seyfried, M. S. (2001). Testing a blowing snow model against distributed snow measurements at Upper Sheep Creek, Idaho, United States of America. *Water Resour. Res.* 37, 1341–1356. doi: 10.1029/2000WR900317

Rambal, S., Ourcival, J. M., Joffre, R., Mouillot, F., Nouvellon, Y., Reichstein, M., et al. (2003). Drought controls over conductance and assimilation of a Mediterranean evergreen ecosystem: scaling from leaf to canopy. *Glob. Chang. Biol.* 9, 1813–1824. doi: 10.1111/j.1365-2486.2003.00687.x

Reichle, R. H., Koster, R. D., Liu, P., Mahanama, S. P. P., Njoku, E. G., and Owe, M. (2007). Comparison and assimilation of global soil moisture retrievals from the Advanced Microwave Scanning Radiometer for the Earth Observing System (AMSR-E) and the Scanning Multichannel Microwave Radiometer (SMMR). *J. Geophys. Res.* 112:D9. doi: 10.1029/2006jd008033

Reynolds, W. D., Elrick, D. E., Youngs, E. G., H., and Booltink, W. G., and, J., Bouma (2002). "Saturated and field saturated water fl ow parameters," in *Methods of Soil Analysis, Part 4. Physical Methods*, SSSA Book Series No. 5., eds J. H. Dane and G. C. Topp (Madison, WI: SSSA), 802–816.

Richards, L. A., and Weaver, L. R. (1944). Moisture retention by some irrigated soils as related to soil-moisture tension. *J. Agric. Res.* 69, 215–235.

Rodriguez-Iturbe, I., D'Odorico, P., Porporato, A., and Ridolfi, L. (1999). On the spatial and temporal links between vegetation, climate, and soil moisture. *Water Resour. Res.* 35, 3709–3722. doi: 10.1029/1999WR900255

Rodriguez-Iturbe, I., Porporato, A., Laio, F., and Ridolfi, L. (2001). Plants in water-controlled ecosystems: active role in hydrologic processes and response to water stress: I. Scope and general outline. *Adv. Water Res.* 24, 695–705. doi: 10.1016/S0309-1708(01)00004-5

Romano, N., and Santini, A. (2002). "Field water capacity," in *Methods of Soil Analysis: Part 4-Physical Methods*, eds J. H. Dane and G. C. Topp (Madison, WI: Soil Science Society of America), 722–738.

Salter, P. J., and Haworth, F. (1961). The available-water capacity of a sandy loam soil. 1. A critical comparison of methods of determining the moisture content of soil at field capacity and at the permanent wilting percentage, *J. Soil Sci.* 12, 326–334. doi: 10.1111/j.1365-2389.1961.tb00922.x

Salvucci, G. D. (1997). Soil and moisture independent estimation of stage-two evaporation from potential evaporation and albedo or surface temperature. *Water Resour. Res.* 33, 111–122. doi: 10.1029/96WR02858

Santanello, J. A., Peters-Lidard, C. D., Garcia, M. E., Mocko, D. M., Tischler, M. A., Moran, M. S., et al. (2007). Using remotely-sensed estimates of soil moisture to infer soil texture and hydraulic properties across a semi-arid watershed. *Remote Sens. Environ.* 110, 79–97. doi: 10.1016/j.rse.2007.02.007

Saxton, K. E., and Rawls, W. J. (2006). Soil water characteristic estimates by texture and organic matter for hydrologic solutions. *Soil Sci. Soc. Am. J.* 70:1569. doi: 10.2136/sssaj2005.0117

Scholes, R. J., and Walker, B. H. (1993). *An African Savanna*. New York, NY: Cambridge University Press.

Seneviratne, S. I., Corti, T., Davin, E. L., Hirschi, M., Jaeger, E. B., Lehner, I., et al. (2010). Investigating soil moisture–climate interactions in a changing climate: a review. *Earth Sci. Rev.* 99, 125–161. doi: 10.1016/j.earscirev.2010.02.004

Seyfried, M., Marks, D., and Chandler, D. G. (2011). Long-term soil water trends across a 1000 m elevation gradient. *Vadose Zone J.* 10, 1276–1286. doi: 10.2136/vzj2011.0014

Seyfried, M. S., and Murdock, M. D. (1996). Calibration of time domain reflectometry for measurement of liquid water in frozen soils. *Soil Sci.* 161, 87–98. doi: 10.1097/00010694-199602000-00002

Seyfried, M. S., Grant, L. E., Marks, D., Winstral, A., and McNamara, J. (2009). Simulated soil water storage effects on streamflow generation in a mountainous snowmelt environment, Idaho, USA. *Hydrol. Process.* 23, 858–873. doi: 10.1002/hyp.7211

Seyfried, M. S., Murdock, M. D., Hanson, C. L., Flerchinger, G. N., and Van Vactor, S. (2001). Long-term soil water content database, reynolds creek experimental watershed, Idaho, United States. *Water Resour. Res.* 37, 2847–2851. doi: 10.1029/2001WR000419

Shokri, N., Lehmann, P., Vontobel, P., and Or, D. (2008). Drying front and water content dynamics during evaporation from sand delineated by neutron radiography. *Water Resour. Res.* 44:W06418. doi: 10.1029/2007wr006385

Shokri, N., and Or, D. (2011). What determines drying rates at the onset of diffusion controlled stage-2 evaporation from porous media? *Water Resour. Res.* 47, W09513. doi: 10.1029/2010WR010284

Smith, T. J., McNamara, J. P., Flores, A. N., Gribb, M. M., Aishlin, P., and Benner, S. (2011). Small Soil storage capacity limits benefit of winter snowpack to upland vegetation. *Hydrol. Process.* 25, 3858–3865. doi: 10.1002/hyp.8340

Sperry, J. S., Hacke, U. G., Oren, R., and Comstock, J. P. (2002). Water deficits and hydraulic limits to leaf water supply. *Plant Cell Environ.* 25, 251–263. doi: 10.1046/j.0016-8025.2001.00799.x

Tesfa, T. K., Tarboton, D. G., Chandler, D. G., and McNamara, J. P. (2009). Modeling soil depth from topographic and land cover attributes. *Water Resour. Res.* 45:W10438. doi: 10.1029/2008WR007474

van Genuchten, M. T. (1980). A closed-form equation for predicting the hydraulic conductivity of unsaturated soils. *Soil Sci. Soc. Am. J.* 44, 892–898. doi: 10.2136/sssaj1980.03615995004400050002x

Veihmeyer, F. J., and Hendrickson, A. H. (1949). Methods of measuring field capacity and permanent wilting percentage of soils. *Soil Sci.* 68, 75–94. doi: 10.1097/00010694-194907000-00007

Vereecken, H., Huisman, J. A., Bogena, H., Vanderborght, J., Vrugt, J. A., and Hopmans, J. W. (2008). On the value of soil moisture measurements in vadose zone hydrology: a review. *Water Resour. Res.* 44:W00D06. doi: 10.1029/2008WR006829

Wang, Z., Feyen, J., Van Genuchten, M. T., and Nielsen, D. R. (1998). Air entrapment effects on infiltration rate and flow instability. *Water Resour. Res.* 34, 213–222. doi: 10.1029/97WR02804

Watson, K. K. (1966). An instantaneous profile method for determining the hydraulic conductivity of unsaturated porous materials. *Water Resour. Res.* 2, 709–715. doi: 10.1029/WR002i004p00709

Western, A. W., Zhou, S.-L., Grayson, R. B., McMahon, T. A., Blöschl, G., and Wilson, D. J. (2004). Spatial correlation of soil moisture in small catchments and its relationship to dominant spatial hydrological processes. *J. Hydrol.* 286, 113–134. doi: 10.1016/j.jhydrol.2003.09.014

Wigmosta, M. S., Vail, L., and Lettenmaier, D. P. (1994). A distributed hydrology-vegetation model for complex terrain. *Water Resour. Res.* 30, 1665–1679. doi: 10.1029/94WR00436

Wood, E. F., Lettenmaier, D. P., and Zartarian, V. G. (1992). A land-surface hydrology parameterization with subgrid variability for general circulation models. *J. Geophys. Res.* 97, 2717–2728. doi: 10.1029/91JD 01786

Conflict of Interest Statement: The authors declare that the research was conducted in the absence of any commercial or financial relationships that could be construed as a potential conflict of interest.

10

Disentangling Natural and Anthropogenic Signals in Lacustrine Records: An Example from the Ilan Plain, NE Taiwan

Jyh-Jaan Huang[1], Chih-An Huh[2†], Kuo-Yen Wei[1], Ludvig Löwemark[1*], Shu-Fen Lin[3], Wen-Hsuan Liao[4,5], Tien-Nan Yang[6], Sheng-Rong Song[1], Meng-Yang Lee[7], Chih-Chieh Su[8] and Teh-Quei Lee[2]

[1] Department of Geosciences, National Taiwan University, Taipei, Taiwan, [2] Institute of Earth Sciences, Academia Sinica, Taipei, Taiwan, [3] Institute of History and Philology, Academia Sinica, Taipei, Taiwan, [4] Research Center for Environmental Changes, Academia Sinica, Taipei, Taiwan, [5] Earth System Science Program, Taiwan International Graduate Program, Academia Sinica, Taipei, Taiwan, [6] Exploration and Development Research Institute, CPC Corporation, Miaoli, Taiwan, [7] Department of Earth and Life Science, University of Taipei, Taipei, Taiwan, [8] Institute of Oceanography, National Taiwan University, Taipei, Taiwan

Edited by:
Gary E. Stinchcomb,
Murray State University, USA

Reviewed by:
Li Wu,
Anhui Normal University, China
Maarten Blaauw,
Queen's University Belfast, UK
Samuel Munoz,
Woods Hole Oceanographic
Institution, USA

*Correspondence:
Ludvig Löwemark
loewemark@gmail.com

†Deceased, Jan. 21st, 2014.

Specialty section:
This article was submitted to
Quaternary Science, Geomorphology
and Paleoenvironment,
a section of the journal
Frontiers in Earth Science

The impact of human activities has been increasing to a degree where humans now outcompete many natural processes. When interpreting environmental and climatic changes recorded in natural archives on historical time scales, it is therefore important to be able to disentangle the relative contribution of natural and anthropogenic processes. Lake Meihua on the Ilan Plain in northeastern Taiwan offers a particularly suitable opportunity to test how human activities known from historical records can be recorded in lacustrine sediment. For this purpose, three cores from Lake Meihua have been studied by a multiproxy approach, providing the first decadal-resolution lacustrine records covering the past 150 years in Taiwan. Profiles of excess ^{210}Pb, ^{137}Cs and ^{239,240}Pu from two short cores (MHL-09-01 and MHL-11-02) allowed a precise chronology to be established. The presence of a yellow, earthy layer with lower levels of organic material coincide with the record of land development associated with the construction of the San-Chin-Gong Temple during AD 1970–1982. Furthermore, in the lower part of the cores, the upwards increasing trend of inc/coh, TOC, TOC/TN, and grain size, coupled with the palynological data (increase of *Alnus, Mallotus, Trema* and herbs) from the nearby core MHL-5A with radiocarbon chronology, suggest that the area surrounding the lake has been significantly affected by agricultural activities since the arrival of Chinese settlers around ~AD 1874. In sum, this study demonstrates that this suite of lacustrine sediments in northeastern Taiwan has recorded human activities in agreement with historical documents, and that different human activities will leave distinct sedimentological, geochemical, and palynological signatures in the sedimentary archives. Therefore, multiproxy reconstructions are important to capture the complex nature of human-environmental interactions. A better understanding of the weathering and erosion response to human activities can also provide useful information for sustainable land-use management.

Keywords: human activities, radionuclides, XRF core scanner, environmental change, organic indicators, pollen

INTRODUCTION

Over the course of the Holocene, the environmental impact of human activities has gradually come to overshadow the impact of natural processes (Montgomery, 2007; Syvitski and Kettner, 2011). These anthropogenic environmental alterations have caused accelerated erosion of land cover to produce anthropogenic sediments (also known as legacy sediments; James, 2013). The erosion, transportation, and re-sedimentation of anthropogenic sediments have attracted increasingly more attention, both because of the need to separate the anthropogenic imprints from natural signals when studying past climatic changes, and because these legacy sediments contain records of the interaction between human society and the environment (James, 2013).

Taiwan, a small island situated on the western margin of the Pacific Ocean covers an area of only 35,980 km^2 with a population over 23 million. The heavy population density (642 people/km^2, 16th in the world) and rapid land-development in recent history should make Taiwan an ideal location to study the influence of anthropogenic processes (Huang and Lin, 2003; Chen et al., 2007a). However, the rapid uplift, steep topography, and high erosion rates up to 3–6 mm yr^{-1} in Taiwan (Dadson et al., 2003) make it difficult to study historical human-environment interactions. Therefore, in Taiwan, millennial to centennial-scale climatic oscillations have only been studied through palynological and geochemical proxies from a limited number of lakes (Wang et al., 2011; Yang et al., 2011b, 2014; Chen et al., 2012; Liew et al., 2014), and even fewer studies have been conducted on human-environment interactions on historical/archaeological time scales due to the scarcity of high resolution material (Lin et al., 2007; Yang et al., 2014).

Unlike other places in Taiwan where uplift is dominating, the Ilan Plain in northeastern Taiwan is characterized by tectonic subsidence which is the result of the westward propagation of the Okinawa Trough (Shyu et al., 2005; **Figure 1A**). Consequently, the subsiding Ilan Plain has acted as a depocenter for sediments delivered by the Lanyang River (the main river system on the Ilan Plain) and has accumulated over 120 m of sediments since last glacial maximum (LGM) (Wei et al., 2003; Chen et al., 2009). Lake Meihua, located at the southern edge of Ilan Plain, has received massive amounts of suspended material from its local drainage and/or nearby local river systems by overbank deposits (**Figures 1A,B**). Previous studies have shown that these sediments can provide exquisite lacustrine records for the study of paleoenvironmental changes during the Holocene (Chen et al., 2012).

Developing reliable proxy records of anthropogenic impact on the environment is difficult for two reasons. First, the changes are often considerably more abrupt and rapid than their natural counterparts, requiring high-resolution techniques and materials, and second, the relation between human activities and the signals recorded in the sedimentary archives are often poorly understood. To overcome these obstacles, we make use of the fast and non-destructive micro X-ray fluorescence (XRF) scanning technique (Croudace et al., 2006; Huang et al., 2016) that can deliver the radiographic images, elemental variations and

organic-related incoherent over coherent ratios (inc/coh) directly from the untreated sediments at sub-mm resolution (Guyard et al., 2007; Rolland et al., 2008; Sáez et al., 2009; Brunschön et al., 2010) to analyze the records of rapid anthropogenic impacts preserved in the sedimentary archives (Miller et al., 2014). Moreover, the well-documented and relatively uncomplicated settlement and development history on the Ilan Plain offers a particularly suitable opportunity to test how human activities known from historical records have been recorded in the lacustrine sediment.

The aim of this study is to assess how known human activities are recorded in the sedimentary archives of a confined lake basin. Therefore, three cores from Lake Meihua have been studied by multi-geochemical and palynological proxies, and address the anthropogenic impacts covering the past 150 years in northern Taiwan. A better understanding of how human activities are recorded in the sedimentary archives is of imperative importance to disentangle natural and anthropogenic signals, and may also provide useful information for land-use management and sustainable environment governance.

STUDY AREA
Geological Background

The lacustrine cores used in this study were retrieved from Lake Meihua (121° 43.57′, 24° 38.35′; 52 ha; 50 m a.s.l.), which is situated on the southern edge of the Ilan Plain, northeastern Taiwan. The Ilan Plain is a triangular-shaped alluvial-fan delta, facing the Pacific Ocean to the east. Two major mountain ranges border the northern and southern flanks, the Hsueshan Range and the Central Range, respectively (**Figure 1A**). The basement rock here is mainly composed of Tertiary argillite, slate and metasandstone (Ho, 1986). The continuous subsidence associated with the westward propagation of the Okinawa Trough has provided over 120 m of sedimentary accommodation space in the Ilan basin since the LGM (Wei et al., 2003; Chen et al., 2009). The high sedimentation rates are the result of intense rain-induced erosion of surrounding mountain areas; the Lanyang River supplies about 17 Mt of suspended sediments per year (Dadson et al., 2003). The position at the southern edge of the Ilan Plain results in Lake Meihua being surrounded by hills, with the only opening toward the Ilan Plain to the north (**Figure 1B**). A normal faulting system relating to the tectonic subsidence along the mountain front (Shyu et al., 2005) together with rapid river aggradation at the northern shore of the lake has allowed the lake to persist for an unusually long time, preserving an exquisite lacustrine record since at least the mid-Holocene (Chen et al., 2012).

Climate and Vegetation

The climate of the Ilan area is humid and subtropical, influenced by both the East Asian summer and winter monsoons, with frequent typhoon activities in summer and autumn. The annual mean temperature is ∼22.2°C and the average annual precipitation over past 50 years is roughly 2760 mm (Central Weather Bureau, www.cwb.gov.tw). The natural vegetation in the study area mainly consists of subtropical evergreen rainforest,

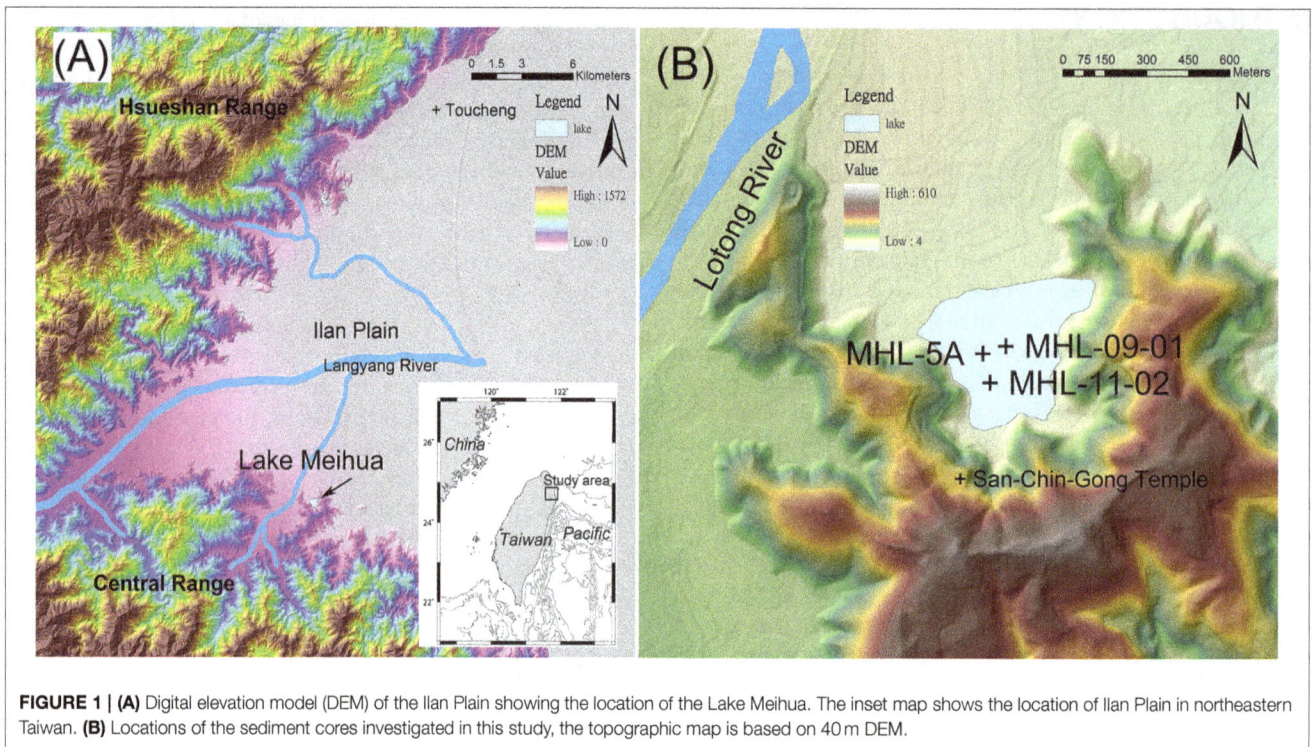

FIGURE 1 | (A) Digital elevation model (DEM) of the Ilan Plain showing the location of the Lake Meihua. The inset map shows the location of Ilan Plain in northeastern Taiwan. **(B)** Locations of the sediment cores investigated in this study, the topographic map is based on 40 m DEM.

although the dominating component of the vegetation is strongly influenced by topography and altitude (Su, 1984). The vegetation of the surrounding mountains consists of subtropical to warm temperate forests dominated by Lauro-Fagaceae elements common in the *Castanopsis-Machilus* Zone (Chen, 2000). Today, much of the plain is cultivated (mostly rice), or covered by herbaceous plants and cultivated trees (Lin et al., 2007).

Human Activities

The history of human settlement in northeastern Taiwan has been reconstructed from archaeological remains (Liu, 1995; Chen et al., 2007b) and historical documents (Yao, 1829; Lu, 1970; Knapp, 1976; Bai, 1991). The colonization history of the Ilan area can be traced back to Neolithic (~5000 yr BP) and Iron age (~1300 yr BP) settlements by aboriginal people of Taiwan (Liu, 1995; Chen et al., 2007b). Before the seventeenth century, Taiwan was primarily populated by aboriginal groups and only a limited number of Chinese migrants due to the isolation policy of the Chinese government (Lu, 1970). However, in the early seventeenth century Chinese settlers mainly from southeastern China started to migrate to Taiwan because of the civil war between the Ming and Qing (or Ching) dynasties on mainland China (Knapp, 1976). Moreover, during the two centuries after the Qing dynasty annexed Taiwan in AD 1683, Chinese settlers established agricultural villages from west to east across the island. Therefore, the settlement history of northeastern Taiwan began quite late compared to western and southern Taiwan due to its isolated geographical location (Lu, 1970; Bai, 1991).

The town of Ilan was first recorded as Kibanuran and occupied by an aboriginal people named Kavalan (meaning

people living on the plain). Chinese settlers reached northern Ilan at ~AD 1768. After fighting with the aboriginal Kavalan people for several years, a settler named Sha Wu and his group finally began their agriculture practices near Toucheng (northern Ilan, **Figure 1A**) in AD 1796 (Lu, 1970). Along with the continuing development of Ilan Plain, Chinese settlers reached the area near Lake Meihua at ~AD 1874 and spread to the Suao area (southern Ilan) in the following decades (Lu, 1970; Bai, 1991).

MATERIALS AND METHODS

Lacustrine Core Sampling

The two short lacustrine cores MHL09-01 and MHL11-02 used for this study were collected on October 1st 2009 and March 9th 2011 from Lake Meihua (**Figure 1B**). To obtain high-quality, undisturbed cores, we used 9-cm diameter transparent acrylic tubes with sharp edges which were inserted directly into the lake sediments on the rubber boat. The tops of the tubes were sealed to maintain the air pressure, and the tubes were pulled up directly to get the lake sediments. Since the tubes were transparent, the interface between water and sediment can be clearly seen to make sure the sediments were well preserved. The recovery length for core MHL-09-01 was 76 cm, and 68 cm for core MHL-11-02. After collection, the cores were kept upright and transported to the laboratory for splitting, description, and subsampling. The archive-halves were used for Itrax-XRF core scanner analysis, and the working-halves were subsampled at a resolution of 0.5 to 1 cm for radionuclides, TOC, and TOC/TN analysis, and 1 to 5 cm for grain size analysis. In addition, the top 205 cm of the 28.63 m long core MHL-5A (**Figure 1B**), which was obtained using hydraulic piston coring in 2005, was used for palynological studies. The sampling intervals for the pollen study were every 10 cm with 1 cm slice thickness, and 2 ml of each slice were used for pollen analysis.

Age Determination

The age models for cores MHL-09-01 and MHL-11-02 were constructed by using excess ^{210}Pb ($^{210}Pb_{ex}$) and were further constrained by the subsurface maximum of ^{137}Cs and $^{239,240}Pu$ (circa A.D. 1963), which are consistent with the history of nuclear fallout. In brief, ^{210}Pb (via ^{210}Po) and $^{239,240}Pu$ in core MHL-09-01 were measured by α-spectrometry, and ^{210}Pb in core MHL-11-02 was measured by γ-spectrometry. ^{137}Cs in both cores was measured by γ-spectrometry. ^{209}Po and ^{242}Pu spikes obtained from ORNL (Oak Ridge National Laboratory), calibrated with NIST-certified SRM-4327 and SRM-996 standards, were used as the yield determinants and added before the digestion of the samples. The details of the radiochemical procedures can be found in Huh and Su (1999) and the references therein. After radiochemical procedures, the Po isotopes were plated onto a silver disc at 80–90°C for 1 h, and the Pu isotopes were electroplated onto a stainless steel disc. ^{210}Pb, ^{214}Pb, and ^{137}Cs in MHL-11-02 were measured by γ-spectrometry based on photon energies around 46.52 keV, 351.99 keV, and 661.62 keV, respectively. Three HPGe detectors were engaged in this study to count over a series of sectioned sediment samples, including one GEM-type (ORTEC GEM-150240) with 150% efficiency (relative to 3 × 3 NaI), one GMX-type (ORTEC GMX-120265) with 100–120% efficiency and one LoAX-type (ORTEC LoAX-70450) with 70% efficiency. The counting efficiencies of the three detectors were calibrated by IAEA reference materials 327 and 375 for sample weight at 100 g as a reference, coupled with an in-house secondary standard for various masses (from 20 to 250 g) to calibrate the effect of sample mass on the attenuation of γ-ray of various energies. Sedimentation accumulation rates were determined from depth profiles of $^{210}Pb_{ex}$. The total ^{210}Pb and the fraction of the total ^{210}Pb that is supported by its precursor, ^{226}Ra. Due to the very short half-lives of all intermediate nuclides between ^{226}Ra and ^{210}Pb, the supported ^{210}Pb is generally in secular equilibrium with all its precursors from ^{226}Ra to ^{214}Pb. The activity of $^{210}Pb_{ex}$ was obtained by subtracting the activity of ^{214}Pb from that of the measured (i.e., total) ^{210}Pb ($^{210}Pb_{ex} = {}^{210}Pb - {}^{214}Pb$). In theory, the activity of $^{210}Pb_{ex}$ is the highest at the core top and decreases down-core because of radioactive decay. Under the assumption of a constant ^{210}Pb flux and a constant sedimentation rate (with the layer of Unit II excluded), the concentration of $^{210}Pb_{ex}$ downcore can be described by:

$$C = C_0 \exp(-\lambda t) \qquad (1)$$

where C is $^{210}Pb_{ex}$ at depth Z, C_0 is $^{210}Pb_{ex}$ at the core top (i.e., Z = 0), λ is the decay constant of ^{210}Pb (0.0311 yr^{-1}), and t is the post-depositional time. Given that t = Z/S, where S indicates sedimentation rate, previous equation can be transformed to

$$C = C_0 \exp[(-\frac{\lambda}{S})Z] \qquad (2)$$

Therefore, we expect to see straight regression lines by plotting C vs. Z on semi-log plots, which have slopes equal to $-\lambda/S$. From the slopes, the sedimentation rate (S) can be determined. The details about the γ-spectrometry method, including the standard materials used to calibrate the detectors and QA/QC of the data, can be found in previous studies (Huh and Su, 1999; Su and Huh, 2002; Huh et al., 2011).

The age model for the uppermost sediments of core MHL-5A is based on the linear interpolation between the date of coring at top, and one AMS (Accelerator Mass Spectrometry) ^{14}C analysis from wood fragments at 194 cm performed by the Rafter Radiocarbon Laboratory, Institute of Geological and Nuclear Sciences, New Zealand. The ^{14}C date was then calibrated using CALIB 7.1.0. (http://calib.qub.ac.uk/calib/). All following ages are given as calibrated years AD.

Multiproxy Sediment Core Analyses

Continuous downcore measurements of elemental variations were done at the Itrax-XRF Core Scanner Lab, Department of Geosciences, National Taiwan University (NTU). Archive halves of core MHL-09-01 and MHL-11-02 were analyzed using the 3 kW Mo target. The XRF measurements were analyzed at 30 kV, 44 mA, 2 mm resolution with 5 s exposure time for MHL-09-01, and 30 kV, 29 mA, 200 μm resolution with 15 s exposure time for MHL-11-02. The original XRF spectrums were processed by the Q-Spec software provided by Cox Analytical Systems to obtain element peak areas as counts. Beside the element counts, the ratio of incoherent to coherent scattering (inc/coh) may be used as an estimate of sedimentary organic contents in lacustrine sediments. In theory, the inc/coh scattering is higher for elements with a low atomic mass than for elements with higher atomic mass (Hodoroaba and Rackwitz, 2014), and therefore has been linked to the mean atomic number and can be largely related to the content of organic matter (Guyard et al., 2007; Rolland et al., 2008; Sáez et al., 2009; Brunschön et al., 2010). Therefore, the inc/coh can become an important reference for the strategy of further sampling, and thus to reduce the measure time and costs on organic experiments. Moreover, optical and radiographic images of the cores were also obtained by the same device.

Samples for total organic carbon (TOC) and total nitrogen (TN) were analyzed at the Department of Earth and Life Science, University of Taipei, following the procedures described in Yang et al. (2011b). In brief, each freeze-dried sample was pre-treated with 1 N HCl to remove carbonate, then washed with deionized water and dried in an oven at 60°C. The samples were then wrapped in a tin capsule and combusted using a ThermoQuest EA1110 elemental analyzer to measure the TOC and TN, and thus the atomic ratio of TOC/TN. The precision of the measurements for TOC and TN is ± 0.4 wt. % and ± 0.6 wt. %, respectively.

Samples for grain size analysis were analyzed at the Institute of Oceanography, National Taiwan University. 1 g of wet sediment was treated sequentially with 10 ml of 15% H_2O_2 and 7.5 ml of 10% HCl to remove organic materials and carbonates, and washed with deionized water. Prior to the analysis, 10 ml of 1% $Na(PO_3)_6$ was added to each sample followed by vigorous shaking. The samples were then ready for analysis using a Beckman Coulter LS-13 320 Laser diffraction particle size analyzer. Reproducibility of replicate analyses of sample and working standard were better than 95%.

For the palynological study, the sample preparation of core MHL-5A followed the classic procedures outlined by Moore et al. (1991), and slightly modified by Lin et al. (2007), involving treatments with hydrochloric acid, potassium hydroxide, sodium pyrophosphate, hydrofluoric acid and acetolysis. For each sample, more than 200 non-aquatic pollen grains were counted, except for samples with very low pollen counts. The taxonomic nomenclature in the pollen diagrams followed the published keys from Huang (1981), Chen and Wang (1999), and the references therein. Percentage calculations and construction of the diagrams were performed using the Tilia and TGView programs (provided by Dr. Eric Grimm, Illinois State Museum). In this study, we will only use the palynological data of the uppermost 205 cm from core MHL-5A.

RESULTS

Lithology

Using sedimentological core descriptions, grain size analysis, and observations of radiographic images, five sedimentary units were defined and correlated between MHL-09-01 and MHL-11-02 (**Figure 2**, **Table 1**). From bottom to top, Unit V (58–76 cm in MHL-09-01 and 66–68 cm in MHL-11-02) consists of a laminated light gray-yellowish silt with gradual changes at the upper boundary. Unit IV (34–58 cm in MHL-09-01 and 34–66 cm in MHL-11-02) is a laminated light gray silt with gradual upper boundary. Unit III (17–34 cm in MHL-09-01 and 18–34 cm in MHL-11-02) is a soupy dark gray silt to very-fine-sand and has a sharp upper boundary especially in MHL-09-01. Unit II (10.5–17 cm in MHL-09-01 and 9–18 cm in MHL-11-02) is made up of a massive yellowish silt with gradual upper boundary, followed by Unit I (0–10.5 cm in MHL-09-01 and 0–9 cm in MHL-11-02), which is a homogenous dark gray silt. In this study, we will focus at Unit III to I. For more details on the lower units, please refer to Huang (2011).

Chronology

Activities of all measured fallout nuclides show anomalously low values in Unit II for both cores (**Figure 2**). However, the $^{210}Pb_{ex}$ profiles of the segments above and below Unit II both show the expected exponential decreases with depths, allowing the sedimentation accumulation rates based on $^{210}Pb_{ex}$ to be estimated. Further chronological constraint comes from the profiles of ^{137}Cs which are characterized by a pronounced peak at the 18.0–18.5 cm interval in MHL-09-01 (also $^{239,240}Pu$ peak) and 17.0–17.5 cm in MHL-11-02, reflecting the history of atmospheric nuclear weapons testing which showed a global fallout maximum in AD 1963 (Pennington et al., 1973).

In MHL-09-01, since the Unit II is covered by 10.5 cm of sediment at the top and commenced ~1.5 cm above the global

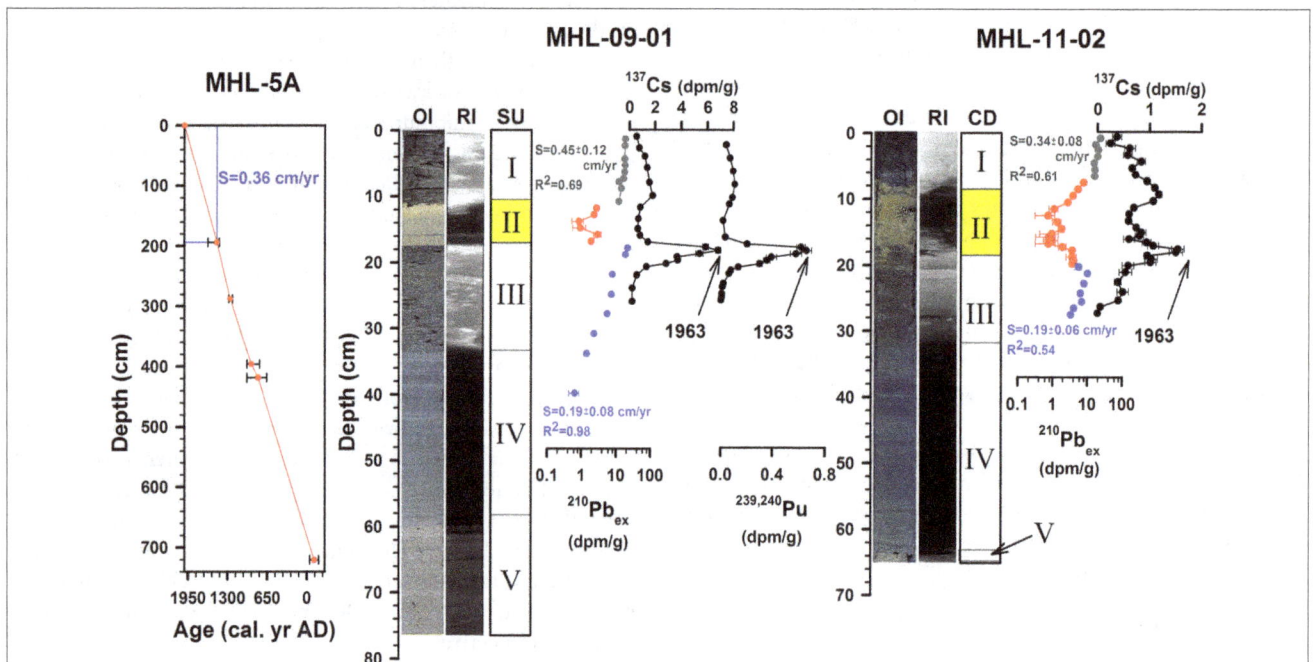

FIGURE 2 | The depth vs. radiocarbon age profile of the upper 720 cm in MHL-5A, and the lithologies and chronologies for core MHL-09-01 and MHL-11-02. The blue box in the left panel shows the segment that was used for palynological reconstruction. This interval has a mean sedimentation rate of 0.36 cm yr^{-1}, which is close both to the average of the age model in MHL-5A, and in agreement with the sedimentation rates reconstructed for the two shorter cores. OI: optical image obtained by Itrax-XRF core scanner; RI: radiographic image obtained by Itrax-XRF core scanner; SU: Sedimentary units. The chronologies are based on the radionuclide profiles of $^{210}Pb_{ex}$, ^{137}Cs and, $^{239,240}Pu$ in core MHL-09-01 and $^{210}Pb_{ex}$ and ^{137}Cs in MHL-11-02. The ^{210}Pb-based sedimentation rates above (in gray color) and below unit II (in blue color) in both cores are also shown next to the $^{210}Pb_{ex}$ profiles.

TABLE 1 | Sedimentary units, its characteristic features and inferred depositional conditions of the Lake Meihua sedimentary sequence in core MHL-09-01 and MHL-11-02.

Sedimentary units	Sedimentological features	Inferred depositional conditions
I	Massive dark gray silt	Modern condition of the lake
II	Massive yellowish silt	Construction Event by San-Chin-Gong Temple
III	Soupy dark gray silt to very fine sand (small plant debris less than 0.2 mm has been found in MHL-11-02)	Impact of Chinese settlers
IV	Laminated light gray silt	Average condition of the lake
V	Laminated light gray-yellowish silt	Impact of flood event (Huang, 2011)

fallout maximum (circa AD 1963), the age of the Unit II can be constrained by ^{210}Pb chronology (**Figure 2**). By least-square fitting of the ^{210}Pb$_{ex}$ data above and below Unit II, we obtained ^{210}Pb-based sedimentation accumulation rates of 0.45 ± 0.12 cm yr^{-1} ($R^2 = 0.69$, $n = 10$) and 0.19 ± 0.08 cm yr^{-1} ($R^2 = 0.98$, $n = 9$), respectively. Thus, the 6.5 cm thick Unit II was deposited $\sim$7 years after the fallout maximum and $\sim$24 years before the core was collected. In other words, Unit II accumulated during AD 1970–1985, with a mean sedimentation accumulation rate of $\sim$0.40 cm/yr. Also, the pattern of radionuclide profiles in MHL-11-02 is similar to those in MHL-09-01 (**Figure 2**). The ^{210}Pb-based sedimentation accumulation rates above and below Unit II are 0.34 ± 0.08 cm yr^{-1} ($R^2 = 0.61$, $n = 7$) and 0.19 ± 0.06 cm yr^{-1} ($R^2 = 0.54$, $n = 8$), while the sedimentation accumulation rate in Unit II is $\sim$0.37 cm yr^{-1}. Therefore, the ages of Unit II in both cores are perfectly consistent with the construction period of San-Ching Gong Temple near the lake (see discussion).

The chronology in the upper 205 cm of MHL-5A is based on the linear interpolation between coring date (i.e., AD 2005) and 1 AMS ^{14}C dating at 194 cm (wood fragments, the dating result is 400 ± 30 yr BP and the median calibrated age is 1476 in the 2σ range of AD 1437–1624, **Table 2**), resulting in a mean sedimentation rate of $\sim$0.36 cm yr^{-1}. The general sedimentation rate indicated by these two age control points is supported by deeper radiocarbon dates in MHL-5A, and also similar to the sedimentation rates that were estimated for the two shorter cores in this study (**Figure 2**, **Table 2**), as well as by previous studies (Huang, 2011; Chen et al., 2012).

Multiproxy Sediment Core Analyses
XRF Core Scanning Measurements

The downcore profiles of inc/coh and Ti in core MHL-09-01 and MHL-11-02 obtained by the Itrax-XRF core scanner will be discussed in this study. The upper 12 cm of XRF measurements in core MHL-09-01 had to be removed because of cracks that formed due to drying during storage. Generally, the inc/coh stays low in lowermost Unit V and IV, increased gradually in Unit III, continuing with an abrupt drop in Unit II and then returning to high values in Unit I. Ti shows an opposite pattern, remaining high in Unit V and IV,

decreasing gradually in Unit III, continuing with an abrupt rise in Unit II and then returning to low values. The relationship between inc/coh and the organic contents is also further confirmed by the positive correlation between inc/coh and TOC measurements (see below) in this study (correlation coefficient, $r = 0.90$ in MHL-09-01 and 0.88 in MHL-11-02) (**Figure 3**).

The inc/coh and Ti profiles in both cores show a strong negative correlation (correlation coefficient, $r = -0.90$ in MHL-09-01 and -0.94 in MHL-11-02) (**Figure 3**). These negative correlations are attributed to the well-known closed-sum effect caused by the dilution of organic materials (Calvert, 1983; Rollinson, 1993). Consequently, the measured variations in the elements (i.e., Ti) will to a large extent mirror the changes in the organic materials (i.e., inc/coh). Some studies have suggested to normalize the elements in the lithogenic component of the sediment against a conservative element (e.g., Al) to allow changes in the input of the elements to be addressed (Kylander et al., 2011; Löwemark et al., 2011). However, it is difficult to remove all the influences generated by variations in organic components (Huang, 2011). Therefore, the interpretation of elemental variations from XRF scanning must be treated with caution especially in organic rich lacustrine sediments, and only element data for Ti is included in this study.

Organic Geochemistry

The general trends of TOC and atomic TOC/TN are similar in both cores (**Figure 3**), while the difference of maximum values may be attributed to the core locations in the lake. The organic content of sediments in Lake Meihua is highly variable, ranging from 0.5% to $\sim$8.0% for TOC (**Figure 3**). Lower values less than 2.0% occur in Units V and IV. A rising trend from the base of Unit III (from 0.5% to $\sim$8.0% in MHL-09-01 and $\sim$0.3% to $\sim$4.0% in MHL-11-02), continues with an abrupt drop to 1.0% in Unit II and then returns to $\sim$4.0% in Unit I of both cores.

The TOC/TN profiles show parallel changes to the TOC (especially in MHL-11-02, **Figure 3**), but vary from 4.0 to 6.5 in MHL-09-01 and 4.0 to 10.0 in MHL-11-02. Lower values occur in Units V and IV, followed by a rising trend up to 5.0 in MHL-09-01 and up to 9.0 in MHL-11-02 in Unit III. An abrupt drop of TOC/TN was found in Unit II, followed by increasing values in Unit I.

Grain Size Analysis

The results of mean grain size analysis are similar in both MHL-09-01 and MHL-11-02 (**Figure 3**). Both cores show finer grain sizes in Unit V, VI, II, and I, ranging from fine-silt to very-fine-silt (Φ:6.0–7.5), while a coarsening upward trend was found in both cores in Unit III. This unit coarsens from silt to very-fine-sand (Φ:6.0–3.2), with the highest peaks at the boundary between Unit III and Unit II. In MHL-11-02, although the highest peak seems to be 3.5 cm below the Unit III/II boundary. This discrepancy is likely caused by mixing or deformation during coring, as can be clearly seen on the optical and radiographic images (**Figure 2**).

TABLE 2 | Radiocarbon dates and corresponding calibrated ages of the upper 720 cm in core MHL-5A from the Lake Meihua.

Depth (cm)	^{14}C age (yr BP)	2σ cal. range (cal. yr AD)	Median Probability (cal. yr AD)	Laboratory code
194 ± 1	400 ± 30	1437–1624	1476	NZA-27808
287.5 ± 0.5	767 ± 30	1218–1282	1255	NZA-27860
396 ± 1	1139 ± 30	777–982	912	NZA-25241
418.5 ± 0.5	1230 ± 90	654–982	800	NTU-4527
720.5 ± 0.5	2098 ± 30	(−194)–(−47)	−121	NZA-25269

All the dating materials are from wood fragments

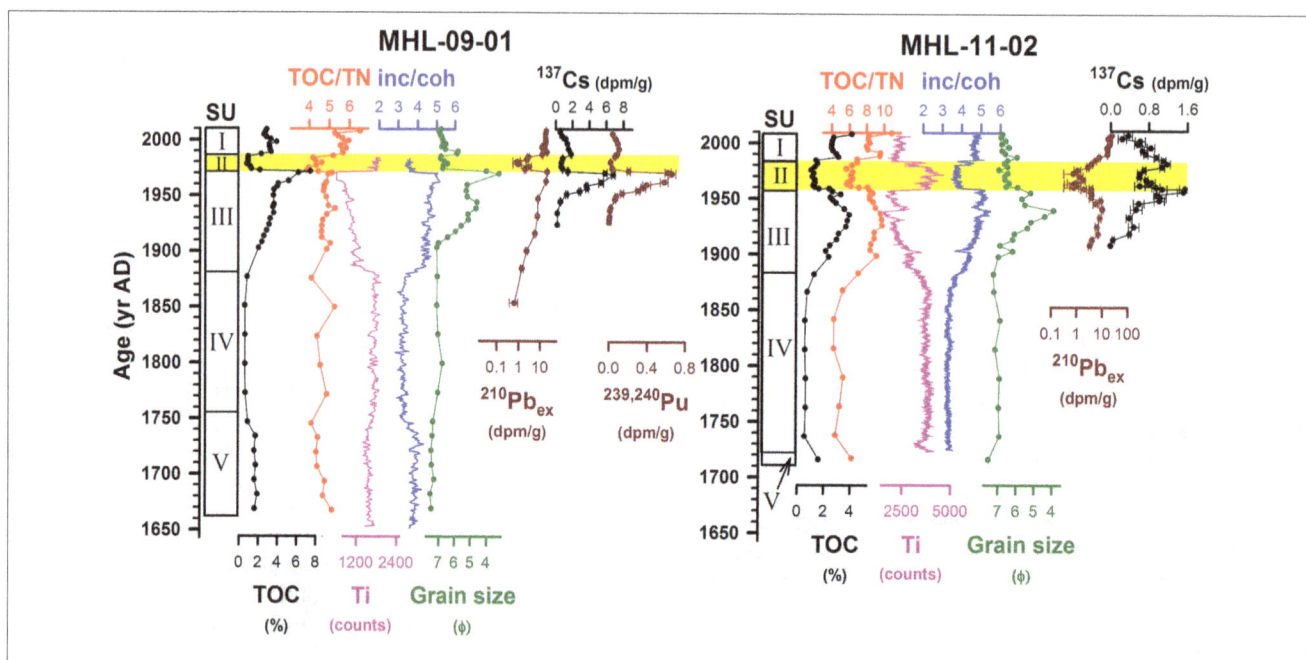

FIGURE 3 | Sediment geochemical records of core MHL-09-01 and MHL-11-02. SU (Sedimentary units), TOC (%, total organic carbon), TOC/TN, inc/coh (organic indicator obtained by Itrax-XRF core scanner), grain size and radionuclides are shown. The interval highlighted by the yellow bars can be related to the erosion caused by the construction of a San-Chin-Gong Temple complex on the hillslope south of the lake from AD 1970 to 1982. The increase in organic matter in Unit III likely represents the land development by Chinese settlers in the late 19th century.

Pollen Assemblages

Pollen of more than 100 taxa, including either families or genera, were identified from the uppermost 205 cm of core MHL-5A. The studied interval represents the time period since AD 1440 and the major pollen taxa can be divided into two zones, as shown in **Figure 5**.

Zone PZ2 (205–55 cm; AD 1476–1860).

This zone is characterized by stable levels of abundant tree and shrub pollen, accounting for over 60% of the terrestrial pollen sum. *Cyclobalanopsis, Quercus, Alnus, Lagerstroemia,* and *Liquidambar* are the predominant contributors, accompanied by *Bischofia, Myrica, Ardisia, Glochidion,* and *Helicia*. The percentages of herbs and ferns are low.

Zone PZ1 (55–0 cm; AD 1860–present).

This zone is marked by a sharp decline in tree pollen, and a large increase in herbaceous and fern pollen. *Lagerstroemia* and *Liquidambar* show dramatic declines, while *Ardisia,*

Engelhardia, Elaeocarpus, Bischofia, Schefflera, and *Calamus* also decrease. Secondary forest taxa, such as *Alnus, Mallotus, Trema* become significant. Herbs, such as Poaceae, Cyperaceae, *Artemisia,* Apiaceae, Amaranthaceae, and Asteraceae rise conspicuously.

DISCUSSION

The Sources of Sediments to Lake Meihua

The TOC/TN can be used to identify the sources of organic materials in sediments as either autochthonous or allochthonous (Meyers, 1994; Lamb et al., 2006; Yang et al., 2011a, 2014). As documented by Meyers (1994) and Lamb et al. (2006), the TOC/TN is usually less than 9 for lacustrine algae (representing the autochthonous primary source) and higher than 12 for vascular plants (representing the allochthonous primary source), respectively. In the Lanyang River catchment, the TOC/TN of primary materials, such as bedrocks, river sediments and soils have been investigated by Kao and Liu (2000) (**Figure 4**), and

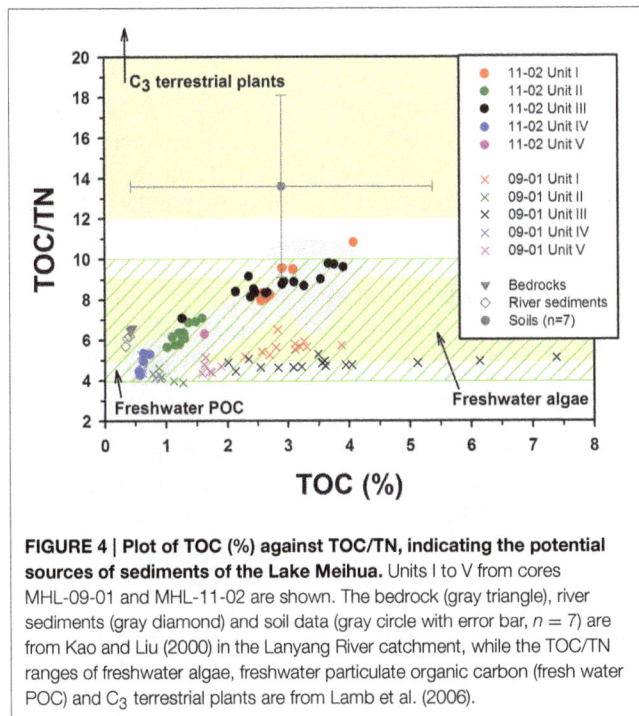

FIGURE 4 | Plot of TOC (%) against TOC/TN, indicating the potential sources of sediments of the Lake Meihua. Units I to V from cores MHL-09-01 and MHL-11-02 are shown. The bedrock (gray triangle), river sediments (gray diamond) and soil data (gray circle with error bar, $n = 7$) are from Kao and Liu (2000) in the Lanyang River catchment, while the TOC/TN ranges of freshwater algae, freshwater particulate organic carbon (fresh water POC) and C_3 terrestrial plants are from Lamb et al. (2006).

also been used to identify the organic sources of lacustrine sediments as reported by Yang et al. (2014) and Yang et al. (2011a).

As suggested by the MHL-5A record in Huang (2011), the TOC and TOC/TN values in Unit IV of both MHL-09-01 and MHL-11-02 (**Figure 4**, blue circle and blue cross) can represent the long-term average conditions in Lake Meihua during the past millennium. The increasing trends of TOC/TN and TOC from Unit IV to Unit III of MHL-11-02 in **Figure 4** (gray arrow) likely indicate an increasing input of carbon-rich terrestrial organic materials in both Unit III and Unit I. However, the bedrocks and river sediments cannot be the major components of these increasing signals because of their low TOC contents (~0.5%, **Figure 4**). Therefore, the soils and the plant debris (found in Unit III of MHL-11-02, **Table 1**) seem to be the main contributors of the increasing organic signals. Conversely, the abrupt drop in Unit II represents the strong dilution by minerogenic materials, indicated also by its low values of TOC, TOC/TN, and radionuclides.

However, the increasing trend of TOC/TN in MHL-09-01 is insignificant (**Figures 3, 4**), although the maximum values of TOC are actually two times higher than those in MHL-11-02. Such observation may be explained as the distance effects of the coring locations in the lake. The terrestrial organic material from the catchment may settle down before it reaches to the center of the lake, where MHL-09-01 was collected. Furthermore, since the TOC/TN in MHL-09-01 remains in the range of freshwater algae, the higher TOC in MHL-09-01 can simply be explained as the lacustrine algae blooming associated with the terrestrial inputs.

The Impact from Construction of San-Chin-Gong Temple

The yellowish Unit II in both cores is the most characteristic unit in the cores. However, the lower values of inc/coh, TOC, TOC/TN, radionuclides in Unit II are very different from the flood layers reported in previous study in Lake Meihua (Huang, 2011; also in the Unit V of this study). There are 8 layers with higher values of inc/coh, TOC, TOC/TN, and grain size in the last 1200 years, and they have been explained as the overbank flooding events and can be linked to the regional climate and earthquake signals (Huang, 2011). Therefore, the natural mechanism cannot explain the presents of Unit II.

The most conspicuous sign of human activities around Lake Meihua is the large temple complex called San-Chin-Gong Temple (**Figure 6**) which is sitting on the slope to the south of the lake (**Figure 1B**). As the headquarters of the Taoism Association in Taiwan, the main building of the temple was constructed over an area of 4 ha during AD 1970–1982, followed by the addition of several annexes in the ensuing years. Such a massive construction in a relatively small catchment must have had a great impact on local environment and the lake just beneath the hill.

The age models for the yellowish Unit II in both cores (**Table 1**, **Figures 2, 3**) are perfectly linked with the historical temple-building record (**Figure 3**). The radionuclide age at the upper boundary of Unit II in MHL-09-01 is AD 1985 (derived from the upper $^{210}Pb_{ex}$ sedimentation rate and the coring date), while the age at the lower boundary is AD 1970 (derived from the ^{137}Cs peak in AD 1963 and the lower $^{210}Pb_{ex}$ sedimentation rate). Thus, the age of Unit II perfectly match with the documented record of temple construction from AD 1970–1982.

The peaks of TOC, inc/coh and grain size at the boundary of Unit III and II also mark the initial stage of the temple building phase, especially in MHL-09-01. The intense deforestation and excavation for the foundation of the temple and surrounding road systems likely is responsible for these changes. Also, Unit II is characterized by relatively lower values of inc/coh, TOC, TOC/TN, radionuclides, and its yellowish color (**Figures 2, 3**). These properties reflect the dilution of organic matter and radionuclides content by increased minerogenic material inputs. The enhanced erosion of the exposed hillslope caused by land development activities introduced radionuclides, organic carbon depleted regolith, and/or even the materials that were used in the temple construction into the lake. These anthropogenic influences are further supported by the increased sediment rates in Unit II (0.40 cm yr^{-1} in MHL-09-01 and 0.37 cm yr^{-1} in MHL-11-02). Therefore, the presence of the yellowish Unit II represents the sedimentary evidence for the construction of the temple complex and surrounding road systems. These observations suggest that locally, human induced changes to the erosional pattern may temporally overshadow the natural signals.

After the construction event of the San-Chin-Gong Temple, the intensity of human activities has remained high in the study area. The construction of a dyke and a new sightseeing area around the lake in recent decades likely have caused the increase in sedimentation rates encountered in Unit I (**Figure 2**, 0.45 cm yr^{-1} in MHL-09-01 and 0.34 cm yr^{-1} in MHL-11-02). Also, the return to high levels of inc/coh, TOC and TOC/TN in

Unit I (**Figures 3, 4**) suggest that runoff delivered more terrestrial organic materials into the lake related to the continuous farming and deforestation in the surrounding area.

Land Development by Chinese Settlers in the Late 19th Century

In the older part of the record, human influence is less dramatic, but still clearly visible from both sedimentary parameters and pollen data. Up till the end of the eighteenth century, the area surrounding Lake Meihua experienced relatively little perturbations from human activities. It was not until AD 1796 that the Chinese settler Sha Wu and his group defeated the aboriginal Kavalan people and invaded the Toucheng area (northeastern Ilan, **Figure 1A**) to extend the frontiers of Chinese settlement and agricultural practices. After that, Chinese settlers started to develop the Ilan Plain, although they were constantly in conflict with the original aboriginal population while extending their territory further, reaching Suao at the southeastern tip of the Ilan Plain in the late nineteenth century (Yao, 1829; Lu, 1970).

The records from Lake Meihua in southern Ilan clearly reflect this history. This study reveals how the increasing trends of inc/coh, TOC, TOC/TN, and grain size in Unit III of both MHL cores (**Figures 3, 4**) record the agronomic activities of Chinese settlers starting from the late nineteenth century. Historical documents suggest that a Chinese settler named Hui-Huang Chen and his group of people started the cultivation works to the south of Lanyang River in AD 1874 (Bai, 1991), at a site close to Lake Meihua. The deforestation, land clearance and following agricultural practices by Chinese settlers likely caused intensified remobilization of terrestrial material, thereby not only increasing the input of TOC and coarser materials into the lake, but also causing algae booming due to the input of

nutrients. These changes in input are also reflected by an increase in plant debris of MHL-11-02, and the less dense nature of the sediments shown in the radiographic images of both cores (**Table 1, Figure 2**).

The change in land use is further supported by the palynological data from core MHL-5A (**Figure 5**, PZ1). The dramatic drop in tree and shrub pollen can be attributed to deforestation, while the dominance of herbs and pioneer species, such as *Alnus, Mallotus,* and *Trema* since around AD 1860 (**Figure 5**, PZ1) likely is the result of anthropogenic vegetation changes due to land clearance, and can therefore be linked to the increasing organic content in the Unit III (**Figure 3**). In the present age model of core MHL-5A, the age for the lower boundary of pollen zone PZ1 is situated at about AD 1860 (**Figure 5**). This is just one decade earlier than the change in agriculture activities indicated by the historical record (Bai, 1991) and the radionuclides ages in the beginning of Unit III in both short cores. However, the low resolution sampling for radiocarbon dating in core MHL-5A makes the exact age of the shift in pollen assemblages uncertain. The sedimentation rates obtained by $^{210}Pb_{ex}$ together with the radionuclides fall out maximum in AD 1963 obtained by ^{137}Cs and $^{239,240}Pu$ in MHL-09-01 and MHL-11-02 suggested that the lower boundary of Unit III can be dated to the 1880s (**Figure 3**). Thus, the evidence of anthropogenic impacts in the lacustrine cores from Lake Meihua may certainly be linked to the agricultural practices by Chinese settlers since AD ~1874, as recorded in historical documents (Bai, 1991).

In contrast to Lake Meihua, where anthropogenic influence can only be clearly detected for the last few 100 years or so, palynological records from Lake Dahu in northern Ilan show a human influence at least since around AD 1400.

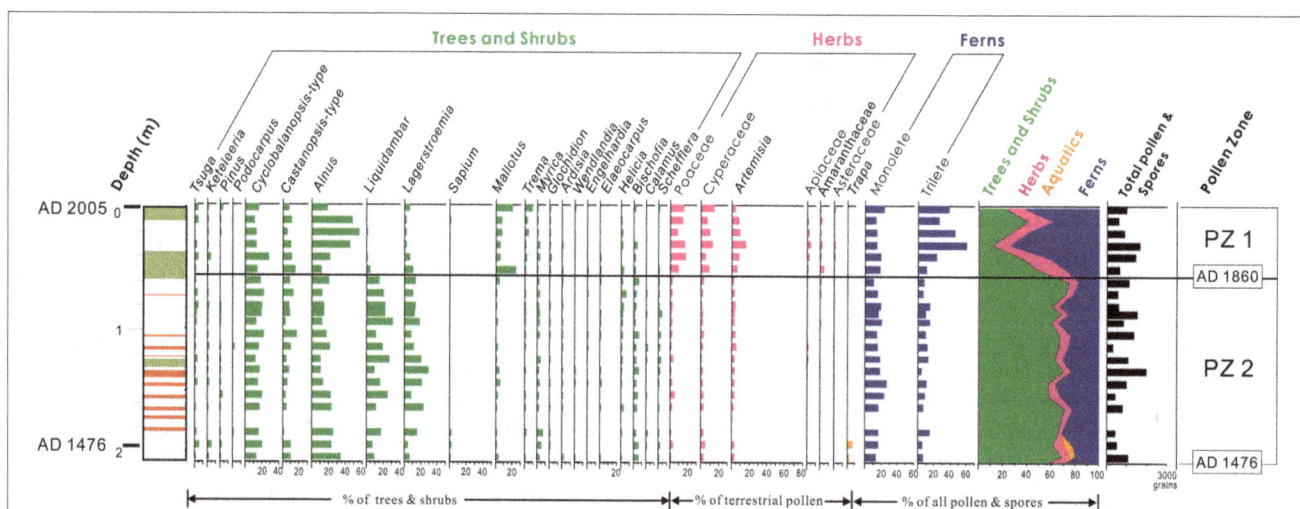

FIGURE 5 | Pollen percentage diagram of selected taxa from core MHL-5A. Pollen percentage diagram of selected taxa from core MHL-5A in the Lake Meihua. In the pollen zone-PZ1, the increasing trends of *Alnus, Mallotus, Trema* and herbs coupled with the increasing trends of organic proxies in Unit III of **Figure 3** suggest that the neighboring area has been significantly affected by agronomic activities starting around AD 1880s and may be attributed to the arrival of Chinese settlers around AD 1874.

FIGURE 6 | View of San-Chin-Gong Temple took from the location of MHL-11-02. As the headquarter of the Taoism Association in Taiwan, the main building of the temple was constructed during AD 1970–1982, followed by the addition of several annexes in the ensuing years.

This suggests that on the northern Ilan Plain, intensive land clearance took place 200 years prior to the arrival of the Chinese settlers (Yang et al., 2014). Consequently, the great drop in arboreal pollen concentration around AD 1400 in Lake Dahu can be attributed to deforestation by aboriginal people to clear land for agriculture. According to the archaeological studies on Ilan Plain (Liu, 1995), the almost 500 years earlier development history in Lake Dahu compared to Lake Meihua can be attributed to the distribution of aboriginal people on Ilan Plain and the later arrival of Chinese settlers in southern Ilan.

This study suggests that while modest land clearance may result in an increase in TOC due to enhanced erosion of the topsoil, a massive intervention, however, such as the temple construction may actually result in a drop in TOC due to dilution of the organic input by massive influx of lithogenic material from the exposed regolith. Where the local end-members of the organic sources are known, TOC/TN can be used to identify the sources and possible transport mechanisms of sediments.

CONCLUSION

Based on sedimentological and geochemical natural archives from Lake Meihua on the Ilan Plain, several general conclusions regarding the interactions between humans and the

environment, and how these interactions are documented in the sedimentary record, may be drawn:

- The presence of a yellowish layer with lower levels of inc/coh, TOC, TOC/TN, and radionuclides can be tied to intensified erosion caused by the land development associated with the construction of the San-Chin-Gong Temple during AD 1970–1982.
- During the temple construction phase, the signals in the sedimentary archives were completely dominated by this event, demonstrating how anthropogenic processes can locally overshadow natural processes in the archives.
- Changes in land use, land clearing, and agriculture are reflected both in sedimentary parameters, such as the TOC, inc/coh, TOC/TN, and grain size, and in pollen counts. The arrival of Chinese settlers, and the following agriculture activities on the southern Ilan Plain, thus can be detected in the sedimentary record since ~AD 1874.

This study demonstrates how lacustrine sediments can be used as recorders of human activities in agreement with historical documents. However, due to the complexity of lake systems in catchments with steep topographies and high erosion rates, a multiproxy approach is necessary to disentangle the relative influence of the different processes.

AUTHOR CONTRIBUTIONS

JH: design of the work, sample taking, geochemical analysis, and drafting the article. CH, WL, and CS: design of the work, sample taking, radionuclide and grain size analysis and interpretation, and drafting the article. KW, LL, SS, and TL: design of the work, funding provider, and drafting the article. SL: Pollen sample analysis and interpretation, and drafting the article. TY, ML: design of the work, sample taking, organic sample analysis and interpretation, and drafting the article.

ACKNOWLEDGMENTS

We are grateful to Miss Yun-Chun Wang, Queenie Chang and Emma Tung for assistance during sample processing and manuscript preparation. Funding for this research was provided by the research project "Environmental Change and Human Activity over the Last 1000 Years in the Lang-Yang River Drainage System: ECHA1000" (NSC 98-2627-M-002-010).

REFERENCES

Bai, Z. C. (1991). "The critical biographies of Chinese settler Hui-Huang Chen (in Chinese)," in *Taiwan Historica,* ed Academia Historica of Taiwan (Taipei: Academia Historica of Taiwan Press), 215–232.

Brunschön, C., Haberzettl, T., and Behling, H. (2010). High-resolution studies on vegetation succession, hydrological variations, anthropogenic impact and genesis of a subrecent lake in southern Ecuador. *Veg. Hist. Archaeobot.* 19, 191–206. doi: 10.1007/s00334-010-0236-4

Calvert, S. E. (1983). Geochemistry of pleistocene sapropels and associated sediments from the eastern Mediterranean. *Oceanol. Acta* 6, 255–267.

Chen, C. W., Kao, C. M., Chen, C. F., and Dong, C. D. (2007a). Distribution and accumulation of heavy metals in the sediments of Kaohsiung Harbor, Taiwan. *Chemosphere* 66, 1431–1440. doi: 10.1016/j.chemosphere.2006.09.030

Chen, H. F., Wen, S. Y., Song, S. R., Yang, T. N., Lee, T. Q., Lin, S. F., et al. (2012). Strengthening of paleo-typhoon and autumn rainfall in Taiwan corresponding to the Southern Oscillation at late Holocene. *J. Quaternary Sci.* 27, 964–972. doi: 10.1002/jqs.2590

Chen, S. H., and Wang, Y. F. (1999). Pollen flora of Yuenyang Lake Nature Preserve, Taiwan (I). *Taiwania* 44, 82–136.

Chen, T. (2000). "Plant ecology of lowland areas in northern Taiwan (in Chinese)," in *Botanical Garden Resources and Its Management*, ed H. Yen. (Taichung: National Museum of Natural Science Press), 9–33.

Chen, W. S., Yang, C. C., and Yang, H. C. (2009). "How to reconstruct the depositional framework of the late Quaternary subsurface sedimentary sequence in the Taiwan coastal plains? (in Chinese)," in *Special Publication of the Central Geological Survey 22*, ed Central Geological Survey (Taipei, Taiwan: Central Geological Survey), 101–114.

Chen, Y. B., Chiu, S. J., and Li, Z. Y. (2007b). *Report of Salvage Excavation on the Site Kiwulan, Ilan County (in Chinese)*. Ilan: Lanyang Museum Press.

Croudace, I. W., Rindby, A., and Rothwell, R. G. (2006). "ITRAX: description and evaluation of a new multi-function X-ray core scanner," in *New Techniques in Sediment Core Analysis*, ed. R. G. Rothwell (London: The Geological Society of London), 51–63.

Dadson, S. J., Hovius, N., Chen, H. G., Dade, W. B., Hsieh, M. L., Willett, S. D., et al. (2003). Links between erosion, runoff variability and seismicity in the Taiwan orogen. *Nature* 426, 648–651. doi: 10.1038/nature02150

Guyard, H., Chapron, E., St-Onge, G., Anselmetti, F. S., Arnaud, F., Magand, O., et al. (2007). High-altitude varve records of abrupt environmental changes and mining activity over the last 4000 years in the Western French Alps (Lake Bramant, Grandes Rousses Massif). *Quaternary Sci. Rev.* 26, 2644–2660. doi: 10.1016/j.quascirev.2007.07.007

Ho, C. S. (1986). *An Introduction to the Geology of Taiwan (in Chinese)*. Taipei: The Ministry of Economic Affairs.

Hodoroaba, V. D., and Rackwitz, V. (2014). Gaining improved chemical composition by exploitation of Compton-to-Rayleigh intensity ratio in XRF analysis. *Anal. Chem.* 86, 6858–6864. doi: 10.1021/ac5000619

Huang, C. Y. (1981). *Spore Flora of Taiwan*. Taipei: Botany Department National Taiwan University.

Huang, J. J. (2011). *Linkage between Natural Disasters and Kiwulan Cultural Hiatus Over the Last 1000 Years in the Lanyang Drainage System (in Chinese)*. Master's thesis. Taipei: National Taiwan University.

Huang, J. J., Löwemark, L., Chang, Q., Lin, T. Y., Chen, H. F., Song, S. R., et al. (2016). Choosing optimal exposure times for XRF core-scanning: suggestions based on the analysis of geological reference materials. *Geochem. Geophys. Geosyst.* 17, 1558–1566. doi: 10.1002/2016gc006256

Huang, K. M., and Lin, S. (2003). Consequences and implication of heavy metal spatial variations in sediments of the Keelung River drainage basin, Taiwan. *Chemosphere* 53, 1113–1121. doi: 10.1016/S0045-6535(03)00592-7

Huh, C. A., Chen, W., Hsu, F. H., Su, C. C., Chiu, J. K., Lin, S., et al. (2011). Modern (<100 years) sedimentation in the Taiwan Strait: rates and source-to-sink pathways elucidated from radionuclides and particle size distribution. *Cont. Shelf Res.* 31, 47–63. doi: 10.1016/j.csr.2010.11.002

Huh, C. A., and Su, C. C. (1999). Sedimentation dynamics in the East China Sea elucidated from ^{210}Pb, ^{137}Cs and ^{239,240}Pu. *Mar. Geol.* 160, 183–196. doi: 10.1016/S0025-3227(99)00020-1

James, L. A. (2013). Legacy sediment: definitions and processes of episodically produced anthropogenic sediment. *Anthropocene* 2, 16–26. doi: 10.1016/j.ancene.2013.04.001

Kao, S. J., and Liu, K. K. (2000). Stable carbon and nitrogen isotope systematics in a human-disturbed watershed (Lanyang-Hsi) in Taiwan and the estimation of biogenic particulate organic carbon and nitrogen fluxes. *Glob. Biogeochem. Cycles* 14, 189–198. doi: 10.1029/1999GB900079

Knapp, R. G. (1976). Chinese frontier settlement in Taiwan. *Ann. Assoc. Am. Geogr.* 66, 43–51. doi: 10.1111/j.1467-8306.1976.tb01071.x

Kylander, M. E., Lind, E. M., Wastegard, S., and Löwemark, L. (2011). Recommendations for using XRF core scanning as a tool in tephrochronology. *Holocene* 22, 371–375. doi: 10.1177/0959683611423688

Lamb, A. L., Wilson, G. P., and Leng, M. J. (2006). A review of coastal palaeoclimate and relative sea-level reconstructions using δ^{13}C and C/N ratios in organic material. *Earth-Sci. Rev.* 75, 29–57. doi: 10.1016/j.earscirev.2005.10.003

Liew, P. M., Wu, M. H., Lee, C. Y., Chang, C. L., and Lee, T. Q. (2014). Recent 4000 years of climatic trends based on pollen records from lakes and a bog in Taiwan. *Quatern. Int.* 349, 105–112. doi: 10.1016/j.quaint.2014.05.018

Lin, S. F., Huang, T. C., Liew, P. M., and Chen, S. H. (2007). A palynological study of environmental changes and their implication for prehistoric settlement in the Ilan Plain, northeastern Taiwan. *Veg. Hist. Archaeobot.* 16, 127–138. doi: 10.1007/s00334-006-0076-4

Liu, Y. C. (1995). "Prehistoric culture type of Ilan region (in Chinese)," in *Proceedings of Ilan Research Symposium*, ed J. Chu (Ilan: Ilan County Government Press), 38–56.

Löwemark, L., Chen, H. F., Yang, T. N., Kylander, M., Yu, E. F., Hsu, Y. W., et al. (2011). Normalizing XRF-scanner data: a cautionary note on the interpretation of high-resolution records from organic-rich lakes. *J. Asian Earth Sci.* 40, 1250–1256. doi: 10.1016/j.jseaes.2010.06.002

Lu, S. X. (1970). *The General History of Ilan County (in Chinese)*. Ilan: Ilan County Government Press.

Meyers, P. A. (1994). Preservation of elemental and isotopic source identification of sedimentary organic matter. *Chem. Geol.* 114, 289–302. doi: 10.1016/0009-2541(94)90059-0

Miller, H., Croudace, I. W., Bull, J. M., Cotterill, C. J., Dix, J. K., and Taylor, R. N. (2014). A 500 year sediment lake record of anthropogenic and natural inputs to Windermere (English Lake District) using double-spike lead isotopes, radiochronology, and sediment microanalysis. *Environ. Sci. Technol.* 48, 7254–7263. doi: 10.1021/es5008998

Montgomery, D. (2007). Soil erosion and agricultural sustainability. *Proc. Natl. Acad. Sci. U.S.A.* 104, 13268–13272. doi: 10.1073/pnas.0611508104

Moore, P. D., Webb, J. A., and Collison, M. E. (1991). *Pollen Analysis*. London: Blackwell Scientific Publications.

Pennington, W., Tutin, T., Cambray, R., and Fisher, E. (1973). Observations on lake sediments using fallout ^{137}Cs as a tracer. *Nature* 242, 324–326. doi: 10.1038/242324a0

Rolland, N., Larocque, I., Francus, P., Pienitz, R., and Laperriere, L. (2008). Holocene climate inferred from biological (Diptera: Chironomidae) analyses in a Southampton Island (Nunavut, Canada) lake. *Holocene* 18, 229–241. doi: 10.1177/0959683607086761

Rollinson, H. R. (1993). *Using Geochemical Data: Evaluation, Presentation, Interpretation*. Upper Saddle River, NJ: Pearson Education.

Sáez, A., Valero-Garcés, B. L., Giralt, S., Moreno, A., Bao, R., Pueyo, J. J., et al. (2009). Glacial to Holocene climate changes in the SE Pacific. The Raraku Lake sedimentary record (Easter Island, 27°S). *Quaternary Sci. Rev.* 28, 2743–2759. doi: 10.1016/j.quascirev.2009.06.018

Shyu, J. B. H., Sieh, K., Chen, Y. G., and Liu, C. S. (2005). Neotectonic architecture of Taiwan and its implications for future large earthquakes. *J. Geophys. Res.* 110:B08402. doi: 10.1029/2004jb003251

Su, C. C., and Huh, C. A. (2002). ^{210}Pb, ^{137}Cs and ^{239,240}Pu in East China Sea sediments: sources, pathways and budgets of sediments and radionuclides. *Mar. Geol.* 183, 163–178. doi: 10.1016/S0025-3227(02)00165-2

Su, H. (1984). Studies on the climate and vegetation types of the natural forests in Taiwan (II): altitudinal vegetation zones in relation to temperature gradient. *Q. J. Chin. Forestry* 17, 57–73.

Syvitski, J., and Kettner, A. (2011). Sediment flux and the Anthropocene. *Philos. Trans. A Math. Phys. Eng. Sci.* 369, 957–975. doi: 10.1098/rsta.2010.0329

Wang, L. C., Wu, J. T., Lee, T. Q., Lee, P. F., and Chen, S. H. (2011). Climate changes inferred from integrated multi-site pollen data in northern Taiwan. *J. Asian Earth Sci.* 40, 1164–1170. doi: 10.1016/j.jseaes.2010.06.003

Wei, K. Y., Chen, Y. G., Chen, W. S., Lai, T. H., Chen, L. C., and Fei, L. Y. (2003). Climate change as the dominant control of the last glacial-Holocene δ^{13}C variations of sedimentary organic carbon in the Lan-Yang Plain, northeastern Taiwan. *West. Pac. Earth Sci.* 3, 57–68.

Yang, T. N., Lee, T. Q., Lee, M. Y., Huh, C. A., Meyers, P. A., Löwemark, L., et al. (2014). Paleohydrological changes in northeastern Taiwan over the past 2 ky inferred from biological proxies in the sediment record of a floodplain lake. *Palaeogeogr. Palaeocl.* 410, 401–411. doi: 10.1016/j.palaeo.2014.06.018

Yang, T. N., Lee, T. Q., Meyers, P. A., Fan, C. W., Chen, R. F., Wei, K. Y., et al. (2011a). The effect of typhoon induced rainfall on settling fluxes of particles and organic carbon in Yuanyang Lake, subtropical Taiwan. *J. Asian Earth Sci.* 40, 1171–1179. doi: 10.1016/j.jseaes.2010.07.016

Yang, T. N., Lee, T. Q., Meyers, P. A., Song, S. R., Kao, S. J., Löwemark, L., et al. (2011b). Variations in monsoonal rainfall over the last 21 kyr inferred from sedimentary organic matter in Tung-Yuan Pond, southern Taiwan. *Quaternary Sci. Rev.* 30, 3413–3422. doi: 10.1016/j.quascirev.2011.08.017

Yao, Y. (1829). Notes for east expedition (in chinese).

Conflict of Interest Statement: The authors declare that the research was conducted in the absence of any commercial or financial relationships that could be construed as a potential conflict of interest.

Millennial-Scale Interaction between Ice Sheets and Ocean Circulation during Marine Isotope Stage 100

Masao Ohno[1]*, Tatsuya Hayashi[2], Masahiko Sato[1,3], Yoshihiro Kuwahara[1], Asami Mizuta[1], Itsuro Kita[1], Tokiyuki Sato[4] and Akihiro Kano[1]

[1] Department of Environmental Changes, Faculty of Social and Cultural Studies, Kyushu University, Fukuoka, Japan, [2] Mifune Dinosaur Museum, Kumamoto, Japan, [3] Geological Survey of Japan, National Institute of Advanced Industrial Science and Technology, Tsukuba, Japan, [4] Faculty of International Resource Sciences, Akita University, Akita, Japan

Edited by:
Qingsong Liu,
Chinese Academy of Sciences, China

Reviewed by:
Hong Ao,
Chinese Academy of Sciences, China
Dario Bilardello,
University of Minnesota, USA

*Correspondence:
Masao Ohno
mohno@scs.kyushu-u.ac.jp

Specialty section:
This article was submitted to
Geomagnetism and Paleomagnetism,
a section of the journal
Frontiers in Earth Science

Waxing/waning of the ice sheets and the associated change in thermohaline circulation have played an important role in global climate change since major continental ice sheets appeared in the northern hemisphere about 2.75 million years ago. In the earliest glacial stages, however, establishment of the linkage between ice sheet development and ocean circulation remains largely unclear. Here, we show new high-resolution records of marine isotope stage 100 recovered from deep-sea sediments on the Gardar Drift, in the subpolar North Atlantic. Results of a wide range of analyses clearly reveal the influence of millennial-scale variability in iceberg discharge on ocean surface condition and bottom current variability in the subpolar North Atlantic during marine isotope stage 100. We identified eight events of ice-rafted debris, which occurred mostly with decreases in sea surface temperature and in current components indicating North Atlantic Deep Water. These decreases are interpreted by weakened deep water formation linked to iceberg discharge, similarly to observations from the last glacial period. Dolomite fraction of the ice-rafted events in early MIS 100 like the last glacial Heinrich events suggests massive collapse of the Laurentide ice sheet in North America. At the same time, our early glacial data suggest differences from the last glacial period: absence of 1470-year periodicity in the interactions between ice sheets and ocean, and northerly shift of the ice-rafted debris belt. Our high-resolution data largely improve the picture of ice-sheet/ocean interactions on millennial time scales in the early glacial period after major Northern Hemisphere glaciation.

Keywords: ice-rafted debris, rock magnetism, North Atlantic deep water, early pleistocene, integrated ocean drilling program

INTRODUCTION

Collapse of continental ice sheets, one of the major threats posed by global warming, made an impact on the global ocean circulation and climate system through the Pleistocene. Lines of evidence have demonstrated that such linkage between the ice sheets and the ocean circulation repeatedly occurred in millennial time scale, as represented by the last glacial Dansgaard-Oeschger (D-O) cycles and Heinrich events (e.g., Hemming, 2004; Clement and Peterson, 2008). It has been argued that the abrupt climate change was caused by collapse of the continental ice sheets and the freshwater discharge into the North Atlantic Ocean, which reduced the North Atlantic Deep Water (NADW) production and weakened the ocean thermohaline circulation (e.g., Broecker, 1991, 1994). Subsequent studies suggested that the millennial-scale changes date back to Marine Isotope Stage (MIS) 100 around the intensification of Northern Hemisphere glaciation (NHG) (Bartoli et al., 2006; Becker et al., 2006; Bolton et al., 2010; Bailey et al., 2012, 2013). For instance at ODP Site 607 on the western flank of the Mid-Atlantic Ridge and Site 981 east of the Rockall Bank (**Figure 1**), Becker et al. (2006) recognized six ice-rafting events during MIS 100 and suggested millennial-scale changes in their ice-rafted debris (IRD) likely by processes similar to D-O cycles. However, the millennial-scale changes were unclear in the carbon isotope records of benthic foraminifer ($\delta^{13}C_{benthic}$) reflecting the bottom water conditions at least partly because of low data resolution. Paleoceanography at the onset of NHG should be improved by multiple data sources with higher time resolution.

Here, we demonstrate the millennial-scale links between iceberg discharge, surface ocean condition, and bottom current variability in the subpolar North Atlantic during the glacial MIS 100, the first pronounced glacial period in the northern hemisphere. Study material was collected at Integrated Ocean Drilling Program (IODP) Site U1314 in the Gardar Drift surrounded by different ice-covered continents. We show new high-resolution data from a wide range of analyses including IRD counts, mineral composition, and abundance of coccoliths. In addition, we focused on rock magnetic properties of sediments as a proxy of bottom current conditions. Rock magnetic properties

are a most potential proxy of bottom current conditions for the sediments poorly containing foraminifera. In the vicinity of the study site, the terrigenous material was largely transported by the northerly bottom current (NADW) in the interglacial periods and by the southerly bottom current (Lower Deep Water: LDW) in the glacial periods (e.g., Oppo and Lehman, 1993; Raymo et al., 2004; Curry and Oppo, 2005; Lynch-Stieglitz et al., 2007; Kissel et al., 2009; Grützner and Higgins, 2010; Alonso-Garcia et al., 2011). Sediments from the north transport have relatively higher magnetic coercivity reflecting the fine grained magnetite of Icelandic sources compared to sediments from the south. The north-south contrast in magnetic property was obviously demonstrated by the ratio of anhysteretic remanent magnetization to initial susceptibility for the last glacial sediments (Kissel, 2005). In our previous study of the IODP Site U1314, Sato et al. (2015) associated magnetic properties of sediments with the NADW production that appeared orbital-scale variability during 2.2–2.9 Ma and abrupt intensification after ~2.68 Ma.

MATERIALS AND METHODS

We analyzed sediment cores at IODP Site U1314 (56°21.9′N, 27°53.3′W in a water depth of 2820 m) in the Gardar Drift, which is surrounded by four different IRD sources; Laurentide, Greenland, Iceland, and European continent (**Figure 1**). A complete spliced section entirely consisting of clay-rich sediments was obtained down to 280 m composite depth (mcd) from the cores of three holes drilled with the advanced piston corer (Expedition 306 Scientists, 2006). Hayashi et al. (2010) constructed an orbital resolution age model for the period of 2.1–2.75 Ma tuned by a hybrid environmental proxy record to the global-standard oxygen isotope curve (Lisiecki and Raymo, 2005); Sato et al. (2015) modified the age model partly by constructing a longer age model to 2.9 Ma. In this study, we have analyzed IRD, X-ray diffraction (XRD), calcareous nannoplankton, and rock magnetic properties of the sediments between 229.5 and 234.5 mcd at a 2-cm interval (250 samples), which correspond to the age between 2.503 and 2.543 Ma with 100~200 years resolution.

In the analysis of IRD, weighed dry sediments (0.5 g) were sieved and the lithic grains (>150 μm) were counted. We excluded authigenic framboidal pyrite in counting.

For XRD measurements, each bulk material was ground and homogenized with an agate mortar. In order to convert X-ray peak intensities to weight percentage data on quartz and calcite and to determine normalized peak intensities on the other minerals, a powdered zincite standard (ZnO) was added in a fixed percentage (5 wt%) to the sample. Relative amounts (wt%) of quartz and calcite were determined by calibration curves measuring artificial standards comprising 5 wt% zincite standard plus matrix consisting of specified weight percentages of quartz and calcite. The XRD data were collected on a Rigaku X-ray Diffractometer RINT 2100 V, using CuKα radiation monochromatized by a curved graphite crystal in a step of 0.02° with a step-counting time of 2 sec. Intensity (area) of each

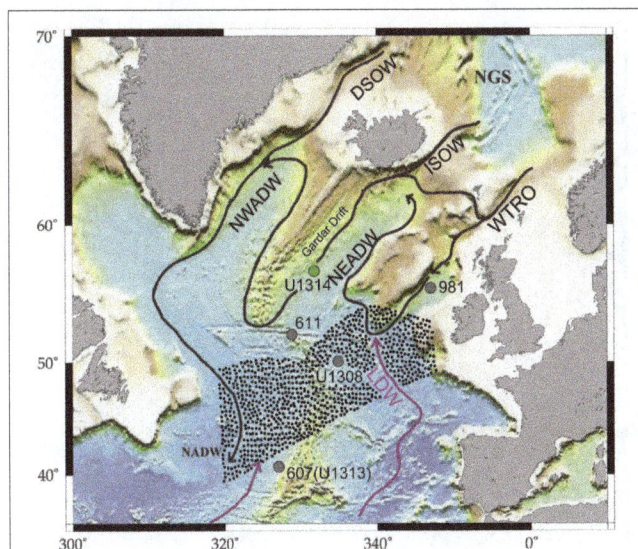

FIGURE 1 | Map of study site. Location of IODP Site U1314 and other sites discussed in the text plotted on bathymetric map of the northern North Atlantic [modified after Raymo et al. (2004)]. Paths of major deep water flows are indicated by arrows. Abbreviations not defined in text are: DSOW, Denmark Strait Overflow Water; NWADW, Northwest Atlantic Deep Water; NEADW, Northeast Atlantic Deep Water; NGS, Norwegian-Greenland Sea. The stippled area represents the Ruddiman IRD belt (Ruddiman, 1977), plotted after Hemming (2004).

elementary peak was calculated following the decomposition (profile fitting) procedure.

The absolute abundance of coccoliths (calcareous nannoplankton), calcareous plates formed by coccolithophores, was measured following the method of Chiyonobu et al. (2006). A powdered sample of 0.02 g was dispersed in 50 ml of water. A portion (0.5 ml) of the solution was then dried and hardened on the cover glass using ultraviolet light. The number of nannoplankton in an 18 mm × 20 μm area was counted through a 1500× microscope and converted into an amount contained in g^{-1} (specimens/g).

Magnetic properties of freeze-dried bulk sediments were measured with an alternating gradient force magnetometer [Princeton Measurements Corporation (Princeton, NY) Micromag 2900]. Hysteresis loops were measured with a maximum field of 1 T to determine the saturation remanence (Mr). We calculated S-ratios as the ratios of isothermal remanent magnetizations (IRMs) acquired in a DC back-field of 0.1 T divided by an initial IRMs acquired in a DC forward-field of 1 T. In addition, IRM acquisition curves were obtained at 29 magnetizing field steps from 1 mT to 1 T after demagnetized in decreasing AF field with the peak field of 1 T. The IRM gradient curves, which are a first derivative of IRM acquisition curve, represent the distribution of magnetic coercivity of each sample.

The observed IRM gradient curves were decomposed into two components with high magnetic and low magnetic coercivity, respectively, using the method in Sato et al. (2015). Rock magnetic experiments of the Site U1314 sediments indicated magnetite as the dominant magnetic mineral on the basis of IRM acquisition curves and S-ratios (Kanamatsu et al., 2009) and low-temperature and high-temperature magnetometry (Zhao et al., 2011).

RESULTS AND DISCUSSION

Ice-Rafting Events and Surface Ocean Condition

Development of continental ice sheets in the circum-North Atlantic region was inferred by IRD records. In the IODP U1314 site, eight ice-rafting events were recognized by the presence of IRD during MIS 100 (labeled as "a" to "h" in order of depth). All the ice-rafting events were accompanied by the deposition of volcanic glasses, which were especially abundant in event (e) at 2523 and event (a) at 2513 ka (**Figure 2B**). Since the volcanic glasses were absent in periods without IRD, the volcanic glass can be regarded as a component of IRD. They deposited largely onto growing ice sheets and/or drifting icebergs, however could have been transported directly by eruption. In this sense, we

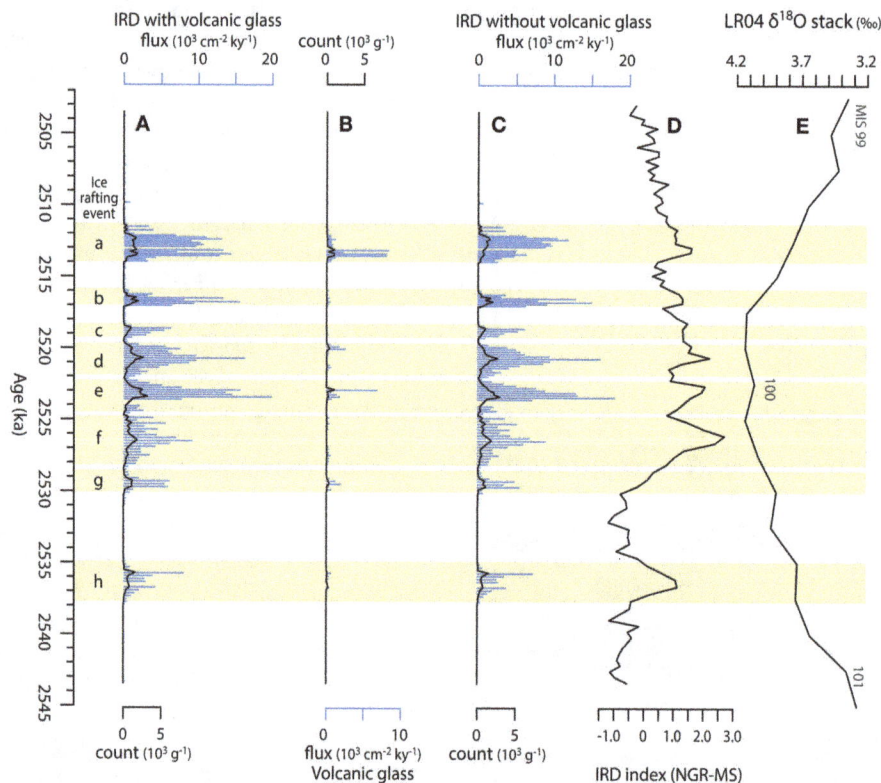

FIGURE 2 | Records of IRD during the MIS 100 glacial period from IODP Site U1314. (A) IRD with volcanic glass (coarse fraction: >150 μm); (B) Volcanic glass (coarse fraction: >150 μm); (C) IRD without volcanic glass (coarse fraction: >150 μm); (D) IRD index by Hayashi et al. (2010) with the age model by Sato et al. (2015); (E) The LR04 stack (Lisiecki and Raymo, 2005). In (A), (B), and (C), black curves and blue bars indicate counts and fluxes of IRD, respectively. Eight ice-rafting events (a, b, c, d, e, f, g, and h) recognized by IRD are shown on the left.

show the IRD in two different ways, including (**Figure 2A**) and excluding the volcanic glass (**Figure 2C**). We found that the IRD excluding the volcanic glass yields better correspondence with the IRD index proposed by Hayashi et al. (2010) for 2.1–2.75 Ma at the same site, which composites natural gamma radiation and initial magnetic susceptibility (**Figure 2D**). As Hayashi et al. (2010) recognized, IRD was conspicuous only when $\delta^{18}O$ values of benthic foraminiferal tests surpassed 3.5‰. This rule is kept in latter glacial stages during the Pleistocene (McManus et al., 1999; Mc Intyre et al., 2001).

Another important aspect is that the IRD record can be divided into three phases (**Figure 2**). The first phase is represented by the single isolated ice-rafting event (h), which occurred during the transitional stage from the MIS 101 interglacial period to the MIS 100 glacial period. Considering volume of ice sheets inferred from $\delta^{18}O$ value 3.7–3.8‰, this phase may be comparable to earlier glacial periods at MIS G4, G2, and 104, each of which is similarly characterized by a single prominent ice-rafting event (Hayashi et al., 2010). The second phase includes six ice-rafting events (g, f, e, d, c, and b) recurred at every two to three thousand years, which started after about 5 kyr from the event (h) in the first phase. Such clear multiple ice-rafting events first appeared in MIS 100 but not prominent in earlier glacial periods. The prolonged event (f) in earlier MIS

100 that has relatively low IRD flux (**Figures 2A,D**) is correlated with a prominent peak the IRD index (Hayashi et al., 2010). In subsequent events IRD increased and showed a saw-tooth oscillation accompanied by long-term decrease in IRD index. The maximum flux of IRD at events e, d, and b was as much as ca. 15,000–18,000 grains/cm²ky, comparable with the IRD flux during the mid-Pleistocene at the same site (Alonso-Garcia et al., 2011). The final phase includes the single ice-rafting event (a) that is separated from the event (b) by about 2-kyr period without IRD, and corresponds to the deglaciation stage between the MIS 100 glacial period and the MIS 99 interglacial period.

The most abundant lithic grains counted as IRD were composed of quartz. In addition, feldspars, amphibole, and mica were commonly observed. The observed IRD composition was consistent with XRD measurements (**Figure 3**), in which quartz shows the strongest correlation with IRD counts (**Figures 3A,B**); mica (**Figure 3C**), and feldspars (**Figures 3D,E**) also increased during ice-rafting events. We did not detect any signal of olivine that indicates contribution of IRD from Icelandic basalt. Signals of a small amount of dolomite were detected associated with the earlier four ice-rafting events (h, g, f, and e; **Figure 3F**). Pyrite was detected in a thin interval between events d and c (**Figure 3G**). As for minerals that are not related to IRD, calcite showed the predominance during interglacial periods and remarkable

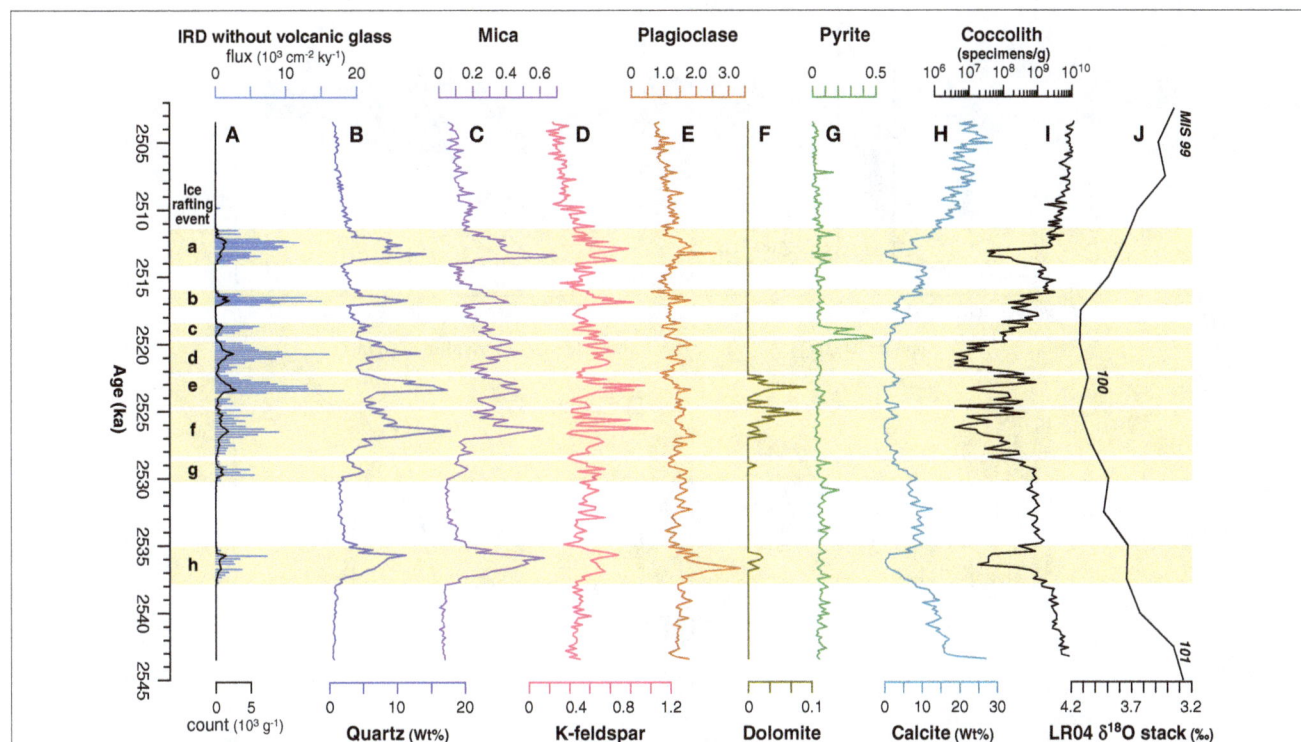

FIGURE 3 | Multi-proxy records of ice-rafting events and ocean surface condition during the MIS 100 glacial period from IODP Site U1314. (A) Ice-rafting events shown by count (black curve) and flux (blue bars) of IRD without volcanic glass; **(B–H)** Mineralogical characteristics of IRD (XRD measurements); **(I)** Ocean surface condition shown by count of coccoliths (logarithmic plot); **(J)** Global ice volume shown by the LR04 stack (Lisiecki and Raymo, 2005). In **(B)** and **(H)**, amounts of quartz and calcite were plotted in weight percent. Amounts of the other minerals were not large enough for estimating weight percent, and were represented by the values of the following representative diffraction peaks relative to that of ZnO (2.81Å): K-feldspar (3.24Å and 3.25Å), mica (10Å), plagioclase (3.21Å and 3.19Å), pyrite (2.71Å), and dolomite (2.89Å).

decline in the first and last ice-rafting events (**Figure 3H**). Total amount of the minerals determined in XRD measurements is estimated to cover up to 50 wt% of the sediments, and the undetermined fraction is likely to be clay minerals.

Surface ocean condition inferred from the absolute abundance of coccoliths (calcareous nannoplankton) showed some relation with ice-rafting events (**Figure 3I**) and also with calcite (**Figure 3H**). Prominent decline in coccolith abundance was recognized in the events h, f, e, d, and a (**Figure 3I**). Coccoliths have been used to interpret the paleoclimate (e.g., Haq, 1980; Sato et al., 2004) and their abundance at the subpolar North Atlantic sites largely depends on the temperature of surface water and the light regime (Baumann et al., 2000). The declined abundance in some ice-rafted events indicates ecological collapses of the coccolithophore community by lowered temperature or expanded sea ice coverage, or both. In turn the three short ice-rafted events g, c, and b were not accompanied by significant change in surface water condition. The record of coccolith abundance of the analyzed section depicts the glacial-interglacial trend as well, mostly reflecting orbital-scale variability in sea surface temperature.

Bottom Current Variability

The bottom current variability was reconstructed from rock magnetic properties of sediments because north currents transport basaltic sediments of Icelandic sources with relatively higher magnetic coercivity compared to acidic sediments transported by south current. **Figure 4** compares typical examples of rock magnetic measurements of sediments with high and low coercivity. The values of coercivity in hysteresis curves ranged from 12 to 19 mT (**Figures 4A,B**). In **Figure 4C**, IRM acquisition curves mostly saturated at 1 T, and therefore, IRM gradient curves (**Figure 4D**), the derivatives of acquisition curves, represent the distribution of coercivity. Using the variability of coercivity distribution, we decomposed the observed IRM gradient curves (**Figure 5D**) into two bottom current components; the interglacial component with high magnetic coercivity (**Figure 5G**), and the glacial component with low coercivity (**Figure 5F**), following the method of Sato et al. (2015). In decomposition, we used the two end-member curves determined by Sato et al. (2015) by averaging the IRM gradient curves of selected samples in 2.2~2.9 Ma: the samples with low S-ratio values (<0.54) for interglacial component and those with

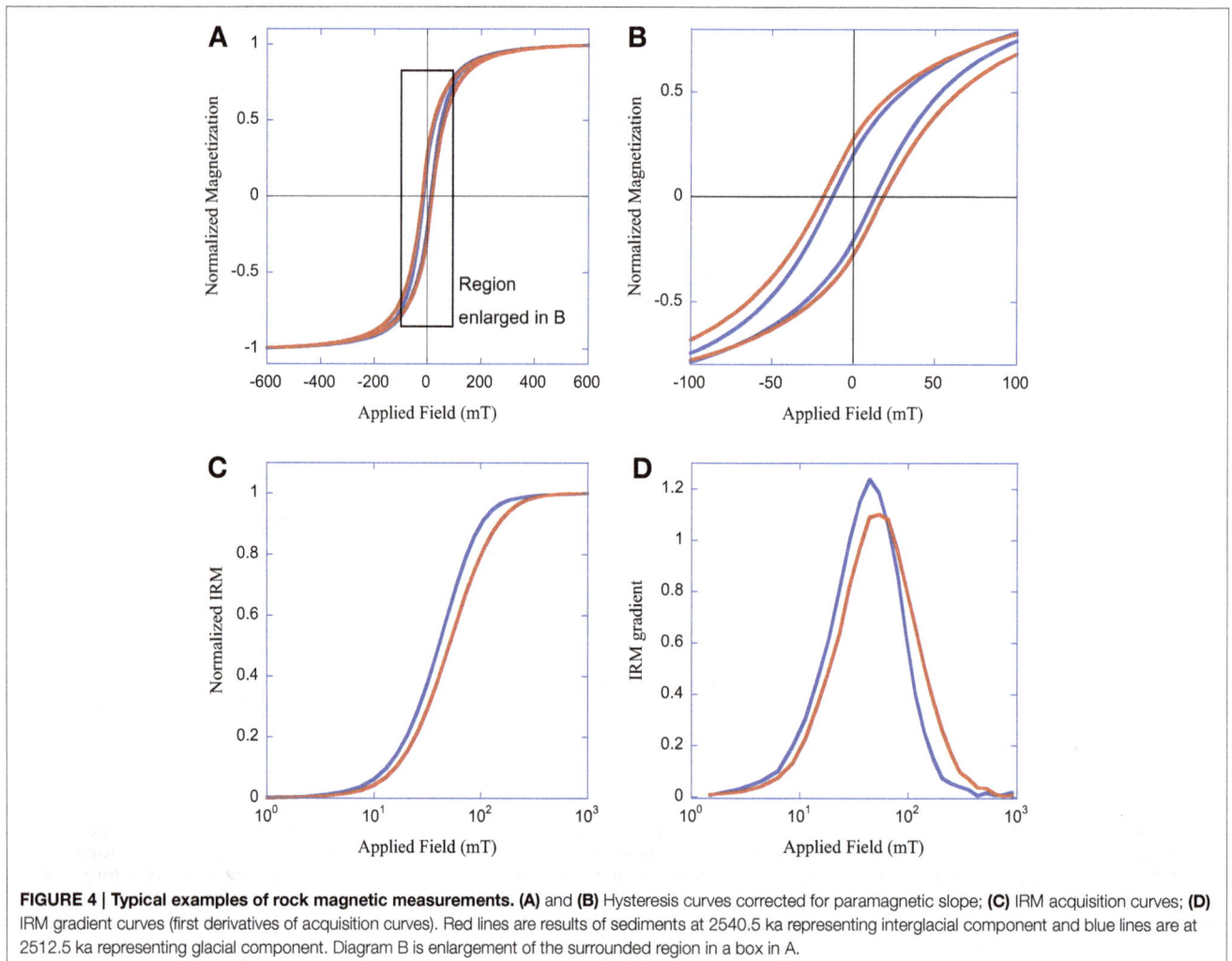

FIGURE 4 | Typical examples of rock magnetic measurements. (A) and **(B)** Hysteresis curves corrected for paramagnetic slope; **(C)** IRM acquisition curves; **(D)** IRM gradient curves (first derivatives of acquisition curves). Red lines are results of sediments at 2540.5 ka representing interglacial component and blue lines are at 2512.5 ka representing glacial component. Diagram B is enlargement of the surrounded region in a box in A.

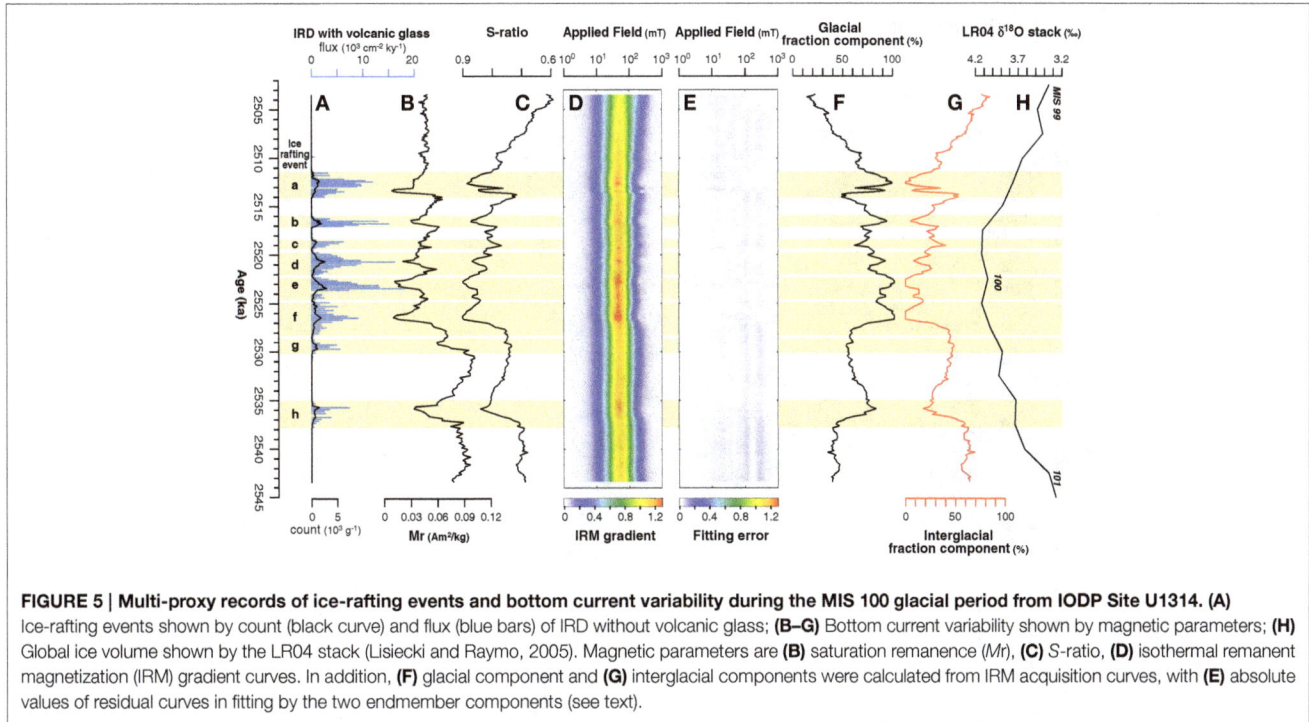

FIGURE 5 | Multi-proxy records of ice-rafting events and bottom current variability during the MIS 100 glacial period from IODP Site U1314. (A) Ice-rafting events shown by count (black curve) and flux (blue bars) of IRD without volcanic glass; **(B–G)** Bottom current variability shown by magnetic parameters; **(H)** Global ice volume shown by the LR04 stack (Lisiecki and Raymo, 2005). Magnetic parameters are **(B)** saturation remanence (*Mr*), **(C)** *S*-ratio, **(D)** isothermal remanent magnetization (IRM) gradient curves. In addition, **(F)** glacial component and **(G)** interglacial components were calculated from IRM acquisition curves, with **(E)** absolute values of residual curves in fitting by the two endmember components (see text).

high S-ratio values (>0.87) for glacial component were chosen, respectively. As a result, the fitting of the observed curve by two components was excellent throughout the study interval; fitting error (integral of the residual curve (R) in Sato et al., 2015) was sufficiently small with an average of 5% (**Figure 5E**). Noise originated from the IRD is negligible because contribution of IRD to the magnetic properties of bulk sediments is small; the geometric mean of the saturation remanence (*Mr*) of bulk sediments (0.051 Am^2/kg) is much higher than *Mr* of sieved (>150 µm) lithic grains (0.0082 Am^2/kg) of 14 representative samples during IRD events. This is confirmed by association between the decreases in *Mr* and the increases in IRD (**Figure 5**).

The bottom current variability reconstructed from rock magnetic properties indicated close connection with ice-rafting events (**Figures 5F,G**). The interglacial component with high magnetic coercivity decreases with ice-rafting events (**Figure 5G**), indicating weakened production of NADW that transports sediments with high magnetic coercivity from the north. Within each ice-rafting event the production of NADW declined rapidly at the commencement of the event, reached a minimum at the peak in the event, and subsequently recovered gradually. Repetition of similar abrupt reduction and gradual recovery of NADW in millennial-scale is reported in the last glacial period by Snowball and Moros (2003). It is noticeable that the gradual recovery of the NADW production continued after the end of the ice-rafting event, which is especially evident in events h, b, and a (**Figure 5G**); these events occurred separately from the succeeding event by more than 2000 years. After the ends of the other ice-rafting events, the NADW production could presumably recover but the recovery was interrupted by the next ice-rafting event that occurred shortly after.

Ice-Sheet/Ocean Interaction

Our high-resolution records in a wide range of analysis reveal a detailed history of ice-sheet/ocean interactions in the earliest Pleistocene subpolar North Atlantic (main results are summarized in **Figure 6**). The vigorous NADW production kept from the MIS 101 interglacial period abruptly reduced associated with the ice-rafting event (h) in the first phase. At that time, however, the production of NADW was able to recover gradually over the next several thousand years because the ice rafting was a single isolated event in the immature glacial stage. These are consistent with abrupt destruction and rapid recovery of surface ocean condition, which occurred synchronous with the event (h). In the full glacial stage (second phase), a series of six ice-rafting events (g–b) recurred in a relatively regular interval of ∼2600 years. These events almost continuously affected ocean surface condition and eventually gave rise to dramatic changes in NADW production. Although the relatively small event (g) gave little impact on NADW production, the next large events (f and e) induced the most serious reductions of NADW production in the MIS 100 glacial period. Subsequently, individual ice-rafting events (d, c, and b) involved some reductions of NADW production, although the long-term trend reversed to gradual recovery of NADW production. In the beginning of the third phase, NADW recovered production to the levels preceding the second phase, but the recovery was again interrupted by the isolated ice-rafting event (a). The impact of this final event was surprisingly large and involved serious destruction of surface ocean condition and the large reduction of NADW production as comparable to the largest ones in the second phase. After that, NADW recovered production vigorously to enter the MIS 99 interglacial period. A previous study (Becker et al., 2006)

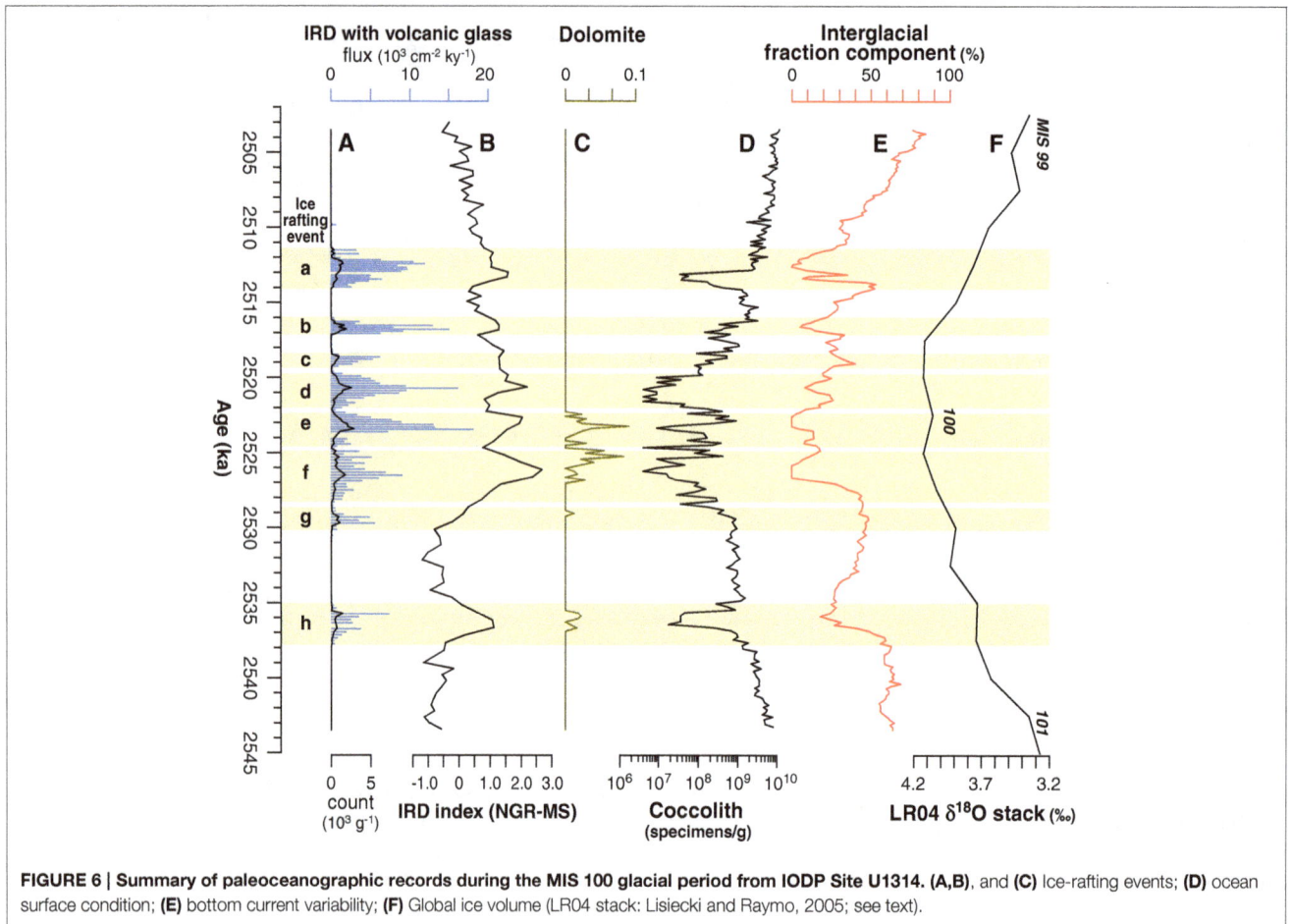

FIGURE 6 | Summary of paleoceanographic records during the MIS 100 glacial period from IODP Site U1314. (A,B), and **(C)** Ice-rafting events; **(D)** ocean surface condition; **(E)** bottom current variability; **(F)** Global ice volume (LR04 stack: Lisiecki and Raymo, 2005; see text).

recognized isotopic signals ($\delta^{13}C_{benthic}$) of weakened NADW production associated with ice-rafting events during the MIS 100 glacial period at ODP Site 607, but they observed no millennial-scale variation, likely because of the low resolution of the carbon isotope record (27 data points during MIS 100). Our new data presented here improve the understanding of millennial-scale activity of the bottom currents in the subpolar North Atlantic that are tightly linked with the development and collapse of circum-North Atlantic ice sheets during MIS 100.

Analogy of MIS 100 to the Last Glacial Period?

As discussed above, our data depicted that millennial-scale abrupt changes in the ice-sheet/ocean system were already established at least in the MIS 100 glacial period. From the viewpoint of the ice-sheet/ocean interaction, they are clearly similar to the last glacial D–O events. On the other hand, with our high-resolution data set, we could not find any evidence supporting that the 1470-year periodicity in the D–O cycles (Schulz, 2002) existed in MIS 100 records at Site U1314. The D–O like periodicity during the MIS G14, G6, and 104 glacial periods was reported from nearby ODP Site 984, but the existence of the last glacial 1470-year periodicity itself has been under debate (e.g., Ditlevsen et al., 2007). Instead, our IRD count data (**Figure 6A**)

show relatively regular spacing between the peaks of the events g–b, which is calculated to be 2600 years.

Among ice-rafting events during MIS 100, the first four (h, g, f, and e) were accompanied by dolomite fractions (**Figure 6C**), which are suggestive of the contribution of icebergs calved from the Laurentide ice sheet in North America (see e.g., Hemming, 2004). In particular, the events f and e may be similar to the last glacial Heinrich event in terms of their relatively large dolomite fractions likely from the Laurentide ice sheet and significantly reduced NADW production (**Figure 6E**). Ice rafting from Laurentide was reduced, and instead relative contribution of IRD from Greenland and European continent increased in the later events during MIS 100.

During the MIS 100 glacial period, the southward expansion of iceberg transportation was obviously more limited than that during the last glacial period. The amount and flux of IRD (>150 μm) (maximum: ca. 2900 grains/g; ca. 18,000 grains/cm²ky) at Site U1314 were comparable to or larger than that within the last glacial Ruddiman IRD belt in N50°–40° (Ruddiman, 1977; Hemming, 2004) [e.g., IODP Site U1308 (Bailey et al., 2010); IODP Site U1313 (Bolton et al., 2010)]. The largest flux of IRD (maximum: 38,000 grains/cm²ky) in the MIS 100 glacial period was reported at DSDP Site 611, which is located 500 km north of the center of the Ruddiman IRD belt (Bailey et al., 2012).

These suggest that the IRD belt shifted northerly in the earliest Pleistocene, likely due to a shorter range of iceberg transportation or northerly-located southern limit of ice-sheets.

CONCLUSIONS

High-resolution data from a wide range of analyses of a sediment core in the subpolar North Atlantic revealed millennial-scale ice-sheet/ocean interactions during MIS 100. Remarkable declines in coccolith abundance and in calcite fraction were observed accompanied by ice-rafting events indicating link between surface ocean condition and ice-rafting events. In addition, link of bottom current variability to ice-rafting events was revealed using rock magnetic properties of sediments. Within each ice-rafting event, the production of NADW declined rapidly at the commencement of the event, reached a minimum at the peak in the event, and subsequently recovered gradually beyond the termination of the event.

The history of ice-sheet/ocean interactions during the MIS 100 glacial period is interpreted as follows. This period is divided into three phases. The first phase (transitional stage from the MIS 101 interglacial period) was characterized by a single isolated ice-rafting event, like earlier ice-rafting events at the MIS G4, G2, and 104 glacial periods. The ice-rafting event involved abrupt destruction of surface ocean condition and rapid reduction of NADW production. After that, the production of NADW recovered gradually over the next several thousand years. In the second phase (full glacial stage), six ice-rafting events recurred at every two to three thousand years with saw-tooth oscillation pattern. The ice-rafting events almost continuously affected ocean surface condition and repeatedly induced serious reductions of the NADW production. In the third phase (transitional stage to the MIS 99 interglacial period), a single isolated ice-rafting event occurred again. The final ice-rafting event involved serious destruction of surface ocean condition and the remarkable reduction of NADW production as comparable to the largest events in the second phase.

Our data indicate that millennial-scale abrupt changes in the ice-sheet/ocean system similar to the last glacial D–O events were already established at least in the MIS 100 glacial period. Among ice-rafting events during MIS 100, two events (f and e) show similarity to the last glacial Heinrich event: massive collapse of the Laurentide ice sheet suggested by relatively large amount of dolomite fractions in addition to serious reduction in NADW production. In contrast, the 1470-year periodicity dominated in last glacial climatic and environmental conditions is absent in our data. In addition, it is likely that the southward expansion of iceberg transportation during the MIS 100 glacial period was more limited than that during the last glacial period.

AUTHOR CONTRIBUTIONS

MO and TH conceived and designed the research. MO and MS performed the rock magnetic experiments and analyses. YK performed the XRD measurements and analyses. TH and MS performed IRD counting. AM performed measurements of calcareous nannoplankton. IK and TS supervised AM. All authors contributed to the interpretation of results. MO and TH prepared the manuscript. All authors commented on and edited the manuscript.

ACKNOWLEDGMENTS

This study used samples provided by the Integrated Ocean Drilling Program (IODP). We are indebted to the staff of the RV JOIDES Resolution and the IODP Bremen Core Repository for their support. This study was performed under the cooperative research program of Center for Advanced Marine Core Research (CMCR), Kochi University (11A021 and 11B019). This study was partly supported by JSPS Grant-in-Aid for Scientific Research (22241006 to MO; 24651011 to IK). We acknowledge comments given by anonymous reviewers.

REFERENCES

Alonso-Garcia, M., Sierro, F. J., Kucera, M., Flores, J. A., Cacho, I., and Andersen, N. (2011). Ocean circulation, ice sheet growth and interhemispheric coupling of millennial climate variability during the mid-Pleistocene (ca 800-400ka). Q. Sci. Rev. 30, 3234–3247. doi: 10.1016/j.quascirev.2011.08.005

Bailey, I., Bolton, C. T., DeConto, R. M., Pollard, D., Schiebel, R., and Wilson, P. A. (2010). A low threshold for North Atlantic ice rafting from "low-slung slippery" late Pliocene ice sheets. Paleoceanography 25, PA1212. doi: 10.1029/2009pa001736

Bailey, I., Foster, G. L., Wilson, P. A., Jovane, L., Storey, C. D., Trueman, C. N., et al. (2012). Flux and provenance of ice-rafted debris in the earliest Pleistocene sub-polar North Atlantic Ocean comparable to the last glacial maximum. Earth Planet. Sci. Lett. 341–344, 222–233. doi: 10.1016/j.epsl.2012.05.034

Bailey, I., Hole, G. M., Foster, G. L., Wilson, P. A., Storey, C. D., Trueman, C. N., et al. (2013). An alternative suggestion for the Pliocene onset of major northern hemisphere glaciation based on the geochemical provenance of North Atlantic Ocean ice-rafted debris. Quat. Sci. Rev. 75, 181–194. doi: 10.1016/j.quascirev.2013.06.004

Bartoli, G., Sarnthein, M., and Weinelt, M. (2006). Late Pliocene millennial-scale climate variability in the northern North Atlantic prior to and after the onset of Northern Hemisphere glaciation. Paleoceanography 21, 4. doi: 10.1029/2005pa001185

Baumann, K.-H., Andruleit, H., and Samtleben, C. (2000). Coccolithophores in the Nordic Seas: comparison of living communities with surface sediment assemblages. Deep Sea Res. II Top. Stud. Oceanogr. 47, 1743–1772. doi: 10.1016/S0967-0645(00)00005-9

Becker, J., Lourens, L. J., and Raymo, M. E. (2006). High-frequency climate linkages between the North Atlantic and the Mediterranean during marine oxygen isotope stage 100 (MIS100). Paleoceanography 21:PA3002. doi: 10.1029/2005PA001168

Bolton, C. T., Wilson, P. A., Bailey, I., Friedrich, O., Beer, C. J., Becker, J., et al. (2010). Millennial-scale climate variability in the subpolar North Atlantic Ocean during the late Pliocene. Paleoceanography 25:PA4218. doi: 10.1029/2010PA001951

Broecker, W. S. (1991). The great ocean conveyor. Oceanography 4, 79–89. doi: 10.5670/oceanog.1991.07

Broecker, W. S. (1994). Massive iceberg discharges as triggers for global climate change. Nature 372, 421–424. doi: 10.1038/372421a0

Chiyonobu, S., Sato, T., Narikiyo, R., and Yamasaki, M. (2006). Floral changes in calcareous nannofossils and their paleoceanographic significance in the equatorial Pacific Ocean during the last 500 000 years. *Island Arc 15*, 476–482. doi: 10.1111/j.1440-1738.2006.00543.x

Clement, A. C., and Peterson, L. C. (2008). Mechanisms of abrupt climate change of the last glacial period. *Rev. Geophys.* 46:RG4002. doi: 10.1029/2006RG000204

Curry, W. B., and Oppo, D. W. (2005). Glacial water mass geometry and the distribution of d 13C of SCO2 in the western Atlantic Ocean. *Paleoceanography 20*, 1–12. doi: 10.1029/2004PA001021

Expedition 306 Scientists (2006). "Site U1314," in *Proceedings Integrated Ocean Drilling Program*, eds J. E. T. Channell, T. Kanamatsu, T. Sato, R. Stein, C. A. Alvarez Zarikian, M. J. Malone, and the Expedition 303/306 Scientists (College Station, TX: Integrated Ocean Drilling Program Management International, Inc.), 1–95.

Ditlevsen, P. D., Andersen, K. K., and Svensson, A. (2007). The DO-climate events are probably noise induced: statistical investigation of the claimed 1470 years cycle. *Clim. Past 3*, 129–134. doi: 10.5194/cp-3-129-2007

Grützner, J., and Higgins, S. M. (2010). Threshold behavior of millennial scale variability in deep water hydrography inferred from a 1.1 Ma long record of sediment provenance at the southern Gardar Drift. *Paleoceanography* 25:PA4204. doi: 10.1029/2009PA001873

Haq, B. U. (1980). Biogeographic history of Miocene calcareous nannoplankton and paleoceanography of the Atlantic Ocean. *Micropaleontology 26*, 414–443. doi: 10.2307/1485353

Hayashi, T., Ohno, M., Acton, G., Guyodo, Y., Evans, H. F., Kanamatsu, T., et al. (2010). Millennialscale iceberg surges after intensification of Northern Hemisphere glaciation. *Geochem. Geophys. Geosyst.* 11:Q09Z20. doi: 10.1029/2010GC003132

Hemming, S. R. (2004). Heinrich events: massive late Pleistocene detritus layers of the North Atlantic and their global climate imprint. *Rev. Geophys.* 42, 1–43. doi: 10.1029/2003rg000128

Kanamatsu, T., Ohno, M., Acton, G., Evans, H., and Guyodo, Y. (2009). Rock magnetic properties of the Gardar Drift sedimentary sequence, Site IODP U1314, North Atlantic: implications for bottom current change through the mid-Pleistocene. *Mar. Geol.* 265, 31–39. doi: 10.1016/j.margeo.2009.06.012

Kissel, C. (2005). Magnetic signature of rapid climatic variations in glacial North Atlantic, a review. *Comptes Rendus Geosci.* 337, 908–918. doi: 10.1016/j.crte.2005.04.009

Kissel, C., Laj, C., Mulder, T., Wandres, C., and Cremer, M. (2009). The magnetic fraction: a tracer of deep water circulation in the North Atlantic. *Earth Planet. Sci. Lett.* 288, 444–454. doi: 10.1016/j.epsl.2009.10.005

Lisiecki, L. E., and Raymo, M. E. (2005). A Pliocene-Pleistocene stack of 57 globally distributed benthic d 18O records. *Paleoceanography 20*, 1–17. doi: 10.1029/2004PA001071

Lynch-Stieglitz, J., Adkins, J. F., Curry, W. B., et al. (2007). Atlantic meridional overturning circulation during the Last Glacial Maximum. *Science 316*, 66–69. doi: 10.1126/science.1137127

Mc Intyre, K., Delaney, M. L., and Ravelo, A. C. (2001). Millennial-scale climate changes and oceanic processes in the late Pliocene and early Pleistocene. *Paleoceanography* 16, 535–543. doi: 10.1029/2000PA000526

McManus, J. F., Oppo, D. W., and, Cullen, J. L. (1999). A 0.5- million-year record of millennial-scale climate variability in the North Atlantic. *Science* 283, 971–975. doi: 10.1126/ science.283.5404.971.

Oppo, D. W., and Lehman, S. J. (1993). Mid-depth circulation of the subpolar North Atlantic during the last glacial maximum. *Science 259*, 1148–1152. doi: 10.1126/science.259.5098.1148

Raymo, M. E., Oppo, D. W., Flower, B. P., et al. (2004). Stability of North Atlantic water masses in face of pronounced climate variability during the Pleistocene. *Paleoceanography* 19, 1–13. doi: 10.1029/2003pa 000921

Ruddiman, W. F. (1977). North Atlantic ice-rafting: a major change at 75,000 years before the present. *Science 196*, 1208–1211. doi: 10.1126/science.196.4295. 1208

Sato, M., Makio, M., Hayashi, T., and Ohno, M. (2015). Abrupt intensification of North Atlantic Deep Water formation at the Nordic Seas during the late Pliocene climate transition. *Geophys. Res. Lett.* 42, 4949–4955. doi: 10.1002/ 2015GL063307

Sato, T., Yuguchi, S., Takayama, T., and Kameo, K. (2004). Drastic change in the geographical distribution of the cold-water nannofossil *Coccolithus pelagicus* (Wallich) Schiller at 2.74 Ma in the late Pliocene, with special reference to glaciation in the Arctic Ocean. *Mar. Micropaleontol.* 52, 181–193. doi: 10.1016/j.marmicro.2004.05.003

Schulz, M. (2002). On the 1470-year pacing of Dansgaard-Oeschger warm events. *Paleoceanography* 17. doi: 10.1029/2000PA000571

Snowball, I., and Moros, M. (2003). Saw-tooth pattern of North Atlantic current speed during Dansgaard-Oeschger cycles revealed by the magnetic grain size of Reykjanes Ridge sediments at 59°N. *Paleoceanography18*, 4–1. doi: 10.1029/2001pa000732

Zhao, M., Ohno, M., Kuwahara, Y., Hayashi, T., and Yamashita, T. (2011). *Magnetic Minerals in Sediments from IODP Site U1314 Determined by Low-Temperature and High-Temperature Magnetism.* Bulletin of the Graduate School of Social and Cultural Studies, Kyushu University, Vol. 17, 77–84.

Conflict of Interest Statement: The authors declare that the research was conducted in the absence of any commercial or financial relationships that could be construed as a potential conflict of interest.

12

Extended-Range Ensemble Predictions of Convection in the North Australian Monsoon Region

Wasyl Drosdowsky and Matthew C. Wheeler *

Research and Development, Bureau of Meteorology, Melbourne, VIC, Australia

Extended-range (<35 day) predictions of area-averaged convection over northern Australia are investigated with the Bureau of Meteorology's Predictive Ocean-Atmosphere Model for Australia (POAMA). Hindcasts from 1980-2011 are used, initialized on the 1st, 11th, and 21st of each month, with a 33-member ensemble. The measure of convection is outgoing longwave radiation (OLR) averaged over the box $120°E$-$150°E$, $5°S$-$17.5°S$. This averaging serves to focus on the intraseasonal and longer time scales, and is an area of interest to users. The raw hindcasts of daily OLR show a strong systematic adjustment away from their initial values during the first week, and then converge to a mean seasonal cycle of similar amplitude and phase to observations. Hence, forecast OLR anomalies are formed by removing the model's own seasonal cycle of OLR, which is a function of start time and lead time, a usual practice for dynamical seasonal prediction. Over all hindcasts, the model forecast root-mean-square (RMS) error is smaller than the RMS error of persistence and climatological reference forecasts for leads 3–35 days. Ensemble spread is less than the forecast RMS error (i.e., under-spread) for days 1–12, but slightly greater than the RMS error for longer leads. Binning the individual forecasts based on ensemble spread shows a generally positive relationship between spread and error. Therefore, greater certainty can be given for forecasts with smaller spread.

Keywords: Australian monsoon, monsoon prediction, tropical prediction, intraseasonal, extended-range, subseasonal, POAMA, dynamical prediction system

INTRODUCTION

As defined by the World Meteorological Organization, extended-range weather forecasts cover the lead-time range of 10–30 days. Stakeholders in agriculture, industry, and the resources sector have continually called out for forecasts on this intermediate range, but few operational products exist. At the Bureau of Meteorology, for example, the only operational product that currently focuses on this range is the Weekly Tropical Climate Note (WTCN: http://www.bom.gov.au/climate/tropnote/tropnote.shtml), which currently provides non-quantitative text-based outlooks of likely large-scale tropical conditions for the coming few weeks.

In recognition of the demand for quantitative extended-range forecast products, the Bureau of Meteorology has sought to provide such quantitative guidance through further development of its dynamical coupled ocean-atmosphere prediction system (Hudson et al., 2011, 2013). Testing of the evolving model/system for skill on the extended range (also known as the intraseasonal or multi-week range) has mostly concentrated on weekly or longer averages of grid-point fields

(Hudson et al., 2011, 2013; Marshall et al., 2014b) or on daily indices of large-scale climate phenomena such as the Madden–Julian Oscillation (MJO; Marshall et al., 2011), Southern Annular Mode (SAM; Marshall et al., 2012), and blocking (Marshall et al., 2014a). Zhu et al. (2014) took the approach of examining the skill of a seamless range of time scales, including the extended range, by using time averages equal in length to the forecast lead time. Recently, Marshall and Hendon (2015) examined the skill of predicting Australian monsoon indices of rain and wind. Together, this research provides encouraging signs of useful extended-range skill in many locations provided there is suitable time-averaging or selection of the intraseasonal climate signals.

For northern Australia, strong intraseasonal variability of tropical convection and rainfall has been long appreciated and documented (see review by Wheeler and McBride, 2011). A frequently-used measure of convection in the Australian monsoon region, and one that is of relevance to the WTCN, is the area-averaged outgoing longwave radiation (OLR) over the box 120°E-150°E, 5°S-17.5°S, as displayed in **Figure 1**. This figure uses dark shading to indicate a negative OLR anomaly, which is indicative of a greater number of cold clouds and/or colder cloud tops than normal in the region, i.e., enhanced convection. Strong intraseasonal variability of convection can be seen in most years. For example, the 2007/08 wet season is made up of about three complete intraseasonal cycles, with monsoon "bursts" occurring in mid-November 2007, late December 2007 to early January 2008, and most of February 2008 (see also Wheeler, 2008). The intraseasonal variability in 2007/08 can be seen to be well correlated to the MJO, as indicated by the times of MJO Phases 4-6 (horizontal thick lines in **Figure 1**, using the definition of MJO phases of Wheeler and Hendon, 2004). It is this empirical relationship that is one of the main inputs to the WTCN. However, in some other years (e.g., 2009/10, 2010/11) any relationship with the MJO is less apparent, and other variability plays an important role (examples provided in Wheeler and McBride, 2011). Noting that the WTCN is currently heavily reliant on the MJO for its outlooks, and that the Bureau's dynamical prediction system attempts to model all of the important sources of variability, it is of interest to see how well the dynamical prediction system performs for this region.

In this work we therefore investigate the quantitative extended-range prediction of the aforementioned area-averaged OLR using an ensemble of hindcasts from the Bureau's operational coupled modeling system, the Predictive Ocean Atmosphere Model for Australia (POAMA) version 2 M (hereafter POAMA-2M). We investigate the model forecast bias, the removal of this bias, the resulting prediction skill, the ensemble spread versus error, and a real-time forecast display.

DATA AND MODEL FORECAST SYSTEM
Observational OLR and MJO Data
The observed OLR data is the NOAA satellite interpolated OLR (Liebmann and Smith, 1996) available from 1974 to the present. Daily MJO index data are the Real-time Multivariate MJO (RMM) index of Wheeler and Hendon (2004) obtained from http://www.bom.gov.au/climate/mjo.

POAMA-2M Forecast System
We analyse POAMA-2M (Hudson et al., 2013) which currently (early 2017) produces the Bureau of Meteorology's operational monthly and seasonal forecasts. This version of POAMA was developed specifically to provide more skilful output on the extended-range time scale (hence the letter "M" for multi-week). Improvements included in this version of POAMA are the use of perturbed atmosphere and ocean initial conditions and a burst ensemble (an ensemble starting from a single initial time as opposed to a lagged ensemble), as well as the use of three different model configurations (using different convective parameterizations or flux correction at the ocean surface) to form a multi-model ensemble (Hudson et al., 2013).

The atmospheric component of POAMA-2M is a spectral model with resolution T47 (~250 km grid) and 17 vertical levels. The ocean component has a zonal resolution of 2° and a varying meridional resolution of 0.5–1.5° with 25 vertical levels. The unperturbed initial conditions are provided by separate data assimilation schemes for the ocean versus the atmosphere and land. For the atmospheric and land initial states, they are generated by nudging of wind, temperature, and humidity toward one of two analysis products that come from different global models (i.e., different to the model used in POAMA-2M). The hindcasts and forecasts are therefore more likely to suffer from "initial shock" than a model that has its own atmospheric data assimilation (Hudson et al., 2011). Perturbations to the initial conditions of the central member are generated using a coupled breeding scheme. Ten perturbed states are produced, providing 11 different initial states that are input to three different configurations of the model, providing a 33-member ensemble (Hudson et al., 2013). This description applies to both the hindcasts and real-time forecasts.

HINDCASTS AND BIAS REMOVAL

We analyse hindcasts from POAMA-2M that have been initialized on the 1st, 11th, and 21st of each month for the period 1980 to 2011. Observations of OLR are not used as part of the model initialization. Instead, the model OLR is computed by the model's radiation scheme and depends critically on the production of convection and clouds by the model's convective parameterization. Therefore, the model OLR is not necessarily the same as observed at the initial condition. Further, since the atmospheric initial conditions are produced by nudging toward a different model (see above), there is high potential for initial shock of the model OLR as the model shifts toward its own attractor. This is explored in **Figure 2**, which shows the annual cycle of observed OLR (black curve) together with the annual cycle of all the day 1 hindcasts (blue) and day 20 hindcasts (green), for the region of interest. Interestingly, the initial (day 1) OLR is close to observed during the wetter months of December-April, but is systematically about 15 Wm^{-2} higher than observed during the drier months of June-October, whereas the day 20 OLR is close to observed during the drier months and systematically too high during the wetter months. Therefore, the initial shock in the OLR field is to increasing OLR (i.e., less

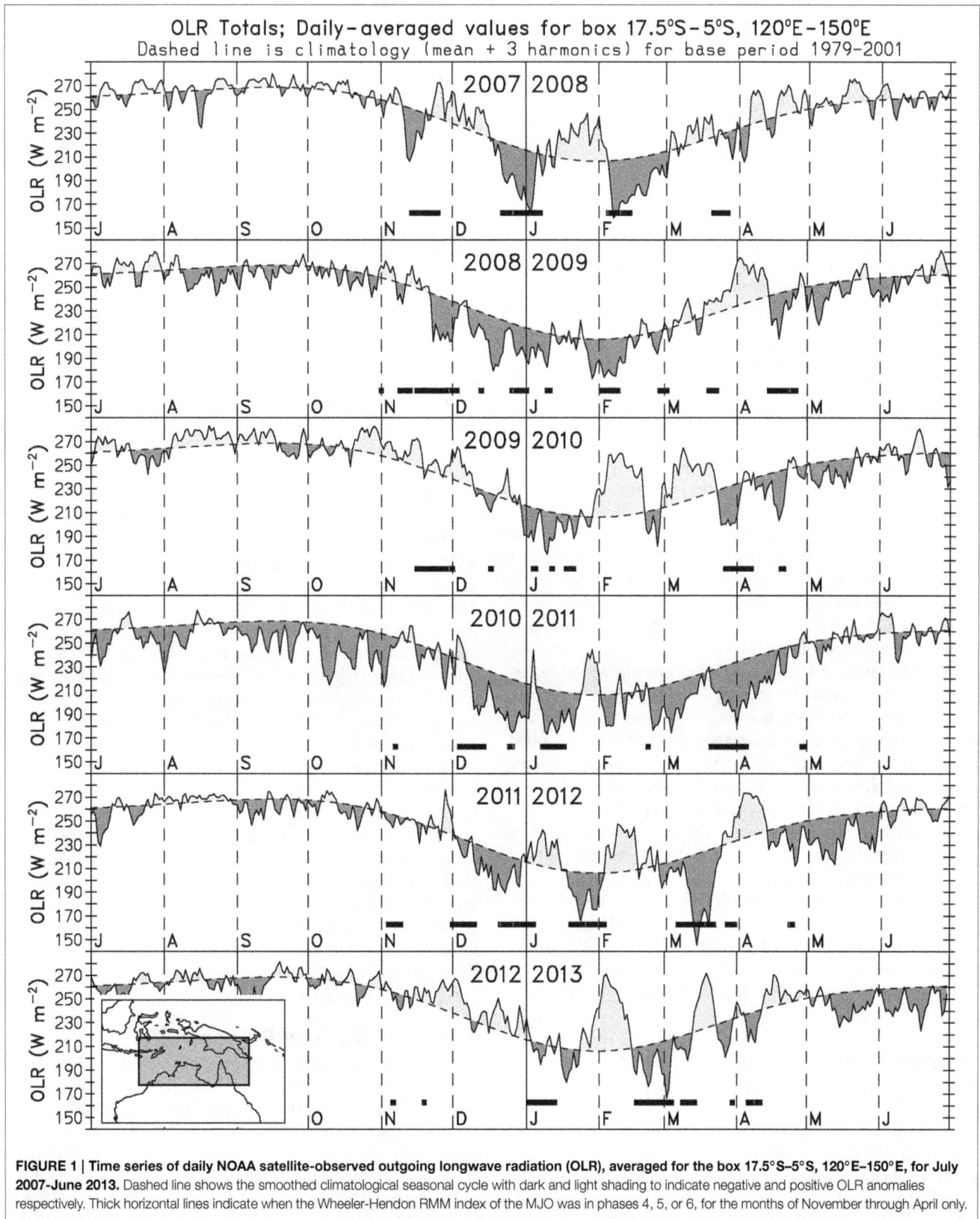

FIGURE 1 | Time series of daily NOAA satellite-observed outgoing longwave radiation (OLR), averaged for the box 17.5°S–5°S, 120°E–150°E, for July 2007–June 2013. Dashed line shows the smoothed climatological seasonal cycle with dark and light shading to indicate negative and positive OLR anomalies respectively. Thick horizontal lines indicate when the Wheeler-Hendon RMM index of the MJO was in phases 4, 5, or 6, for the months of November through April only.

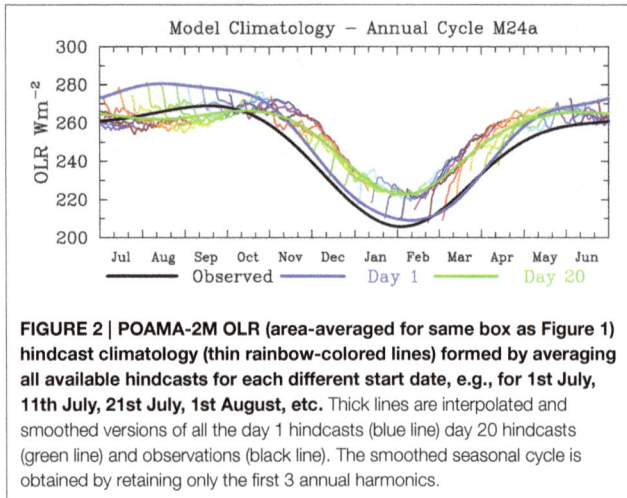

FIGURE 2 | POAMA-2M OLR (area-averaged for same box as Figure 1) hindcast climatology (thin rainbow-colored lines) formed by averaging all available hindcasts for each different start date, e.g., for 1st July, 11th July, 21st July, 1st August, etc. Thick lines are interpolated and smoothed versions of all the day 1 hindcasts (blue line) day 20 hindcasts (green line) and observations (black line). The smoothed seasonal cycle is obtained by retaining only the first 3 annual harmonics.

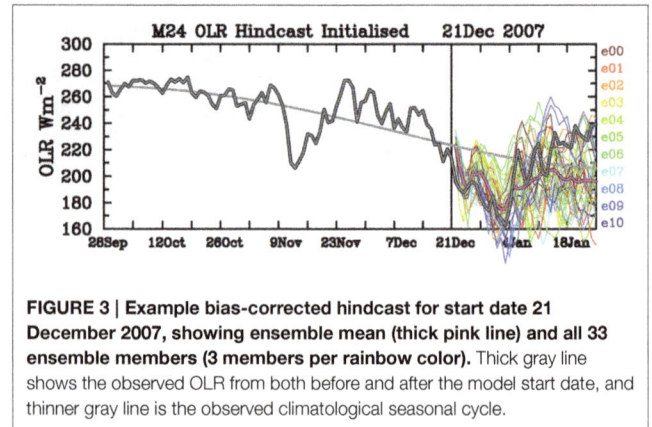

FIGURE 3 | Example bias-corrected hindcast for start date 21 December 2007, showing ensemble mean (thick pink line) and all 33 ensemble members (3 members per rainbow color). Thick gray line shows the observed OLR from both before and after the model start date, and thinner gray line is the observed climatological seasonal cycle.

convection) in the climatologically-wet months but decreasing OLR (i.e., more convection or colder surface temperatures) in the dry months. Further, the individual hindcast climatologies from each start date (thin colored lines) demonstrate that most of this initial shock occurs in the first few days of the hindcasts. This figure is a strong demonstration for the need to compute anomalies for the model with respect to the lead-time dependent hindcast climatology, a procedure that is now common for dynamical seasonal prediction (Stockdale, 1997; Hudson et al., 2011). This is a first-order linear correction for the initial shock, model drift, and model mean bias.

Separate lead-dependent hindcast climatologies are calculated for each of the three model configurations making up the multi-model ensemble. Resulting individual ensemble anomalies, and the multi-model ensemble mean anomaly, may then be plotted relative to the observed climatology, as shown for an example hindcast in **Figure 3**. In this example the forecast ensemble mean (thick pink line) is shown to track the verifying observations quite well in the first couple of weeks, with the ensemble members gradually spreading around it. Note that the same color is used for ensemble members using the same initial condition, as input to the three different model configurations, resulting in three lines of each color. Interestingly, there is a grouping of these ensemble members for short lead times, showing that for the first few days it is the initial condition that most determines the forecast trajectory, rather than the model physics. We will return to this issue in the next section.

HINDCAST PERFORMANCE AND SKILL-SPREAD RELATIONSHIP

Model OLR vs. Observed OLR

Model performance is evaluated for the bias-corrected multi-model ensemble mean OLR anomalies in comparison with the verifying observed OLR anomalies. Two metrics have been calculated; the correlation (shown later) and the root mean square (RMS) error for lead times from 1 to 35 days. These measures are calculated over all the hindcasts, and over various subsets. **Figure 4** shows the RMS error for the ensemble mean

from all hindcasts (upper panel) and for the summer monsoon months of December through March (DJFM; lower panel). Also shown are the RMS error for forecasts computed using climatology (i.e., a zero anomaly), the RMS error for persistence of the initial daily anomaly, and the hindcast "skill" computed as a percentage improvement of POAMA over climatology (using the same numerical scale as the error, i.e., from 0 to 40%).

The RMS error of POAMA is smaller than that of both persistence and climatology for the range 2–35 days for both the all season and the summer monsoon (i.e., DJFM) season. Although the POAMA hindcast RMS error is greater in DJFM, the hindcasts are more skilful relative to climatology during this season, with a percentage improvement of about 36% at a lead of 1 day compared to 30% for the all-seasons case, and a percentage improvement of about 16% compared to 9% at a lead of 10 days. We contend that these percentage improvements represent a useful level of skill.

The spread (pink lines in **Figure 4**) for the all-season case is less than the model RMS error for days 1 to 12, i.e., indicating that the ensemble is under-spread. Between days 12 and 16 the spread is very close to the RMS error for both the model forecasts and climatological forecasts. Beyond day 16 the spread of the ensemble exceeds the RMS error of a climatological forecast, which implies that the POAMA forecasts have slightly greater variance than observed. For the summer months only, the ensemble appears to be under-spread for lead times up to about 20 days, and the spread stays below the RMS error of a climatological forecast for all leads.

Impact of MJO on Forecast Skill

As noted in the introduction, some, but not all, seasons show a strong relationship between the OLR over northern Australia and the phase of the MJO. It has also been previously shown that prediction of the RMM index of the MJO is somewhat more skilful when the MJO is strong at the initial time (Rashid et al., 2011). It therefore seems a reasonable hypothesis that the prediction of OLR over northern Australia should be more skilful when the MJO is strong in the initial conditions. We test this hypothesis by examining the impact of the MJO on the forecast skill when stratifying the hindcasts according to the presence or absence of a strong MJO signal (**Figure 5A**), and also by the phase

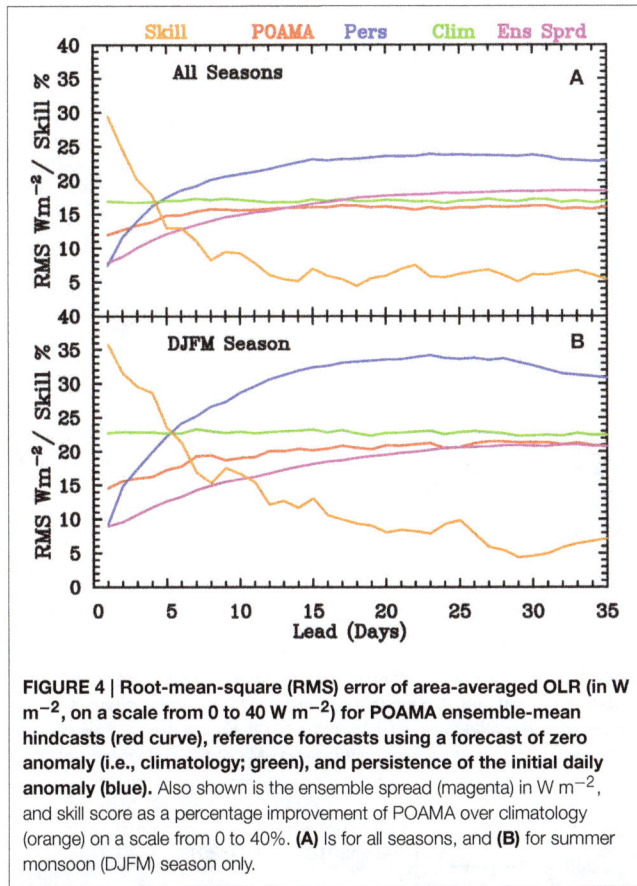

FIGURE 4 | Root-mean-square (RMS) error of area-averaged OLR (in W m^{-2}, on a scale from 0 to 40 W m^{-2}) for POAMA ensemble-mean hindcasts (red curve), reference forecasts using a forecast of zero anomaly (i.e., climatology; green), and persistence of the initial daily anomaly (blue). Also shown is the ensemble spread (magenta) in W m^{-2}, and skill score as a percentage improvement of POAMA over climatology (orange) on a scale from 0 to 40%. (A) Is for all seasons, and (B) for summer monsoon (DJFM) season only.

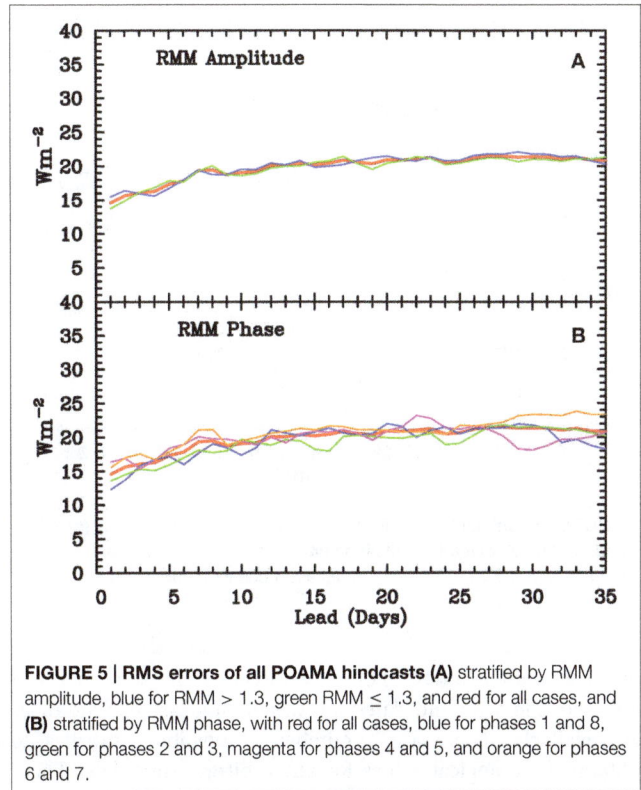

FIGURE 5 | RMS errors of all POAMA hindcasts (A) stratified by RMM amplitude, blue for RMM > 1.3, green RMM ≤ 1.3, and red for all cases, and (B) stratified by RMM phase, with red for all cases, blue for phases 1 and 8, green for phases 2 and 3, magenta for phases 4 and 5, and orange for phases 6 and 7.

of the MJO (**Figure 5B**) at the initial time. However, there is no evident impact of the amplitude or phase of the MJO on the hindcast RMS error. For the correlation skill (not shown), there appears a weak increase in correlation between days 2 and 9 for hindcasts initialized with a strong MJO, but it is not statistically significant. Thus we cannot confirm our hypothesis above. This appears consistent with the result of Marshall et al. (2011) who found that although POAMA was able to correctly simulate and predict the relationship between the MJO and rainfall (i.e., convection) over most of the tropical Indo-Pacific, it was not able to do this over the Maritime Continent and northern Australia, indicating that there is still room for improvement in these extended-range forecasts for northern Australia.

Ensemble Spread vs. Observed Uncertainty and Error

The relationship between the spread and the underlying observed variability is further highlighted by the rank histogram or "Talagrand diagram" (Talagrand et al., 1997; Hamill, 2001), shown in **Figure 6** using all hindcasts. This histogram is constructed by counting where the verifying observation lies amongst all the ensemble members for each hindcast. Ideally the distribution should be flat, which occurs when the ensemble spread matches the observed variability, whereas a U-shaped distribution indicates insufficient spread with many verifying observations falling near the extremes or outside the range of

the ensemble members, and a domed distribution indicates too much ensemble spread. All three types of distributions can be seen in **Figure 6**, which shows too little spread at 3 days lead, too much spread at 20 days lead, and about the right level of spread at 10 days lead. This result is consistent with **Figure 4A** which suggested that ensemble spread was insufficient at short leads (up to ~12 days) and over spread beyond about 16 days. The lack of ensemble spread at short lead times is exacerbated by the use of the same initial conditions for each of the three model configurations. This shows up in **Figure 6** in the strong peak found at every third bin, due to the very small separation between the three ensemble members with identical initial conditions. As discussed by Hudson et al. (2013), this clearly signifies the need for using different initial conditions for each of the three model configurations.

Binning the individual forecasts based on ensemble spread shows a generally positive relationship between spread and error (**Figure 7**). The spread-error correlations are 0.55 for the 3-day lead, 0.57 for the 10-day lead, and 0.90 for the 20-day lead. These correlations are more encouraging of a positive relationship between spread and error than others that have been listed in the literature (as reviewed by Grimit and Mass, 2007), especially for the 20-day lead. Therefore, especially for the 20-day lead, greater certainty can be given to forecasts with smaller spread.

REAL-TIME FORECASTS

Real-time forecasts have been performed once a week since August 2011 and twice a week since February 2013, and run

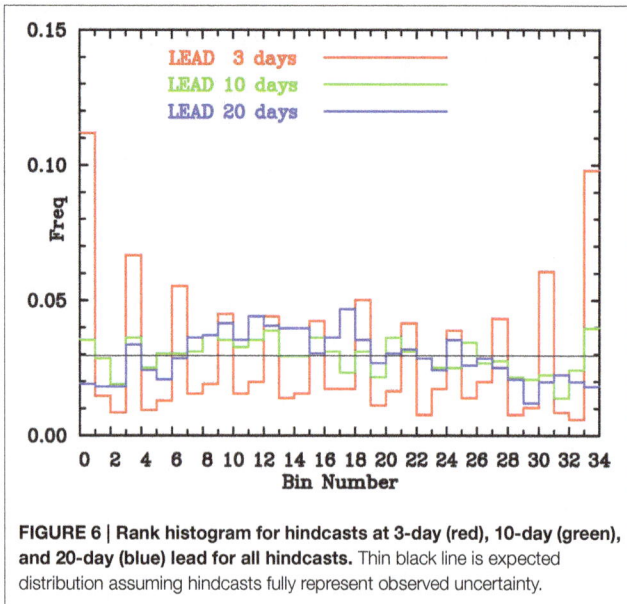

FIGURE 6 | Rank histogram for hindcasts at 3-day (red), 10-day (green), and 20-day (blue) lead for all hindcasts. Thin black line is expected distribution assuming hindcasts fully represent observed uncertainty.

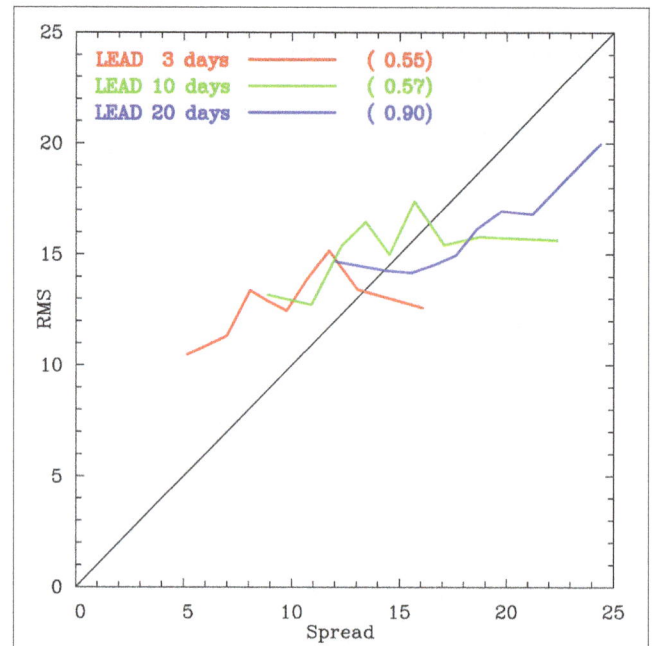

FIGURE 7 | Spread vs. RMS error for hindcasts at a lead of 3 days (red) 10 days (green) and 20 days (blue). Binning of the hindcasts is based on the ensemble spread, using 9 bins containing 128 hindcasts each. Numbers in brackets are correlation between spread and RMS error for each lead time.

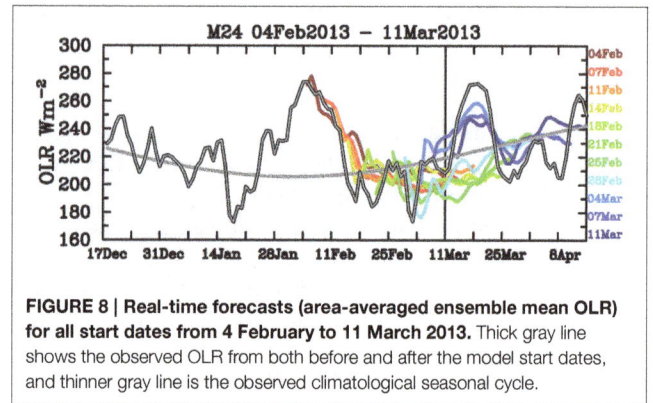

FIGURE 8 | Real-time forecasts (area-averaged ensemble mean OLR) for all start dates from 4 February to 11 March 2013. Thick gray line shows the observed OLR from both before and after the model start dates, and thinner gray line is the observed climatological seasonal cycle.

out to at least 120 days. For most forecasts, the start date will not coincide with the dates for which hindcast climatologies are available. There are a number of possible strategies to obtain climatological values for any arbitrary start date. These include simply using the nearest available climatology or linearly interpolating between the two closest dates. The approach adopted here is to fit a smoothed annual cycle to each lead time, similar to the annual cycles for the observed and initialized data, and then obtain a climatological value for any start date and lead time by interpolation from this smoothed annual cycle. Smoothed annual cycles for lead times of day 1 (blue) and day 20 (green) are shown in **Figure 2**.

Real-time anomalies are then calculated in a similar manner to that used for the hindcasts. The consistency of the forecasts over a run of forecasts is demonstrated in **Figure 8**, which shows the ensemble mean of a sequence of 11 forecasts initialized between 04 Feb 2013 and 11 Mar 2013. This example shows a general agreement between one start time and the next, although with a notable exception to this for forecasts initialized before 1st March vs. those initialized after 1st March, showing that the model had a "change of heart" at that time. Comparison with the observed OLR anomalies shows that the forecast of positive OLR anomalies for 12–25 March by those forecasts initialized after 1st March, indicating suppressed convection, verified well. Overall, verification of all the available real-time forecasts (**Figure 9**) shows that the hindcast skill is generally maintained by independent forecasts. The limited real-time POAMA forecasts have lower RMS errors than persistence beyond the first day of the forecast, which is better than was the case for the hindcasts (**Figure 4**), which took 2–3 days to beat persistence. However, compared to climatological forecasts, the POAMA real-time forecasts lose skill more swiftly than the hindcasts, with the POAMA RMS error slightly exceeding that of climatology at lead times greater than about 15 days. We cannot think of a difference in the model prediction system between

the hindcasts and real-time forecasts that could explain these skill differences (e.g., the model physics and initialisation remain the same), so we assume that the skill differences are a result of the rather short period (<5 years) of real-time data, and the associated statistical uncertainty this causes.

CONCLUSIONS

We examine the ability of the latest version of the POAMA system (POAMA-2M) to simulate the seasonal cycle of OLR over the north Australian monsoon region, and to predict its variability on timescales out to 35 days. After the period of initial shock is over (i.e., after the first few days), the model's seasonal cycle shows that it does not have enough convection

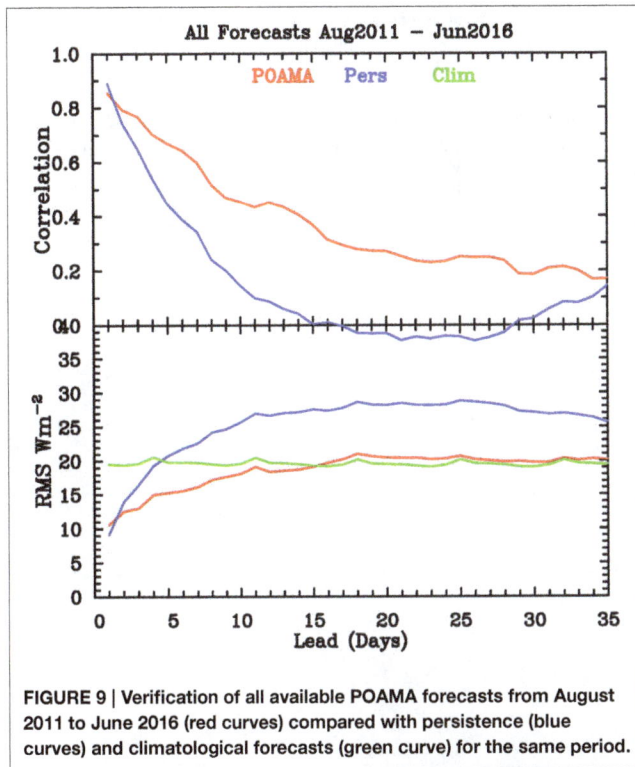

FIGURE 9 | Verification of all available POAMA forecasts from August 2011 to June 2016 (red curves) compared with persistence (blue curves) and climatological forecasts (green curve) for the same period.

during the wet season and consequently has smaller amplitude than observed. After correcting for this bias, model hindcasts show skill, over persistence and climatological forecasts, at lead times beyond 2 days. These hindcasts show that the model has greater skill during the summer monsoon season from December to March, but this skill is largely independent of the state of the MJO. Examination of the spread of hindcasts suggests that the model has insufficient spread at short lead times (<12 days), and slightly too much spread at longer lead times (>16 days). Real-time forecasts are constructed in a similar manner to the hindcasts, with the required bias correction obtained by interpolating the model seasonal cycle to give values at each real-time forecast start time. Verification of all available forecasts from August 2011 through to June 2016 suggests that the hindcast skill relative to persistence and climatology is mostly maintained in completely independent forecasts.

AUTHOR CONTRIBUTIONS

WD and MW jointly designed the research, interpreted the results, and wrote the paper. WD analyzed the prediction system output and made **Figures 2–9**. MW made **Figure 1**, undertook the milestone reporting to funding bodies and managers, formatted the manuscript, and submitted the manuscript.

FUNDING

This work was partially supported during 2011–13 by the Northern Australia/Monsoon Prediction project of the Managing Climate Variability Program managed by the Grains Research and Development Corporation.

ACKNOWLEDGMENTS

We thank the POAMA and related teams for their dedication to producing and maintaining POAMA, and for supporting its products. In particular, we thank Griffith Young for the POAMA web pages and computing support. Greg Browning and Andrew Marshall kindly provided internal reviews.

REFERENCES

Grimit, E. P., and Mass, C. F. (2007). Measuring the ensemble spread–error relationship with a probabilistic approach: stochastic ensemble results. *Mon. Weather Rev.* 135, 203–221. doi: 10.1175/MWR 3262.1

Hamill, T. M. (2001). Interpretation of rank histograms for verifying ensemble forecasts. *Mon. Weather Rev.* 129, 550–560. doi: 10.1175/1520-0493(2001)129<0550:IORHFV>2.0.CO;2

Hudson, D., Alves, O., Hendon, H. H., and Wang, G. (2011). The impact of atmospheric initialisation on seasonal prediction of tropical Pacific SST. *Clim. Dyn.* 36, 1155–1171. doi: 10.1007/s00382-010-0763-9

Hudson, D., Marshall, A. G., Yin, Y. H., Alves, O., and Hendon, H. H. (2013). Improving intraseasonal prediction with a new ensemble generation strategy. *Mon. Weather Rev.* 141, 4429–4449. doi: 10.1175/MWR-D-13-00059.1

Liebmann, B., and Smith, C. A. (1996). Description of a complete (interpolated) outgoing longwave radiation dataset. *Bull. Am. Met. Soc.* 77, 1275–1277.

Marshall, A. G., and Hendon, H. H. (2015). Subseasonal prediction of Australian summer monsoon anomalies. *Geophys. Res. Lett.* 42, 10913–10919. doi: 10.1002/2015GL067086

Marshall, A. G., Hudson, D., Hendon, H. H., Pook, M. J., Alves, O., and Wheeler, M. C. (2014a). Simulation and prediction of blocking in the Australian region and its influence on intra-seasonal rainfall

in POAMA-2. *Clim. Dyn.* 42, 3271–3288. doi: 10.1007/s00382-013-1974-7

Marshall, A. G., Hudson, D., Wheeler, M. C., Alves, O., Hendon, H. H., Pook, M. J., et al. (2014b). Intra-seasonal drivers of extreme heat over Australia in observations and POAMA-2. *Clim. Dyn.* 43, 1915–1937. doi: 10.1007/s00382-013-2016-1

Marshall, A. G., Hudson, D., Wheeler, M. C., Hendon, H. H., and Alves, O. (2011). Assessing the simulation and prediction of rainfall associated with the MJO in the POAMA seasonal forecast system. *Clim. Dyn.* 37, 2129–2141. doi: 10.1007/s00382-010-0948-2

Marshall, A. G., Hudson, D., Wheeler, M. C., Hendon, H. H., and Alves, O. (2012). Simulation and prediction of the Southern Annular Mode and its influence on Australian intra-seasonal climate in POAMA. *Clim. Dyn.* 38, 2483–2502. doi: 10.1007/s00382-011-1140-z

Rashid, H. A., Hendon, H. H., Wheeler, M. C., and Alves, O. (2011). Prediction of the Madden-Julian oscillation with the POAMA dynamical prediction system. *Clim. Dyn.* 36, 649–661. doi: 10.1007/s00382-010-0754-x

Stockdale, T. N. (1997). Coupled ocean–atmosphere forecasts in the presence of climate drift. *Mon. Weather Rev.* 125, 809–818. doi: 10.1175/1520-0493(1997)125<0809:COAFIT>2.0.CO;2

Talagrand, O., Vautard, R., and Strauss, B. (1997). "Evaluation of probabilistic prediction systems," in *Proceedings, ECMWF Workshop on Predictability*. Reading: ECMWF, 1–25.

Wheeler, M. C. (2008). Seasonal climate summary southern hemisphere (summer 2007-08): mature La Nina, an active MJO, strongly positive

SAM, and highly anomalous sea-ice. *Aust. Meteorol. Mag.* 57, 379–393.

Wheeler, M. C., and Hendon, H. H. (2004). An all-season real-time multivariate MJO index: development of an index for monitoring and prediction. *Mon. Weather Rev.* 132, 1917–1932. doi: 10.1175/1520-0493(2004)132<1917:AARMMI>2.0.CO;2

Wheeler, M. C., and McBride, J. L. (2011). "Australasian monsoon," in *Intraseasonal Variability in the Atmosphere-Ocean Climate System, 2nd Edn*, eds W. K. M. Lau and D. E. Waliser (Berlin: Springer), 147–198.

Zhu, H., Wheeler, M. C., Sobel, A. H., and Hudson, D. (2014). Seamless precipitation prediction skill in the tropics and extratropics from a global model. *Mon. Weather Rev.* 142, 1556–1569. doi: 10.1175/MWR-D-13-00222.1

Conflict of Interest Statement: The authors declare that the research was conducted in the absence of any commercial or financial relationships that could be construed as a potential conflict of interest.

The reviewer SW and the handling Editor declared their shared affiliation, and the handling Editor states that the process nevertheless met the standards of a fair and objective review.

13

Challenges of Quantifying Meltwater Retention in Snow and Firn: An Expert Elicitation

Dirk van As*, Jason E. Box and Robert S. Fausto

Department of Glaciology and Climate, Geological Survey of Denmark and Greenland, Copenhagen, Denmark

Thirty-four experts took part in a survey of the most important and challenging topics in the field of meltwater retention in snow and firn, to reveal those topics that present the largest potential for scientific advancement. The most important and challenging topic to the expert panel is spatial heterogeneity of percolation, both in measurement and model studies. Studying percolation blocking by ice layering, particularly in modeling, also provides large potential for science advancement, as well as hydraulic conductivity and capillary forces in snow/firn. Model studies can benefit from improved initialization, and improved calculation of accumulation and liquid water at the surface. Firn coring should be performed more often, though we argue that also data that are relatively simple to collect, but of great importance to retention such as surface accumulation, density and temperature, are too sparse due to the high logistical expenses involved in field campaigns. Generally speaking, retention changes are expected to be of importance to the climatic (surface) mass balance and thus ice loss in coming decades, more so for Greenland than Antarctica or ice masses elsewhere.

Keywords: snow, firn, melt, percolation, retention, refreezing, spatial heterogeneity, ice layer blocking

Edited by:
Martin Hoelzle,
University of Fribourg, Switzerland

Reviewed by:
Shad O'Neel,
USGS Alaska Science Center, USA
Christoph Schneider,
Humboldt University of Berlin,
Germany

*Correspondence:
Dirk van As
dva@geus.dk

Specialty section:
This article was submitted to
Cryospheric Sciences,
a section of the journal
Frontiers in Earth Science

INTRODUCTION

The mass balances of ice sheets, ice caps, and glaciers worldwide include one large mass input, namely snow accumulation. Yet depending on the location, several large mass loss components exist such as ice-dynamic mass transfer to the oceans, basal melting, surface meltwater runoff, and sublimation. Globally, the loss components are found to out-compete the gains in the current climate, which is largely attributed to increased melting at the air-ice and ocean-ice interfaces, also impacting ice dynamic discharge (IPCC, 2013).

Surface runoff occurs when melting, rainfall, and condensation exceed evaporation and retention in snow, firn, porous ice, and supraglacial lakes (Cuffey and Paterson, 2010). Our ability to determine surface runoff thus depends on our ability to determine several entirely independent processes, each with their own complexities. Out of these processes, retention of water in snow, firn or ice may not be the largest contributor averaged over an entire glacier, ice cap or ice sheet, yet regionally (in the *percolation area*) it can be large enough to accommodate all liquid water that percolates into the surface layer (e.g., Benson, 1962). At lower elevations, the snow and firn (or even porous ice) periodically gets overwhelmed by liquid water in the *runoff area*. Challenges in determining runoff are largest due to the uncertainties in quantifying retention in snow and firn, as this is the only component that takes place at depth (in the order of meters), thus out of sight and reach of basic, direct observational methods (Harper et al., 2012). Indirect methods have existed for decades, such as determining changes in density/stratigraphy from repeat snow pits (e.g., Techel and Pielmeier, 2011) or firn cores (e.g., Vallelonga et al., 2014), continuous subsurface temperature measurements for capturing the release of latent heat from refreezing (Humphrey et al., 2012; Charalampidis et al., 2016), or radar systems capable of identifying strong reflectors in porous snow or firn as ice layers (e.g., Koenig et al., 2016). Such methods are excellent for determining the location and quantity of the retained mass, but do not allow us to track and comprehend all physical processes involved in percolation and refreezing. This then limits our ability to model these processes with accuracy.

Whereas part of the difficulty in studying retention stems from not being able to track water movement through snow/firn with detail without altering the medium itself by digging pits or drilling cores, another aspect is the memory of the system. Namely, when meltwater refreezes it will impact future percolation due to changes in snow/firn density and temperature (Bezeau et al., 2013). This continues until the layer and the changes it caused in the snow/firn, migrated downward far enough to escape the direct influence of surface processes, i.e., the maximum percolation depth (Harper et al., 2012). Depending on the net accumulation rate (Mosley-Thompson et al., 2001), this can take many years. So if a model is forced by inaccurate surface energy quantities, or initialized with inaccurate snow/firn densities and temperatures, or struggles representing physical processes such as gravitational densification (Arthern et al., 2010; Morris and Wingham, 2014), thermal conduction (Sturm et al., 1997), and movement of liquid water (Colbeck, 1974; Hirashima et al., 2010), then not only the first retention event may be calculated inaccurately, but in a cascade effect all retention events may be affected. The impact on the climatic mass balance can be substantial for instance when consecutive high melt seasons occur, with ice layer formation shifting conditions in favor of surface runoff as opposed to percolation (Machguth et al., 2016).

In this paper, we aim to identify the most crucial topics in meltwater retention in order to reveal the largest potential for scientific advancement in this field. We apply expert elicitation, which is common practice in investigations with high uncertainty due to the lack of data, for instance in case of rare events or future predictions. In glaciology the expert survey methodology was applied e.g., in estimating future sea level contributions from the ice sheets (Bamber and Aspinall, 2013). Our approach differs in that we will not attempt quantification of meltwater retention in snow and firn through expert elicitation.

METHODS

The questionnaire used to conduct the expert elicitation on meltwater retention in snow and firn, consisted of 18 entries listing snow/firn processes and properties (**Table 1**). On a scale of 1–10, 10 being highest, the expert was asked to rate each topic in terms of: (1) importance to meltwater retention, (2) difficulty level in measuring with accuracy, and (3) difficulty level in modeling with accuracy (providing ratings R_{ret1-3} for each topic/question). Also, the expert was asked to rate "the

importance of the change in meltwater retention in snow and firn in a warming climate on decadal time scales for mass loss from" (1) Antarctica, (2) Greenland, and (3) other glaciated regions (ratings R_{reg1-3}). In case of substantial doubt the expert was asked not to provide an answer. In order to be able to assess the expert levels of the sources, the questionnaire also asked for the self-assessed (1) expert level on the topic of meltwater retention in snow and firn, (2) experience level performing firn measurements, (3) experience level modeling meltwater retention in snow and firn (ratings R_{exp1-3}), and (4) the years (Y) since (or until) obtaining a Ph.D. degree.

We invited those present at the "Workshop on observing and modeling meltwater retention processes in snow and firn on ice sheets and glaciers" hosted by the Geological Survey of Denmark and Greenland (GEUS) on 1–3 June 2016, and other known topical experts to take part in the survey. With the aim of publishing the results indicated on the questionnaire, all subjects gave their written consent by returning the document. The expert panel of 34 excludes the first author of this study. On average, the retention expert level is found to be 6.4 ($R_{exp1,av}$). Experience levels are lower for modeling of retention ($R_{exp2,av} = 5.6$), and particularly for measuring ($R_{exp3,av} = 4.9$), the latter of which is likely related to the high cost associated with field expeditions.

We apply expert weighting to each answer by multiplication with an individual expert factor, calculated as:

$$F_{exp} = \left(\frac{R_{exp}}{R_{exp,av}} + \frac{Y + 10}{Y_{av} + 10} \right) / 2 \tag{1}$$

Note, that the self-assessed expert level and scientific seniority weigh equally in calculating one's expert level. Ratings R_{ret1} and R_{reg1-3} are weighted using expert level rating R_{exp1}, ratings R_{ret2} using R_{exp2}, and ratings R_{ret3} using R_{exp3}. Expert factors vary between 0.29 and 2.18, a factor of 7.4 difference.

Likewise, we apply importance weighting by multiplication with the importance factor of each snow/firn topic:

$$F_{imp} = \frac{R_{ret1,EW}}{R_{ret1,EW,av}} \tag{2}$$

where $R_{ret1,EW}$ is the expert-weighted importance averaged over all experts. Importance factors range between 0.67 and 1.26.

Using these methods, we obtain averages and expert-weighted (EW) averages for every surveyed snow/firn topic, in terms of their importance to meltwater retention, and their difficulty to quantify by means of measuring and modeling (**Table 1**). We also present expert- and importance-weighted (EIW) averages of the difficulties in measuring and modeling. Note, that after the weighting values can exceed the 1–10 range.

RESULTS

The 34 experts attributed average values between 4.4 and 8.5 with an overall average of 6.7 to the importance of snow/firn topics (**Table 1**). Expert weighting has little impact on the average (6.8), nor on the ranking of the importance of processes/properties. Surface accumulation (8.6) and the availability of liquid water at

TABLE 1 | Expert judgment assessment of snow/firn processes and properties relevant to retention.

Topic	Importance to snow/firn processes			Difficulty level in measuring with accuracy				Difficulty level in modeling with accuracy			
	Av	SD	EW	Av	SD	EW	EIW	Av	SD	EW	EIW
Surface water (melt+rain)	8.4[2]	1.4	8.4[2]	5.6	2.4	5.5	6.7	5.8	2.1	5.7	7.0
Surface accumulation	8.5[1]	1.3	8.6[1]	4.6	1.8	4.7	5.9	5.8	1.7	5.8	7.3[4]
Surface accumulation density	6.8	1.5	6.9	5.4	2.7	5.3	5.3	5.7	1.9	5.9	5.9
Surface temperature	7.2	1.9	7.2	3.0	1.9	3.3	3.5	4.5	1.6	4.5	4.7
Surface albedo (melt-albedo feedback)	7.3[5]	1.6	7.3[5]	4.0	2.0	4.3	4.6	5.9	1.3	5.9	6.3
Near-surface solar radiation penetration	5.4	2.1	5.6	7.2	1.7	7.4	6.1	6.3	2.0	6.6	5.4
Near-surface ventilation by wind	4.4	1.9	4.6	7.3[4]	2.0	7.8[4]	5.2	7.3[4]	1.9	7.3[4]	4.9
Grain size	6.4	1.5	6.5	5.1	1.9	5.2	5.0	6.7	2.0	6.8	6.5
Gravitational densification	6.2	1.6	6.3	5.6	1.9	5.8	5.3	4.6	1.7	4.7	4.3
Heat conduction	6.5	1.5	6.6	5.3	1.8	5.5	5.3	4.8	2.0	5.0	4.8
Capillary forces acting on liquid water	6.4	1.6	6.7	7.7[2]	1.6	8.2[2]	8.0[4]	7.0[5]	2.1	7.3[5]	7.2[5]
Hydraulic conductivity	6.8	1.6	7.1	7.4[3]	1.5	8.1[3]	8.5[2]	7.3[3]	1.6	7.7[3]	8.0[3]
Blocking of meltwater percolation by ice layering	7.5[4]	1.5	7.7[4]	7.2	1.9	7.3	8.2[3]	8.3[2]	1.3	8.5[2]	9.5[2]
Spatial heterogeneity of meltwater percolation	7.6[3]	1.6	7.9[3]	7.9[1]	2.1	8.6[1]	9.9[1]	8.6[1]	1.7	9.1[1]	10.5[1]
Deep firn processes / lower boundary conditions	4.9	2.0	5.3	7.3[5]	2.1	7.6[5]	5.9	6.5	2.4	6.8	5.3
Spatial (vertical) resolution	–	–	–	5.9	2.2	6.6	6.6	5.3	2.1	5.7	5.7
Time resolution (capability of resolving relevant cycles)	–	–	–	6.4	2.2	6.7	6.7[5]	5.2	1.9	5.7	5.7
Model initialization	–	–	–	–	–	–	–	6.6	1.7	7.0	7.0
Average	6.7	1.6	6.8	6.1	2.0	6.3	6.3	6.2	1.8	6.4	6.4

Listed are the arithmetic-mean averages (Av), standard deviations (SD), expert-weighted averages (EW), and expert- and importance-weighted averages (EIW) of the expert ratings on a scale of 1–10. Ranks 1–5 indicating highest values are in superscript.

the surface (8.4) are considered most important, followed by the spatial heterogeneity of meltwater percolation (7.9), percolation blocking by ice layering (7.7), and surface albedo through the melt-albedo feedback (7.3). Least important of the selected topics is considered to be near-surface ventilation by wind (4.6).

EW-values of the difficulty level in measuring the snow/firn processes/properties vary considerably (**Table 1**). Whereas measuring surface temperature (or in fact any surface variable) is considered relatively straightforward (3.3), measuring the spatial heterogeneity of meltwater percolation is judged to be the largest challenge (8.6). Other large challenges are found in measuring the effect of capillary forces on liquid water (8.2), hydraulic conductivity (8.1), ventilation by wind (7.8), and processes occurring in the bottom half of the firn (7.6).

Also in modeling, surface temperature is considered least difficult to determine with accuracy (4.5), and spatial heterogeneity most difficult (9.1, **Table 1**). The top five modeling challenges align well with the high ranking observational challenges, except that the modeling of percolation blocking by ice layering ranks second (8.5). On average, measuring and modeling retention-related topics (6.4) are judged to be equally complicated (6.3 vs. 6.4).

In **Figure 1A**, we identify four quadrants in a plot of importance vs. difficulty of the survey topics. Most topics are located in the quadrants of high importance and low difficulty, or

low importance and high difficulty. Three topics are judged to be of above-average importance and difficulty: spatial heterogeneity in, and blocking of, meltwater percolation, and the hydraulic conductivity of the snow/firn.

EIW-values confirm that these three are top ranking both in measuring and modeling (**Table 1, Figure 1B**). Also the topic of capillary forces acting on liquid water ranks high (8.0 and 7.2). The top five in measuring is completed by temporal resolution, i.e., the difficulty in resolving relevant firn process cycles from repeat cores. In modeling, surface accumulation has a top-ranking EIW-value (7.3).

Finally, we turn to the question on how important meltwater retention in snow/firn is to the change in the mass balance of ice masses around the globe in coming decades (R_{reg1-3}). This expert judgment assessment indicates that, on these timescales, retention may be an important factor for ice loss from Antarctica (5.2), Greenland (8.4), and glaciated regions elsewhere (6.5).

DISCUSSION

On some topics nearly all experts agree as shown by a small standard deviation (**Table 1**). For instance, all but one expert assess the modeling of meltwater percolation blocking by ice layers to be difficult (7–10), resulting in a small standard deviation of 1.3. In the other extreme, in judging the difficulty

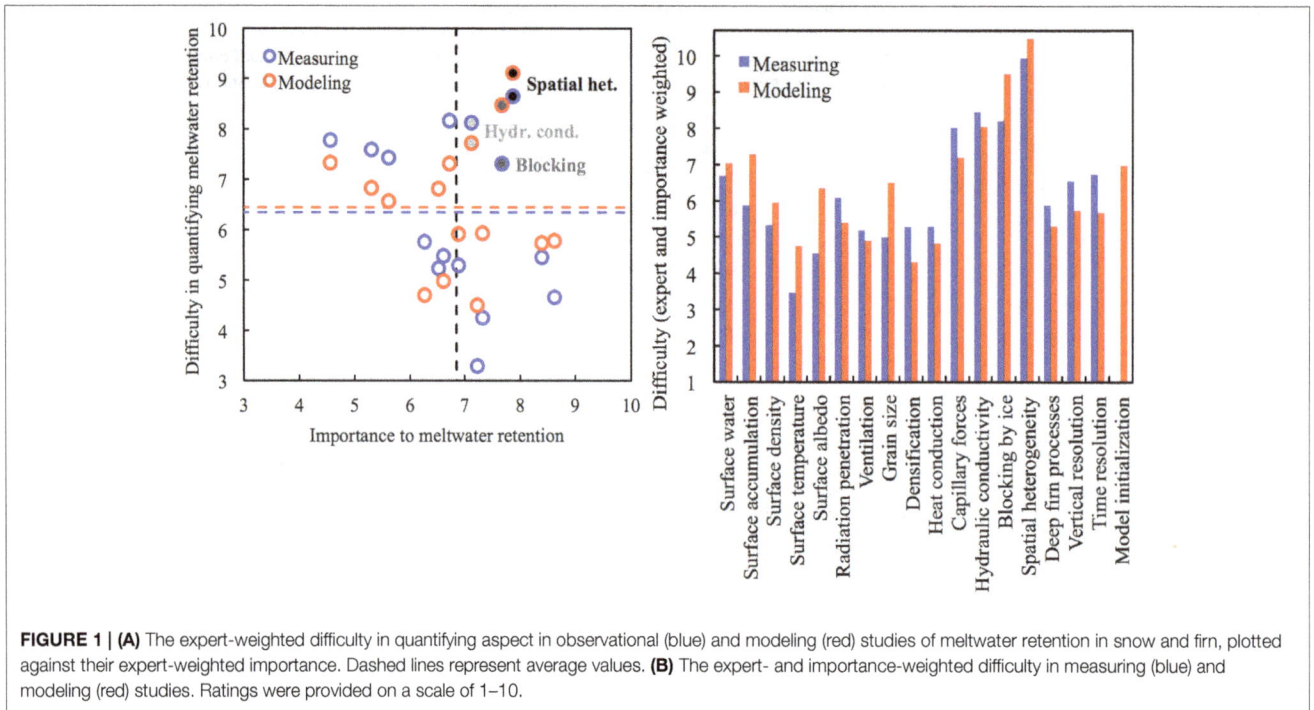

FIGURE 1 | (A) The expert-weighted difficulty in quantifying aspect in observational (blue) and modeling (red) studies of meltwater retention in snow and firn, plotted against their expert-weighted importance. Dashed lines represent average values. **(B)** The expert- and importance-weighted difficulty in measuring (blue) and modeling (red) studies. Ratings were provided on a scale of 1–10.

level of measuring surface accumulation density, a considerable disagreement is found with entries over the entire 1–10 range, resulting in a larger standard deviation of 2.7. To remedy the situation in which the less experienced potentially increase disagreement, we added weight to the opinions of those with more experience in the field of retention by calculating expert factors for all survey partakers, based on the expert's seniority level and self-assessed expert levels (see Methods Section). The calculation is arguably arbitrary, and therefore kept simple. However, we find that expert weighting does not significantly change the average topic score; expert weighting contributes between −0.16 and +0.71 with an average of +0.22 to the R_{ret} questions. Therefore, the results are expert-level insensitive, because the survey part takers with a below-average expert level generally agree with their more experienced peers. The adjustment through importance weighting is larger with adjustment between −2.59 (snow/firn ventilation being difficult to measure, but not important for retention studies) and +1.49 (surface accumulation being somewhat difficult to model, but very important for retention studies), yet the average importance-adjustment is negligible at −0.04.

If a topic related to meltwater retention in snow/firn is considered to be both important and difficult to assess, we regard this as a topic in which the largest scientific advances can be made. These topics have an above-average EIW score (**Table 1**). Yet, we also need to consider the topics that are important, but less difficult to quantify. In retention modeling, these topics leave little room for advancement. On the other hand, in measuring, an additional challenge exists in reaching the often remote and inaccessible accumulation areas of glaciers and ice caps/sheets. So while the measurement may be relatively simple, observational data remain sparse due to high logistical expense. A data shortage

is also reflected in the experts' opinion that relevant cycles in retention are not well-resolved by repeat firn coring (**Table 1**). Therefore, we argue that also the relatively simple, yet important, measurements should be targeted for scientific advancement in retention. Strikingly, these are all the measurements of surface variables listed in **Table 1** (surface water, accumulation, density, temperature, and albedo) for which automated solutions exist or can be developed, such as equipping weather stations (e.g., Mölg et al., 2008; Van As et al., 2011) with radiometers, sonic rangers, and snow-water-equivalent sensors.

With sparsity of observational data, it becomes of primary importance that existing data are available to everyone. Excellent examples of publicly available databases containing firn data exist (e.g., the SUMup database; Koenig et al., 2016), and efforts should be made by the global glaciological community to expand those databases. Likewise, only few models are capable of detailed calculations of meltwater retention in snow and firn. The more the model codes become available to the entire research community, the more researchers can build on previous efforts, and the largest the scientific advancement will be.

CONCLUSION

In this study, we used the results of 34 experts that took part in an expert elicitation of meltwater retention in snow and firn, to identify those snow/firn processes and properties that are both important and challenging to quantify, thus providing the largest potential for scientific advancement. We applied expert- and importance-weighting, only the latter of which proved influential.

We find the following topics to present the largest potential for scientific advancement, both in measuring and modeling:

- spatial heterogeneity of meltwater percolation in snow/firn,
- blocking of percolating meltwater by ice layers, especially in model studies,
- hydraulic conductivity,
- and capillary forces acting on liquid water.

Additionally, in modeling the following topics are worthy of further emphasis:

- quantifying surface water,
- quantifying accumulation,
- and model initialization.

Based on the survey results, we also argue for:

- obtaining more firn cores and to sample them in more detail,
- and measuring surface variables (liquid water, accumulation, density, temperature, and albedo) at more sites, since data sparsity is not caused by the difficulty of the measurement itself, but by the difficulty in getting to the accumulation areas of glaciers and ice caps/sheets.

In a final question on the future change of retention due to climate change, the expert panel answered that in the coming decades, meltwater retention is expected to be of importance to mass loss in glaciated regions worldwide, especially in Greenland.

AUTHOR CONTRIBUTIONS

DV, RF, and JB conceived the study. DV compiled the questionnaire, analyzed the data, and wrote the text with contributions by JB and RF.

ACKNOWLEDGMENTS

We are grateful to the anonymous experts taking part in this study. Support for this study was provided by Denmark's Nature and Universe grant DFF 4002-00234 through the Retain project (retain.geus.dk) and the Danish Energy Agency (www.ENS.dk) through the Programme for Monitoring of the Greenland Ice Sheet (www.PROMICE.dk). This study with human subjects was carried out in accordance with the ethical guidelines by the Geological Survey of Denmark and Greenland (GEUS).

REFERENCES

Arthern, R. J., Vaughan, D. G., Rankin, A. M., Mulvaney, R., and Thomas, E. R. (2010). *In situ* measurements of Antarctic snow compaction compared with predictions of models. *J. Geophys. Res.* 115:F03011. doi: 10.1029/2009JF001306

Bamber, J. L., and Aspinall, W. P. (2013). An expert judgement assessment of future sea level rise from the ice sheets. *Nat. Clim. Change* 3, 424–427. doi: 10.1038/nclimate1778

Benson, C. S. (1962). *Stratigraphic Studies in the Snow and Firn of the Greenland Ice Sheet*. Research Report 70, SIPRE.

Bezeau, P., Sharp, M., Burgess, D., and Gascon, G. (2013). Firn profile changes in response to extreme 21st-century melting at Devon Ice Cap, Nunavut, Canada. *J. Glaciol.* 59, 981–991. doi: 10.3189/2013JoG12J208

Charalampidis, C., Van As, D., Colgan, W. T., Fausto, R. S., MacFerrin, M., and Machguth, H. (2016). Thermal tracing of retained meltwater in the lower accumulation area of the southwestern Greenland ice sheet. *Ann. Glaciol.* 57. doi: 10.1017/aog.2016.2. Available online at: https://www.igsoc.org/annals/57/72/accepted.html

Colbeck, S. C. (1974). The capillary effects on water percolation in homogeneous snow. *J. Glaciol.* 13, 85–97.

Cuffey, K. M., and Paterson, W. S. B. (2010). *The Physics of Glaciers*. Burlington, MA: Academic Press, 704.

Harper, J., Humphrey, N., Pfeffer, W. T., Brown, J., and Fettweis, X. (2012). Greenland ice-sheet contribution to sea-level rise buffered by meltwater storage in firn. *Nature* 491, 240–243. doi: 10.1038/nature11566

Hirashima, H., Yamaguchi, S., Sato, A., and Lehning, M. (2010). Numerical modeling of liquid water movement through layered snow based on new measurements of the water retention curve. *Cold Regions Sci. Technol.* 64, 94–103. doi: 10.1016/j.coldregions.2010.09.003

Humphrey, N. F., Harper, J. T., and Pfeffer, W. T. (2012). Thermal tracking of meltwater retention in Greenland's accumulation area. *J. Geophys. Res.* 117:F01010. doi: 10.1029/2011JF002083

IPCC (2013). "Climate change 2013: the physical science basis," in *Contribution of Working Group I to the Fifth Assessment Report of the Intergovernmental Panel on Climate Change*, eds T. F. Stocker, D. Qin, G.-K. Plattner, M. Tignor, S. K. Allen, J. Boschung, A. Nauels, Y. Xia, V. Bex, and P. M. Midgley (Cambridge; New York, NY: Cambridge University Press), 1535.

Koenig, L. S., Ivanoff, A., Alexander, P. M., MacGregor, J. A., Fettweis, X., Panzer, B., et al. (2016). Annual Greenland accumulation rates (2009–2012) from airborne snow radar. *Cryosphere* 10, 1739–1752. doi: 10.5194/tc-10-1739-2016

Machguth, H., MacFerrin, M., Van As, D., Box, J. E., Charalampidis, C., Colgan, W., et al. (2016). Greenland meltwater storage in firn limited by near-surface ice formation. *Nat. Clim. Change* 6, 390–393. doi: 10.1038/nclimate2899

Mölg, T., Cullen, N. J., Hardy, D. R., Kaser, G., and Klok, L. (2008). Mass balance of a slope glacier on Kilimanjaro and its sensitivity to climate. *Int. J. Climatol.* 28, 881–892. doi: 10.1002/joc.1589

Morris, E. M., and Wingham, D. J. (2014). Densification of polar snow: measurements, modeling, and implications for altimetry. *J. Geophys. Res.* 119, 349–365. doi: 10.1002/2013JF002898

Mosley-Thompson, E., McConnell, J. R., Bales, R. C., Lin, P. N., Steffen, K., Thompson, L. G., et al. (2001). Local to regional-scale variability of annual net accumulation on the Greenland ice sheet from PARCA cores. *J. Geophys. Res.* 106, 33839–33851. doi: 10.1029/2001JD900067

Sturm, M., Holmgren, J., Konig, M., and Morris, K. (1997). The thermal conductivity of seasonal snow. *J. Glaciol.* 43, 26–41.

Techel, F., and Pielmeier, C. (2011). Point observations of liquid water content in wet snow–investigating methodical, spatial and temporal aspects. *Cryosphere* 5, 405–418. doi: 10.5194/tc-5-405-2011

Vallelonga, P., Christianson, K., Alley, R. B., Anandakrishnan, S., Christian, J. E. M., Dahl-Jensen, D., et al. (2014). Initial results from geophysical surveys and shallow coring of the Northeast Greenland Ice Stream (NEGIS). *Cryosphere* 8, 1275–1287. doi: 10.5194/tc-8-1275-2014

Van As, D., Fausto, R. S., and PROMICE Project Team (2011). Programme for Monitoring of the Greenland Ice Sheet (PROMICE): first temperature and ablation records. *Geol. Surv. Den. Greenland Bull.* 23, 73–76.

Conflict of Interest Statement: The authors declare that the research was conducted in the absence of any commercial or financial relationships that could be construed as a potential conflict of interest.

14

Carbon Leaching from Tropical Peat Soils and Consequences for Carbon Balances

Tim Rixen[1,2]*, Antje Baum[1], Francisca Wit[1] and Joko Samiaji[3]

[1] Leibniz Center for Tropical Marine Ecology, Bremen, Germany, [2] Department of Biogeochemistry, Institute of Geology, University of Hamburg, Hamburg, Germany, [3] Faculty of Fishery and Marine Science, University of Riau, Pekanbaru, Indonesia

Drainage and deforestation turned Southeast (SE) Asian peat soils into a globally important CO_2 source, because both processes accelerate peat decomposition. Carbon losses through soil leaching have so far not been quantified and the underlying processes have hardly been studied. In this study, we use results derived from nine expeditions to six Sumatran rivers and a mixing model to determine leaching processes in tropical peat soils, which are heavily disturbed by drainage and deforestation. Here we show that a reduced evapotranspiration and the resulting increased freshwater discharge in addition to the supply of labile leaf litter produced by re-growing secondary forests increase leaching of carbon by ∼200%. Enhanced freshwater fluxes and leaching of labile leaf litter from secondary vegetation appear to contribute 38 and 62% to the total increase, respectively. Decomposition of leached labile DOC can lead to hypoxic conditions in rivers draining disturbed peatlands. Leaching of the more refractory DOC from peat is an irrecoverable loss of soil that threatens the stability of peat-fringed coasts in SE Asia.

Keywords: tropical peat soil, degradation, secondary vegetation, carbon loss, Sumatra, Indonesia

Edited by:
Francien Peterse,
Universiteit Utrecht, Netherlands

Reviewed by:
William Patrick Gilhooly III,
Indiana University Purdue University
Indianapolis, USA
Chris Evans,
Centre for Ecology and Hydrology, UK

*Correspondence:
Tim Rixen
tim.rixen@leibniz-zmt.de

Specialty section:
This article was submitted to
Biogeoscience,
a section of the journal
Frontiers in Earth Science

INTRODUCTION

Pristine peat swamp forests are rare with only circa 10% left on the islands of Borneo and Sumatra (Miettinen and Liew, 2010). The remaining disturbed peat soils are drained, covered by plantations, shrubs and secondary forests, and are characterized by a heavily altered hydrological cycle. As seen in a river catchment on Borneo, deforestation decreases evapotranspiration and thereby increases freshwater fluxes as well as the riverine export of DOC (Moore et al., 2013). Depending on the soil types in their catchments, DOC concentrations in rivers differ in their response to changes in freshwater fluxes, which are mostly linked to precipitation. In rivers draining organo-mineral soils, enhanced freshwater fluxes raise DOC concentrations, whereas DOC concentrations in high-latitude rivers draining peat soils do not respond to changing precipitation rates until flooding occurs at which point the surface runoff dilutes the DOC concentration in the river (Moore and Jackson, 1989; Clark et al., 2008).

Processes controlling carbon leaching from tropical peat soils have so far not been studied. Approximately half of all tropical peat soils are located in Indonesia (0.207×10^{12} m^2) and mostly in the coastal plains of the islands of Irian Jaya, Borneo, and Sumatra (Page et al., 2011). This study quantifies carbon leaching from Indonesian peat soils based on data obtained during nine expeditions to the Siak and five other major rivers draining peatlands in Central Sumatra. Furthermore, we developed a mixing model in order to simulate processes responsible for carbon leaching and applied our results to Indonesian peatlands (**Figure 1, Table 1,** Figure S1).

METHODS

Study Area and Expeditions

The investigated rivers were the Rokan, Siak, Kampar, Indragiri, Batang Hari, and Musi (**Figure 1**). All rivers originate in the Barisan mountains, pass through the Northeastern lowlands and discharge into the Malacca Strait. On their way to the ocean all rivers cut through peat soils, which are located in the coastal flat plains and cover 3.5–30.2% of their catchments (**Table 1**). The Siak river, which was investigated most intensively over the years, originates at the confluence of its two headstreams, S. Tapung Kanan and S. Tapung Kiri (**Figure 1B**). Its main tributary is the Mandau river. The S. Tapung Kiri drains non-peatlands, whereas the S. Tagung Kanan, the Mandau, and the Siak cut through peatlands covering in total 21.9% of the Siak catchment (**Table 1**).

DOC Sampling and Analysis

In the rivers DOC samples were taken using a Niskin bottle at a water-depth of ~1 m. After sampling DOC samples were filtered through 0.45 μm cellulose-acetate filters and immediately acidified with phosphoric acid to a pH ≤ 2 and stored cool and dark until analysis. DOC samples were analyzed by means of high temperature catalytic oxidation using a Dohrmann DC-190 and a Shimadzu TOC-V$_{CPH}$ total Organic Carbon Analyzer. Before injection into the furnace, the samples were decarbonated by purging with oxygen. The evolving CO_2 was purified, dried and detected by a non-dispersive infrared detection system.

Groundwater samples for the determination of DOC concentrations were taken with a pore-water-lance during the expedition in March 2008. On top of the 4 mm thick pipe a syringe was attached to extract pore water from the soil. The obtained samples were preserved and analyzed as the river samples.

DOC Age Determination

The age determination of peat and water samples was conducted at the Leibniz Laboratory for Radiometric Dating and Isotope Research in Kiel, Germany, in 2007. Organic fragments of the peat sample were inspected and collected under a light microscope and pre-treated based on an acid-alkali-acid cleaning with diluted hydrochloric (1%) and sodium hydroxide (1%). The extracted organic material (humin acid fraction) was precipitated, washed, and dried. The river water was pre-treated for dating by filtering and freeze-drying the sample. River water and peat samples were combusted in evacuated, flame sealed quartz tubes containing copper oxide at 900°C. The evolved CO_2 was reduced to graphite with H_2 on an iron catalyst. The resulting graphite-iron powder was pressed into aluminum target holders for the ion sputter source with the help of a pneumatic press. The ^{14}C concentration of the sample is calculated by comparing the $^{14}C/^{12}C$ ratio of each sample, determined by AMS, with those of an international standard (NIST Oxalic Acid standard 2–OxII). Radiocarbon concentrations are reported in percent Modern Carbon (pMC) with ±1-σ measurement uncertainty. 100 pMC is defined as

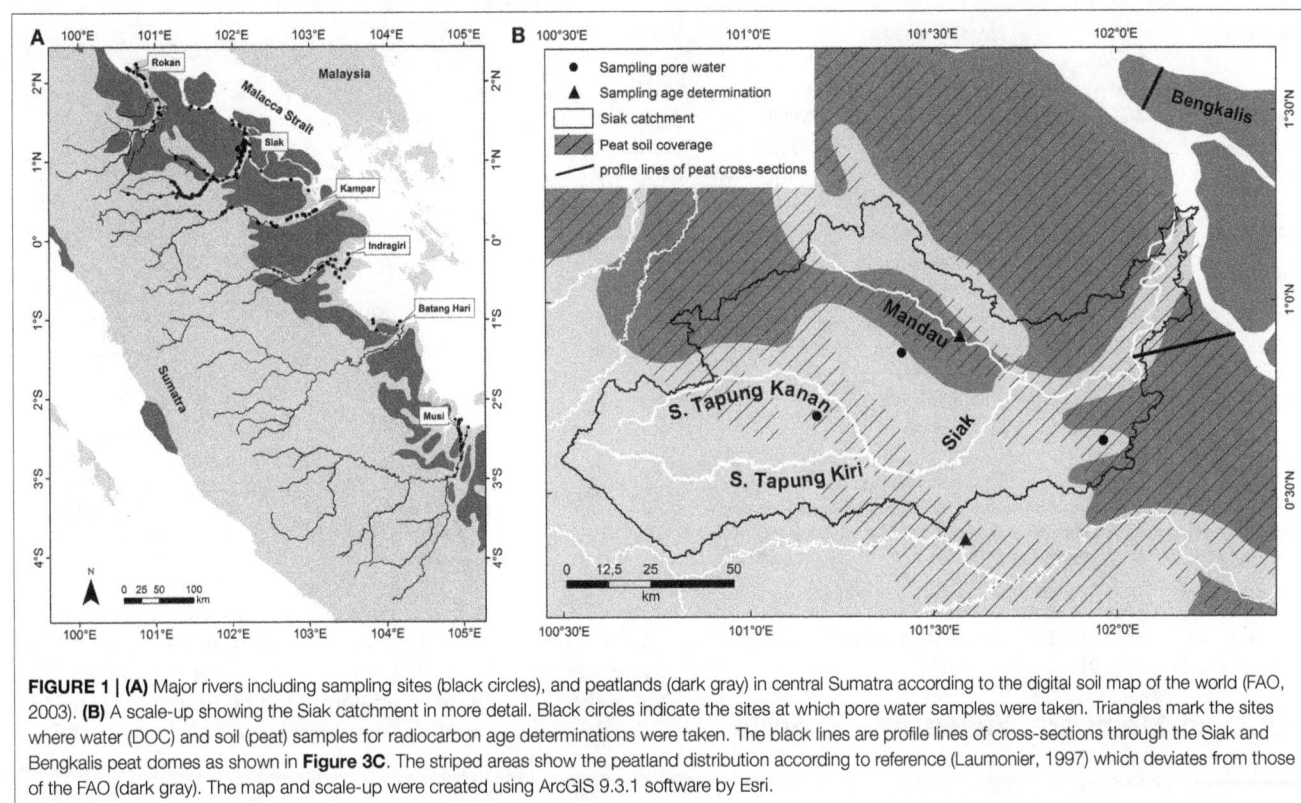

FIGURE 1 | (A) Major rivers including sampling sites (black circles), and peatlands (dark gray) in central Sumatra according to the digital soil map of the world (FAO, 2003). **(B)** A scale-up showing the Siak catchment in more detail. Black circles indicate the sites at which pore water samples were taken. Triangles mark the sites where water (DOC) and soil (peat) samples for radiocarbon age determinations were taken. The black lines are profile lines of cross-sections through the Siak and Bengkalis peat domes as shown in **Figure 3C**. The striped areas show the peatland distribution according to reference (Laumonier, 1997) which deviates from those of the FAO (dark gray). The map and scale-up were created using ArcGIS 9.3.1 software by Esri.

TABLE 1 | Rivers and expeditions as well as riverine DOC end-member concentrations, monthly precipitation rates (Schneider et al., 2014) averaged for the catchment area of the respective rivers (Table S2), and the DOC yields calculated according to Equation (1).

River	Expedition		DOC conc. (µM)	Precipitation (mm)	DOC yield (g C m^{-2} year^{-1})
	Month	Year			
Siak	9	2004	2187 ± 40	250 ± 42	49 ± 15
Siak	8	2005	2159 ± 136	426 ± 55	82 ± 24
Siak	3	2006	1633 ± 55	216 ± 48	32 ± 11
Siak	11	2006	1849 ± 22	244 ± 67	40 ± 15
Siak	3	2008	2205 ± 56	409 ± 43	81 ± 22
Siak	10	2009	2632 ± 164	317 ± 42	75 ± 22
Siak	4	2013	633 ± 88	205 ± 38	12 ± 4
Peat coverage 21.9%	Mean (expedition)		1900 ± 593	295 ± 50	53 ± 25
	1986–2013			232 ± 88	39 ± 22
Rokan	4	2006	833 ± 50	300 ± 4	22 ± 6
Rokan	3	2008	728 ± 54	438 ± 15	29 ± 7
Peat coverage 30.2%	Mean (expedition)		781 ± 53	369 ± 69	25 ± 8
	1986–2013			250 ± 96	17 ± 0
Kampar	4	2006	1236 ± 51	290 ± 57	32 ± 10
Kampar	11	2008	1325 ± 39	281 ± 37	33 ± 9
Peat coverage 22.4%	Mean (expedition)		1280 ± 44	285 ± 5	33 ± 8
	1986–2013			248 ± 97	28 ± 13
Indragiri	3	2008	846 ± 159	366 ± 47	28 ± 9
Indragiri	10	2009	774 ± 71	272 ± 83	19 ± 8
Peat coverage 11.9%	Mean (expedition)		810 ± 36	319 ± 47	23 ± 7
	1986–2013			209 ± 82	15 ± 7
Batang Hari	10	2009	377 ± 10	214 ± 58	7 ± 3
Batang Hari	10	2012	241 ± 1	249 ± 54	5 ± 2
Batang Hari	4	2013	*314 ± 00	190 ± 42	5 ± 2
Peat coverage 5%	Mean (expedition)		311 ± 55	218 ± 25	6 ± 2
	1986–2013			199 ± 74	6 ± 3
Musi	11	2008	423 ± 21	436 ± 075	17 ± 5
Musi	10	2012	*264 ± 00	249 ± 46	6 ± 2
Peat coverage 3.5%	Mean (expedition)		344 ± 80	343 ± 93	11 ± 5
	1986–2013			223 ± 105	7 ± 4
		Mean precipitation rate		227 ± 19	

*Results obtained for each river during all expeditions are presented as "mean expedition." Additionally DOC yields were calculated by using the mean riverine DOC end-member concentration and the precipitation rates averaged for the entire period of observations between 1987 and 2013. The "Mean precipitation rate" is the average of all precipitation rates determined within the catchments of all six rivers. *Samples were taken at a salinity of approximately zero; no samples were taken in the estuary.*

the radiocarbon concentration of the atmosphere in 1950 AD (Stuiver and Polach, 1977).

DOC End-Members, Yields, and Exports

In order to estimate DOC exports from the rivers into the ocean, DOC end-member concentrations were determined. These DOC end-member concentrations were derived from the correlation between DOC concentrations measured in the river estuary and salinity, wherein the zero salinity y-intercept defined the riverine DOC end-member concentration (**Table 1**, Figure S2). As DOC respiration and its release from e.g., phytoplankton could lead to deviations from the mixing line, which affect the regression equation and thus the calculated DOC end-member concentrations, the standard deviation of the y-intercept was obtained from a least square fit in order to achieve a quantitative error estimate (Bevington and Robinon, 1992). The calculated DOC end-member concentrations were subsequently multiplied by the river discharge in order to obtain the riverine DOC exports into the ocean. The DOC export normalized to the respective catchment area is the DOC yield, which was calculated as the product of the freshwater flux (FW$_{flux}$) and the DOC end-member concentration (Equation 2):.

$$FW_{flux}\,[L\,m^{-2}]\ = (Precipitation\,^*$$
$$(1 - (Evapotranspiration/100)))(1)$$
$$DOC\,yield\,[g\,m^{-2}\,yr^{-1}]\ = FW_{flux}\,^*\,DOC_{end\text{-}member}\qquad (2)$$

Precipitation, Evapotranspiration, and Freshwater Flux

The 1 × 1 degree gridded GPCC-rainfall data (Global Precipitation Climatology Centre, Landsurface Monitoring Product 1.0; Schneider et al., 2011) were used to calculate the precipitation rates and their standard deviations in the river catchments during our expeditions (Table 1, Table S2). The full GPCC data covering the period between 1986 and 2013 was used to calculate the mean precipitation rate for the individual catchments. Precipitation rates and river discharges measured on the island of Borneo showed an annual mean evapotranspiration (ET) of 37.9% of disturbed and 67.7% of pristine peatlands (Moore et al., 2013). These values fall almost in the range of 41–72% determined for Indonesia (Kleinhans, 2003; Kumagai

et al., 2005; Baum et al., 2007) suggesting a mean ET of 56.5 ± 15.5%. Since already 90% of all peatlands on Sumatra and Borneo were disturbed in 2008 (Miettinen and Liew, 2010) we used an evapotranspiration of 37.9% and considered an error range of 15.5% for calculating the freshwater flux.

River Catchments and Peat Coverage

An elevation model based on 30 arc-second data obtained from the shuttle radar topographic mission (SRTM, USGS, 2004) was established by using the geographical information system ArcGis with the extension ArcHydro in order to identify the individual river catchments and calculate the respective catchment areas. The peat area was obtained from the "Digital soil map of the world" (version 3.6, scale 1:5,000,000, resolution 3′ × 3′; FAO, 2003) and combined with the derived catchments to quantify the coverage in percentage (Table 1).

Mixing Model Principle

In order to better understand the processes controlling the DOC concentrations in rivers, a mixing model was produced (Figure 2). Since the majority of the available data was collected from the Siak, the mixing model was first of all developed based on data from the Siak. Later on the derived results were extrapolated and compared to data obtained from the other studied rivers.

In the model, the simulated DOC concentrations in the Siak were calculated as the product of mixing of three different water types: (i) groundwater from peatlands, (ii) surface runoff from peatlands, and (iii) freshwater discharges from parts of the

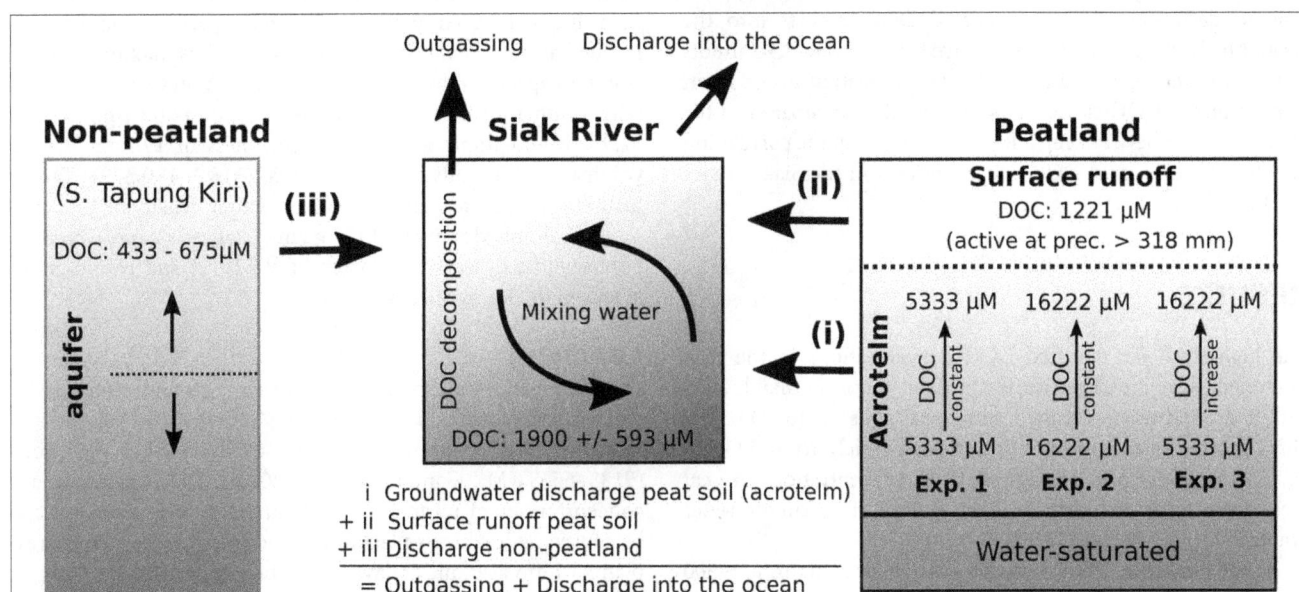

FIGURE 2 | Schematic of the model structure with the three water types (i, ii, and iii) that mix within the Siak. The source waters originate from the non-peatlands, which are to a large extent drained by the S. Tapung Kiri (See Figure 1), and from the peatlands. In peatlands it was further distinguished between surface runoff and groundwater discharge from the acrotelm. Groundwater DOC concentrations were varied during the three experiments. Based on our observations the DOC concentration of the surface runoff was set to 1221 μM and those in S. Tapung Kiri tended to increase from 433 to 675 μM with increasing precipitation rates (Table S3 and Figure S1). DOC decomposition and outgassing in addition to DOC discharges into the ocean balance DOC inputs along with the source waters. More detailed information is given in the Supplementary Material.

catchment that are not covered by peatlands (non-peatlands), which are largely drained by the S. Tapung Kiri (**Figures 1B, 2**). Freshwater discharges were calculated separately for the peat area and non-peat area within the catchment. The freshwater discharge from the peat area was divided into groundwater discharge and surface runoff based on the water storage capacity of the acrotelm, which is the upper, partly water-saturated soil horizon in which groundwater levels respond to varying precipitation rates (Holden, 2005). Acrotelm overflow and factors influencing DOC concentrations in the groundwater, as well as the DOC concentrations for the water types will be discussed in more detail in the following sections.

Besides the mixing of the different water types, the decomposition of DOC is the second factor controlling DOC concentrations in the Siak and was studied during a DOC decomposition experiment (Rixen et al., 2008). The results of this experiment, which included the photochemical and bacterial degradation of DOC, showed that roughly 27% of the DOC was degradable within a period of 2 weeks, whereas the remaining larger fraction was refractory on that respective time scale. The obtained equation, describing the DOC decomposition as a function of DOC concentration, was implemented into a box diffusion model that satisfactorily reproduced the oxygen concentrations in the Siak river, which are mainly controlled by the respiration of the labile DOC. In addition to the DOC concentration, the DOC decomposition equation within our mixing model also depends on time. The residence time of water within the Siak was derived from freshwater discharges and a fixed river profile. Accordingly, the residence time decreased with an increasing freshwater discharge, which in turn lowers the DOC decomposition in the rivers because there was less time to decompose DOC. Simulated DOC exports into the ocean finally result from the difference between the DOC inputs from the water types and the DOC decomposition in the river. The simulated DOC concentrations were then compared to the DOC end-member concentrations obtained from the correlation between the measured DOC concentrations in the Siak estuary and the salinity.

RESULTS

The highest so far reported DOC concentrations in the Siak catchment were measured during our expedition in March 2008 within groundwater obtained from peat soils (**Figure 1**). The DOC groundwater concentrations varied widely from 5333 to 16,222 μM (Table S3). The lowest DOC concentrations in peat areas were measured with 1221.6 μM in surface runoff water (Table S3).

In the Siak the DOC concentration increased from ~500 to 1300 and from 1300 to 1900 μM around the S. Tapung Kanan/Kiri and Mandau junctions because S. Tapung Kanan and the Mandau are rivers draining relatively large peat areas and thus are enriched in DOC (Baum et al., 2007; Rixen et al., 2008). During our expeditions the DOC concentrations in the S. Tapung Kiri, which drains the non-peat areas within the Siak catchment, varied between 344 and 675 μM. DOC

concentrations <675 μM were only measured during expeditions at which the precipitation was <300 mm, which might indicate that DOC concentrations decrease with decreasing precipitation rates as observed in many other rivers draining mineral soils (Figure S1). In the estuary, decreasing DOC concentrations correlating to increasing salinities indicate mixing between DOC-rich Siak water and DOC-poor ocean water (Figure S2). During our expeditions in the Siak catchment, the DOC end-member concentrations varied between 636 ± 88 and 2632 ± 164 μM and increased with increasing precipitation rates up to 318 mm (**Figure 3A**). After precipitation rates exceeded 318 mm the DOC concentrations dropped. The lowest DOC end-member concentrations were measured during our expedition in April 2013. Although this value is within the range of DOC end-member concentrations determined in the other studied rivers (**Table 1**) it represents a real exception for the Siak. The low DOC concentrations in the Siak estuary which finally lead to the low DOC end-member concentrations were assumed to be caused by enhanced DOC respiration within an senescent plankton bloom that occurred in the estuary during our expedition (Wit et al., 2015).

DISCUSSION

Acrotelm Thickness

Due to the low DOC concentrations in the surface runoff water, flooding with a resulting acrotelm overflow is one explanation for decreasing DOC concentrations in Siak at high precipitation rates (**Figure 3A**). A precipitation rate of 318 mm, beyond which the DOC concentrations decrease in the Siak, indicates accordingly the maximum water storage capacity of the acrotelm in the Siak catchment. In order to convert this maximum water storage capacity into acrotelm depth, evapotranspiration and pore volume need to be considered. The evapotranspiration of 37.9% would imply an acrotelm thickness of 19.7 cm if pore volume in peat soils would be 100% (318 * 1–0.379). Taking furthermore into account an organic matter density of 1 g cm^{-3} and a peat bulk density of 0.127 g cm^{-3} suggests a more realistic pore volume of 87.3 % (Warren et al., 2012) and results in an acrotelm thickness of 22.4 cm.

Leaching

In the Siak catchment the lower groundwater DOC concentrations of about 5333 μM (Table S3) fall in the range of DOC concentrations measured in other peat soils (5183–6658 μM; Gandois et al., 2013). DOC groundwater concentrations of up to 16,222 μM are extremely high but can be caused by leaching of leaf litter produced by secondary forest plants within a couple of days (Yule and Gomez, 2009).

Contrary to secondary plants, leaves of endemic peat plants are more resistant to degradation (Treutter, 2006; Lim et al., 2014) and therefore less DOC is leached from endemic leaf litter. In DOC pore water profiles, such a reduced leaching is reflected in the absence of a pronounced DOC gradient (Gandois et al., 2013). DOC pore water profiles in soils with secondary vegetation are expected to show a strong gradient with higher

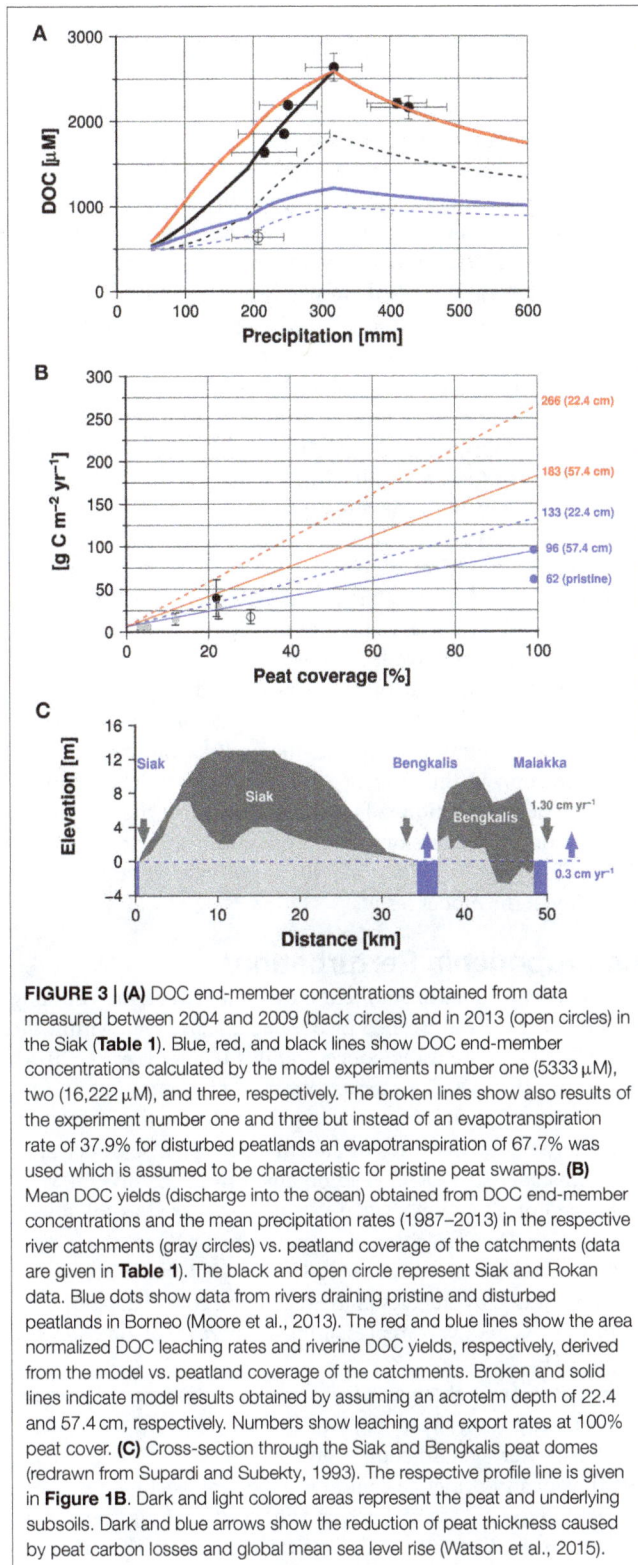

FIGURE 3 | (A) DOC end-member concentrations obtained from data measured between 2004 and 2009 (black circles) and in 2013 (open circles) in the Siak (**Table 1**). Blue, red, and black lines show DOC end-member concentrations calculated by the model experiments number one (5333 µM), two (16,222 µM), and three, respectively. The broken lines show also results of the experiment number one and three but instead of an evapotranspiration rate of 37.9% for disturbed peatlands an evapotranspiration of 67.7% was used which is assumed to be characteristic for pristine peat swamps. **(B)** Mean DOC yields (discharge into the ocean) obtained from DOC end-member concentrations and the mean precipitation rates (1987–2013) in the respective river catchments (gray circles) vs. peatland coverage of the catchments (data are given in **Table 1**). The black and open circle represent Siak and Rokan data. Blue dots show data from rivers draining pristine and disturbed peatlands in Borneo (Moore et al., 2013). The red and blue lines show the area normalized DOC leaching rates and riverine DOC yields, respectively, derived from the model vs. peatland coverage of the catchments. Broken and solid lines indicate model results obtained by assuming an acrotelm depth of 22.4 and 57.4 cm, respectively. Numbers show leaching and export rates at 100% peat cover. **(C)** Cross-section through the Siak and Bengkalis peat domes (redrawn from Supardi and Subekty, 1993). The respective profile line is given in **Figure 1B**. Dark and light colored areas represent the peat and underlying subsoils. Dark and blue arrows show the reduction of peat thickness caused by peat carbon losses and global mean sea level rise (Watson et al., 2015).

DOC concentrations in the acrotelm top due to leaching of labile leaf litter and lower DOC concentrations at the acrotelm base where peat leaching dominates. This is however an expected observation that in the future needs to be proved by data, but

in order to test whether such a soil DOC gradient helps to better explain the DOC end-member concentrations in the Siak, sensitivity experiments were carried out with our mixing model.

Numerical Leaching Experiments for the Siak

Possible effects of leaf litter and peat leaching on the DOC concentrations in the Siak were studied by performing three model experiments (**Figure 2**) and comparing the results with the riverine DOC end-members (**Figure 3A**). In the first two experiments, groundwater DOC concentrations were considered to be constant as generally assumed for peatlands. With a DOC concentration of 5333 µM, the first experiment reflected leaf litter leaching from pristine peat soils, whereas the second experiment mimicked leaf litter leaching from secondary forest plants with a DOC concentration of 16,222 µM. During the third experiment, DOC concentrations increased from 5333 µM at the base to 16,222 µM at the top of the acrotelm. The resulting linear DOC concentration gradient within the acrotelm corresponds to the assumption that endemic leaf litter leaching dominates at the base and secondary leaf litter leaching controls the DOC supply at the soil surface. DOC leaching became accordingly a function of the water level within the acrotelm and therewith also of the precipitation rates.

The model experiments 1 and 2 showed increasing DOC concentrations with increasing precipitation rates prior to the acrotelm overflow at 318 mm precipitation, after which surface runoff diluted DOC concentrations (**Figure 3A**). In the model the increasing DOC concentrations prior to the overflow were caused by reduced DOC decomposition as a consequence of a lower residence time of water at higher precipitation rates. The first experiment (groundwater DOC concentration of 5333 µM) results in simulated DOC concentrations lower than the measured end-members, whereas the second experiment (DOC groundwater concentration of 16,222 µM) leads to simulated DOC concentrations similar to those measured in the Siak during our expeditions. This suggests that leaf litter leaching from secondary vegetation could strongly affect the DOC concentration in the Siak, which is also obvious because peat soils in the Siak catchment are heavily disturbed with hardly any endemic vegetation left (Miettinen and Liew, 2010). The consistency between simulated and end-member DOC concentrations was further improved during the third experiment, indicating that leaching of leaf litter from secondary forest plants dominates at the top of the acrotelm and leaf litter leaching from endemic peat vegetation gains importance toward the base of the acrotelm.

Considering an evapotranspiration rate of 67.7%, as assumed to be characteristic for pristine peat swamps, instead of 37.9%, would not affect the trend but the values, which means that then all simulated DOC concentrations would fall below the measured DOC end-members. This implies that the evapotranspiration rates derived from the experiments carried out on Borneo of 37.9% apply also to Sumatra.

The concept of two DOC sources (leaf litter and peat) also agrees quite well with the radiocarbon ages of peat and DOC in rivers and the results derived from our DOC decomposition experiment, which showed that DOC in the rivers consists of a labile and a more refractory fraction (Rixen et al., 2008). The radiocarbon age of a peat sample obtained at a soil depth of 5 cm (**Figure 1**) revealed an age of 890 ± 25 years BP. At high latitudes DOC can be younger than the surrounding peat because of vertical water movements (Clymo and Bryant, 2008), but since at our study site younger peat was already eroded a DOC age of 890 ± 25 years represents a minimum estimate. DOC ages in the Mandau, a main tributary of the Siak, were with 575 ± 30 years BP similar to those measured in the disturbed channel on Borneo (Moore et al., 2013). Assuming that the secondary leaf-litter DOC is of modern age and DOC leached from peat is as old as the peat suggests that 65 ± 5% of the DOC in the river could be derived from peat and 35 ± 5% might originate from labile leaf litter. In combination with the results from the decomposition experiments and in line with the leaching characteristics of leaf litter this furthermore implies that a major share (65–73%) of the peat-derived DOC in the Siak is old and refractory on time scales of days to weeks. The smaller remaining fraction is young, labile, and originated from leaching of leaf litter produced by secondary forest plants. Since the decomposition of labile DOC was the main factor controlling the consumption of dissolved oxygen, the hypoxic conditions in the Siak can be seen as a consequence of the re-growing secondary forest and its production of labile leaf litter.

If groundwater DOC concentrations in the Siak are controlled by two DOC sources in the acrotelm, the acrotelm thickness becomes an important factor for the leaching of carbon from soils. The deeper the acrotelm, the greater the distance between the groundwater and secondary leaf litter at the soil surface and the lower the mean DOC groundwater concentration. This reduction should be limited by the DOC groundwater concentrations of 5333 μM caused by peat leaching at the bottom of the acrotelm. However, as indicated in experiment 1, such a low DOC groundwater concentration can only explain the DOC end-member concentration measured during our last expedition in April 2013 if one assumes an evapotranspiration of 67.7% as observed in pristine peat swamps (**Figure 3A**). Considering however the overall status of Sumatran peat lands this appears unlikely which in turn supports the hitherto held assertion that an enhanced DOC respiration associated with the occurrence of a plankton bloom lowered the DOC concentrations in the Siak estuary in April 2013 (Wit et al., 2015).

Extrapolation of Siak Results

So far we used the mixing model to calculate DOC end-member concentrations depending on precipitation rates and varying DOC concentrations in the acrotelm wherein the best fit between simulated and measured DOC end-member concentrations in the Siak could be obtained by assuming two DOC sources in the acrotelm: peat and secondary leaf litter. The next step was to study the impact of the peat coverage on the DOC end-member concentration. Therefore, all model settings remained unchanged except the precipitation rate and the peat coverage. The precipitation rate was set at 227 mm, which is the monthly mean precipitation of all studied river catchments covering the period between 1986 and 2013 (**Table 1**), and instead of using the

Siak peat coverage of 21.9% (**Table 1**) the peat coverage was set in a range from 0 to 100% (**Figure 3B**). The resulting simulated DOC concentrations were multiplied by the freshwater flux to obtain the DOC yields (See Equations 2, 3). Since the majority of DOC originates from peatlands, DOC yields are expected to increase with increasing peat coverage, which indeed is evident from the DOC yields obtained from the studied rivers. However, the Rokan river deviates from this trend with a lower DOC yield because coastal erosion, favored by mangrove deforestation (Butcher, 1996), increased suspended matter concentrations and DOC adsorption to suspended clay. Apart from that the simulated DOC yields agree quite well with the data-based end-members. This agreement can further be improved by using an acrotelm depth of 57.4 cm instead of 22.4 cm, which is well within the depth-range of acrotelms found in drained forests and plantations (Hooijer et al., 2012).

By increasing the acrotelm depth from 22.4 to 57.4 cm, the DOC yield was reduced from 133 to 96 g C m^{-2} yr^{-1} at a peat soil coverage of 100% (**Figure 3B**). This is consistent with the observed DOC yield in a channel draining disturbed peatlands in Borneo (Moore et al., 2013) and suggests that these model results are representative for disturbed SE Asian peatlands. The modeled DOC soil leaching rate associated with the riverine DOC export yield of 96 g C m^{-2} yr^{-1} at the acrotelm depth of 57.4 cm amounts to 183 g C m^{-2} yr^{-1}, implying a DOC decomposition in the river of 87 g C m^{-2} yr^{-1}. This estimate for DOC decomposition falls within the range of the data-based estimate on CO$_2$ emissions from SE Asian peat lands of 105 ± 27.5 g C m^{-2} yr^{-1} (Wit et al., 2015) and further supports the reliability of the model results.

Anthropogenic Perturbations

DOC concentrations of 5661 μM measured in a channel draining pristine peat soils in Borneo (Moore et al., 2013) fall within the range of DOC pore water concentrations (Gandois et al., 2013), suggesting that DOC decomposition is of minor importance under natural conditions, as expected from the refractory characteristics of leaf litter of endemic peat plants. Under such a circumstance, the DOC yield obtained from the pristine river in Borneo of 62 g C m^{-2} yr^{-1} can be considered as the lowest estimate on DOC leaching. Compared to DOC leaching rates derived from the model of 183 g C m^{-2} yr^{-1}, this implies that leaching caused by changes in the vegetation cover has risen from 62 to 183 g C m^{-2} yr^{-1} representing an increase of almost 200% (**Figure 4**).

In Borneo, changes in vegetation cover decreased evapotranspiration from 67.7 to 37.9%. This nearly doubles the freshwater flux fraction from 0.32 to 0.62 (Moore et al., 2013), which in principle could raise the peat leaching rate from 62 to ~118 g C m^{-2} yr^{-1}. Such an enhanced leaching rate is close to that of 108 g C m^{-2} yr^{-1} derived from our model by assuming a constant groundwater DOC concentration of 5333 μM at a peat coverage of 100%. However, a peat leaching rate of 108 or 118 g C m^{-2} yr^{-1} contributes ~60% to the total leaching of 183 g C m^{-2} yr^{-1} (**Figure 4**). If deforestation of pristine peat swamp forests and the reduced evapotranspiration raised leaching of DOC from peat from 62 to 108 g C m^{-2} yr^{-1}, the increase from 108 to 183 g C m^{-2} yr^{-1} appears to be caused by leaching of

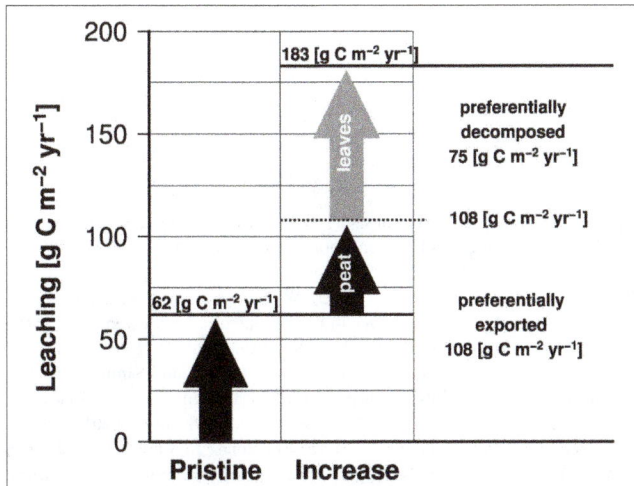

FIGURE 4 | Leaching of organic carbon from pristine peat soils and the additional leaching caused by human impacts subdivided according to DOC sources (peat soil and leaf litter).

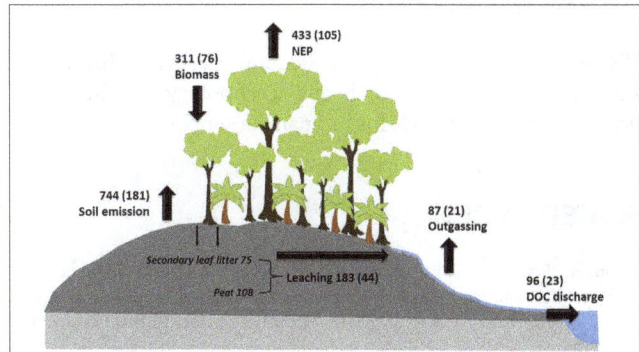

FIGURE 5 | Carbon fluxes in disturbed peatlands. Net ecosystem carbon losses representing a negative net ecosystem production (NEP; Hirano et al., 2007), CO_2 emissions from disturbed peat soils due to aerobic peat decomposition (Miettinen and Liew, 2010) and fires (van der Werf et al., 2008), carbon leaching, outgassing, and DOC discharge into the ocean. The carbon uptake by the growing biomass results from the difference between soil emission and NEP. River outgassing is the difference between leaching and DOC discharge into the ocean. The numbers and the numbers in brackets are fluxes in g C m^{-2} yr^{-1} and Tg C yr^{-1}, respectively.

leaf litter produced by secondary forest plants (**Figure 4**). This in turn implies that changes in the hydrological cycle and the regrowth of secondary forest plants explain 38% and 62% of the increase in DOC leaching from degraded peatlands, respectively.

Carbon Budgets

In order to emphasize the role of DOC leaching it needs to be seen in the context of CO_2 losses from soils caused by aerobic peat respiration and the CO_2 uptake by the re-growing vegetation, which reduces carbon losses from the ecosystem (**Figure 5**). Estimates of CO_2 emissions from disturbed peat soils due to aerobic peat decomposition (Miettinen and Liew, 2010) and fires (van der Werf et al., 2008) of 53 and 128 Tg C yr^{-1}, respectively, and a disturbed peatland area of 0.24 10^{12} m^2 suggest a peat carbon loss of 744 g C m^{-2} yr^{-1}. The net ecosystem carbon loss from disturbed peatlands covered by secondary forest amounts to a carbon loss of 433 g C m^{-2} yr^{-1} (Hirano et al., 2007), indicating that re-growing biomass could reduce CO_2 emissions from disturbed peatlands into the atmosphere by 311 g C m^{-2} yr^{-1}. A leaching rate of 183 g C m^{-2} yr^{-1} increases carbon losses from the ecosystem by 42% from 433 to 616 g C m^{-2} yr^{-1} of which 48% (87 g C m^{-2} yr^{-1}) returns back into the atmosphere and 52% (96 g C m^{-2} yr^{-1}) is exported to the ocean. Since the leached DOC originates from two different sources its role for the ecosystem varies: DOC leaching of labile leaf litter of 75 g C m^{-2} yr^{-1} reduces the carbon uptake by the re-growing biomass by 24% and its preferential decomposition in the river can lead to hypoxia as seen e.g., in the Siak river (Rixen et al., 2008, 2010). The leaching of DOC from peat of 108 g C m^{-2} yr^{-1} represents in turn an irrecoverable loss of soil.

Assuming a peat carbon density of 65.35 kg m^{-3} (Warren et al., 2012) peat carbon losses caused by oxidation (744 g C m^{-2} yr^{-1}) and leaching (108 g C m^{-2} yr^{-1}) lower the peat thickness by $\sim$1.30 cm yr^{-1} [(0.744 + 0.108)/65.35]. Such a fast shrinking of coastal peat domes, that partly even form the coast as seen on the Island of Bengkalis (**Figure 3C**), is a serious threat to the stability of the coastal peat plains that cover 10% of the Indonesian land mass.

CONCLUSION

In order to study processes controlling carbon leaching from tropical peat soils a mixing model was developed and validated by DOC concentrations measured in the groundwater, surface runoff water, and river in the Siak catchment. This model was subsequently used to quantify carbon leaching from tropical peat soils wherein the obtained results were compared with data obtained from other Sumatran rivers. Since these rivers reveal peat soil coverages of <30.2%, also data obtained from the literature were used. Our results show that a reduced evapotranspiration and a resulting increased freshwater discharge in addition to the supply of labile leaf litter produced by re-growing secondary forests increase leaching of carbon by $\sim$200%. Since the leached carbon originates from two different sources, namely peat and secondary vegetation, the resulting ecological consequences differ: Leached peat carbon is an irrecoverable loss of land that in addition to peat oxidation weakens the stability of peat-fringed coasts. This calls for mitigating strategies but reforestation by secondary forest plants bears an ecological threat. Leaching of their labile leaf litter supplies DOC of which the decomposition can lead to oxygen deficiencies in peat draining rivers.

AUTHOR CONTRIBUTIONS

TR, AB, and JS conceived and led the research conducted in Sumatra. Data were collected by TR, AB, FW, and JS. AB and FW performed chemical analyses. TR and FW conducted online measurements. TR and AB designed the study; FW and JS were involved in study design. TR led the writing of the paper. All authors discussed results and commented on the manuscript.

ACKNOWLEDGMENTS

We would like to thank the Federal German Ministry of Education, Science, Research and Technology (BMBF Bonn, grant number 03F0392C) for its financial support.

REFERENCES

Baum, A., Rixen, T., and Samiaji, J. (2007). Relevance of peat draining rivers in central Sumatra for the riverine input of dissolved organic carbon into the ocean. *Estuar. Coast. Shelf Sci.* 73, 563–570. doi: 10.1016/j.ecss.2007.02.012

Bevington, P. R., and Robinon, K. D. (1992). *Data Reduction and Error Analysis for the Physical Sciences.* New York, NY: McGraw-Hill Companies.

Butcher, J. G. (1996). *The Salt Farm and the Fishing Industry of Bagan Si Api Api.* Ithaca, NY: Southeast Asia Program Publications at Cornell University.

Clark, J. M., Lane, S. N., Chapman, P. J., and Adamson, J. K. (2008). Link between DOC in near surface peat and stream water in an upland catchment. *Sci. Total Environ.* 404, 308–315. doi: 10.1016/j.scitotenv.2007.11.002

Clymo, R. S., and Bryant, C. L. (2008). Diffusion and mass flow of dissolved carbon dioxide, methane, and dissolved organic carbon in a 7-m deep raised peat bog. *Geochim. Cosmochim. Acta* 72, 2048–2066. doi: 10.1016/j.gca.2008.01.032

FAO (2003). *Soil Map of the World*, Vol. 5. Rome: FAO/UNESCO.

Gandois, L., Cobb, A. R., Hei, I. C., Lim, L. B. L., Salim, K. A., and Harvey, C. F. (2013). Impact of deforestation on solid and dissolved organic matter characteristics of tropical peat forests: implications for carbon release. *Biogeochemistry* 114, 183–199. doi: 10.1007/s10533-012-9799-8

Hirano, T., Segah, H., Harada, T., Limin, S., June, T., Hirata, R., et al. (2007). Carbon dioxide balance of a tropical peat swamp forest in Kalimantan, Indonesia. *Glob. Chang. Biol.* 13, 412–425. doi: 10.1111/j.1365-2486.2006.01301.x

Holden, J. (2005). Peat land hydrology and carbon release: why small-scale process matters. *Philos. Trans. R. Soc.* 363, 2891–2913. doi: 10.1098/rsta.2005.1671

Hooijer, A., Page, S., Jauhiainen, J., Lee, W. A., Lu, X. X., Idris, A., et al. (2012). Subsidence and carbon loss in drained tropical peatlands. *Biogeosciences* 9, 1053–1071. doi: 10.5194/bg-9-1053-2012

Kleinhans, A. (2003). *Einfluss der Waldkonversion auf den Wasserhaushalt Eines Tropischen Regenwaldeinzugsgebietes in Zentral Sulawesi (Indonesien);(Influence of Forest Conversion on the Water Balance of a Tropical Rainforest Catchment in Central Sulawesi (Indonesia)).* Göttingen: Georg-August-Universität Göttingen.

Kumagai, T. O., Saitoh, T. M., Sato, Y., Takahashi, H., Manfroi, O. J., Morooka, T., et al. (2005). Annual water balance and seasonality of evapotranspiration in a Bornean tropical rainforest. Agricultural and forest. *Meteorology* 128, 81–92. doi: 10.1016/j.agrformet.2004.08.006

Laumonier, Y. (1997). *The Vegetation and Physiography of Sumatra.* Dordrecht: Kluwer Academic Publishers.

Lim, T. Y., Lim, Y. Y., and Yule, C. M. (2014). Bioactivity of leaves of *Macaranga* species in tropical peat swamp and non-peat swamp environments. *J. Trop. For. Sci.* 26, 134–141. Available online at: http://www.jstor.org/stable/23617022

Miettinen, J., and Liew, S. C. (2010). Degradation and development of peatlands in peninsular Malaysia and in the islands of Sumatra and Borneo since 1990. *Land Degradation Dev.* 21, 285–296. doi: 10.1002/ldr.976

Moore, S., Evans, C. D., Page, S. E., Garnett, M. H., Jones, T. G., Freeman, C., et al. (2013). Deep instability of deforested tropical peatlands revealed by fluvial organic carbon fluxes. *Nature* 493, 660–663. doi: 10.1038/nature11818

Moore, T. R., and Jackson, R. J. (1989). Dynamics of dissolved organic carbon in forested and disturbed catchments, Westland, New Zealnd 2. Larry River. *Water Resour. Res.* 25, 1331–1339. doi: 10.1029/WR025i006p01331

Page, S. E., Rieley, J. O., and Banks, C. J. (2011). Global and regional importance of the tropical peatland carbon pool. *Glob. Chang. Biol.* 17, 798–818. doi: 10.1111/j.1365-2486.2010.02279.x

Rixen, T., Baum, A., Pohlmann, T., Balzer, W., Samiaji, J., and Jose, C. (2008). The Siak, a tropical black water river in central Sumatra on the verge of anoxia. *Biogeochemistry* 90, 129–140. doi: 10.1007/s10533-008-9239-y

Rixen, T., Baum, A., Sepryani, H., Pohlmann, T., Jose, C., and Samiaji, J. (2010). Dissolved oxygen and its response to eutrophication in a tropical black water river. *J. Environ. Manage.* 91, 1730–1737. doi: 10.1016/j.jenvman.2010.03.009

Schneider, U., Becker, A., Finger, P., Meyer-Christoffer, A., Rudolf, B., and Ziese, M. (2011). "GPCC monitoring product: near real-time monthly land-surface precipitation from rain-gauges based on SYNOP and CLIMAT data," in *Global Precipitation Climatology Centre (GPCC)*, Deutscher Wetterdienst. Available online at: http://gpcc.dwd.de/

Schneider, U., Becker, A., Finger, P., Meyer-Christoffer, A., Rudolf, B., and Ziese, M. (2014). *GPCC Monitoring Product: Near Real-Time Monthly Land-Surface Precipitation from Rain-Gauges Based on SYNOP and CLIMAT Data.* Offenbach: Deutscher Wetterdienst.

Stuiver, M., and Polach, H. A. (1977). Discussion: reporting of 14C data. *Radiocarbon* 19, 355–363.

Supardi and Subekty, A. D. (1993). General geology and peat resources oh the Siak Kanan and Bengkalis Island peat deposits, Sumatra, Indonesia. *Geol. Soc. Am. Special Paper* 286, 45–61.

Treutter, D. (2006). Significance of flavonoids in plant resistance: a review. *Environ. Chem. Lett.* 4, 147–157. doi: 10.1007/s10311-006-0068-8

USGS (2004). *Global Land Cover Facility (GLCF) 30 Arc Second SRTM Elevation Data.* Maryland, MD: University of Maryland.

van der Werf, G. R., Dempewolf, J., Trigg, S. N., Randerson, J. T., Kasibhatla, P. S., Giglio, L., et al. (2008). Climate regulation of fire emissions and deforestation in equatorial Asia. *Proc. Natl. Acad. Sci. U.S.A.* 105, 20350–20355. doi: 10.1073/pnas.0803375105

Warren, M. W., Kauffman, J. B., Murdiyarso, D., Anshari, G., Hergoualc'h, K., Kurnianto, S., et al. (2012). A cost-efficient method to assess carbon stocks in tropical peat soil. *Biogeosciences* 9, 4477–4485. doi: 10.5194/bg-9-4477-2012

Watson, C. S., White, N. J., Church, J. A., King, M. A., Burgette, R. J., and Legresy, B. (2015). Unabated global mean sea-level rise over the satellite altimeter era. *Nat. Clim. Change* 5, 565–568. doi: 10.1038/nclimate2635

Wit, F., Muller, D., Baum, A., Warneke, T., Pranowo, W. S., Muller, M., et al. (2015). The impact of disturbed peatlands on river outgassing in Southeast Asia. *Nat. Commun.* 6:10155. doi: 10.1038/ncomms10155

Yule, C., and Gomez, L. (2009). Leaf litter decomposition in a tropical peat swamp forest in Peninsular Malaysia. *Wetlands Ecol. Manag.* 17, 231–241. doi: 10.1007/s11273-008-9103-9

Conflict of Interest Statement: The authors declare that the research was conducted in the absence of any commercial or financial relationships that could be construed as a potential conflict of interest.

15

Influence of Stress Field Changes on Eruption Initiation and Dynamics

Roberto Sulpizio * and Silvia Massaro

Dipartimento di Scienze della Terra e Geoambientali, Università degli Studi di Bari, Bari, Italy

We review here three main (first-order) mechanisms of stress variation able to influence the triggering of volcanic eruptions and the possible impact on eruption dynamics. They are short- and long-term unloading, seismic energy effects, and changes in far field stress due to geodynamic processes. We present an equilibrium equation for rupture of magma chamber and opening of a dyke up to the surface, taking into account the contribution of each mechanism within the equation. The equation considers the effect of possible superimposition of the three mechanisms with internal processes to the magmatic system, and it is also used for discussing the possible influence on eruption dynamics. The different possible contribution to the eruption triggering are discussed for each mechanism, highlighting how, in many cases, a single mechanism alone is not sufficient for driving eruptive activity if the magmatic system is not close to eruptive conditions.

Edited by:
Thorvaldur Thordarson,
University of Iceland, Iceland

Reviewed by:
Agust Gudmundsson,
University of London, UK
Sonia Calvari,
Istituto Nazionale di Geofisica e
Vulcanologia, Italy

*Correspondence:
Roberto Sulpizio
roberto.sulpizio@uniba.it

Specialty section:
This article was submitted to
Volcanology,
a section of the journal
Frontiers in Earth Science

Keywords: stress change, volcanic eruptions, eruptive dynamics, unloading, seismic energy, far field stress

INTRODUCTION

Increasing evidence supports the idea that stress changes play a fundamental role in triggering volcanic eruptions and in controlling their dynamics (Hill et al., 2002; Manga and Brodsky, 2006).

Stress changes in volcanic areas may vary in origin due to short- or long-term processes (Gudmundsson and Philipp, 2006; Andrew and Gudmundsson, 2007; Watt et al., 2008; Plateaux et al., 2014). The first includes earthquakes and landslides (Stein, 1999; Hill et al., 2002; Harris and Ripepe, 2007; Walter, 2007; Walter et al., 2007; Watt et al., 2008; De la Cruz-Reyna et al., 2010), while the second comprises unloading due to erosion and deglaciation (Davydov et al., 2005; Sigmundsson et al., 2010), tidal effects (Sohn, 2004; Cazaneve and Chen, 2010), or changes in the tectonic regime (Ventura and Vilardo, 1999; Waite and Smith, 2004; Diez et al., 2005; Miura and Wada, 2007; Lehto et al., 2010; Carbone et al., 2014). These processes superimpose to the possible local stress variations related to internal dynamics of a volcano, such as pressure increase in the magma chamber due to magma influx from depth or buoyancy induced by magma differentiation processes (Massol and Jaupart, 1999; Gudmundsson, 2006, 2016; Cañon-Tapia, 2014). Although usually claimed for explaining eruptive style transitions (i.e., from effusive to explosive, or from no-activity to eruption, Hasabe et al., 2001; Adams et al., 2006; Ida, 2007; Di Traglia et al., 2009; Schneider et al., 2012; Ripepe et al., 2013; Kereszturi et al., 2014), the magmatic processes internal to a volcano alone are sometimes not sufficient for equalling the elastic energy due to lithostatic loading (e.g., Gudmundsson, 2016; Sulpizio et al., 2016). In other cases the physical and chemical characteristics of the deposits do not support triggering mechanisms like magma mixing or bubble nucleation. For example, arrival of gas rich magma in a magmatic system or magma evolution within the chamber itself are usually claimed for explaining transitions from effusive to explosive eruptions, even in cases in which the geological evidences are lacking (i.e., the erupted material is poorly vesicular, as in the case of many basaltic eruptions; Fink et al., 1992; Wylie et al., 1999). In other cases, the arrival of fresh magma into a magma chamber is postulated as trigger magma chamber rupture and eventually fed an eruption, even in the absence of petrological evidences (i.e., mingling and mixing; e.g., Davì et al., 2011). All these considerations claim for discussion about the state of the art and perspectives about the interplay between volcanic activity and changes in the stress field. This review has not the presumption of being exhaustive of all the knowledge on stress changes and volcanic eruptions, but we will critically review the main mechanisms inducing short- and long-term stress changes at volcanoes, and their possible influence on eruption initiation and its dynamics. The review is intended to focus on the first order effects of stress change. In particular, the changing strength and strain energy due to not homogeneous lithosphere or different volcano edifice (e.g., Gudmundsson, 2012a, 2016) is not explicitly discussed, although they are implicitly contained in the equations describing the driving/resisting pressures. The review is organized in four main chapters: stress changes due to unloading, effects of seismic energy, changes in regional stress field (far field), and influence of stress change on eruption dynamics.

STRESS CHANGES DUE TO UNLOADING

The unloading processes are the most common way to change the lithostatic load. This may induce fracture initiation/propagation, which changes the lithostatic component of the stress at any point in the lithosphere and, ultimately, may result in eruption initiation. The importance of unloading processes on volcanic activity is testified by the long-term eruptive histories of many volcanoes, which reveal that changes in eruption rate and/or magma composition follows partial destruction of the edifice (Presley et al., 1997; Hildenbrand et al., 2004; Hora et al., 2007; Longprè et al., 2009; Boulesteix et al., 2012).

The unloading can be a short- or long-term process, and the different mechanisms will be reviewed following the temporal scale of action.

Long-Term Processes

Many surface load variations occurring over a long time scale (such as deglaciation at mid high latitudes) have been suggested to have a significant impact on eruptive behavior (Jellinek et al., 2004; Sinton et al., 2005; Sigmundsson et al., 2010; Geyer and Bindeman, 2011; Hooper et al., 2011). A retreating ice cap of limited dimensions and thickness (e.g., radius of only a few kilometers) will affect only the shallowest parts of a magmatic system. Conversely, a retreating ice cap with a radius of tens of kilometers or more may influence the generation of melt down to the mantle (Gudmundsson, 1986; Andrew and Gudmundsson, 2007; Sigmundsson et al., 2010).

This can be expressed in a simple way considering the expression of pressure in the elastic Earth:

$$P = p_0(\sigma_{xx} + \sigma_{yy} + \sigma_{zz}) \tag{1}$$

or, in cylindrical coordinates:

$$P = \frac{1}{3}p_0(\sigma_{xx} + \sigma_{yy} + \sigma_{zz}) \tag{2}$$

where $p_0 = rgh$, σ_{rr}, $\sigma_{\theta\theta}$, and σ_{zz} the radial, tangential and vertical stress, respectively (**Table 1**).

Considering a disc load, the vertical stress at a depth z in the Earth crust and distance R from the load center can be expressed as Davis and Selvadurai (2001) and Pinel and Jupart (2004):

$$\sigma_{zz} = p_0 \left[1 - \frac{z^3}{\left(R^2 + z^2\right)^{3/2}} \right] \tag{3}$$

The other two horizontal stress components are equal to:

$$\sigma_{rr} = \sigma_{\theta\theta} = \frac{p_0}{2} \left[(1 + 2v) - \frac{2(1 + v)z}{\sqrt{(R^2 + z^2)}} + \frac{z^3}{\left(R^2 + z^2\right)^{3/2}} \right] \tag{4}$$

where v the Poisson ratio here equal to 0.5. Taking into account equations (2) to (4) the pressure under a disc load overlying an elastic space is:

$$P = \frac{2}{3}p_0(1 + v) \left[1 - \frac{z}{\sqrt{(R^2 + z^2)}} \right] \tag{5}$$

It is evident that for $R \to \infty$ both stress and pressure simplify to lithostatic. The influence of disc load is greater in the upper crust, while it decreases with depth (increasing z).

Using these equations Sigmundsson et al. (2010) calculated the influence of unloading due to melting of an ice cap (**Figure 1**). The calculations were performed for two different ice models, both with radius 50 km and constant thinning rate during 110 years. The first model has a uniform thinning rate of 50 cm year^{-1} (corresponding to surface pressure change of 4.5 kPa year^{-1}). The second model thins by 25 cm year^{-1} between 0 and 30 km, and by 62 cm year^{-1} between 30 and 50 km (**Figure 1**). It can be seen that the pressure decrease is, in average, around 4–5 kPa year^{-1} in the first 10 years, increasing up to 6–7 kPa year^{-1} after 110 years. It means an average reduction of pressure of 0.5–0.6 MPa in about one century. It is worth nothing that the main part of pressure decrease is located in the upper 10–15 km (**Figure 2**), which is also the location of shallow magma chambers and magmatic feeding/conduit systems.

The effects on a shallow magma chamber can be numerically simulated considering it as a cavity of an idealized shape (sphere or ellipsoid) within an elastic homogeneous crust and filled

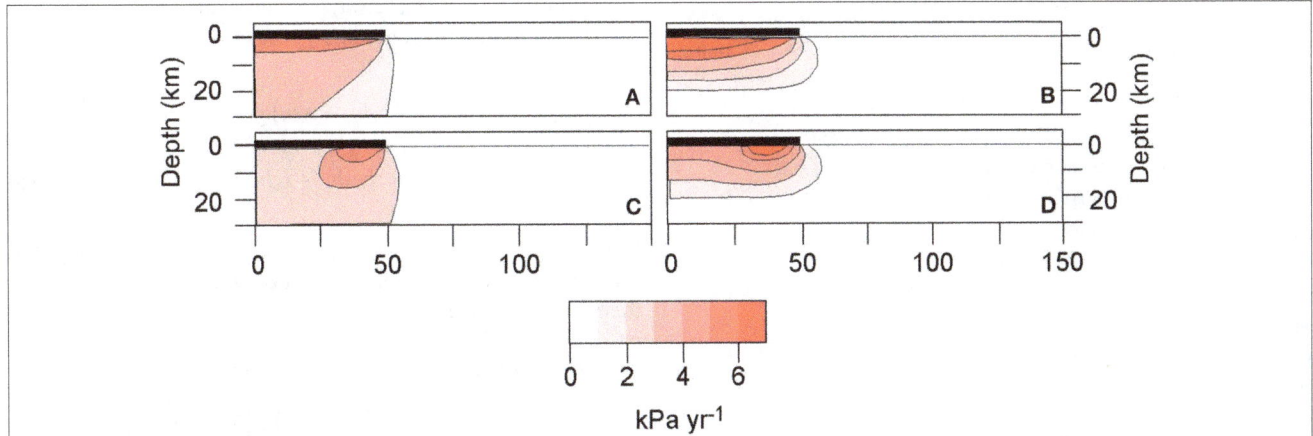

FIGURE 1 | Pressure decrease below a melting ice cap (modified after Sigmundsson et al., 2010). Results shows two different ice caps, both with radius 50 km and constant thinning rate during 110 years. **(A,B)** Uniform thinning rate of 50 cm year^{-1} (corresponding to surface pressure change of 4.5 kPa year^{-1}). **(C,D)** Thinning rate of 25 cm year^{-1} between 0 and 30 km, and of 62 cm year^{-1} between 30 and 50 km. The final volume reduction is the same in both models. **(A,C)** Average yearly stress change in the initial 10 years after thinning begins. **(B,D)** Average yearly stress changes 100–110 years after the beginning of thinning.

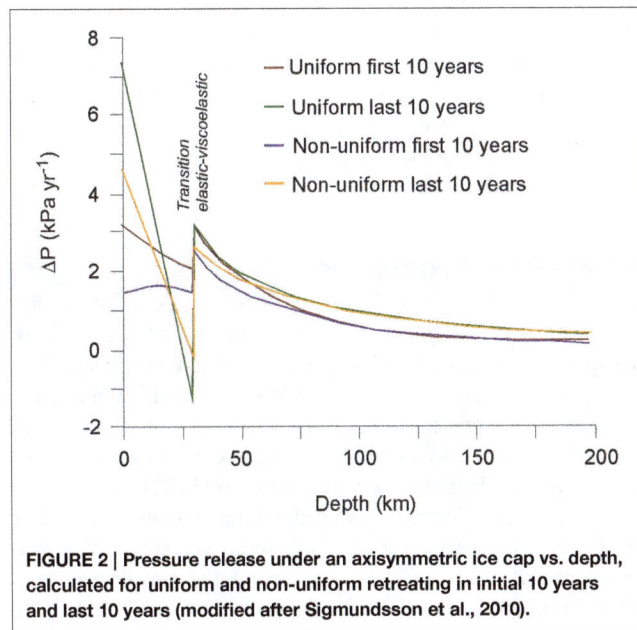

FIGURE 2 | Pressure release under an axisymmetric ice cap vs. depth, calculated for uniform and non-uniform retreating in initial 10 years and last 10 years (modified after Sigmundsson et al., 2010).

with an inviscid fluid. The magma has the same density of the surrounding crust and the reference state is lithostatic. Although strong, these assumptions can provide a first order picture of the unloading effects on a shallow reservoir (Gudmundsson, 2006).

Surface load variation induces a magma pressure change (ΔP_m) and a modification of the excess magma pressure required for dyke initiation P_e (Gudmundsson, 2012b). The failure of the chamber wall that marks the dyke initiation occurs when the minimum compressive deviatoric stress reaches the tensile strength (T_0) of the host rocks (Pinel and Jaupart, 2005; Gudmundsson, 2012b). Applying this rupture criterion in three dimensions, the ΔP required for dyke initiation can be defined

(Albino et al., 2010). However, to allow dyke propagation we need a sufficient magma overpressure (P_o, also named driving pressure and net pressure, values up to several tens of MPa), which is the driving mechanism of a hydrofracture (a fluid-driven extension fracture; Gudmundsson, 2012b). Overpressure is the result of the combined effects of the initial excess pressure in the magma chamber, the (eventual) magma buoyancy, and the lithostatic load (Gudmundsson, 2006). It also acts against the normal stress applied on the potential dyke fracture before magma emplacement, and it coincides with the minimum principal compressive stress σ_3. A general form to express overpressure is (Gudmundsson, 1990, 2012b):

$$P_0 = P_e + (\rho_r - \rho_m)\,gh_1 + \sigma_d + R_f \qquad (6)$$

where deviatoric stress $\sigma_d = \sigma_1 - \sigma_3$, and ρ_r is the rock density and ρ_m the magma density (**Table 1**). To allow a dyke to reach the surface and feed an eruption a minimum overpressure (ΔP_{0m}) is required, in order to maintain the dyke open (Anderson, 1936; Costa et al., 2007). Taking also into account the viscous and frictional resisting forces per unit area (R_f) the Equation (6) changes into:

$$P_0 - \Delta P_{0m} = P_e + (\rho_r - \rho_m)\,gh_1 + \sigma_d + R_f \qquad (7)$$

where h_1 indicates the different height of magma column during dike propagation to the surface.

Equation (6) considers the condition for dyke initiation, while the Equation (7) highlights constrains for dyke to reach the surface and feed an eruption. Because lithospheric inhomogeneity is not here considered, Equation (7) does not contain some important constraints for dyke propagation like stress barriers, elastic mismatch, and Cook-Gordon delamination (Gudmundsson, 2011). Defining ΔF_g as the difference in gravitational force at chamber rupture and at an arbitrary time during dyke propagation [$\Delta F_g = (\rho_r$

TABLE 1 | List of symbols used in the text and equations.

Notation	Description	Unit
σxx, σyy, σzz	Stress component in the Cartesian coordinates	Pa
σrr, $\sigma\theta\theta$, σzz	Stress component in the Cylindrical coordinates	Pa
σ_1	Maximum stress component	Pa
σ_3	Minimum stress component	Pa
σ_d	Deviatoric stress	Pa
σ_l	Lithostatic stress	Pa
σ_e	Seismic stress	Pa
σ_{ff}	Far field stress	Pa
σ_{tec}	Homogeneous horizontal tensile stress	Pa
σ_t	Total stress field	Pa
ν	Poisson's ratio	
g	Gravitational acceleration	m/s^2
h	Height	m
h_1	Height of magma column during dike propagation	m
h_i	Height of the ice cup	m
h_r	Rock thickness	m
R	Distance from the disc load center	m
R_f	Resisting force per unit area	Pa
ρ	Crustal density	kg/m^3
ρ_m	Magma density	kg/m^3
ρ_r	Rock density	kg/m^3
ρ_i	Ice density	kg/m^3
P	Pressure	Pa
P_l	Lithostatic pressure	Pa
P_0	Driving pressure	Pa
P_e	Excess magma pressure required for dyke initiation	Pa
P_u	Unloading pressure	Pa
ΔP	Pressure variation	Pa
ΔP_m	Magma pressure change	Pa
ΔP_{0m}	Minimum magma pressure change	Pa
$\Delta P(K)$	Pressure reduction within the magma chamber induced by the removal of a surface conical load	Pa
ΔF_g	Difference in gravitational force between chamber rupture and an arbitrary time during dyke propagation	N
E	Young Modulus	Pa
K	Bulk Modulus	Pa
T_0	Tensile strength of the host rocks	Pa
V	Initial volume of the reservoir	m^3
ΔV	Volume variation of the reservoir	m^3
V_e	Erupted volume in presence of edifice collapse	m^3
V_n	Erupted volume in absence of edifice collapse	m^3
D	Effective depth accounting for the total deficit of mass with respect to before rifting	m
W	Graben width	m
z	Depth	m
z_c	Depth below the rift	m
z_{in}	Depth of the crustal reservoir	m
z_1, z_2	Depth of the dyke trajectories	m

$- \rho_m)g(h_1 - h)]$, it is possible to descend that for $\Delta P_{0m} < \Delta F_g + R_f$ only dyke injection is possible but not eruption.

An unloading event always reduces lithostatic load, and therefore it induces changes in σ_d because:

$$\sigma_d = \sigma_1 - \sigma_3 = (\rho_r h_r - \rho_i h_i)g - \frac{(\rho_r h_r - \rho_i h_i)g}{\nu - 1}$$
$$= (\rho_r h_r - \rho_i h_i)g\left[1 - \frac{1}{\nu - 1}\right] \quad (8)$$

where ν is the Poisson's ratio, and ρ_i and h_i the density and thickness of ice cap, respectively (**Table 1**). Equation (7) can therefore be written as:

$$P_0 - \Delta P_{0m} = P_e + (\rho_r - \rho_m)gh$$
$$+ (\rho_r h_r - \rho_i h_i)g\left[1 - \frac{1}{\nu - 1}\right] + R_f \quad (9)$$

Magma pressure changes strongly depend on the chamber shape as well as on its depth. As a general rule, dyke propagation is favored for spherical and oblate shapes of magma chambers, whereas it is inhibited for prolated ones (Gudmundsson, 2012b).

In any case, models and simple calculations show that the reduction of stress and pressure may range between a few kPa (10 years' time span) up to less than 1 MPa, about three and one orders of magnitude less than the tensile strength of rocks. This means that, in general, the ice thinning effect on the failure of shallow magma chambers is minimal (Andrew and Gudmundsson, 2007; Sigmundsson et al., 2010), and can be decisive only if the magma batch is close to the rupture conditions.

Short-Term Processes

Numerical modeling have linked short time scale redistribution of surface loads, such as partial destruction of edifices or flank collapse events, to eruption triggering and changes in eruption style (e.g., Pinel and Jaupart, 2005; Manconi et al., 2009). Large flank collapses are common phenomena in the evolution of volcanic edifices, and sometimes these events trigger explosive eruptions (Le Friant et al., 2003; Roverato et al., 2011).

The pressure decrease induced within a magma chamber by the partial destruction of a sub-aerial volcanic edifice can be quantified using an elastic model for the two-dimensional plane strain approximation (Pinel and Jaupart, 2005). Using the same approach, Pinel and Albino (2013) calculated the effect of unloading of a conical edifice over an elastic lithosphere, obtaining similar results than the removal of the ice cap. In particular, they considered a very shallow, elliptical magma batch (top 1 km of depth) filled with fluid of the same density of the surrounding rocks and bulk modulus K. Removing a conical load of 2 km radius, 1 km height, and density 2,800 kg/m^3, induces a change in the magma chamber related to:

$$\Delta P = -K\frac{\Delta V}{V} \quad (10)$$

with V being the initial volume of the reservoir (Pinel and Albino, 2013).

In the vicinity of the reservoir the pressure variation within the crust differs from the homogeneous case (**Figure 3**), being

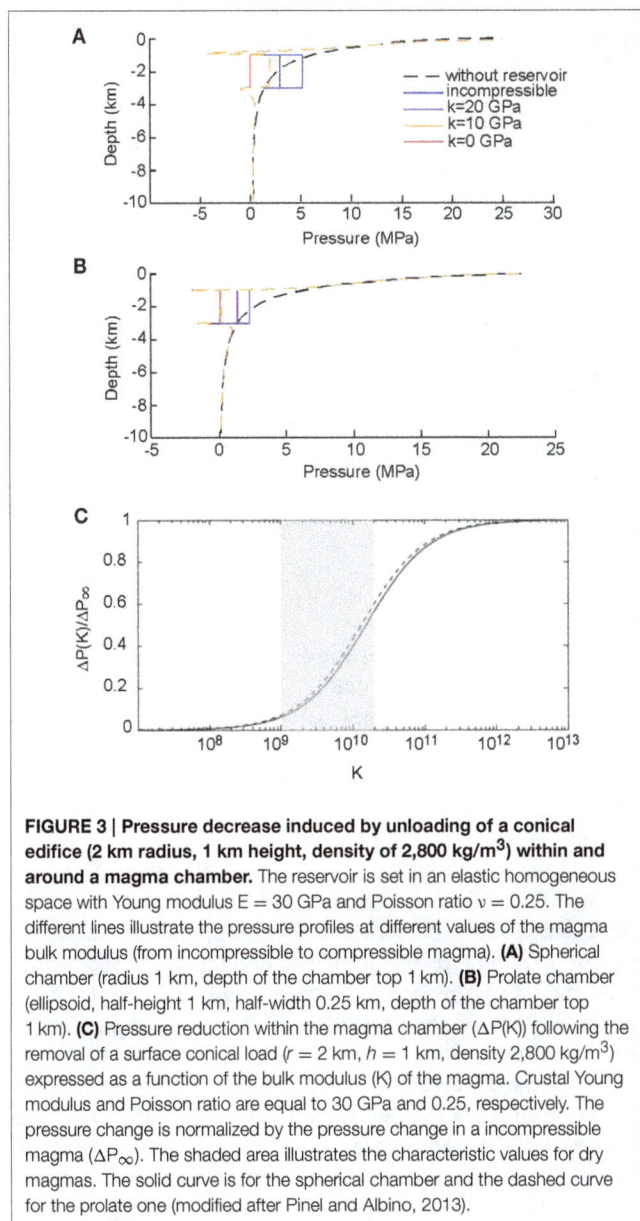

FIGURE 3 | Pressure decrease induced by unloading of a conical edifice (2 km radius, 1 km height, density of 2,800 kg/m³) within and around a magma chamber. The reservoir is set in an elastic homogeneous space with Young modulus E = 30 GPa and Poisson ratio ν = 0.25. The different lines illustrate the pressure profiles at different values of the magma bulk modulus (from incompressible to compressible magma). (A) Spherical chamber (radius 1 km, depth of the chamber top 1 km). (B) Prolate chamber (ellipsoid, half-height 1 km, half-width 0.25 km, depth of the chamber top 1 km). (C) Pressure reduction within the magma chamber (ΔP(K)) following the removal of a surface conical load (r = 2 km, h = 1 km, density 2,800 kg/m³) expressed as a function of the bulk modulus (K) of the magma. Crustal Young modulus and Poisson ratio are equal to 30 GPa and 0.25, respectively. The pressure change is normalized by the pressure change in a incompressible magma (ΔP∞). The shaded area illustrates the characteristic values for dry magmas. The solid curve is for the spherical chamber and the dashed curve for the prolate one (modified after Pinel and Albino, 2013).

higher for spherical shape than for the prolate one. Pressure also increases at the chamber margins, and is most extreme at the chamber top. This is because the deformation of the magma chamber walls due to unloading is partially counterbalanced by pressure partition within the magma chamber.

The amount of the magma pressure reduction increases with the value of the bulk modulus. This is because for incompressible magmas (larger value of k) no reservoir volume change occurs, and only pressure lowering within the chamber compensates the reduction induced by the unloading event. The effect of compressibility is shown in **Figure 3C**.

Figure 4 shows the pressure reduction within a spherical reservoir with a top at 1 km depth, induced by the removal of the upper 20% volume of the volcano edifice (mean value based on field observations; Voight and Elsworth, 1997). The

erupted volume is larger than that in the absence of edifice collapse ($V_e > V_n$; **Table 1**; **Figure 4**) when the small edifices are considered. As the edifice size increases the V_e/V_n ratio decreases. When large strato-volcanoes are partially destroyed by flank collapse this volume reduces to zero, possibly resulting in the abortion of any incipient eruption. Shallow magma batches require smaller edifice size to reach the point of aborted eruption, (**Figure 4A**), whereas deep chambers reduce any effect of edifice collapse on erupted magma volume. This is because any edifice collapse reduces the lithostatic load on the magma batch, and the magma volume required for reaching the eruptive conditions is smaller than in the case of larger edifices or deeper magma chambers (Manconi et al., 2009).

The magma reservoir shape also influences the possibility of eruption following an edifice collapse. **Figure 5** shows that the influence of the collapse is smaller for a prolate reservoir, and a larger edifice size is required than for a spherical reservoir at the same depth. Having a prolate chamber with top at depth of 1 km, eruption is only aborted when the edifice radius is greater than 6 km.

EFFECT OF SEISMIC ENERGY

Earthquakes can stress magmatic systems either through static stresses (the offset of the fault which generates a permanent deformation in the crust) or through dynamic stresses from the seismic waves (Manga and Brodsky, 2006). Both stresses increase with the seismic moment of the earthquake, but they decay in different way with distance r from the generation area. In particular, static stresses decreases as $1/r^3$, whereas dynamic stresses fall off more gradually (as $1/r^{1.66}$) and are proportional to the seismic wave amplitude (e.g., Lay and Wallace, 1995).

The stress transfer due to regional earthquakes may be of great importance in reawakening a dormant system. Previous works suggested a statistical correlation among large earthquakes and eruptions in time and space (Linde and Sacks, 1998; Hill et al., 2002; Marzocchi, 2002; Walter and Amelung, 2007; Walter et al., 2007). However, not all the large earthquakes trigger eruptions, and this is compelling evidence that the magmatic system needs to be ready to erupt under a new energetic equilibrium. This implies that the eruption triggering depends on the initial state of the magmatic system prior to the earthquake (magma composition, volatiles, chamber overpressure, strength of the host rocks, and type, size, and distance of the foci; Hill et al., 2002). In this framework, an important event is the unclamping of previous faults, which is the reduction in normal stress due to earthquake energy.

Dynamic and static deformation due to an earthquake may increase volcanic activity (Hill et al., 2002; Walter and Amelung, 2007). Seismic body and surface waves induce dynamic deformation, whereas displacement across a fault and subsequent viscoelastic relaxation of the crust account for permanent static deformation. A statistically significant response immediately after the earthquake (Linde and Sacks, 1998) has been observed for volcanoes at 750 km or more from the epicenters, suggesting they are triggered by dynamic deformation (Brodsky et al., 1998; Manga and Brodsky, 2006). The effect of static deformation

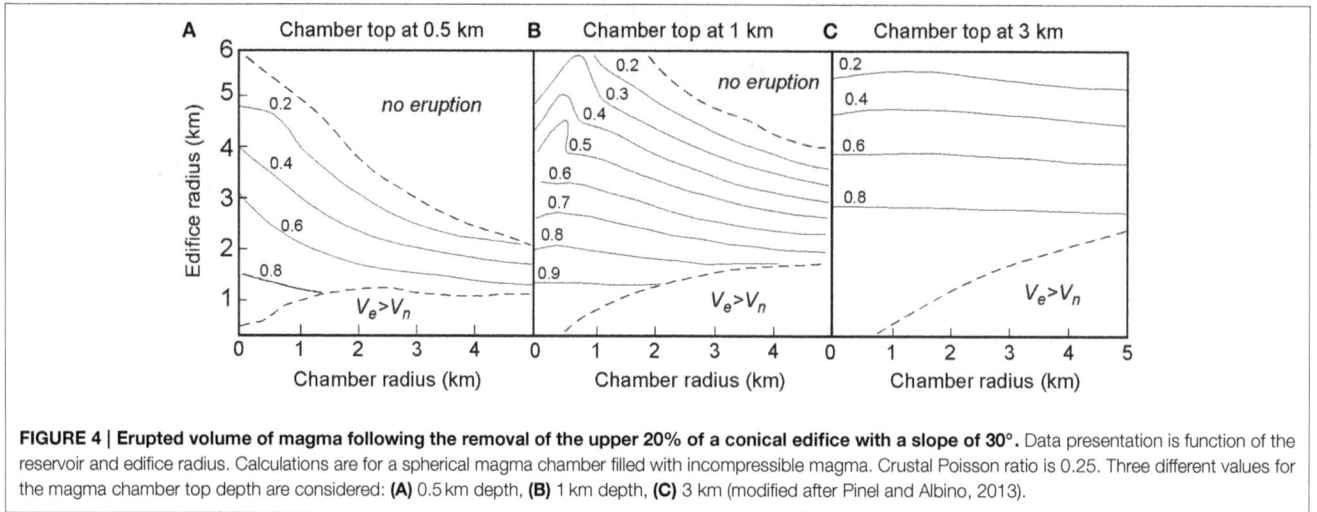

FIGURE 4 | Erupted volume of magma following the removal of the upper 20% of a conical edifice with a slope of 30°. Data presentation is function of the reservoir and edifice radius. Calculations are for a spherical magma chamber filled with incompressible magma. Crustal Poisson ratio is 0.25. Three different values for the magma chamber top depth are considered: **(A)** 0.5 km depth, **(B)** 1 km depth, **(C)** 3 km (modified after Pinel and Albino, 2013).

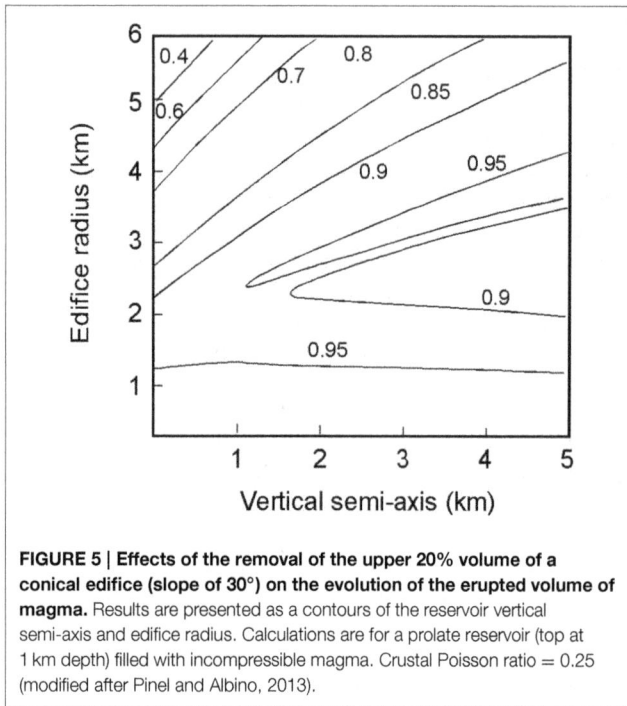

FIGURE 5 | Effects of the removal of the upper 20% volume of a conical edifice (slope of 30°) on the evolution of the erupted volume of magma. Results are presented as a contours of the reservoir vertical semi-axis and edifice radius. Calculations are for a prolate reservoir (top at 1 km depth) filled with incompressible magma. Crustal Poisson ratio = 0.25 (modified after Pinel and Albino, 2013).

in triggering eruptions remains poorly understood and it is unclear whether it is the most effective type of deformation in promoting eruptions (Marzocchi et al., 2002; Selva et al., 2004). The amplitude of static deformation decays more rapidly with distance than the seismic waves (Hill et al., 2002). Follows that to have eruption triggering from static deformation is most likely at volcanoes located in proximity to an earthquake rupture plane.

Classical examples of interaction between earthquakes and volcanic eruptions are the Kamchatka 1952 (M 9.0, followed by renewal of activity at Karpinsky and Maly Semiachik volcanoes, and at the Tao-Rusyr Caldera), Chile 1960 (M 9.5, followed by renewal of activity at Cordón-Caulle Planchón-Peteroa, Tupungatito and Calbuco volcanoes), Alaska 1964 (M 9.2,

followed by renewal of activity at Trident and Redoubt volcanoes), Sumatra-Andaman 2004–2005 (M 9.3 and M 8.7, followed by renewal of activity at Talang and Barren Island volcanoes; Sepulveda et al., 2005; Walter and Amelung, 2007). All these examples are from subduction zones, which most of the time are partially locked and accumulate stress that is released during earthquakes (**Figure 6A**). Walter and Amelung (2007) related the triggering of the eruption listed above to the change in volumetric strain, which is the sum of the normal components of the strain tensor. Negative volumetric strain corresponds to volumetric contraction (compressing the rock), and positive volumetric strain corresponds to volumetric expansion (decompressing the rock). Earthquakes in subduction areas are associated with volumetric contraction in the near-trench portion of the forearc and volumetric expansion in the far-trench portion, which is where the volcanic arc is usually located (**Figure 6B**). The main observation is that all the erupted volcanoes underwent volumetric expansion induced by the earthquake. A direct mechanical effect of stress change due to volumetric expansion may be the unclamping of the fissure system. A pre-existing network of cracks may be connected, nucleate and thereby facilitate preferred paths for magma ascent. Unclamping of fracture system was claimed for the earthquake occurred on Kamchatka peninsula on January 1st, 1996 along a SW–NE trending fracture system, which triggered the twin-eruption at the volcanoes Karymsky and Akademia Nauk (Walter, 2007). The earthquake is hypothesized to have prompted dilatation of the magmatic system together with extensional normal stress at intruding N–S trending dykes, allowing magma to propagate to the surface.

Taking into account Equation (10) and adding the contribution of seismic stress, it can be written:

$$P_0 - \Delta P_{0m} = P_e + (\rho_r - \rho_m) gh$$
$$+ (\rho_r h_r - \rho_i h_i) g \left[1 - \frac{1}{\nu - 1} \right] + \sigma_e + R_f$$

$$(11)$$

NON-SEISMIC CHANGES IN REGIONAL STRESS FIELD

The change in tectonic stress has been claimed as trigger of large ignimbrite eruptions or for controlling the eruptive style of explosive eruptions (Korringa, 1973; Aguirre-Díaz and Labarthe-Hernandez, 2003; Miller and Wark, 2008; Costa et al., 2011).

The first order influence of far-field stress (σ_{ff}) on eruption triggering was investigated using numerical simulations, which

demonstrated how the combined effect of crustal extension and magma chamber overpressure can sustain linear dyke-fed explosive eruptions with mass fluxes in excess of 10^{10} kg/s from shallow-seated (4–6 km depth) chambers affected by extensional stress regime (Costa et al., 2011). The model shows that for a far-field stress above the value able to counterbalance the lithostatic pressure at the fragmentation depth (**Figure 7**), a dyke of any length remains opened, and the Mass Eruption Rate (MER) is strongly controlled by the 3D geometry and

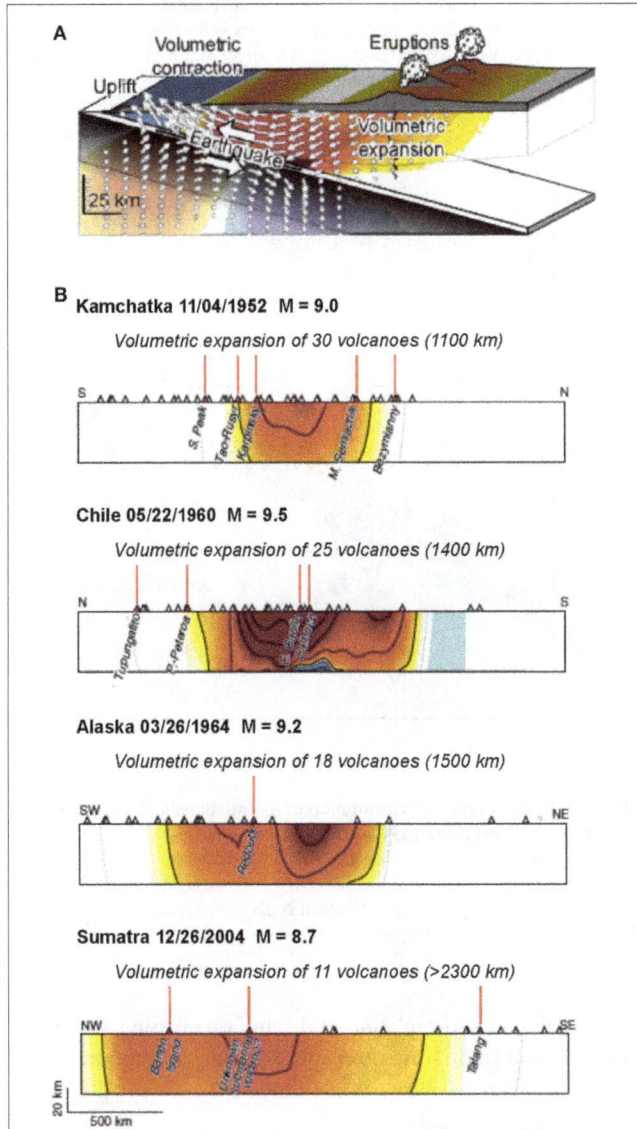

FIGURE 6 | (A) Schematic cross sections of subduction zone. Arrows indicate the displacement field, while colors rank the volumetric deformation (red colors for volumetric expansion, blue colors for volumetric contraction) associated with a megathrust earthquake. (B) Volumetric deformation associated with megathrust earthquakes. Triangles indicate the historically active volcanoes. Red lines highlight the volcanoes that erupted within 3 years from earthquakes. Red colors indicate positive strain (volumetric expansion), blue colors indicate negative strain (volumetric contraction). Contours represent 0 μ strain (dashed line) and 5 μ strain increments (solid black lines) (modified after Walter and Amelung, 2007).

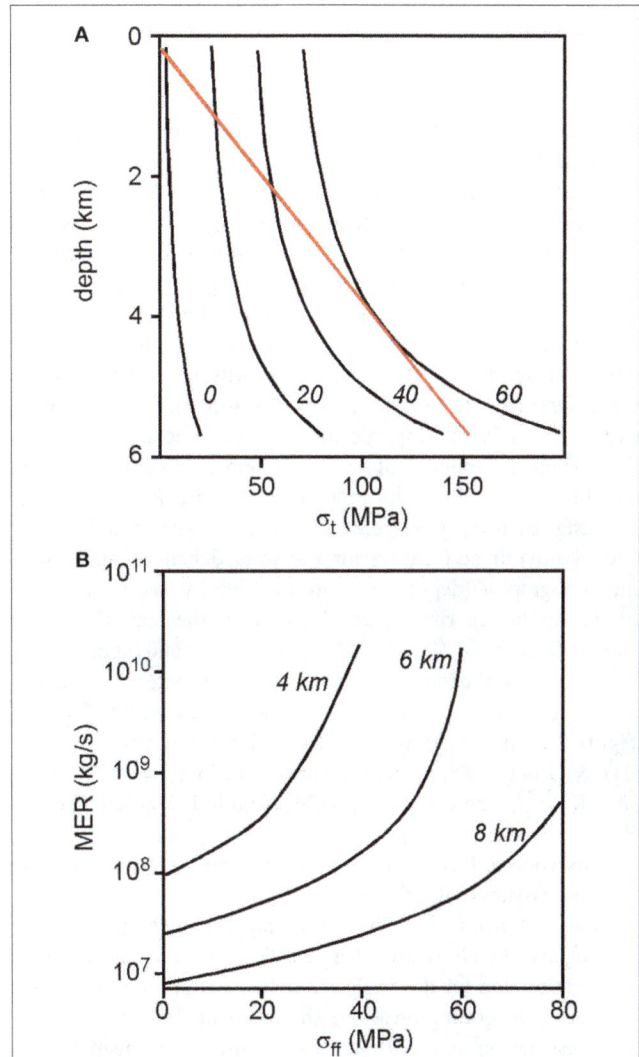

FIGURE 7 | (A) Profile of dyke tensile stress σ_t along the vertical axis obtained using the analytical solution for a pressurized magma chamber under the effect of different far-field extensional stresses. Values between 0 and 60 MPa of the far-field stresses σ_{ff} were considered for a magma chamber with circular cross-section. Red line represents the lithostatic pressure. For σ_{ff} of 40 MPa or larger the dyke remains open throughout its length and magma flow dynamics is mainly controlled by the 3D geometry and extension of the system. (B) Maximum Eruption Rate (MER) as a function of extensional stress σ_{ff} for a dyke thickness of 5 m for magma chambers at 4, 6, and 8 km depths and overpressures above lithostatic of 20 MPa, respectively (modified after Costa et al., 2011).

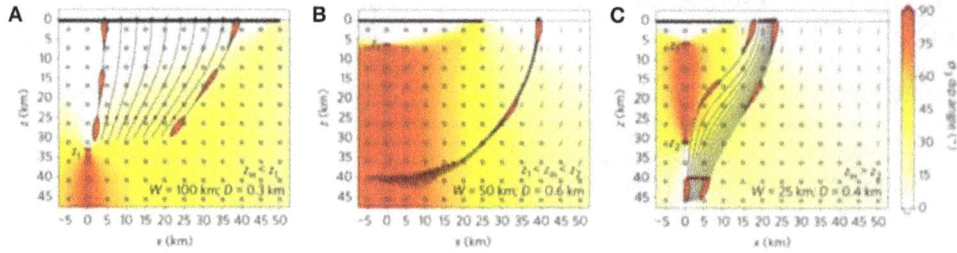

FIGURE 8 | Dyke trajectories for z_{in} above (A), within (B), and below (C) the stress barrier zone. Red squares indicate the upper tip of the dyke at injection; red circles indicate that a dyke arrested and formed a sill; red triangles indicate the arrival location at the surface. Black and gray segments show the directions of σ_1 and σ_3, respectively (a circle indicates direction perpendicular to the page), for three nominal sets of graben width (W) and depth (D). The dip angle of σ_3 is color-shaded. Where the dip angle is subvertical (reddish color), a stress barrier to vertical ascent of dykes is acting (modified after Maccaferri et al., 2014).

extension of the system. It is worth noting that the requested value of σ_{ff} is as high as 40–60 MPa, which is not easily matched during normal geodynamic processes. As an example, a homogeneous horizontal tensile stress $\sigma_{tec} = 5$ MPa was used by Maccaferri et al. (2014) for modeling the dyke trajectories in rifting areas. In this model the dyke opens under assigned normal and shear stress given by the internal overpressure and by the shear component of the tectonic plus unloading stresses, respectively. The overpressure within the dyke is set as the difference between the magma pressure and the confining stress, which is the superposition of the lithostatic pressure, the normal component of the topographic unloading and the tectonic stress. When the unloading pressure $P_u = \rho\, gD$ (ρ is crustal density, g acceleration due to gravity, and D is the effective depth accounting for the total deficit of mass from the topographic depression and low-density sediments with respect to before rifting) dominates over the tectonic tensile stress ($D > \pi\sigma_{tec}/(2\rho g)$, 250 m for $\rho = 3{,}000$ kgm^{-3}), σ_3 becomes vertical beneath the rift in a volume centered at a depth $z_c = \rho g DW/(\pi\sigma_{tec})$, where W is the graben width (**Table 1**; **Figure 8**). The upper limit of the volume is given by $z_1 = (W/2K)(1-(1-K^2)^{1/2})$ and the lower one by $z_2 = (W/2K)(1 + (1-K^2)^{1/2})$, where $K = \pi\sigma_{tec}/(2P_0)$ (**Table 1**; Maccaferri et al., 2014).

This volume forms a stress barrier zone, which deflects the ascending dykes to the rift sides.

Sideways from the rift center, σ_3 becomes first inward dipping and then horizontal (**Figure 8**). Three scenarios for dyke propagation and for the final surface distribution of magmatism can occur, depending on where the dykes nucleate relatively to the stress barrier zone. When $z_{in} < z_1$ in-rift volcanism occurs, while off-rift volcanism occurs for $z_1 < z_{in} < z_2$ and $z_{in} > z_2$ (**Table 1**; **Figure 8**).

Incorporating the far-field tectonic stress in Equation (11) we have:

$$P_0 - \Delta P_{0m} = P_e + (\rho_r - \rho_m)\, gh$$
$$+ \left(\rho_r h_r - \rho_i h_i\right) g \left[1 - \frac{1}{\nu - 1}\right] + \sigma_e + \sigma_{tec} + R_f$$

$$\tag{12}$$

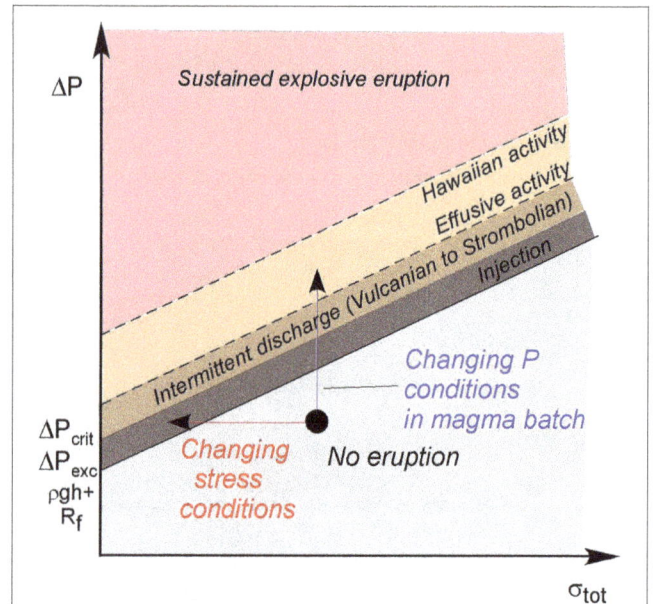

FIGURE 9 | Qualitative representation of the relationships between magmatic overpressure (ΔP) and lithostatic stress (σ_3). Eruption initiation may be triggered by an increase of ΔP at above the critical value ΔP_{crit} (the magma overpressure required to maintain a conduit open from the magma chamber to the surface) or reduction of σ_3 (after Sulpizio et al., 2016).

which is the final formulation of driving vs. resisting forces that drives the transition from no eruption to full on eruption.

INFLUENCE ON ERUPTION DYNAMICS

With the only exception of unloading due to ice cap retreat, the above discussed changes in stress field may play a role also in short term variation of eruption style. Complex transitions between effusive and explosive eruptive styles are frequently described in volcanic activity (e.g., Jaupart and Allegre, 1991; Villemant and Boudon, 1998; Adams et al., 2006; Platz et al., 2007) and the alternation of pyroclastic deposits and lavas is common in almost all stratovolcanoes. Shifts in eruptive style have been related to many complex sub-surface processes such as decompression-induced crystallization (Hammer et al., 2000; Blundy and Cashman, 2005), increase in magma viscosity due to groundmass crystallization caused by volatile loss and temperature gradients (Stevenson et al., 1996; Manga, 1998; Melnik and Sparks, 2002; Cashman and Sparks, 2013), and time-dependent release of overpressure due to the contrasting effects of magma viscosity and elastic energy released from country rocks deformation (Wylie et al., 1999).

All these processes can for sure participate to changes in eruptive style, but sharp changes in local or far-field stress may sometimes play a similar role in driving eruptive activity. This is especially true when dealing with changing eruptive style in eruptions or eruptive cycles with similar magmatic composition, which do not account for any petrologic or textural trigger of the changing eruptive behavior. For instance, the interplay between magma overpressure and stress acting on the volcanic system was claimed for explaining the eruptive style transitions of Monte dei Porri (Salina Island, Italy; Sulpizio et al., 2016), and effusive eruptions following local stress decrease due to spreading of the volcanic edifice were repeatedly observed at Mount Etna volcano (Borgia et al., 1992; Froger et al., 2001; Lundgren et al., 2004; Neri et al., 2004).

The contribution of stress lowering to the change of eruptive style can be easily explained using the Equation (12) in the ΔP vs. σ_{tot} space (**Figure 9**; Sulpizio et al., 2016). It shows how a transition from no-eruption to eruption or from a given eruptive style to another is allowed through the superimposition of internal magmatic pressure (increase of ΔP) and changing in the total stress field (σ_t), defined by the sum of all the defined partial stresses defined early.

SUMMARY AND CONCLUSIONS

Understanding the interplay between crustal stress and volcanic activity and its dynamics is essential for comprehension of a number of natural phenomena and for mitigating the related hazards and risk. Significant evidence of coupling between

stress change and volcanic events emerges from investigation of tectonic earthquakes, flank collapses, and also long-term processes such as erosion and landslides. The effect of these processes superimposes on changes in magma overpressure, including the growth of gas bubbles and input of new magma in the chamber. This is because, although dyke initiation and propagation to the surface is usually governed by the depth-dependent magma parameters, the source location is also subject to the stress field conditions that vary from one point to another in the crust and that can promote or prevent brittle failures.

During last decades many authors provided precious contributions to this topic, and this review presented the state of the art of the knowledge about some of the main mechanisms inducing stress change and able to influence eruption initiation and dynamics. In particular, we reviewed three main pivotal issues correlated to stress: the unloading and its long- and short-term effects, the seismic energy, and the regional (or far-field) stress changes. Their occurrence alone was used as a preliminary guide in this study.

The contribution of each mechanism has been analyzed, and an equilibrium equation for magma chamber rupture and dyke opening to the surface has been presented. The equation was also used for interpreting the possible changes in eruptive style of single eruptions or eruptive cycles. The three mechanisms can have different impact on magmatic systems, and can influence or not the triggering of volcanic eruptions. However, it emerges clearly from this review how a single mechanism is hardly responsible for eruption initiation, but the concur of internal processes is usually necessary. It emerges how internal (magmatic processes) end external (stress field variations) processes concur in modulating eruptive activity.

AUTHOR CONTRIBUTIONS

All authors listed, have made substantial, direct and intellectual contribution to the work, and approved it for publication.

REFERENCES

Adams, N., Houghton, B. F., Fagents, S. A., and Hildreth, W. (2006). The transition from explosive to effusive eruptive regime: the example of the 1912 Novarupta eruption, Alaska. *Geol. Soc. Am. Bull.* 118, 620–634. doi: 10.1130/B25768.1

Aguirre-Díaz, G., and Labarthe-Hernández, G. (2003). Fissure ignimbrites: fissure-source origin for voluminous ignimbrites of the Sierra Madre Occidental and its relationship with Basin and Range faulting. *Geology* 31, 773–776. doi: 10.1130/G19665.1

Albino, F., Pinel, V., and Sigmundsson, F. (2010). Influence of surface load variations on eruption likelihood: application to two Icelandic subglacial volcanoes, Grimsvötn and Katla. *Geophys. J. Int.* 181, 1510–1524. doi: 10.1111/j.1365-246x.2010.04603.x

Anderson, E. M. (1936). The dynamics of formation of cone sheets, ring dykes and cauldron subsidences. *Proc. R. Soc. Edinb.* 56, 128–163. doi: 10.1017/S0370164600014954

Andrew, R. E. B., and Gudmundsson, A. (2007). Distribution, structure, and formation of Holocene lava shields in Iceland. *J. Volcan. Geotherm. Res.* 168, 137–154. doi: 10.1016/j.jvolgeores.2007.08.011

Blundy, J., and Cashman, K. (2005). Rapid decompression-driven crystallization recorded by melt inclusions from Mount St. Helens volcano. *Geology* 33, 793–796. doi: 10.1130/G21668.1

Borgia, A., Ferrari, L., and Pasquarè, G. (1992). Importance of Gravitational Spreading in the Tectonic and Volcanic Evolution of Mt. Etna. *Nature* 357:231. doi: 10.1038/357231a0

Boulesteix, T., Hildenbrand, A., Gillot, P. Y., and Soler, V. (2012). Eruptive response of oceanic islands to giant landslides: new insigths from the geomorphologic evolution of the Teide-Pico Viejo volcanic complex (Tenerife, Canary). *Geomorphology* 138, 61–73. doi: 10.1016/j.geomorph.2011.08.025

Brodsky, E. E., Sturtevant, B., and Kanamori, H. (1998). Earthquakes, volcanoes, and rectified diffusion. *J. Geophy. Res.* 103, 827–838. doi: 10.1029/98jb02130

Cañon-Tapia, E. (2014). Volcanic eruption triggers: a hierarchical classification. *Earth Sc. Rev.* 129, 100–119. doi: 10.1016/j.earscirev.2013.11.011

Carbone, D., Aloisi, M., Vinciguerra, S., and Puglisi, G. (2014). Stress, strain and mass changes at Mt. Etna during the period between the 1991–93 and 2001 flank eruptions. *Earth Sci. Rev.* 138, 454–468. doi: 10.1016/j.earscirev.2014.07.004

Cashman, K. V., and Sparks, R. S. J. (2013). How volcanoes work: a 25 year perspective. *Geol. Soc. Am. Bull.* 125, 664–690. doi: 10.1130/B30720.1

Cazaneve, A., and Chen, J. (2010). Time-variable gravity from space and present-day mass redistribution in the Earth system. *Earth Planet. Sci. Lett.* 298, 263–274. doi: 10.1016/j.epsl.2010.07.035

Costa, A., Gottsmann, J., Melnik, O., and Sparks, R. S. J. (2011). A stress-controlled mechanism for the intensity of very large magnitude explosive eruptions. *Earth Planet. Sci. Lett.* 310, 161–166. doi: 10.1016/j.epsl.2011.07.024

Costa, A., Melnik, O., and Sparks, R. S. J. (2007). Controls of conduit geometry and wallrock elasticity on lava dome eruptions. *Earth Planet. Sci. Lett.* 260, 137–151. doi: 10.1016/j.epsl.2007.05.024

Davì, M., De Rosa, R., Donato, P., and Sulpizio, R. (2011). The Lami pyroclastic succession (Lipari, Aeolian Islands): a clue for unravelling the eruptive dynamics of the Monte Pilato rhyolitic pumice cone. *J. Volcan. Geotherm. Res.* 201, 285–300. doi: 10.1016/j.jvolgeores.2010.09.010

Davis, R. O., and Selvadurai, A. P. S. (2001). *Elasticity and Geomechanics.* Cambridge, UK: Cambridge University Press 1996.

Davydov, M. N., Kedrinskii, V. K., Chernov, A. A., and Takayama, K. (2005). Generation and evolution of cavitation in magma under dynamic unloading. *J. Appl. Mechan. Technic. Physics* 46, 208–215. doi: 10.1007/s10808-005-0036-2

De la Cruz-Reyna, S., Tárraga, M., Ortiz, R., and Martínez-Bringas, A. (2010). Tectonic earthquakes triggering volcanic seismicity and eruptions. Case studies at Tungurahua and Popocatépetl volcanoes. *J. Volcan. Geotherm. Res.* 193, 37–48. doi: 10.1016/j.jvolgeores.2010.03.005

Diez, M., La Femina, P. C., Connor, C. B., Strauch, W., and Tenorio, T. V. (2005). Evidence for static stress changes triggering the 1999 eruption of Cerro Negro Volcano, Nicaragua and regional aftershock sequences. *Geoph. Res. Lett.* 32, L04309. doi: 10.1029/2004gl021788

Di Traglia, F., Cimarelli, C., de Rita, D., and Gimeno Torrente, D. (2009). Changing eruptive styles in basaltic explosive volcanism: examples from Croscat complet cone, Garrotxa Volcanic Field (NE Iberian Peninsula). *J. Volcan. Geotherm. Res.* 180, 89–109. doi: 10.1016/j.jvolgeores.2008.10.020

Fink, J., Anderson, S. W., and Manley, C. R. (1992). Textural constraints on effusive silicic volcanism: beyond the permeable foam model. *J. Geophys. Res.* 97, 9073–9083. doi: 10.1029/92JB00416

Froger, J. L., Merle, O., and Briole, P. (2001). Active spreading and regional extension at Mount Etna imaged by SAR interferometry. *Earth Planet. Sci. Lett.* 187, 245–258. doi: 10.1016/S0012-821X(01)00290-4

Geyer, A., and Bindeman, I. (2011). Glacial influence on caldera-forming eruptions. *J. Vol. Geotherm. Res.* 202, 127–142. doi: 10.1016/j.jvolgeores.2011.02.001

Gudmundsson, A. (1986). Mechanical aspects of postglacial volcanism and tectonics of the Reykjanes Peninsula, southwest Iceland. *J. Geophys. Res.* 91, 711–721. doi: 10.1029/jb091ib12p12711

Gudmundsson, A. (1990). Emplacement of dikes, sills and crustal magma chambers at divergent plate boundaries. *Tectonophysics* 176, 257–275. doi: 10.1016/0040-1951(90)90073-H

Gudmundsson, A. (2006). How local stresses control magma-chamber ruptures, dyke injections, and eruptions in composite volcanoes. *Earth Sci. Rev.* 79, 1–31. doi: 10.1016/j.earscirev.2006.06.006

Gudmundsson, A. (2011). Deflection of dykes into sills at discontinuities and magma chamber formation. *Tectonophysics* 500, 50–64. doi: 10.1016/j.tecto.2009.10.015

Gudmundsson, A. (2012a). Strengths and strain energies of volcanic edifices: implications for eruptions, collapse calderas, and landslides. *Nat. Hazards Earth Syst. Sci.* 12, 2241–2258. doi: 10.5194/nhess-12-2241-2012

Gudmundsson, A. (2012b). Magma chambers: formation, local stresses, excess pressures, and compartments. *J. Volcanol. Geotherm. Res.* 237–238, 19–41. doi: 10.1016/j.jvolgeores.2012.05.015

Gudmundsson, A. (2016). The mechanics of large volcanic eruptions. *Earth Sci. Rev.* 163, 72–93. doi: 10.1016/j.earscirev.2016.10.003

Gudmundsson, A., and Philipp, S. J. (2006). How local stress fields prevent volcanic eruptions. *J. Vol. Geotherm. Res.* 158, 257–268. doi: 10.1016/j.jvolgeores.2006.06.005

Hammer, J. E., Cashman, K. V., and Voight, B. (2000). Magmatic processes revealed by textural and compositional trends in Merapi dome lavas. *J. Volcanol. Geotherm. Res.* 100, 165–192. doi: 10.1016/S0377-0273(00)00136-0

Harris, A. J. L., and Ripepe, M. (2007). Regional earthquakes as a trigger for enhanced volcanic activity: evidence from MODIS thermal data. *Geoph. Res. Lett.* 34, L02304. doi: 10.1029/2006GL028251

Hasabe, N., Fukutani, A., Sudo, M., and Tagami, T. (2001). Transition of eruptive style in an arc-arc collision zone: K-Ar dating of Quaternary monogenetic and polygenetic volcanoes in the Higashi-Izu region, Izu peninsula, Japan. *Bull. Volcanol.* 63, 377–386. doi: 10.1007/s004450100158

Hildenbrand, A., Gillot, P., and Le Roy, I. (2004). Volcano-tectonic and geochemical evolution of an oceanic intra-plate volcano: Tahiti-Nui (french Polynesia). *Earth Planet. Sci. Lett.* 217, 349–365. doi: 10.1016/S0012-821X(03)00599-5

Hill, D. P., Pollitz, F., and Newhall, C. (2002). Earthquake-volcano interactions. *Phys. Today* 55, 41–47. doi: 10.1063/1.1535006

Hooper, A., Ofeigsson, B., Sigmundsson, F., Lund, B., Einarsson, P., Geirsson, H., et al. (2011). Increased crustal capture of magma at volcanoes with retreating ice cap. *Nat. Geosci.* 4, 783–786. doi: 10.1038/ngeo1269

Hora, J. M., Singer, B. S., and Worner, G. (2007). Volcano evolution and eruptive flux on the thick crust of the Andean Central Volcanic Zone: 40Ar/39Ar constraints from Volcan Parinacota, Chile. *GSA Bull.* 119, 343–362. doi: 10.1130/B25954.1

Ida, Y. (2007). Driving force of lateral permeable gas flow in magma and the criterion of explosive and effusive eruptions. *J. Volcanol. Geotherm. Res.* 162, 172–184. doi: 10.1016/j.jvolgeores.2007.03.005

Jaupart, C., and Allegre, C. (1991). Gas content, eruption rate and instabilities of eruption regime in silicic volcanoes. *Earth Planet. Sci. Lett.* 102, 413–429. doi: 10.1016/0012-821X(91)90032-D

Jellinek, A. M., Manga, M., and Saar, M. O. (2004). Did melting glaciers cause volcanic eruptions in eastern California? Probing the mechanics of dike formation. *J. Geophys. Res.* 109:B09206. doi: 10.1029/2004JB002978

Keressturi, G., Nemeth, K., Cronin, S. J., Procter, J., and Augustìn-Flores, J. (2014). Influences on the variability of eruption sequences and style transitions in the Aukland Volcanic Field, New Zeland. *J. Volcanol. Geotherm. Res.* 286, 101–115. doi: 10.1016/j.jvolgeores.2014.09.002

Korringa, M. K. (1973). Linear vent area of the Soldier Meadow Tuff, an ash-flow sheet in northwestern Nevada. *Geol. Soc. Am. Bull.* 84, 3849–3866. doi: 10.1130/0016-7606(1973)84<3849:LVAOTS>2.0.CO;2

Lay, T., and Wallace, T. (1995). Modern *Global Seismology,* Vol. 58. London: Academic Press.

Le Friant, A., Boudon, G., Deplus, C., and Villemant, B. (2003). Large-scale flank collapse events during the activity of Montagne Pelè, Martinique, Lesser Antilles, *J. Geophys. Res.* 108, 2055. doi: 10.1029/2001JB001624

Lehto, H. L., Roman, D. C., and Moran, S. C. (2010). Temporal changes in stress preceding the 2004–2008 eruption of Mount St. Helens, Washington. *J. Volcanol. Geotherm. Res.* 198, 129–142. doi: 10.1016/j.jvolgeores.2010.08.015

Linde, A. T., and Sacks, I. S. (1998). Triggering of volcanic eruptions. *Nature* 395, 888–890. doi: 10.1038/27650

Longprè, M.-A., Troll, V. R., Walter, T. R., and Hansteen, T. H. (2009). Volcanic and geochemical evolution of the Teno massif, Tenerife, Canary Islands: some repercussions of giant landslides on ocean island magmatism. *Geochem. Geophys. Geosyst.* 10, Q12017. doi: 10.1029/2009GC002892

Lundgren, P., Casu, F., Manzo, M., Pepe, A., Berardino, P., Sansosti, E., et al. (2004). Gravity and magma induced spreading of Mount Etna volcano revealed by satellite radar interferometry. *J. Geophys. Res.* 31, L04602. doi: 10.1029/2003gl018736

Maccaferri, F., Rivalta, E., Keir, D., and Acocella, V. (2014). Off-rift volcanism in rift zones determined by crustal unloading. *Nat. Geosci. Lett.* 7, 297–300. doi: 10.1038/ngeo2110

Manconi, A., Longprè, M. A., Walter, T. R., Troll, V. R., and Hansteen, T. H. (2009). The effects of flank collapses on volcano plumbing systems. *Geology* 460, 1.099–1.102. doi: 10.1130/g30104a.1

Manga, M. (1998). Rheology of bubble-bearing magmas. *J. Volcanol. Geotherm. Res.* 87, 15–28. doi: 10.1016/S0377-0273(98)00091-2

Manga, M., and Brodsky, E. E. (2006). Seismic triggering of eruptions in the far field: volcanoes and geysers. *Annu. Rev. Earth Planet. Sci.* 34, 263–291. doi: 10.1146/annurev.earth.34.031405.125125

Marzocchi, W. (2002). Remote seismic influence on large explosive eruptions. *J. Geophys. Res.* 107, 2018. doi: 10.1029/2001jb000307

Marzocchi, W., Casarotti, E., and Piersanti, A. (2002). Modeling the stress variations induced by great earthquakes on the largest volcanic eruptions of the 20th Century. *J. Geophy. Res.* 107, 2320. doi: 10.1029/2001jb001391

Massol, H., and Jaupart, C. (1999). The generation of gas overpressure in volcanic eruptions. *Earth Planet. Sc. Lett.* 166, 57–70. doi: 10.1016/S0012-821X(98)00277-5

Melnik, O., and Sparks, R. S. J. (2002). "Dynamics of magma ascent and lava extrusion at Soufrière Hills Volcano, Montserrat," in *The Eruption of Soufrière Hills Volcano, Montserrat, From 1995 to 1999*, Vol. 21, eds T. H. Druitt and B. P. Kokelaar (London: Geological Society), 153–171.

Miller, C., and Wark, D. A. (2008). Supervolcanoes and their explosive supereruptions. *Elements* 4, 11–16. doi: 10.2113/GSELEMENTS.4.1.11

Miura, D., and Wada, Y. (2007). Effects of stress in the evolution of large silicic magmatic systems: an example from the Miocene felsic volcanic field at Kii Peninsula, SW Honshu, Japan. *J. Volcanol. Geotherrm Res.* 167, 300–319. doi: 10.1016/j.jvolgeores.2007.05.017

Neri, M., Acocella, V., and Behncke, B. (2004). The role of the Pernicana Fault System in the spreading of Mt. Etna (Italy) during the 2002–2003 eruption. *Bull. Volcanol.* 66, 417–430. doi: 10.1007/s00445-003-0322-x

Pinel, V., and Albino, F. (2013). Consequences of volcano sector collapse on magmatic storage zones: insights from numerical modeling. *J. Volcanol. Geotherm. Res.* 252, 29–37. doi: 10.1016/j.jvolgeores.2012.11.009

Pinel, V., and Jaupart, C. (2005). Some consequences of volcanic edifice destruction for eruption conditions. *J. Volcanol. Geotherm. Res.* 145, 68–80. doi: 10.1016/j.jvolgeores.2005.01.012

Pinel, V., and Jupart, C. (2004). Magma storage and horizontal dyke injection beneath a volcanic edifice. *Earth Planet. Sci. Lett.* 221, 245–262. doi: 10.1016/S0012-821X(04)00076-7

Plateaux, R., Béthoux, N., Bergerat, F., and Lépinay, B. M. (2014). Volcano-tectonic interactions revealed by inversion of focal mechanisms: stress field insight around and beneath the Vatnajokull ice cap in Iceland. *Front. Earth Sci.* 2:9. doi: 10.3389/feart.2014.00009

Platz, T., Cronin, S. J., Cashman, K. V., Stewart, R. B., and Smith, I. E. M. (2007). Transition from effusive to explosive phases in andesite eruptions: a case-study from the AD1655 eruption of Mt.Taranaki, New Zealand. *J. Volcanol. Geotherm. Res.* 161, 1534. doi: 10.1016/j.jvolgeores.2006.11.005

Presley, T. K., Sinton, J. M., and Pringle, M. (1997). Postshield volcanism and catastrophic mass wasting of the Waianae Volcano, Oahu, Hawaii. *Bull. Volcanol.* 58, 597–616. doi: 10.1007/s004450050165

Ripepe, M., Marchetti, E., Ulivieri, G., Harris, A., Dehn, J., Burton, M., et al. (2013). Effusive to explosive transition during the 2003 eruption of Stromboli volcano. *Geology* 33, 341–344. doi: 10.1130/G21173.1

Roverato, M., Capra, L., Sulpizio, R., and Norini, G. (2011). Stratigraphic reconstruction of two debris avalanche deposits at Colima Volcano (Mexico): insights into pre-failure conditions and climate influence. *J. Volcanol. Geotherm. Res.* 207, 33–46. doi: 10.1016/j.jvolgeores.2011.07.003

Schneider, A., Rempel, A. W., and Cashman, K. V. (2012). Conduit degassing and thermal controls on eruptive styles at Mt. St. Helens. *Earth Planet. Sci. Lett.* 357, 347–354. doi: 10.1016/j.epsl.2012.09.045

Selva, J., Marzocchi, W., Zencher, F., Casarotti, E., Piersanti, A., and Boschi, E. (2004). A forward test for interaction between remote earthquakes and volcanic eruptions: the case of Sumatra (June 2000) and Denali (November 2002) earthquakes. *Earth Planet. Sci. Lett.* 226, 383–395. doi: 10.1016/j.epsl.2004.08.006

Sepulveda, F., Lahsen, A., Bonvalot, S., Cembrano, J., Alvarado, A., and Letelier, P. (2005). Morpho-structural evolution of the Cordón Caulle geothermal region, Southern Volcanic Zone, Chile: insights from gravity and 40Ar/39Ar dating. *J. Volcanol. Geotherm. Res.* 148, 207–233. doi: 10.1016/j.jvolgeores.2005.03.020

Sigmundsson, F., Pinel, V., Lund, B., Albino, F., Pagli, C., Geirsson, H., et al. (2010). Climate effects on volcanism: Influence on magmatic systems of loading and unloading from ice mass variations with examples from Iceland. *Phil. Trans. R. Soc. Lond.* 368, 2.519–2.534. doi: 10.1098/rsta.2010.0042

Sinton, J., Gronvold, K., and Saemundsson, K. (2005). Postglacial eruptive history of the western volcanic zone, Iceland. *Geochem. Geophys. Geosyst.* 6:Q1200. doi: 10.1029/2005GC001021

Sohn, R. A. (2004). Microearthquake patterns following the 1998 eruption of Axial Volcano, Juan de Fuca Ridge: mechanical relaxation and thermal strain. *J. Geophys. Res.* 109, B01101. doi: 10.1029/2003jb002499

Stein, R. S. (1999). The role of stress transfer in earthquake occurrence. *Nature* 402, 605–609. doi: 10.1038/45144

Stevenson, R. J., Dingwell, D. B.,Webb, S. L., and Sharp, T. G. (1996). Viscosity of microlite-bearing rhyolitic obsidians: an experimental study. *Bull. Volcanol.* 58, 298–309. doi: 10.1007/s004450050141

Sulpizio, R., Lucchi, F., Forni, F., Massaro, S., and Tranne, C. (2016). Unravelling the effusive-explosive transitions and the construction of a volcanic cone from geological data: the example of Monte dei Porri eruptive cycle, Salina, Aeolian Islands (Italy). *J. Volcanol. Geotherm. Res.* 327, 1–22. doi: 10.1016/j.jvolgeores.2016.06.024

Ventura, G., and Vilardo, G. (1999). Slip tendency analysis of the Vesuvius faults: implications for the seismotectonic and volcanic hazard assessment. *Geophy. Res. Lett.* 26, 3229–3232. doi: 10.1029/1999GL005393

Villemant, B., and Boudon, G. (1998). Transition from dome-forming to plinian eruptive styles controlled by H_2O and Cl degassing. *Nature* 392, 65–69. doi: 10.1038/32144

Voight, B., and Elsworth, D. (1997). Failure of volcano slopes. *Geotechnique* 47, 535, 1–31. doi: 10.1680/geot.1997.47.1.1

Waite, G. P., and Smith, R. B. (2004). Seismotectonics and stress field of the Yellowstone volcanic plateau from earthquake first-motions and other indicators. *J. Geophys. Res.* 109, B02301. doi: 10.1029/2003JB002675

Walter, T. R. (2007). How a tectonic earthquake may wake up volcanoes: stress transfer during the 1996 earthquake–eruption sequence at the Karymsky Volcanic Group, Kamchatka. *Earth Planet. Sci. Lett.* 264, 347–359. doi: 10.1016/j.epsl.2007.09.006

Walter, T. R., and Amelung, F. (2007). Volcanic eruptions following M≥9 megathrust earthquakes: implications for the Sumatra-Andaman volcanoes. *Geology* 35, 539–542. doi: 10.1130/G23429A.1

Walter, T. R., Wang, R., Zimmer, M., Grosser, H., Luhr, B., and Ratdomopurbo, A. (2007). Volcanic activity influenced by tectonic earthquakes: static and dynamic stress triggering at Mt. Merapi. *Geophys. Res. Lett.* 34, L05304. doi: 10.1029/2006gl028710

Watt, S. F. L., Pyle, D. M., and Mather, T. A. (2008). The influence of great earthquakes on volcanic eruption rate along the Chilean subduction zone. *Earth Planet. Sci. Lett.* 277, 399–407. doi: 10.1016/j.epsl.2008.11.005

Wylie, J. J., Voight, B., and Whiteheed, J. A. (1999). Instability of magma flow from volatile dependent viscosity. *Science* 285, 1883–1885. doi: 10.1126/science.285.5435.1883

Conflict of Interest Statement: The authors declare that the research was conducted in the absence of any commercial or financial relationships that could be construed as a potential conflict of interest.

16

Stress Field Control during Large Caldera-Forming Eruptions

Antonio Costa[1]* and Joan Martí[2]

[1] Istituto Nazionale di Geofisica e Vulcanologia, Bologna, Italy, [2] CSIC, Institute of Earth Sciences "Jaume Almera," Barcelona, Spain

Crustal stress field can have a significant influence on the way magma is channeled through the crust and erupted explosively at the surface. Large Caldera Forming Eruptions (LCFEs) can erupt hundreds to thousands of cubic kilometers of magma in a relatively short time along fissures under the control of a far-field extensional stress. The associated eruption intensities are estimated in the range 10^9–10^{11} kg/s. We analyse syn-eruptive dynamics of LCFEs, by simulating numerically explosive flow of magma through a shallow dyke conduit connected to a shallow magma (3–5 km deep) chamber that in turn is fed by a deeper magma reservoir (>~10 km deep), both under the action of an extensional far-field stress. Results indicate that huge amounts of high viscosity silicic magma (>10^7 Pa s) can be erupted over timescales of a few to several hours. Our study provides answers to outstanding questions relating to the intensity and duration of catastrophic volcanic eruptions in the past. In addition, it presents far-reaching implications for the understanding of dynamics and intensity of large-magnitude volcanic eruptions on Earth and to highlight the necessity of a·future research to advance our knowledge of these rare catastrophic events.

Keywords: super-eruptions, magma ascent dynamics, extensional stress, volcanic conduit model, fissure eruptions

Edited by:
Agust Gudmundsson,
Royal Holloway, University of London,
UK

Reviewed by:
Hiroaki Komuro,
Shimane University, Japan
Antonio M. Álvarez-Valero,
University of Salamanca, Spain

*Correspondence:
Antonio Costa
antonio.costa@ingv.it

Specialty section:
This article was submitted to
Volcanology,
a section of the journal
Frontiers in Earth Science

INTRODUCTION

There is compelling evidence that Large Caldera-Forming Eruptions (LCFEs) are characterized by extremely large intensities. Estimations of Mass Eruption Rates (MERs) obtained with different independent methods (Wilson and Walker, 1981; Hildreth and Mahood, 1986; Wilson and Hildreth, 1997; Baines and Sparks, 2005; Costa et al., 2014; Martí et al., 2016; Roche et al., 2016) indicate MERs of the orders 10^9–10^{11} kg/s (e.g., Bishop Tuff, Campanian Ignimbrite, Oruanui eruption, Taupo eruption, Peach Spring Tuff, Young Toba Tuff), implying durations of few to several hours only to evacuate even thousands of km^3 of magma.

Most LCFEs occur in both subduction zone and extensional environments characterized by relatively low rates of magma production (see Jellinek and De Paolo, 2003 and references therein) implying that the thousand km^3 volume magma chambers feeding those events have to accumulate over long periods (>10^5 years; Jellinek and De Paolo, 2003).

In order to erupt, magmas stored in relatively shallow chambers (3–8 km; e.g., Smith et al., 2005, 2006; Matthews et al., 2011; Chesner, 2012) normally have to overcome critical overpressures up to ~50 MPa for nucleating new fractures and up to ~10 MPa for propagating magma up to the surface (Rubin, 1995; Jellinek and De Paolo, 2003). Whereas for small magma chambers such overpressures can be easily achieved, for very large chamber volumes it is more problematic to reach such overpressures, and dyke formation and propagation are, as a consequence, inhibited (Jellinek and De Paolo, 2003). In some cases there is clear evidence of new injection of magma (and associated oversaturation of volatiles) as main cause to achieve the required overpressure to open the magma chamber (e.g., Sparks et al., 1977; Pallister et al., 1992; Self, 1992; Folch and Martí, 1998). However, in most large calderas this is not so clear. In contrast, tectonic triggers (i.e., decrease of ambient pressure due to tectonic—earthquake—activity) would be a plausible mechanism (see Aguirre-Díaz and Labarthe-Hernañdez, 2003; Martí et al., 2009), despite they have not been sufficiently explored yet. In the case of tectonic triggers, the magma chamber would evacuate the magma through the pre-existing faults or newly formed fractures without needing any over-pressurization of the magma chamber (Martí et al., 2009).

Irrespectively of the mechanism that leads to the rupture of the magma chamber during caldera eruptions, syn-eruptive dynamics of magma ascent in high intensity eruptions are not clear and there have not been many attempts to quantitatively describe them (Costa et al., 2011). Dykes feeding these eruptions have to be long enough and remain open over much of their length throughout the entire explosive activity. The mechanics of feeding explosive silicic ignimbrite eruptions through a linear fissure (Korringa, 1973; Aguirre-Díaz and Labarthe-Hernañdez, 2003) or from multiple vents along a fissure (Suzuki-Kamata et al., 1993; Wilson, 2001; Smith et al., 2005, 2006; Folch and Martí, 2009) are largely unexplored.

Magma emplacement through dykes and the capability of magma to reach the surface strongly depend on the local stresses across the different layers that constitute the volcano (Gudmundsson, 2006). As explained for instance by Gudmundsson (2006), a dyke propagated upward from a magma chamber can reach the surface only if the stress field along all its path is favorable to magma-fracture propagation. This implies that the stress field has to promote extension-fracture formation as well as keep the stress field homogenized along the entire path of the dyke to the surface.

Moreover, once that a critical magma chamber pressure is reached (e.g., by intrusion of new magma or by evolution of volatiles or both or by external triggering because a fracture can reach the chamber roof) and a dyke can propagate in the surrounding rocks, in order to produce an explosive eruption, magma has to fragment. Because of the typical silicic compositions (e.g., Chesner, 1998, 2012; Matthews et al., 2012) and high crystal contents (e.g., Gottsmann et al., 2009; Costa et al., 2011) effective viscosity of those magmas is very high ($>10^7$ Pa s; e.g., Costa et al., 2011). Since the fragmentation depth is controlled by effective magma viscosity, in order to be able to keep dykes open at deep fragmentation levels it is necessary that local magma overpressure counterbalances the lithostatic load at that depth (Costa et al., 2009, 2011). Hence extremely larger overpressures should be attained.

Concerning this point, Costa et al. (2011) showed that coupling of magma overpressure with the effects of a far-field extensional stress can play a pivotal role. As we mentioned above, most LCFEs have been recorded in extensional environments (e.g., Jellinek and De Paolo, 2003; Sobradelo et al., 2010), but even where LCFEs occur in convergent regions they appear to be associated with local extension (Miller et al., 2008; Acocella and Funiciello, 2010).

Besides tectonic stress, local extension can be produced by the growth of magma chambers and reservoirs exceeding several hundreds of cubic kilometers in volume due to a "magmatic" stress field on local and regional scales. Either counteracts or adds to dominant tectonic stresses depending on the sign and intensity of the far-field stress and on the magma chamber shape and orientation (Gudmundsson, 1988, 1998; Gudmundsson et al., 1997).

In this contribution, we start briefly reviewing the general tectonic settings of LCFEs and other evidence of stress field control during LCFE. Then we summarize the model of Costa et al. (2011) for LCFEs adapted in order to account for the effects of a pressurized magma reservoir (**Figure 1**). Finally, we apply the Costa et al. (2011) model to show how, for magma chamber and reservoir depths and magma properties typical of a LCFE similar to the Young Toba Tuff (YTT), the local stress field, due to the combined effects of relatively low pressurizations of magma chamber and reservoir, and far-field stress, can promote large MERs.

TECTONIC SETTINGS OF LCFEs AND ERUPTION CONDITIONS

Collapse calderas are volcanic subsidence structures that can be recognized in many volcanic systems and may form in any geodynamic environment (Gudmundsson, 1988). However, the largest caldera eruptions, those that erupt hundreds to thousands of cubic kilometers of magma, are invariably associated with silicic magmas and extensional structures occurring mostly in subduction zones, continental rifts or extensional environments of Basin and Range type, always related to a thick continental crust (Geyer and Martí, 2008; Cole et al., 2010; Reyners, 2010; Rowland et al., 2010). This fact suggests that this type of extensional stress is a requisite (or a favorable factor) for the formation of large magma chambers responsible for calderas (see also e.g., Gudmundsson, 1998, 2006; Hughes and Mahood, 2008).

It is generally accepted that there is a positive linear relationship between the area of the caldera and the volume of material extruded during the eruption (Smith, 1979; Spera and Crisp, 1981; Geyer and Martí, 2008). So, large calderas are related to the eruption of large volumes of magma. The tendency of large collapse calderas to form in areas with extensional tectonics and in relatively thick (≥ 30 km) continental crusts suggests that these conditions are the most favorable to accumulate large volumes of silicic magmas at shallow depths (Sobradelo et al., 2010). Another feature that characterizes large collapse calderas is their association with shallow (3–5 km deep) magma chambers, thus suggesting that mechanical conditions to form such calderas are only achieved under stress configurations related to small aspect ratios, i.e., depth vs. extent of the chamber (Martí et al., 2009; Geyer and Martí, 2014).

The mechanisms by which a magma chamber opens to the surface and then evolves into a caldera-forming event are still

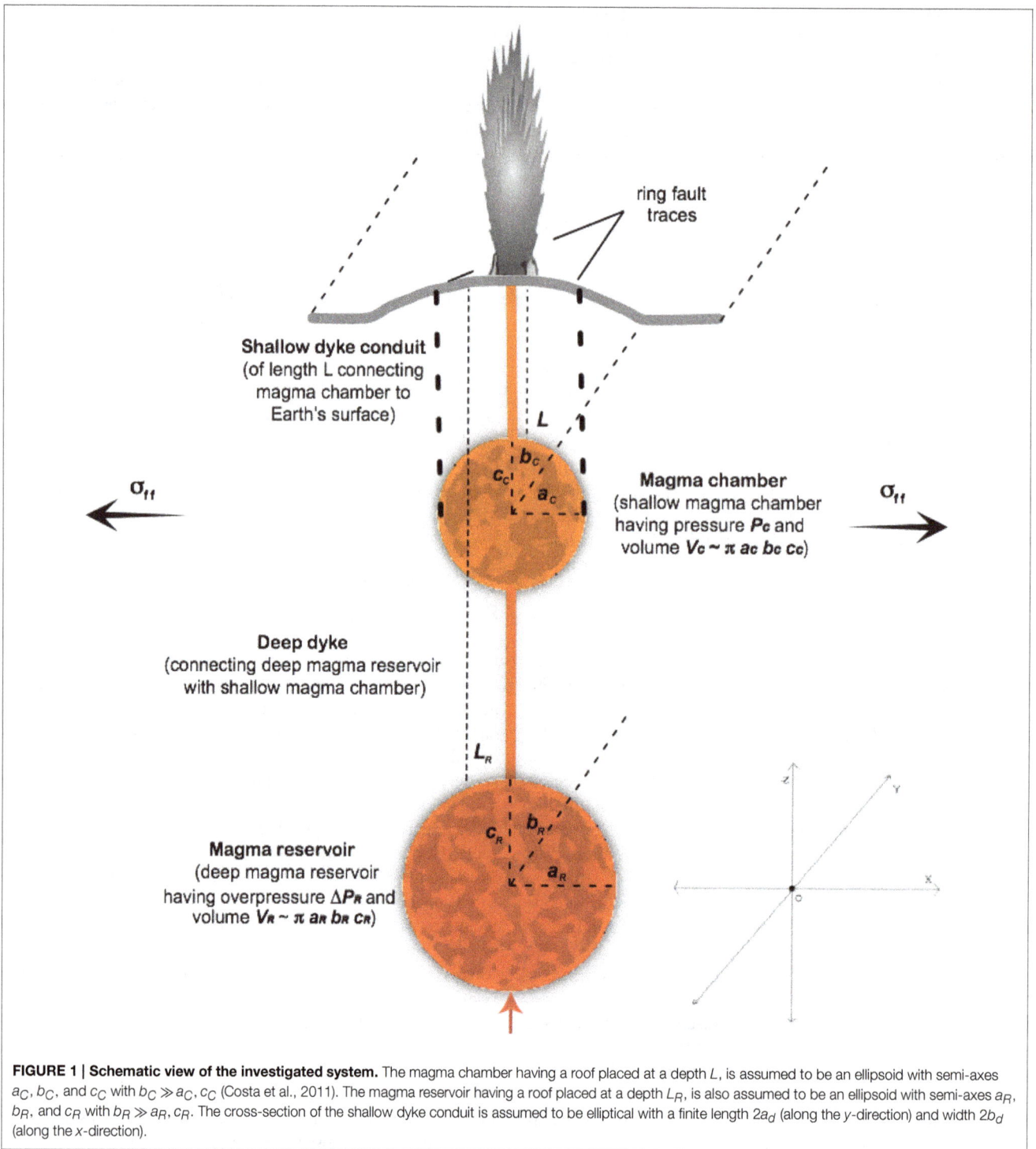

FIGURE 1 | Schematic view of the investigated system. The magma chamber having a roof placed at a depth L, is assumed to be an ellipsoid with semi-axes a_C, b_C, and c_C with $b_C \gg a_C$, c_C (Costa et al., 2011). The magma reservoir having a roof placed at a depth L_R, is also assumed to be an ellipsoid with semi-axes a_R, b_R, and c_R with $b_R \gg a_R$, c_R. The cross-section of the shallow dyke conduit is assumed to be elliptical with a finite length $2a_d$ (along the y-direction) and width $2b_d$ (along the x-direction).

not fully understood. In any case, there is general consensus that collapse calderas require very specific stress conditions to form, which will be defined by the stress field, size, shape, and depth of the magma chamber, magma rheology and gas content, and state of deformation (e.g., presence of local and regional faults) of the host rock (see Acocella, 2007; Martí et al., 2009; Acocella et al., 2015). Under these circumstances, the rupture of

a magma chamber may be an intrinsic cause by an increase of magma pressure due to injection of new magma into the chamber and/or oversaturation of volatiles due to crystallization, or may be favored by outside by reducing stresses through a tectonic event (e.g., earthquake). The mechanisms by which the eruption will progress and caldera will collapse may be different depending on each scenario (see Martí et al., 2009), but it is not the aim of this study to discuss in detail such differences. On the contrary, we will concentrate on common aspects related to magma flow through fractures and on the conditions needed for having large MERs characteristics of LCFEs, regardless on how these fractures have opened.

According to the stratigraphic features shown by caldera forming deposits we can differentiate two main caldera types or end members. One comprises the caldera forming episode preceded by a Plinian eruption that may erupt a considerable volume of magma. This type of calderas, named underpressure calderas by Martí et al. (2009), are characterized in the field by the presence of relatively thick Plinian deposits underlying the caldera-forming ignimbrites, and would correspond to those calderas in which the initiation of caldera collapse requires a substantial decompression of the magma chamber (Druitt and Sparks, 1984). Examples of such calderas are Crater Lake (e.g., Bacon, 1983), Katmai (e.g., Hildreth, 1991), Santorini (e.g., Druitt and Francaviglia, 1992). The other end-member corresponds to those calderas in which caldera collapse starts at the beginning of the eruption, without any Plinian phase preceding it. The succession of deposits that characterize these calderas, called overpressure calderas by Martí et al. (2009), is composed only of the caldera forming ignimbrites, with occasional some minor pyroclastic surge deposits at the base of the caldera-forming succession. In this case, it is interpreted that these calderas start forming since the beginning of the eruption without needing (significant) decompression of the magma chamber (Gudmundsson et al., 1997; Martí et al., 2009). Examples of this type of caldera are La Pacana (e.g., Gardeweg and Ramirez, 1987), Cerro Galán (Folkes et al., 2011), Aguas Calientes (Petrinovic et al., 2010); El Abrigo (Pittari et al., 2008), Bolaños graben caldera (Aguirre-Díaz et al., 2008).

BASIC MODEL OF MAGMA ASCENT DURING LCFEs

Previous numerical models of LCFE were used to study syn-eruptive dynamics of magma ascent of these eruptions. In particular, Folch and Martí (2009) presented simulations of LCFE based on Macedonio et al. (2005) conduit model, considering an initial eruption phase from a central-vent conduit, a transition to peripheral fissure-vent conduits, and a final phase controlled by piston-like subsidence. In line with their results, during the eruption phase from peripheral ring fissures, MERs would increase by almost an order of magnitude. During the piston-like subsidence, pressure increases back to lithostatic and the MER tends to stabilize near these larger values and does not change significantly after the subsidence starts. Because of the limitations of the model and the assumed rigid conduit geometry, the simulations by Folch and Martí (2009) were limited to relatively small MERs, $<10^9$ kg/s. Such low MERs would imply long eruption durations, in the range of few tens to hundreds hours, or longer, even for erupting relatively small volumes of magma.

Contrarily to the simplifying assumptions of most volcanic conduit models, rock mechanics imply the most efficient way of moving magma through cold lithosphere is via dykes (Rubin, 1995), and this is supported from field evidence (e.g., Gudmundsson, 2002) and geophysical analysis (e.g., Hautmann et al., 2009; Sigmundsson et al., 2010). Because the complexity in describing coupled magma-rock dynamics, explosive volcanic eruptions have been commonly modeled in terms of multiphase flows through rigid conduits of a fixed cross-section. Costa et al. (2009) and Costa et al. (2011) generalized Macedonio et al. (2005) model considering magma flow from an elastic dyke with elliptical cross-section emanating from magma chamber reaching the surface as a linear dyke or as dyke evolving to a cylinder at shallower depths. Costa et al. (2011) showed that MERs of the order of 10^9–10^{11} kg/s can be obtained for realistic input parameters if the effect of an extensional stress field on the dynamics of magma ascent is accounted for.

Here we summarize the model proposed by Costa et al. (2011) that is based on the following mass and momentum equations:

$$\frac{\partial}{\partial z}(\rho A U) = 0 \qquad (1)$$

and

$$U\frac{\partial U}{\partial z} = -\frac{1}{\rho}\frac{\partial P}{\partial z} - g - f_{ft} \qquad (2)$$

where z denotes the vertical coordinate along the dyke axis, $A = \pi a_d b_d$, is the cross-section area of an elliptical dyke having semi-axes a_d and b_d, U is the vertical mixture velocity, g is the gravity acceleration, and f_{ft} is the friction term calculated for an elliptical cross-section (Costa et al., 2009). The model assumes steady-state conditions. This is justified because eruption durations and time-scales of pressure variations at base of the conduit are of the order of hours, much longer than magma travel times in the conduit that are of the order of minutes (e.g., Wilson et al., 1980; Folch et al., 1998).

The model considers that fragmentation occurs when the gas volume fraction, α, reaches a critical value of 0.75 (Sparks, 1978). Despite this simplification, the results are in line with other fragmentation criteria proposed by e.g., Melnik (1999), or Papale (1999).

The dyke semi-axes a_d and b_d depend on the difference between magmatic pressure and normal stress in host rocks ΔP in accord to the following relationships (e.g., Muskhelishvili, 1963; Sneddon and Lowengrub, 1969; Costa et al., 2009):

$$a_d(z) = a_{d0}(z) + \frac{\Delta P}{2G}\left[2(1-\upsilon)b_{d0}(z) - (1-2\upsilon)a_{d0}(z)\right] \quad (3a)$$

$$b_d(z) = b_{d0}(z) + \frac{\Delta P}{2G}\left[2(1-\upsilon)a_{d0}(z) - (1-2\upsilon)b_{d0}(z)\right] \quad (3b)$$

$$\Delta P = P - (\rho_r g z - \sigma_t) \qquad (4)$$

where, G is the rigidity of wallrock, υ is Poisson's ratio, a_{d0} and b_{d0} are the unpressurized values of the semi-axes, σ_t is the tensile stress along the axis of the dyke conduit due to the presence of magma chamber under the effect of an extensional far-field stress σ_{ff} acting on the plane y–z (see **Figure 1**).

As in Costa et al. (2011), as first-order approximation, the tensile stress along the dyke conduit due the effects of magma chamber and reservoir under the effect of an extensional far-field stress is calculated using the general analytical solutions by Gao (1996) obtained in the limit of a plane 2D geometry. Such a first-order approximation can capture general large-scale features (Costa et al., 2011). We would like to remark that, in this way, we account for the effects of magma chamber and reservoir on the tensile stress during magma transport in the shallow dyke conduit but we do not investigate their internal dynamics (out of the scope of this study). Magma chamber pressure is used as boundary condition at the base of the shallow dyke conduit and assumed chamber volumes to estimate durations once we calculated MERs.

The main limitations of the magma transport model presented above and the solving methodology are discussed in Costa et al. (2009) and Costa et al. (2011).

CONTROL OF LOCAL STRESS FIELD ON ERUPTION DYNAMICS AND INTENSITIES

In the framework of the model described in Section Basic Model of Magma Ascent during LCFE, we consider a relatively shallow magma chamber connected to the surface through a shallow dyke conduit. Internal pressures of the shallow chamber range from over- to under-pressure conditions. In terms of stress distribution we also consider the effect of a deeper reservoir that can be in neutral conditions or over-pressurized with respect to the lithostatic loading (**Figure 1**, terminology as in Gudmundsson, 2012). Irrespectively of the process that formed the fracture (magma chamber overpressure or tectonic events), we assume that dyke is already opened and we study syn-eruptive magma transport.

In the approximation of elastic deformation, valid because the short time scales (from few to several hours) characterizing LCFEs, the dyke will tend to open or close as function of the local magmatic pressure with respect the local loading. Besides the lithostatic loading we need to account for the contribution σ_t to the tensile stress along the axis of the shallow dyke conduit, due to the presence of a more or less pressurized magma chamber and reservoir under the effect of an extensional far-field stress. The contribution of an extensional stress on keeping open the base of the dyke was discussed by Costa et al. (2011) who, however, have not considered the effect due to the presence of a deep pressurized reservoir. They described a critical extensional stress that produces a tensile stress at the base of the dyke able to counterbalance the lithostatic loading. Nevertheless, because fragmentation levels during LCFEs are very deep due to the typical large magma viscosities (Costa et al., 2011), the critical extensional stress was very high ($\sigma_{ff} \approx 50$–60 MPa) and close to rifting regime (Turcotte and Schubert, 2002).

Here, our results show that even the contribution of a deep magma reservoir with an over-pressure of about 10 MPa is able to halve critical extensional stresses allowing the dyke to remain open until the magma pressure goes back to sub-neutral conditions. Once formed, a long dyke can remain open even for magma chamber pressures from ~10 MPa above the lithostatic

loading (considered a typical value to propagate a dyke; e.g., Gudmundsson, 2006, 2012) to ~10 MPa below the lithostatic loading. Once the pressure at the base of the dyke decreases below a critical value, the eruption stops and the system has to recover again large magmatic pressure before it can erupt, i.e. the dyke can act as a valve.

Although, Jellinek and De Paolo (2003) show the difficulty to overpressurize a large magma chamber with typical magma rate production in subduction zone and extensional environments, overpressures of ~10 MPa could be easily achieved because of magma crystallization (and as we discussed above, magma associated to LCFEs are typically characterized by high crystallinity; Costa et al., 2011). Such overpressures can be sufficient to trigger an eruption, as the roof of the shallow magma chamber can undergo to intensive heating that can significantly weaken the strength of the overlying rocks (Gregg et al., 2012, 2015). Another end-member for triggering the eruption could be when a fracture opens from surface or all through the host rock above the magma chamber if there is a pre-existing fault.

Concerning deep magma reservoirs, overpressures of ~10 MPa may be generated by the contribution of CO_2 exsolution (e.g., Folch and Martí, 1998; Gudmundsson, 2015).

However, we would like to remark that here we focus on the syn-eruptive dynamics of magma ascent inside the shallow dyke conduit during LCFEs and not on magma chamber dynamics and mechanics able to trigger such eruptions that are likely related to magma chamber roof failure and have been explored by other authors (e.g., Burov and Guillou-Frottier, 1999; Folch and Martí, 2004; Gray and Monaghan, 2004; Gregg et al., 2012, 2015; Geyer and Martí, 2014). For the purposes of this study, aimed at describing the first-order features only, chamber and reservoir shapes are not crucial and for simplicity, as in Costa et al. (2011), we assume them as elongated bodies with circular cross-sections. This choice is consistent with the observation that common geometries are general oblate-ellipsoidal chambers (Gudmundsson, 2012 and references therein). Moreover, as discussed in (Gudmundsson, 2012) long-lived magma chambers cannot have very irregular shapes as their surfaces can be assumed smooth (Jaeger, 1961, 1964; Gudmundsson, 2012). On this basis, under isotropic conditions, on very long time-scales, chamber geometries should tend to become sub-circular.

We also need to consider that the aspect ratio of the caldera produced during those eruptions may not be indicative of the chamber shape because as magma overpressure decreases below a critical value, the shallow dyke cannot anymore be kept opened around the fragmentation depth and tends to collapse forming a local restriction that would stop the eruption (Costa et al., 2009, 2011) even if a large fraction of the stored magma is still in the chamber. Caldera shapes can reflect only chamber cross-sections and the fraction of magma volumes evacuated. However, caldera eruptions tend to erupt all eruptible magma (all magma that may vesiculate and have a density lower than the host rock while the crustal block in collapsing) likely because the control of the piston-like subsidence able to maintain the necessary excess pressure (Martí et al., 2000; Folch and Martí, 2009; Gudmundsson, 2015; Geshi and Miyabuchi, 2016). Also it is common to see in large silicic calderas the emplacement of degassed magma in the form of extrusive dome along the ring faults immediately after caldera formation (e.g., Williams, 1941; Lipman, 1984). This implies that we should not expect a large volume of magma remaining in the chamber. In fact many calderas (see Geyer and Martí, 2008) exhibit post-caldera volcanism of mafic composition, thus indicating that nothing or very little was remaining in the magma chamber after caldera collapse, so deeper mafic magmas can cross it without being trapped (mixed) by the resident magma.

In our model the solution for the stress field is calculated using the general analytical solutions by Gao (1996) valid for a pressurized elliptical hole obtained in the limit of a plane 2D geometry (approximation valid for $c_C \gg a_C, b_C$, and $c_R \gg a_R, b_R$, **Figure 1**). Such solutions also assume that the medium is homogeneous and purely elastic. The elastic rheology is however a reasonable approximation because the time scales of the eruption, that are of the order of few to several hours, are much shorter than the viscous time of the magmatic system, of the order of hundreds of years or longer (Jellinek and De Paolo, 2003). However, concerning assumption of medium homogeneity we need keep in mind that rock stress distribution can be affected by presence of pore fluids, temperature, and alteration of the different layers (Gudmundsson, 2006). Moreover, active faults, block boundaries are neglected. In addition the solution is valid for an unbounded domain, neglects the effect of topography, and the far field stress is assumed to be homogeneous. However, the analytical solution is able to capture first-order, general large-scale features even with all these limitations (Costa et al., 2011).

We applied the model described above to a LCFE similar to YTT for which magma physical parameters, erupted volumes, tectonic settings are known and independent estimations of MERs are available (Costa et al., 2014). In particular, we studied a magma chamber located at 3–5 km depth under extensional far-field stresses, σ_{ff}, ranging from 0 to -50 MPa (**Figure 1**). The upper limit of this spectrum (comparable with stresses needed for nucleating new fractures) can be considered representative of the transition toward an active extensional setting (Turcotte and Schubert, 2002). The stress field perturbation due to a magma reservoir having a top at 10 km depth was also considered.

Concerning magma chamber and reservoir volumes, we assumed a chamber volume, $V_C = \frac{4}{3}\pi a_C b_C c_C$, of $\sim$5000 km^3 (considering a chamber extension c_C of 100 km, consistent with Toba caldera geometry) and a similar reservoir volume, $V_R = \frac{4}{3}\pi a_R b_R c_R$, of $\sim$5000 km^3.

Before to proceed, it is useful to summarize some basic effects due to the combination of different magma chamber geometries under the action of a far-field extensional stress. For an elongated magma chamber with a circular cross-section (i.e., with an aspect ratio $a_C/c_C \approx 1$) near neutral pressure conditions, the maximum tensile stress is at the base of the dyke ($x = 0$, $z = c_C$; **Figure 1**) and in this case is $\sigma_t \approx 3\sigma_{ff}$ (e.g., Gudmundsson, 1988). For a prolate cross-section $\sigma_t \approx (1 + 2a_C/c_C)\sigma_{ff}$ with $a_C \geq c_C$ and $b_C \gg a_C$ (e.g., Gudmundsson, 1988). For an oblate cross-section the maximum stress is at the two lateral tips of the ellipse. The stress along the dyke,

connecting the magma chamber to the surface, depends on the intensity of extensional stress, magma chamber pressure, and magma chamber aspect ratio a_C/c_C, magma reservoir depth, magma reservoir aspect ratio a_R/c_R, and magma reservoir overpressure ΔP_R.

Considering the above estimation for the magma chamber volume, and the geometry of the Toba caldera ($\sim$100 $\times$ 30 km), the chamber would be roughly approximated by an oblate ellipsoid having an elongation of $\sim$100 km, a width of $\sim$30 km, and a height of $\sim$3 km. However, for the sake of simplicity, consistently with our basic magma conduit flow model and the analytical solution for the stress field, we assumed a chamber (and reservoir) with a circular cross section, $b_C \gg a_C = c_C$ with $2a_C = D_C$ being D_C the equivalent diameter (in our case $D_C \sim 10$ km). In our approximation the maximum effect on the tensile stress is along the dyke conduit at the center of the circular cross section of the chamber, whereas, considering the more realistic case of an oblate ellipsoid, it would be around the lateral tips of the ellipse with an intensity factor an order of magnitude larger due to the different geometry aspect ratio ($\sim$10). We would like to remark that we aim to estimate the order of magnitude control of the stress along the dyke conduit due to magma chamber and reservoir and not investigate their detailed magma-rock mechanics. Our approximation represents a minimum bound for the actual effects on the tensile stress that can be even an order of magnitude larger in case of elongated geometries.

Concerning chemical and physical magma properties of the shallow chamber, we considered a magma composition like that of YTT (e.g., Chesner, 2012; Matthews et al., 2012) characterized by a high SiO$_2$ content ($\sim$70–75%), a water content of about 5–6%, temperatures of 700–780°C and high crystallinity (up to 40%). These properties are very similar to those characterizing magmas of other LCFEs such as, for example, those associated to the Aira Caldera which erupted more than 300 km^3 of magma (Aramaki, 1984) and to the fissure eruptions of ignimbrite from Southern Sierra Madre Occidental, Mexico (Aguirre-Díaz and Labarthe-Hernández, 2003; Gottsmann et al., 2009).

Since our model uses cross-section averaged variables only, magma properties are treated in an approximate way. This includes equilibrium water exsolution, absence of gas overpressure with respect to magma pressure, and constant viscosity assumptions. A more realistic description of the effective viscosity (out of the scope of this paper) should account for the coupling with dissolved water, heat loss, viscous dissipation, crystal resorption, and the associated local effects (Costa and Macedonio, 2005; Costa et al., 2007).

Considering those properties and approximations, using the models of Giordano et al. (2008) for melt viscosity and Costa et al. (2009) and Cimarelli et al. (2011) for accounting for the effects of crystals, we estimated a reference effective magma viscosity of 10^8 Pa s. In our simplified model, the main effect of magma viscosity is on magma fragmentation depth as high viscosities tend to move fragmentation level at greater depths, because the critical volume fraction of bubbles is attained earlier upon magma ascent. Initial dyke thickness is assumed to be ∼100 m, consistent with typical values estimated for such magma viscosity (Wada, 1994).

Other simplifications are related to rock properties that, for the sake of simplicity, are assumed constant with depth. Although, the variations with depth of some properties such as the Young modulus are evident (e.g., Paulatto, 2010; Costa et al., 2013), the dependence of others, such as the variation of fracture toughness with confining pressure, is not very clear (Rubin, 1993). For example Abou-Sayed (1977) found a 50% increase for limestone toughness at confining pressure of 7 MPa, whereas Schmidt and Huddle (1977) found almost no increase at a confining pressure of 7 MPa but an increase up to a factor 4 at 60 MPa.

All the model input parameters are reported in **Table 1**.

The effect of the far-field extensional stresses is shown in **Figure 2** where we reported the profiles of the tensile stress, σ_t, along the vertical axis of the shallower dyke conduit for both an unpressurized magma reservoir and for a magma reservoir with 10 MPa overpressure (for each condition three different magma chamber pressures are considered). **Figure 2** shows that, for an unpressurized magma reservoir, σ_t can counterbalance the lithostatic loading at the dyke base only if the far-field stress is about −40 MPa or larger. Whereas when the reservoir has an

overpressure of 10 MPa such condition is reached for a far-field stress around −30 MPa or lower.

As shown in **Figure 3**, the increased tensile stress affects the local pressure difference (given as the magma pressure minus the effects of lithostatic loading and tensile stress). In order to get almost neutral conditions at the base of the dyke, in absence of a pressurized reservoir, a far-field stress of about −40 MPa or larger is needed. Whereas, considering a pressurized reservoir a far-field stress of about −30 MPa or lower is enough.

The maximum sustainable length of the shallow dyke conduit (Costa et al., 2009, 2011) and the MER as function of the extensional far-field stress are reported in **Figure 4**. We can see that maximum sustainable lengths of the dyke can range from a few to a few tens kms depending on the extensional far-field stress and magma reservoir overpressure. Similarly the maximum MER can span from 10^9 to about 10^{11} kg/s as function of the extensional stress, magma chamber depth, and magma reservoir overpressure. As it is shown for comparison in **Figure 4** and discussed by Costa et al. (2011), a much shallower magma chamber (e.g., 3 km depth) would be able to erupt higher MER for the same far-field stress.

DISCUSSION AND OPEN PROBLEMS

We have shown that in order to produce extremely large eruption intensities of highly silicic magmas, such as those characterizing LCFEs, it is necessary to consider the effects of the local stress field resulting from the combination of an extensional far-field stress and a pressurized magma reservoir. Our simulations indicate that MERs of $10^{10}-10^{11}$ kg/s are promoted during moderate to high extensional far field stress (20–40 MPa), a pressurized magma reservoir (∼10 MPa), and relatively shallow magma chambers (3–5 km).

We need to remark that these calculations represent a first-order description aimed at capturing some general features and the proper order of magnitude of the estimated quantities. For a more accurate description other factors should be considered besides the approximations described above. A more correct description of a dual magma chamber system (Melnik and Costa, 2014) should consider not only the effect of a pressurized magma reservoir on the tensile stress within the shallow dyke conduit, but even the control of the magma reservoir on the dynamics of the shallow magma chamber. This is important to characterize how the response of the deeper dyke to local pressure variation can alter magma feeding into the chamber on time-scales comparable to the eruption duration. Dealing with such complex dynamics is out the scope of this work and is the subject of ongoing research.

Our study analyses the conditions required to keep magma conduits open during caldera formation and these will be the same irrespectively on how these eruption pathways have been opened. Because the high magma viscosity, fragmentation level is typically very deep (basically at the roof of the chamber). The effect of a far-field extensional stress is needed in order to permit magma pressure can counterbalance lithostatic loading.

TABLE 1 | Parameters used in the simulations (estimated from Costa et al., 2011, 2014; Chesner, 2012).

Symbol	Parameter	YTT
x_{tot}	Concentration of dissolved gas	6 wt%
T	Magma temperature	1053 K
x_c	Magma crystal fraction	40 wt%
μ	Magma viscosity	10^8 Pa s
E_D	Dynamic rock Young modulus	40 GPa
G	Static host rock rigidity	6 GPa
υ	Poisson ratio	0.3
β	Bulk modulus of melt/crystal	10 GPa
ρ_{lo}	Density of the melt phase	2300 kg m^{-3}
ρ_{co}	Density of crystals	2800 kg m^{-3}
ρ_r	Host rock density	2600 kg m^{-3}
S	Solubility coefficient	$4.1 \cdot 10^{-6}$ Pa$^{-1/2}$
N	Solubility exponent	0.5
L	Depth of magma chamber roof	5 km
P_C	Magma chamber pressure	115–140 MPa
$2a_C$ (D_C)	Magma chamber width (Equivalent diameter)	30 km (10 km)
$2c_C$ (D_C)	Magma chamber height (Equivalent diameter)	3 km (10 km)
$2b_C$	Magma chamber elongation	100 km
V_C	Magma chamber volume	5000 km^3
L_R	Depth of magma reservoir roof	10 km
$2a_R$	Magma reservoir width	10 km
$2c_R$	Magma reservoir height	10 km
$2b_R$	Magma reservoir elongation	100 km
V_R	Magma reservoir volume	5000 km^3
ΔP_R	Magma reservoir overpressure	0–20 MPa

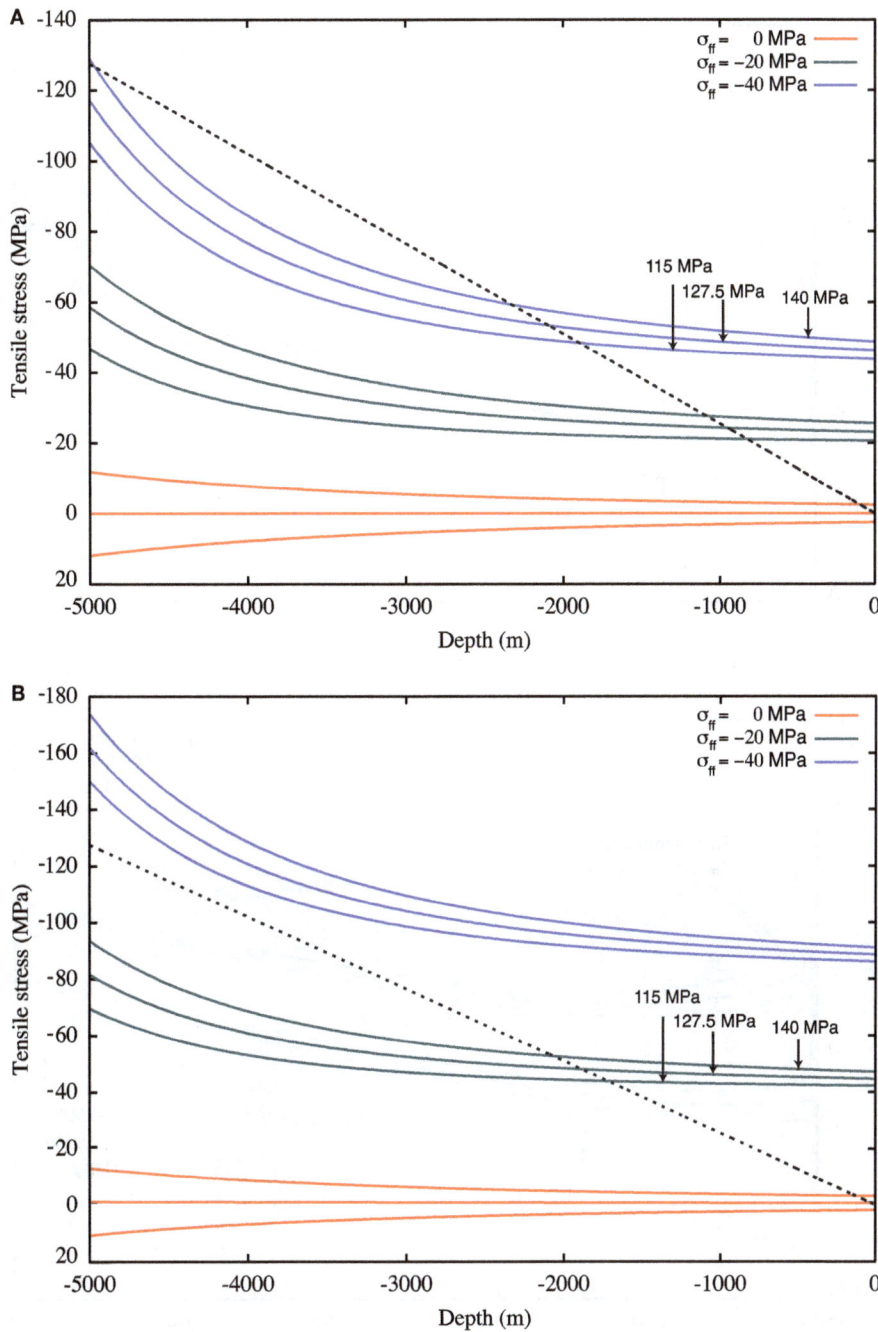

FIGURE 2 | Dyke tensile stress σ_t profile along the vertical axis of the shallower dyke conduit obtained using the analytical solution presented by Gao (1996) for a pressurized magma chamber at 5 km depth (at lithostatic pressure and ±12.5 MPa, as indicated) under the effect of different far-field extensional stresses (from 0 to −40 MPa, as indicated in the figures), for an unpressurized magma reservoir (A) and a pressurized (+10 MPa) reservoir at 10 km depth (B). Dashed line represents the lithostatic pressure. Note that the critical extensional stress that will produce a tensile stress at the dyke base able to counterbalance the lithostatic pressure is different in the two cases: in **(A)** for σ_{ff} of −40 MPa or stronger the dyke remains open, whereas in **(B)**, due to the presence of a pressurized magma reservoir, this critical stress is decreased at 20–30 MPa (for more realistic geometries these effects will be much larger, see Section Discussion and Open Problems).

However, the current model does not consider rock failure occurring in over- and under-pressure conditions (Costa et al., 2009). In our case significant rock failure should occur below and across fragmentation depth (**Figure 3**), implying the partial destruction of the roof of the magma chamber likely moving the fragmentation inside the chamber producing a stable system

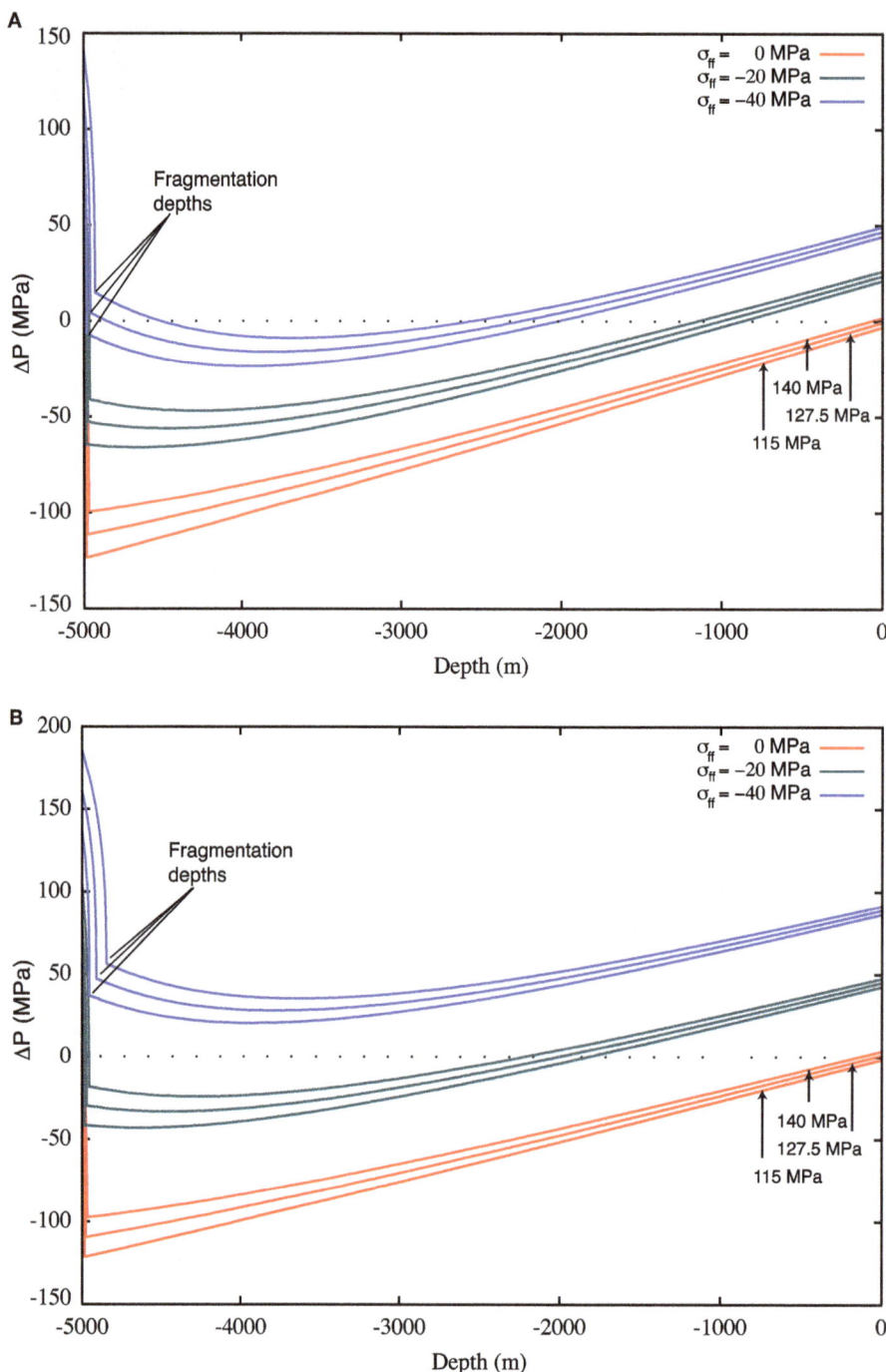

FIGURE 3 | Pressure difference (magma pressure−lithostatic loading−tensile stress) profile along the vertical axis of the shallow dyke conduit obtained using the analytical solution presented by Gao (1996) for a pressurized magma chamber at 5 km depth (at lithostatic pressures and ±12.5 MPa, as indicated) under the effect of different far-field extensional stresses (from 0 to −40 MPa, as indicated in the figures), for a unpressurized magma reservoir (A) and a pressurized (+10 MPa) reservoir at 10 km depth (B). On the top **(A)**, for an unpressurized magma reservoir σ_{ff} moduli larger then 40 MPa are needed to reach neutral conditions along the dyke. On the bottom **(B)**, the presence of a pressurized magma reservoir reduces the critical extensional stress needed to reach neutral conditions at 20–30 MPa.

more suitable to maintain the dyke opened. This implies that, in caldera eruptions that are preceded by a Plinian phase (i.e., MER $\lesssim 10^9$ kg/s; Koyaguchi et al., 2010) for getting a transition to larger MER, the caldera collapse phase must be accompanied by a change in the mechanical conditions of the system that allow magma to fragment much deeper when flowing up through

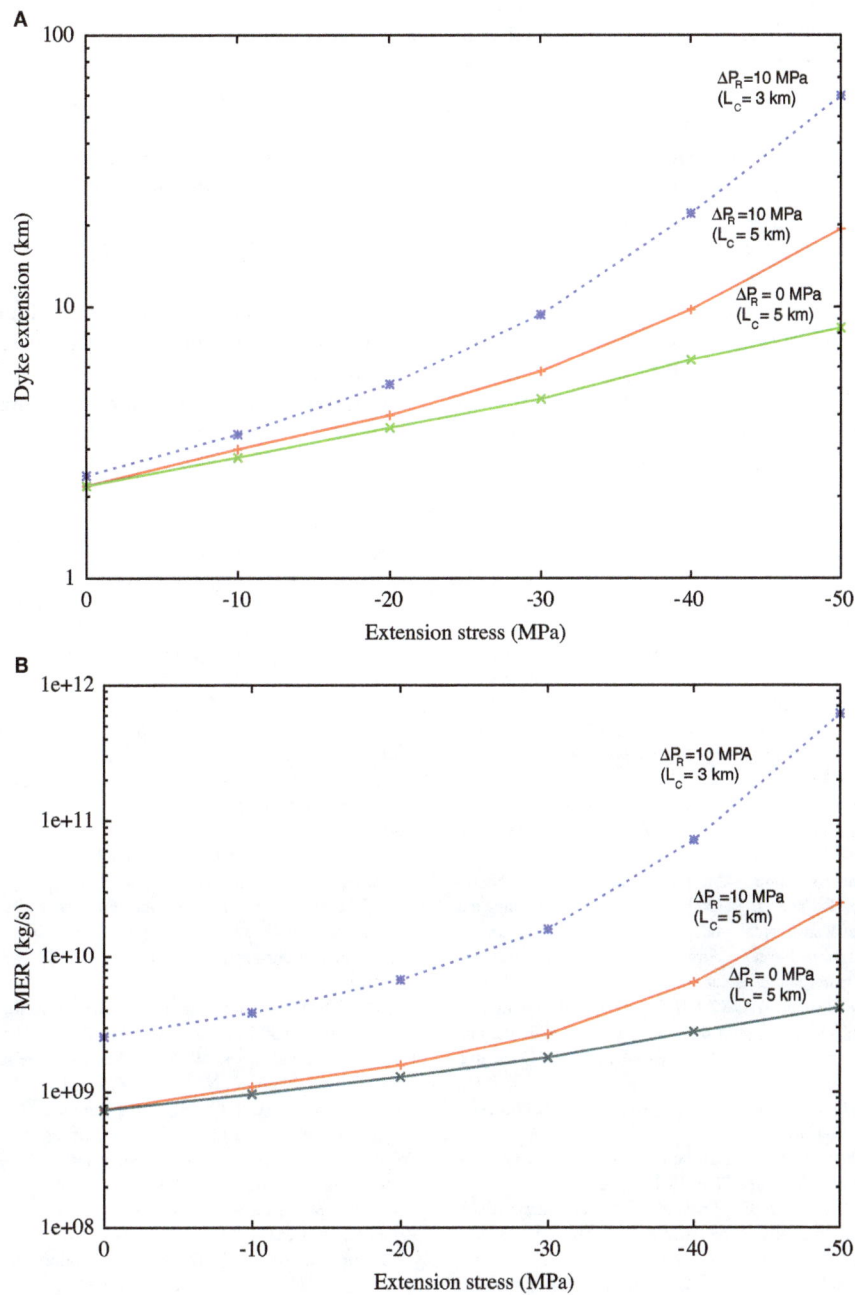

FIGURE 4 | Effects of extensional far-field stress on (A) maximum sustainable length of the shallow dyke conduit and (B) MER. (A) Maximum sustainable lengths of the shallow dyke conduit depend on the extensional far-field stress and magma reservoir overpressure, ranging from a few to a few tens kms. **(B)** Similarly maximum MERs are a function of the extensional stress and magma reservoir overpressure, ranging from 10^9 to 10^{11} kg/s.

the ring fault. This agrees with previous models (e.g., Williams, 1941; Druitt and Sparks, 1984; Martí et al., 2000) that assume a considerable decompression on the magma chamber as a necessary requisite to permit caldera collapse. In the case where a part of the ring fault was open by a tectonic event, the conditions for magma fragmentation to occur very deep in the conduit would be achieved early in the eruption. In such situations it would not be necessary to have a significant decompression

of the chamber to allow caldera collapse, as it is suggested by the lack of pre-caldera deposits in many LCFE (Martí et al., 2009).

CONCLUSIONS

Our study indicates that in order to erupt large volumes of silicic magmas during LCFE in relatively short times (e.g., large

MER estimated in the range 10^9-10^{11} kg/s), it is necessary to account for the combined effect of extensional far-field stress and pressurization conditions of magma chambers and reservoirs. In presence of a pressurized deep reservoir even intermediate extensional crustal stresses (20–30 MPa) facilitate an efficient evacuation of large magmatic chambers through a shallow dyke conduit. Largest MERs are promoted in system having a shallow magma chamber (3–5 km). Large MERs are maintained even for under-pressurized magma chamber. Simulation results are consistent with geological observations of LCFE. However, the model assumes that the dyke is already formed and does not account for rock failure that could change drastically the geometry of the system (e.g., chamber roof collapse) as the eruption proceeds. Our results help to address future research aimed to advance our knowledge on the dynamics of these rare catastrophic events.

AUTHOR CONTRIBUTIONS

AC developed the new code version and ran the simulations. AC, JM analyzed the results and wrote the manuscript.

ACKNOWLEDGMENTS

Two reviewers and the editor Agust Gudmundsson are warmly thanked for their constructive feedback. AC is grateful to T. Koyaguchi and P. Gregg for fruitful discussion during his stay at the Earthquake Research Institute, the University of Tokyo.

REFERENCES

Abou-Sayed, A. S. (1977). "Fracture toughness KIC of triaxially loaded Indiana limestone," in *Proc. 17th U.S. Symposium Rock Mechanics* (Keystone).

Acocella, V. (2007). Understanding caldera structure and development: An overview of analogue models compared to natural calderas. *Earth Sci. Rev.* 85, 125–160. doi: 10.1016/j.earscirev. 2007.08.004.

Acocella, V., Di Lorenzo, R., Newhall, C., and Scandone, R. (2015). An overview of recent (1988 to 2014) caldera unrest: knowledge and perspectives. *Rev. Geophys.* 53, 896–955. doi: 10.1002/2015RG000492

Acocella, V., and Funiciello, F. (2010). Kinematic setting and structural control of arc volcanism, Earth Planet. *Sci. Lett.* 289, 43–53. doi: 10.1016/j.epsl.2009.10.027

Aguirre-Díaz, G. J., Labarthe-Hernández, G., Tristán-González, M., Nieto-Obregón, J., and Gutiérrez-Palomares, I. (2008). "Ignimbrite Flare-up and graben-calderas of the Sierra Madre Occidental, Mexico," in *Caldera Volcanism: Analysis, Modelling and Response, Developments in Volcanology*, Vol. 10., eds J. Martí and J. Gottsmann (Amsterdam: Elsevier), 492.

Aguirre-Díaz, G., and Labarthe-Hernández, G. (2003). Fissure ignimbrites: fissure-source origin for voluminous ignimbrites of the Sierra Madre Occidental and its relationship with Basin and Range faulting. *Geology* 31, 773–776. doi: 10.1130/G19665.1

Aramaki, S. (1984). Formation of the Aira Caldera, Southern Kyushu, 22,000 Years Ago. *J. Geophys. Res.* 89, 8485–8501.

Bacon, C. R. (1983). Eruptive history of Mount Mazama and Crater Lake Caldera, Cascade Range, U.S.A. *J. Volcanol. Geother. Res.* 18, 57–115.

Baines, P. G., and Sparks, R. S. J. (2005). Dynamics of giant volcanic ash clouds from supervolcanic eruptions. *Geophys. Res. Lett.* 32, L24808. doi: 10.1029/2005GL024597

Burov, E. B., and Guillou-Frottier, L. (1999). Thermomechanical behaviour of large ash flow calderas. *J. Geophys. Res.* 104, 23081–23109. doi: 10.1029/1999JB900227

Chesner, C. A. (1998). Petrogenesis of the Toba Tuffs, Sumatra, Indonesia. *J. Petrol.* 39, 397–438.

Chesner, C. A. (2012). The toba caldera complex. *Quat. Int.* 258, 5–18. doi: 10.1016/j.quaint.2011.09.025

Cimarelli, C., Costa, A., Mueller, S., and Mader, H. (2011). Rheology of magmas with bimodal crystal size and shape distributions: insights from analogue experiments. *Geochem. Geophys. Geosyst.* 12, Q07024. doi: 10.1029/2011GC003606

Cole, J. W., Spinks, K. D., Deering, C. D., Nairn, I. A., and Leonard, G. S. (2010). Volcanic and structural evolution of the Okataina Volcanic Centre; dominantly silicic volcanism associated with the Taupo Rift, New Zealand. *J. Volcanol. Geother. Res.* 190, 123–135. doi: 10.1016/j.jvolgeores.2009. 08.011

Costa, A., Gottsmann, J., Melnik, O., and Sparks, R. S. J. (2011). A stress-controlled mechanism for the intensity of very large magnitude explosive eruptions. *Earth Planet. Sci. Lett.* 310, 161–166. doi: 10.1016/j.epsl.2011.07.024

Costa, A., and Macedonio, G. (2005). Viscous heating effects in fluids with temperature-dependent viscosity: triggering of secondary flows. *J. Fluid Mech.* 540, 21–38. doi: 10.1017/S0022112005006075

Costa, A., Melnik, O., and Vedeneeva, E. (2007). Thermal effects during magma ascent in conduits. *J. Geophys. Res.* 112:B12205. doi: 10.1029/2007JB004985

Costa, A., Smith, V., Macedonio, G., and Matthews, N. (2014). The magnitude and impact of the Youngest Toba Tuff super-eruption. *Front. Earth Sci.* 2:16. doi: 10.3389/feart.2014.00016

Costa, A., Sparks, R. S. J., Macedonio, G., and Melnik, O. (2009). Effects of wall-rock elasticity on magma flow in dykes during explosive eruptions. *Earth Planet. Sci. Lett.* 288, 455–462. doi: 10.1016/j.epsl.2007.05.024

Costa, A., Wadge, G., Stewart, R., and Odbert, H. (2013). Coupled sub-daily and multi-week cycles during the lava dome eruption of Soufriere Hills Volcano, Montserrat. *J. Geophys. Res.* 118, 1895–1903. doi: 10.1002/jgrb. 50095

Druitt, T. H., and Francaviglia, V. (1992). Caldera formation on Santorini and the physiography of the islands in the late Bronze Age. *Bull. Volcanol.* 54, 484–493.

Druitt, T. H., and Sparks, R. S. J. (1984). On the formation of calderas during ignimbrite eruptions. *Nature* 310, 679–681.

Folch, A., and Martí, J. (1998). The generation of overpressure in felsic magma chambers by replenishment. *Earth Planet. Sci. Lett.* 163, 301–314.

Folch, A., and Martí, J. (2004). Geometrical and mechanical constraints on the formation of ring-fault calderas. *Earth Planet. Sci. Lett.* 221, 215–225. doi: 10.1016/S0012-821X(04)00101-3

Folch, A., and Martí, J. (2009). Time-dependent chamber and vent conditions during explosive caldera-forming eruptions. *Earth Planet. Sci. Lett.* 280, 246–253. doi: 10.1016/j.epsl.2009.01.035

Folch, A., Martí, J., Codina, R., and Vazquez, M. (1998). A numerical model for temporal variations durino explosive central vent eruptions. *J. Geophys. Res.* 103, 20883–20899.

Folkes, C. B., Wright, H. M., Cas, R. A. F., de Silva, S. L., Lesti, C., and Viramonte, J. G. (2011). A re-appraisal of the stratigraphy and volcanology of the Cerro Galan volcanic system, NW Argentina. *Bull. Volcanol.* 73:1427–1454. doi: 10.1007/s00445-011-0459-y

Gao, X.-L. (1996). A general solution of an infinite elastic plate with an elliptic hole under biaxial loading. *Int. J. Pres. Ves. Piping* 67, 95–104.

Gardeweg, M., and Ramirez, C. F. (1987). La Pacana caldera and the Atana Ignimbrite - A major ash-flow and resurgent caldera complex in the Andes of northern Chile. *Bull. Volcanol.* 49, 547–566.

Geshi, N., and Miyabuchi, Y. (2016). Conduit enlargement during the precursory Plinian eruption of Aira Caldera, Japan. *Bull. Volcanol.* 78, 63. doi: 10.1007/s00445-016-1057-9

Geyer, A., and Martí, J. (2008). The new worldwide Collapse Caldera Database (CCDB): a tool for studying and understanding caldera processes. *J. Volcanol. Geotherm. Res.* 175, 334–354. doi: 10.1016/j.jvolgeores.2008.03.017

Geyer, A., and Martí, J. (2014). A short review of our current understanding of the development of ring faults during collapse caldera formation. *Front. Earth Sci.* 2:22, doi: 10.3389/feart.2014.00022

Giordano, D., Russel, J. K., and Dingwell, D. B. (2008). Viscosity of magmatic liquids: a model, Earth Planet. *Sci. Lett.* 271, 123–134, doi: 10.1016/j.epsl.2008.03.038

Gottsmann, J., Lavallée, Y., Martí, J., and Aguirre-Díaz, G. (2009). Magma–tectonic interaction and the eruption of silicic batholiths. *Earth Planet. Sci. Lett.* 284, 426–434. doi: 10.1016/j.epsl.2009.05.008

Gray, J. P., and Monaghan, J. J. (2004). Numerical modelling of stress fields and fracture around magma chambers. *J. Volcanol. Geotherm. Res.* 135, 259–283. doi: 10.1016/j.jvolgeores.2004.03.005

Gregg, P. M., de Silva, S. L., Grosfils, E. B., and Parmigiani, J. P. (2012). Catastrophic caldera-forming eruptions: thermomechanics and implications for eruption triggering and maximum caldera dimensions on Earth. *J. Volcanol. Geotherm. Res.* 241–242, 1–12. doi: 10.1016/j.jvolgeores.2012.06.009

Gregg, P. M., Grosfils, E. B., and de Silva, S. L. (2015). Catastrophic caldera-forming eruptions II: the subordinate role of magma buoyancy as an eruption trigger. *J. Volcanol. Geotherm. Res.* 305, 100–113. doi: 10.1016/j.jvolgeores.2015.09.022

Gudmundsson, A. (1988). Effect of tensile stress concentration around magma chambers on intrusion and extrusion frequencies. *J. Volcanol. Geotherm. Res.* 35, 179–194.

Gudmundsson, A. (1998). Formation and development of normal-fault calderas and the initiation of large explosive eruptions. *Bull. Volcanol.* 60, 160–170.

Gudmundsson, A. (2002). Emplacement and arrest of sheets and dykes in central volcanoes. *J. Volcanol. Geotherm. Res.* 116, 279–298. doi: 10.1016/S0377-0273(02)00226-3

Gudmundsson, A. (2006). How local stresses control magma-chamber ruptures, dyke injections, and eruptions in composite volcanoes. *Earth Sci. Rev.* 79, 1–31. doi: 10.1016/j.earscirev.2006.06.006

Gudmundsson, A. (2012). Magma chambers: Formation, local stresses, excess pressures, and compartments. *J. Volcanol. Geotherm. Res.* 237–238, 19–42. doi: 10.1016/j.jvolgeores.2012.05.015

Gudmundsson, A. (2015). Collapse-driven eruptions. *J. Volcanol. Geotherm. Res.* 304, 1–10. doi: 10.1016/j.jvolgeores.2015.07.033

Gudmundsson, A., Martí, J., and Turon, E. (1997). Stress fields generating ring faults in volcanoes. *Geophys. Res. Lett.* 24, 1559–1562. doi: 10.1029/97GL01494

Hautmann, S., Gottsmann, J., Sparks, R. S. J., Costa, A., Melnik, O., and Voight, B. (2009). Modelling ground deformation caused by oscillating overpressure in a dyke conduit at Soufriere Hills Volcano, Montserrat. *Tectonophysics* 471, 87–95. doi: 10.1016/j.tecto.2008.10.021

Hildreth, W. (1991). The timing of caldera collapse at Mount Katmai in response to magma withdrawal toward Novarupta. *Geophys. Res. Lett.* 18, 1541–1544.

Hildreth, W., and Mahood, G. A. (1986). Ring-fracture eruption of the Bishop Tuff. *Geol. Soc. Am. Bull.* 97, 396–403.

Hughes, G. H., and Mahood, G. A. (2008). Tectonic controls on the nature of large silicic calderas in volcanic arcs. *Geology* 36, 627–630. doi: 10.1130/G24796A.1

Jaeger, J. C. (1961). The cooling of irregularly shaped igneous bodies. *Am. J. Sci.* 259, 721–734.

Jaeger, J. C. (1964). Thermal effects of intrusions. *Rev. Geophys.* 2, 443–466.

Jellinek, A. M., and De Paolo, D. J. (2003). A model for the origin of large silicic magma chambers: precursors of caldera-forming eruptions. *Bull. Volcanol.* 65, 363–381. doi: 10.1007/s00445-003-0277-y

Korringa, M. K. (1973). Linear vent area of the Soldier Meadow Tuff, an ash-flow sheet in northwestern Nevada. *Geol. Soc. Am. Bull.* 84, 3849–3866.

Koyaguchi, T., Suzuki, Y. J., and Kozono, T. (2010). Effects of the crater on eruption column dynamics. *J. Geophys. Res.* 115, B07205. doi: 10.1029/2009JB007146

Lipman, P. W. (1984). The roots of ash-flow calderas in North America: windows into the tops of granitic batholiths. *J. Geophys. Res.* 89, 8801–8841.

Macedonio, G., Neri, A., Martí, J., and Folch, A. (2005). Temporal evolution of flow conditions in sustained magmatic explosive eruptions. *J. Volcanol. Geotherm. Res.* 143, 153–172. doi: 10.1016/j.jvolgeores.2004.09.015

Martí, J., Folch, A., Macedonio, G., and Neri, A. (2000). Pressure evolution during caldera forming eruptions. *Earth Planet. Sci. Lett.* 175, 275–287. doi: 10.1016/S0012-821X(99)00296-4

Martí, A., Folch, A., Costa, A., and Engwell, A. (2016). Reconstructing the plinian and co-ignimbrite sources of large volcanic eruptions: a novel approach for the Campanian Ignimbrite. *Nat. Sci. Rep.* 6:21220. doi: 10.1038/srep21220

Martí, J., Geyer, A., and Folch, A. (2009). "A genetic classification of collapse calderas based on field studies, analogue and theoretical modelling," in *Volcanology: the Legacy of GPL Walker*, eds T. Thordarson and S. Self (London: IAVCEI-Geological Socety of London), 249–266.

Matthews, N. E., Pyle, D. M., Smith, V. C., Wilson, C. J. N., Huber, C., and van Hinsberg, V. (2011). Quartz zoning and the pre-eruptive evolution of the ~340 ka Whakamaru magma systems, New Zealand. *Contr. Mineral. Petrol.* 163, 87–107. doi: 10.1007/s00410-011-0660-1

Matthews, N. E., Smith, V. C., Costa, A., Pyle, D. M., Durant, A. J., and Pearce, N. J. G. (2012). Ultra-distal tephra deposits from super-eruptions: examples from Toba, Indonesia and Taupo Volcanic Zone, New Zealand. *Toba Super Erupt. Impact Ecosys. Hominin* 258, 34–79. doi: 10.1016/j.quaint.2011.07.010

Melnik, O. (1999). Fragmenting magma. *Nature* 397, 394–395.

Melnik, O., and Costa, A. (2014). "Dual chamber-conduit models of non-linear dynamics behaviour at Soufriere Hills volcano," in *The Eruption of Soufriere Hills Volcano, Montserrat from 2000 to 2010*, eds G. Wadge, R. E. A. Robertson and B. Voight (London: Memoir of the Geological Society of London, The Geological Society of London), 501. doi: 10.1144/M39.3

Miller, C., Wark, D., Self, S., Blake, S., and John, D. (2008). (Potentially) Frequently asked questions about supervolcanoes and supereruptions. *Elements* 4, 16.

Muskhelishvili, N. (1963). *Some Basic Problems in the Mathematical Theory of Elasticity*. Leiden: Noordhof.

Pallister, J. S., Hoblitt, R. P., and Reyes, A. G. (1992). A basalt trigger for the 1991 eruptions of Pinatubo volcano. *Nature* 356, 426–428.

Papale, P. (1999). Strain-induced magma fragmentation in explosive eruptions. *Nature* 397, 425–428.

Paulatto, M., Minshull, T. A., Baptie, B., Dean, S., Hammond, J. O. S., Henstock, T., et al. (2010). Upper crustal structure of an active volcano from refraction/reflection tomography, Montserrat, Lesser Antilles. *Geophys. J. Int.* 180, 685–696. doi: 10.1111/j.1365-246X.2009.04445.x

Petrinovic, I., Martí, J., Aguirre-Díaz, G., Guzmán, S., Geyer, A., and Salado Paz, N. (2010). The Cerro Aguas Calientes caldera, NW Argentina: an example of a tectonically controlled, polygenetic collapse caldera, and its regional significance. *J. Volcanol. Geotherm. Res.* 194, 15–26. doi: 10.1016/j.jvolgeores.2010.04.012

Pittari, A., Cas, R. A. F., Wolff, J. A., Nichols, H. J., and Martí, J. (2008). The use of lithic clast distributions in pyroclastic deposits to understand pre- and syn-caldera collapse processes: a case study of the Abrigo Ignimbrite, Tenerife, Canary Islands," in *Caldera Volcanism: Analysis, Modelling and Response: Developments in Volcanology*, eds J. Gottsmann, and J. Martí (Amsterdam: Elsevier), 97–142.

Reyners, M. E. (2010). Stress and strain from earthquakes at the southern termination of the Taupo Volcanic Zone, New Zealand. *J. Volcanol. Geotherm. Res.* 190, 82–88. doi: 10.1016/j.jvolgeores.2009.02.016

Roche, O., Buesch, D. C., and Valentine, G. A. (2016). Slow-moving and far-travelled dense pyroclastic flows during the Peach Spring super-eruption. *Nat. Communic.* 7:10890. doi:10.1038/ncomms10890

Rowland, J. V., Wilson, C. J. N., and Gravley, D. M. (2010). Spatial and temporal variations in magma-assisted rifting, Taupo Volcanic Zone, New Zealand. *J. Volcanol. Geotherm. Res.* 190, 89–108. doi: 10.1016/j.jvolgeores.2009.05.004

Rubin, A. M. (1993). Tensile fracture of rock at high confining pressure: implications for dike propagation. *J. Geophys. Res.* 98, 15919–15935.

Rubin, A. M. (1995). Propagation of magma-filled cracks. *Annu. Rev. Earth Planet. Sci.* 23, 287–336.

Schmidt, R. A., and Huddle, C. W. (1977). Effect of confining pressure on fracture toughness of Indiana limestone. *Int. J. Rock Mech. Min. Sci. Geomech. Abstr.* 14, 289–293.

Self, S. (1992). Krakatau revisited: the course of events and interpretation of the 1883 eruption. *Geol. J.* 28, 109–121.

Sigmundsson, F., Hreinsdóttir, S., Hooper, A., Arnadottir, T., Pedersen, R., Roberts, M. J., et al. (2010). Intrusion triggering of the 2010 Eyjafjallajokull explosive eruption. *Nature* 468, 426–430. doi: 10.1038/nature09558

Smith, R. L. (1979). Ash flow magmatism. *Geol. Soc. Am. Spec. Papers* 180, 5–27.

Smith, V. C., Shane, P. A., Nairn, I. A., and Williams, C. M. (2006). Geochemistry and magmatic properties of eruption episodes from Haroharo Linear Vent Zone, Okataina Volcanic Centre, Taupo Volcanic Zone, New Zealand during the last 10 kyr. *Bull. Volcanol.* 69, 57–88. doi: 10.1007/s00445-006-0056-7

Smith, V. C., Shane, P., and Nairn, I. A. (2005). Trends in rhyolite geochemistry, mineralogy, and magma storage during the last 50 kyr at Okataina and Taupo volcanic centres, Taupo Volcanic Zone, New Zealand. *J. Volcanol. Geotherm. Res.* 148, 372–406. doi: 10.1016/j.jvolgeores.2005.05.005

Sneddon, I. N., and Lowengrub, M. (1969). *Crack Problems in the Classical Theory of Elasticity.* New York, NY: Wiley & Sons.

Sobradelo, R., Geyer, A., and Martí, J. (2010). Statistical data analysis of the CCDB (Collapse Caldera Database): Insights on the formation of caldera systems. *J. Volcanol. Geother. Res.* 198, 241–252. doi: 10.1016/j.jvolgeores.2010.09.003

Sparks, R. S. J. (1978). The dynamics of bubble formation and growth in magmas: a review and analysis. *J. Volcanol. Geotherm. Res.* 3, 1–37.

Sparks, R. S. J., Sigurdsson, H., and Wilson, L. (1977). Magma mixing: a mechanism for triggering acid explosive eruptions. *Nature* 267, 315–318.

Spera, F., and Crisp, J. A. (1981). Eruption volume, periodicity, and caldera area: relationships and inferences on development of compositional zonation in silicic magma chambers. *J. Volcanol. Geother. Res.* 11, 169–187.

Suzuki-Kamata, K., Kamata, H., and Bacon, C. R. (1993). Evolution of the caldera-forming eruption at Crater Lake, Oregon, indicated by component analysis of lithic fragments. *J. Geophys. Res.* 98, 14059–14074.

Turcotte, D. L., and Schubert, G. (2002). *Geodynamics, 2nd Edn.* Cambridge: Cambridge University Press.

Wada, Y. (1994). On the relationship between dike width and magma viscosity. *J. Geophys. Res.* 99, 17743–17755. doi: 10.1029/94JB00929

Williams, H. (1941). "Calderas and their origin," in *Bulletin of the Department of Geological Sciences*, Vol. 25, eds University of California Press (Berkeley, CA: University of California Press), 239–346.

Wilson, C. J. N. (2001). The 26.5 ka Oruanui eruption, New Zealand: an introduction and overview. *J. Volcanol. Geotherm. Res.* 112, 133–174. doi: 10.1016/S0377-0273(01)00239-6

Wilson, C. J. N., and Hildreth, W. (1997). The Bishop Tuff: new insights from eruptive stratigraphy. *J. Geol.* 105, 407–439.

Wilson, C. J. N., and Walker, G. P. L. (1981). "Violence in pyroclastic flow eruptions," in *Tephra Studies,* eds S. Self and R. S. J. Sparks (Dordrecht: D. Reidel), 441–448.

Wilson, L., Sparks, R. S. J., and Walker, G. P. L. (1980). Explosive volcanic eruptions: IV. The control of magma properties and conduit geometry on eruptive column behaviours. *Geophys. J. R. Astron. Soc.* 63, 117–148.

Conflict of Interest Statement: The authors declare that the research was conducted in the absence of any commercial or financial relationships that could be construed as a potential conflict of interest.

17

Role of Sediment Size and Biostratinomy on the Development of Biofilms in Recent Avian Vertebrate Remains

Joseph E. Peterson[1]*, Melissa E. Lenczewski[2], Steven R. Clawson[3] and Jonathan P. Warnock[4]

[1] Department of Geology, University of Wisconsin Oshkosh, Oshkosh, WI, USA, [2] Department of Geology and Environmental Geosciences, Northern Illinois University, DeKalb, IL, USA, [3] Department of Entomology, University of Wisconsin-Madison, Madison WI, USA, [4] Department of Geoscience, Indiana University of Pennsylvania, Indiana, PA, USA

Edited by:
Michel Laurin,
UMR7207 Centre de Recherche sur la Paléobiodiversité et les Paléoenvironnements (CR2P), France

Reviewed by:
Subir Bera,
University of Calcutta, India
Jürgen Kriwet,
University of Vienna, Austria
Louise Zylberberg,
Centre National de la Recherche Scientifique (CNRS), France

*Correspondence:
Joseph E. Peterson
petersoj@uwosh.edu

Specialty section:
This article was submitted to Paleontology,
a section of the journal Frontiers in Earth Science

Microscopic soft tissues have been identified in fossil vertebrate remains collected from various lithologies. However, the diagenetic mechanisms to preserve such tissues have remained elusive. While previous studies have described infiltration of biofilms in Haversian and Volkmann's canals, biostratinomic alteration (e.g., trampling), and iron derived from hemoglobin as playing roles in the preservation processes, the influence of sediment texture has not previously been investigated. This study uses a Kolmogorov Smirnov Goodness-of-Fit test to explore the influence of biostratinomic variability and burial media against the infiltration of biofilms in bone samples. Controlled columns of sediment with bone samples were used to simulate burial and subsequent groundwater flow. Sediments used in this study include clay-, silt-, and sand-sized particles modeled after various fluvial facies commonly associated with fossil vertebrates. Extant limb bone samples obtained from *Gallus gallus domesticus* (Domestic Chicken) buried in clay-rich sediment exhibit heavy biofilm infiltration, while bones buried in sands and silts exhibit moderate levels. Crushed bones exhibit significantly lower biofilm infiltration than whole bone samples. Strong interactions between biostratinomic alteration and sediment size are also identified with respect to biofilm development. Sediments modeling crevasse splay deposits exhibit considerable variability; whole-bone crevasse splay samples exhibit higher frequencies of high-level biofilm infiltration, and crushed-bone samples in modeled crevasse splay deposits display relatively high frequencies of low-level biofilm infiltration. These results suggest that sediment size, depositional setting, and biostratinomic condition play key roles in biofilm infiltration in vertebrate remains, and may influence soft tissue preservation in fossil vertebrates.

Keywords: taphonomy, biofilm, biostratinomy, vertebrate paleontology, sedimentology

INTRODUCTION

The presence of non-biomineralized osteocytes and blood vessels in vertebrate fossils and sub-fossils from various fluvial deposits has been well-established in literature (Pawlicki, 1978; Schweitzer et al., 2005, 2007, 2016; Asara et al., 2007; Bertazzo et al., 2015; Lee et al., 2017). Particles of aggregated hematite have been suggested to play a role in the preservation of collagen in such fossils (Schweitzer et al., 2005, 2007, 2016; Lee et al., 2017). Crucially, growth of microbial biofilms, prokaryotic organisms in an exopolymeric matrix (Schweitzer et al., 2016), on the surface of bone and the subsequent sealing of natural osteological vectors like pores of blood vessels and Haversian and Volkmann's canals have been suggested to facilitate the preservational mechanisms that occur during the early phases of diagenesis; biofilm infiltration leads to mineralization, retarding further metabolization of organic materials and promoting soft tissue preservation (Trinajstic et al., 2007; Peterson et al., 2010; Schweitzer et al., 2016).

Some previous analyses of early diagenetic biofilm growth on archosaur bones used the Extant Phylogenetic Bracket (EPB) method of inference after Witmer (1995) to focus on the influence of higher order phylogenetic variation and biostratinomic processes (e.g., trampling and subsequent fracturing before burial) with extant analogs (Peterson et al., 2010). Crushing of a bioclast (biostratinomy), which can stem from particle transport in high-energy fluvial systems, diagenetic alteration, or macrovertebrate bioturbation, produces new vectors for microbial infiltration of bone pore spaces in the form of cracks and fractures, providing open conditions for decomposition of internal organic materials in bones. Bone alteration, such as crushing and fracturing, has previously been illustrated to have a statistically significant impact on biofilm formation and infiltration (Peterson et al., 2010).

Another critical factor in the preservation of more labile tissues in fossils is the sediment in which they are buried (Allison and Briggs, 1991; Schweitzer et al., 2007; Piñeiro et al., 2012). From site to site, sedimentary deposits will vary in texture, porosity, permeability, groundwater geochemistry, and microbial composition (Briggs, 2003; Maier et al., 2009). As the growth rate of microbial biofilm depends on the nutrients available at nucleation sites, factors that influence the transport of nutrients, such as burial media, must be taken into account (Allison and Briggs, 1991; Briggs, 2003; Trinajstic et al., 2007; Piñeiro et al., 2012).

The hypothesis of this study was that the burial media (sands, silts, and clays) and biostratinomic condition (whole or crushed bone) influence the development and infiltration of biofilms and authigenic mineralization in extant theropod bones. By further understanding the role of biofilms in the early diagenetic phases of fossilization, predictions can be made regarding the conditions necessary for likely soft tissue preservation in fossilized bone.

METHODOLOGY

Sample Preparation

To test the hypothesis of this study, controlled column experiments were utilized to simulate groundwater flow during early diagenesis (Leal-Batista and Lenczewski, 2006; Lenczewski et al., 2007; Greenhagen et al., 2014; Waska, 2014). The water flowed through a variety of sediments with textures representative of common fluvial settings containing extant theropod bones, which were then analyzed for degree of biofilm formation, defined as a series of Classes (after Peterson et al.,

2010). To appropriately model the taphonomy of fossil theropod specimens previously reported to possess primary soft-tissues that may be influenced by biofilms, such as subadult tyrannosaurs (Peterson et al., 2010), eight femora of the extant theropod *Gallus gallus domesticus* (Domestic Chicken) were selected and carefully stripped of bulk muscle and connective tissues with a sterilized scalpel. Samples were obtained from commercial sources (i.e., organic grocery stores in Northern Illinois). The handling of all bone samples was performed with nitrile gloves to avoid contamination to the microflora in this experiment. No permits were required for this study.

Column Construction

To explore the role of burial media on the development of microbial biofilms in extant theropod bone, control column experiments were used to simulate groundwater flow after burial in different sediments (Greenhagen et al., 2014; Waska, 2014). Three sedimentary groups were developed and modeled after fluvial facies commonly associated with vertebrate fossils; channel sands, crevasse splays, and distal flood-basin deposits (**Table 1**). Sediment mass ratios were based on particle size analysis values for common late Cretaceous fluvial depositional environments (Peterson et al., 2011). Sand and clay was obtained from commercial sources (quartz masonry sand and bentonite commonly used in installation of groundwater wells) and silt was obtained from a local Peoria loess deposit in northern Illinois. All sediments were dried prior to use.

Columns were constructed from 10 cm wide PVC pipes and each was approximately 30 cm long with a final volume of approximately 2,700 cm^3 (**Figure 1**). The upper and lowermost 50 mL of each column were filled with high density polyethylene (HDPE) spheres to aid in equal distribution of water inflow and outflow. The lower layer of HDPE spheres was covered with 500 mL of coarse silica sand to further aid in water flow, followed by 1,000 mL of sediment modeling specific depositional environments (**Table 1**), with each column containing a different fluvial analog and the bone samples. The columns were then filled with 250 mL of coarse silica sand, followed by an additional 500 mL of HDPE spheres. To simulate pre-burial biostratinomic variation, two bone samples were reserved for placement in each sediment column; one sample was placed intact, and the second sample was crushed. Crushing was conducted in a sterile Whirl-Pak bag perpendicular to the long axis of the bone to simulate bones that have been fractured prior to burial, as often observed in the vertebrate fossil record (e.g., Behrensmeyer et al., 1986). The column was subsequently filled with sediment to within 2–3 cm of the top of the column, where a final layer of HDPE spheres were added.

TABLE 1 | Sample columns, particle size ratios, and depositional models based on particle size analysis from Peterson et al. (2011).

Column	Sand:silt:clay ratio	Depositional model
A	2:1:1	Channel sand
B	1:2:1	Crevasse splay
C	1:1:2	Flood basin

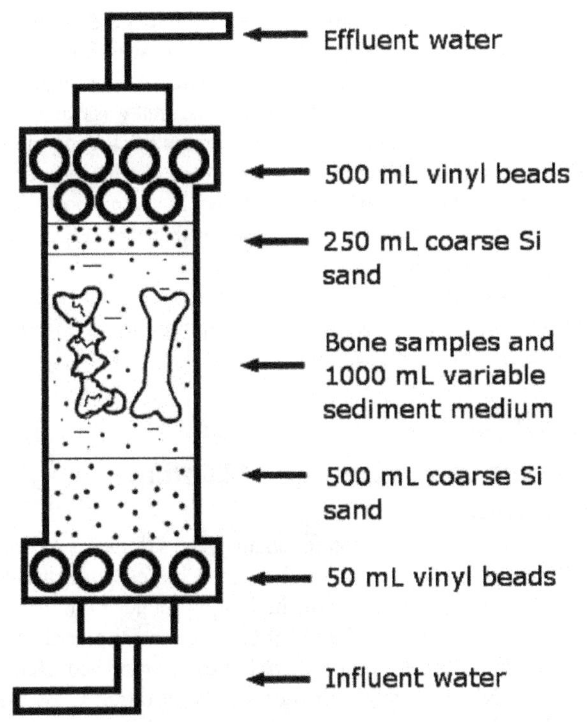
Effluent water
500 mL vinyl beads
250 mL coarse Si sand
Bone samples and 1000 mL variable sediment medium
500 mL coarse Si sand
50 mL vinyl beads
Influent water

FIGURE 1 | Schematic of PVC flow-through columns used in experiment.

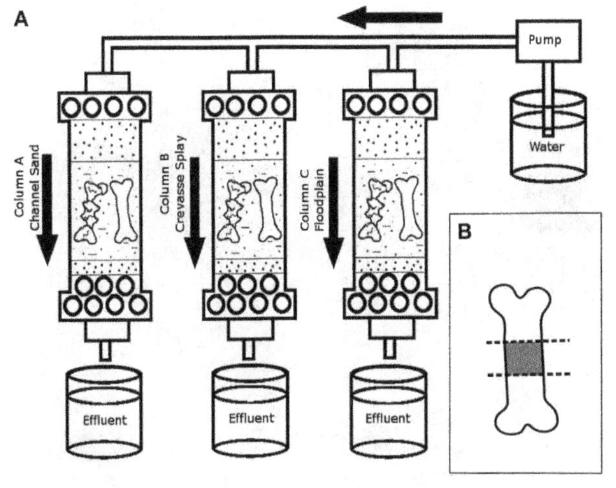
A
Pump
Water
Column A Channel Sand
Column B Crevasse Splay
Column C Floodplain
Effluent
Effluent
Effluent
B

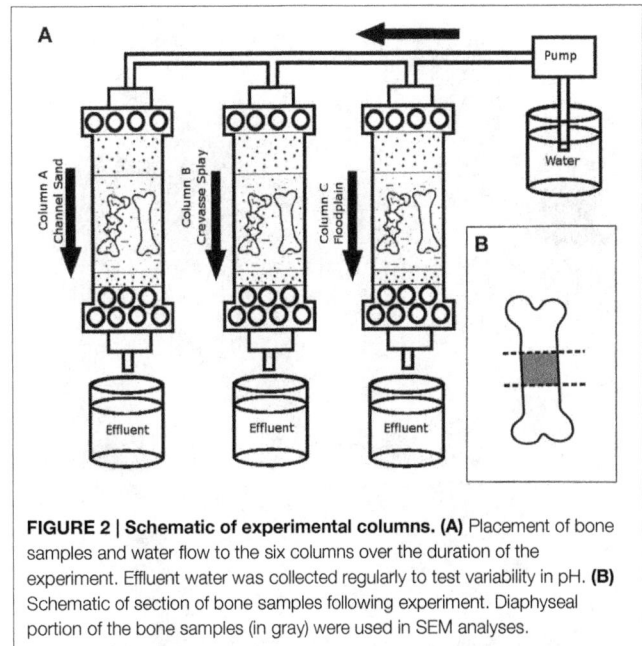

FIGURE 2 | Schematic of experimental columns. (A) Placement of bone samples and water flow to the six columns over the duration of the experiment. Effluent water was collected regularly to test variability in pH. **(B)** Schematic of section of bone samples following experiment. Diaphyseal portion of the bone samples (in gray) were used in SEM analyses.

Columns were then placed on a laboratory bench and connected to a low-flow peristaltic pump that bottom fed groundwater directly into the columns. The groundwater was sourced from a well-located near Northern Illinois University campus, however the pump was not connected directly to the well. Instead, an influent carboy, filled with well-water, was connected to the pump. The carboy was connected to the peristaltic pump, and the tubing was flushed with three times its internal volume, before being connected to the bottom influents of the columns.

The time zero for each column started on the first day of observing an effluent from the column. The experiment maintained continuous flushing of the columns for 180 days with short shut-off times (up to 10 min) for water source replacement (i.e., refilling the influent carboy with well-water). Source water was kept the same throughout the experiment to maintain a similar microbial community. Effluent from the columns was monitored for pH and temperature every 3–4 days. The effluent from the columns indicated that the temperature was around room temperature (25°C) and the pH varied from 6.5 to 7.9 throughout the experiment (**Figure 2A**).

The experiment ran for 180 days at a continuous flow rate (500 mL/day), after which the bones and sediment samples in contact with the bones were collected. All handling of specimens was conducted while wearing nitrile gloves. Each bone sample was cut into three equal transverse sections for scanning electron microanalysis (SEM) (**Figure 2B**), which focused on the diaphyseal portion of each studied limb element. Sectioning was performed using a laminar flow hood, a set of sterilized forceps, and 20 flame-sterilized cutting plates. A new flame-sterilized blade was used for each sample and the forceps were re-sterilized prior to cutting each specimen. All specimens were placed in labeled sterile Whirl-Pak bags and stored in a −70°C freezer to preserve the biofilms until preparation for SEM analyses.

Scanning Electron Microanalysis

Prior to SEM analysis, each sample was removed from its Whirlpak bag using nitrile gloves and lypholized. Samples were then fixed with Canemco carbon tape to a standard SEM aluminum sample stub. Using an SPI magnetron sputter coater, the specimens were Au/Pd coated to a thickness of ~40 nm to ensure optimal adhesion and conduction in preparation for analysis. Specimens were imaged at 300x magnification utilizing secondary electron imaging (SEI) via a JEOL 5900 Low Vacuum Scanning Electron Microscope with Oxford INCA EDS at Beloit College. All images were taken with a 40 spotsize, 20 keV accelerating voltage at an average working distance of 10 mm. For each specimen, 25–30 images of Haversian and Volkmann's canals were collected. To confirm that the substances encrusting pore spaces in *Gallus gallus domesticus* study specimens were biofilms, micrographs of colonial bacteria within the material were collected at ultra-high magnifications for each specimen (450x–5,000x) (**Figure 3**), revealing microbial communities surrounded by an undulatory extracellular matrix [e.g., extracellular polymeric substance (EPS)].

Statistical Analysis

In order to quantify the degree of biofilm infiltration in bone samples, natural vectors in the form of Volkmann's and Haversian canals were identified at the bone surfaces at a magnification of 300x and tabulated for degree of biofilm infiltration (**Table 2, Figure 4**). In order to standardize

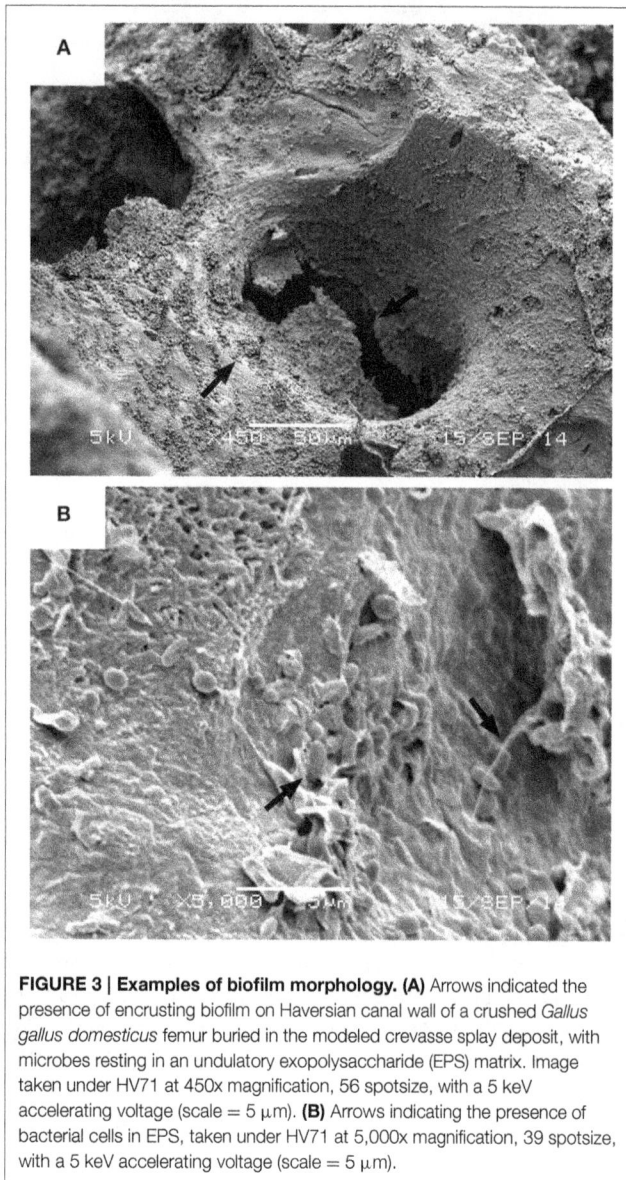

FIGURE 3 | Examples of biofilm morphology. (A) Arrows indicated the presence of encrusting biofilm on Haversian canal wall of a crushed *Gallus gallus domesticus* femur buried in the modeled crevasse splay deposit, with microbes resting in an undulatory exopolysaccharide (EPS) matrix. Image taken under HV71 at 450x magnification, 56 spotsize, with a 5 keV accelerating voltage (scale = 5 μm). (B) Arrows indicating the presence of bacterial cells in EPS, taken under HV71 at 5,000x magnification, 39 spotsize, with a 5 keV accelerating voltage (scale = 5 μm).

TABLE 2 | Degree of biofilm infiltration into observed natural vectors in bone samples of *G. gallus domesticus*, defined as 4 distinct nominal Classes (after Peterson et al., 2010).

Class	Pore Infiltration
0	No biofilm infilling of vector.
1	Biofilm coating inner vector wall.
2	Thick biofilm coating or infilling.
3	Complete/near complete biofilm infilling.

comparisons among Haversian and Volkmann's canals, canals and other vectors were chosen based on their diameter, with only vectors between 50 and 100 μm being chosen for analysis.

In order to quantify the degree of modern biofilm infiltration in extant theropod bone, natural osteological vectors were identified on the surface of SEM samples at a magnification of

300×. Following Peterson et al. (2010), a series of four nominal classes were established to categorize the degree of biofilm infiltration (**Table 2**; **Figures 4A–D**). A natural vector with no visible biofilm was classified as Class 0 (**Figure 4A**), a vector with a thin biofilm coating the inner wall of the canal was classified as Class 1 (**Figure 4B**), a Class 2 vector exhibits thickening biofilm that is beginning to close the aperture of the pore aperture of the vector (**Figure 4C**), and a Class 3 vector possesses a nearly complete infilling of the canal by biofilm (**Figure 4D**). Kolmogorov-Smirnov Goodness-of-fit tests were performed on the pore size/infiltration matrix to identify variation in biofilm infiltration among samples in different taphonomic conditions at a 0.05 significance level ($\alpha = 0.05$).

RESULTS

Variance in the Degree of Biofilm Infiltration

Degree of biofilm infiltration for all analyzed samples is recorded in **Tables 3A,B**. Both whole and crushed bones from modeled channel sediments displayed the highest frequency of infiltration Class 3 (whole −50%; crushed −40%). Frequencies of infiltration observed for whole and crushed bones in modeled channel sediments decrease with each successive infiltration Class, such that Class 2 is the second most frequently observed, followed by Class 1 and Class 0. For bone samples in modeled crevasse splay sediments, whole bones display a similar pattern to channel sediments, with Class 3 being the most frequent (44%), followed by Classes 2 and 1 (26% each) and finally Class 0. However, crushed bones from modeled crevasse splay sediments displayed the highest proportion of infiltration Class 0 (36%), followed in order of decreasing frequency by Class 1, 2, and 3. Results from modeled flood basin sediments follow those of the channel sediments, with Class 3 being the most frequently recorded for both whole (81%) and crushed (62%) bones. Each subsequent infiltration Class is less frequent for whole bone in modeled flood basin sediments. Crushed bones revealed fewer Class 2 than 3, and equal frequencies of Classes 1 and 0 (8% each).

Burial Media

Bone samples buried in the modeled flood-basin deposit (clay) exhibited a trend toward high frequencies of Class 3 biofilm infiltration and very low frequencies of lower-class biofilm infiltration (**Figure 5**). Samples buried in modeled channel deposits (sand) and crevasse splays (silt) exhibited similar trends of moderate frequencies of mid-level biofilm infiltration (Classes 1 and 2).

Biostratinomic Condition

Crushed-bone samples consistently exhibited lower frequencies of maximum biofilm infiltration (Class 3) than their whole-bone counterparts (**Figure 5**) in channel and flood-basin deposits. Crushed- and whole-bone samples buried in modeled crevasse splay deposits exhibit considerable variability. Alternatively, whole- and crushed-bone samples in channel sands follow comparable trends, exhibiting moderate-levels (Classes 1 and 2) of biofilm infiltration. Whole- and crushed-bone samples buried

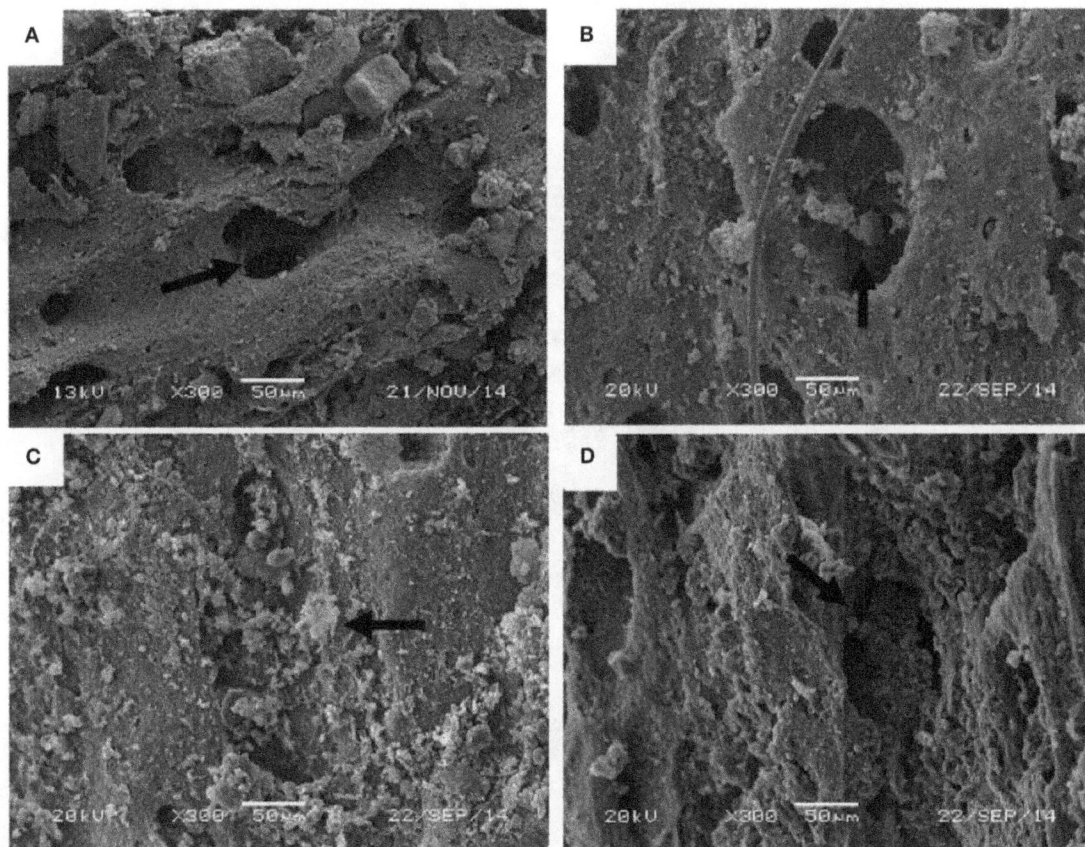

FIGURE 4 | Examples of Vector Infiltration Classes for statistical analysis: (A) Class 0—no biofilm visible inside of canal. Arrow indicating the biofilm-free canal wall; **(B)** Class 1—some biofilm coating inner pore wall. Arrow indicating the presence of early-stage biofilm development; **(C)** Class 2—biofilm thickening and beginning to seal pore aperture of a whole femur. Arrow indicating the increasing abundance of biofilm; **(D)** Class 3—near complete infilling of the canal. Arrow indicating the presence of nearly complete covering of biofilms. All images are SEI, taken under HV71 at 300x magnification, 40 spotsize, with a 20 keV accelerating voltage and a working distance of 8–12 mm.

TABLE 3 | Kolmogorov Smirnov Goodness of Fit results: (A) number of natural osteological vectors tallied with Class of biofilm infiltration; **(B)** results of significant difference comparisons.

A. Number of Vectors Per Infiltration Class

Class	Channel (Whole)	Channel (Crushed)	Crevasse (Whole)	Crevasse (Crushed)	Flood basin (Whole)	Flood basin (Crushed)
0	1 (6%)	4 (14%)	1 (4%)	8 (36%)	0 (0%)	2 (8%)
1	4 (25%)	6 (21%)	7 (26%)	7 (32%)	1 (3%)	2 (8%)
2	3 (19%)	7 (25%)	7 (26%)	5 (23%)	5 (16%)	6 (23%)
3	8 (50%)	11 (39%)	12 (44%)	2 (9%)	25 (81%)	16 (61%)
Total	16	28	27	22	31	26

B. Kolmogorov Smirnov Test Results (0.05)

	Channel (Whole)	Channel (Crushed)	Crevasse (Whole)	Crevasse (Crushed)	Flood basin (Whole)	Flood basin (Crushed)
Channel (Crushed)	x					
Crevasse (Whole)	x	x				
Crevasse (Crushed)	0.05	0.05	0.05			
Flood basin (Whole)	0.05	0.05	0.05	0.05		
Flood basin (Crushed)	x	0.05	x	0.05	0.05	

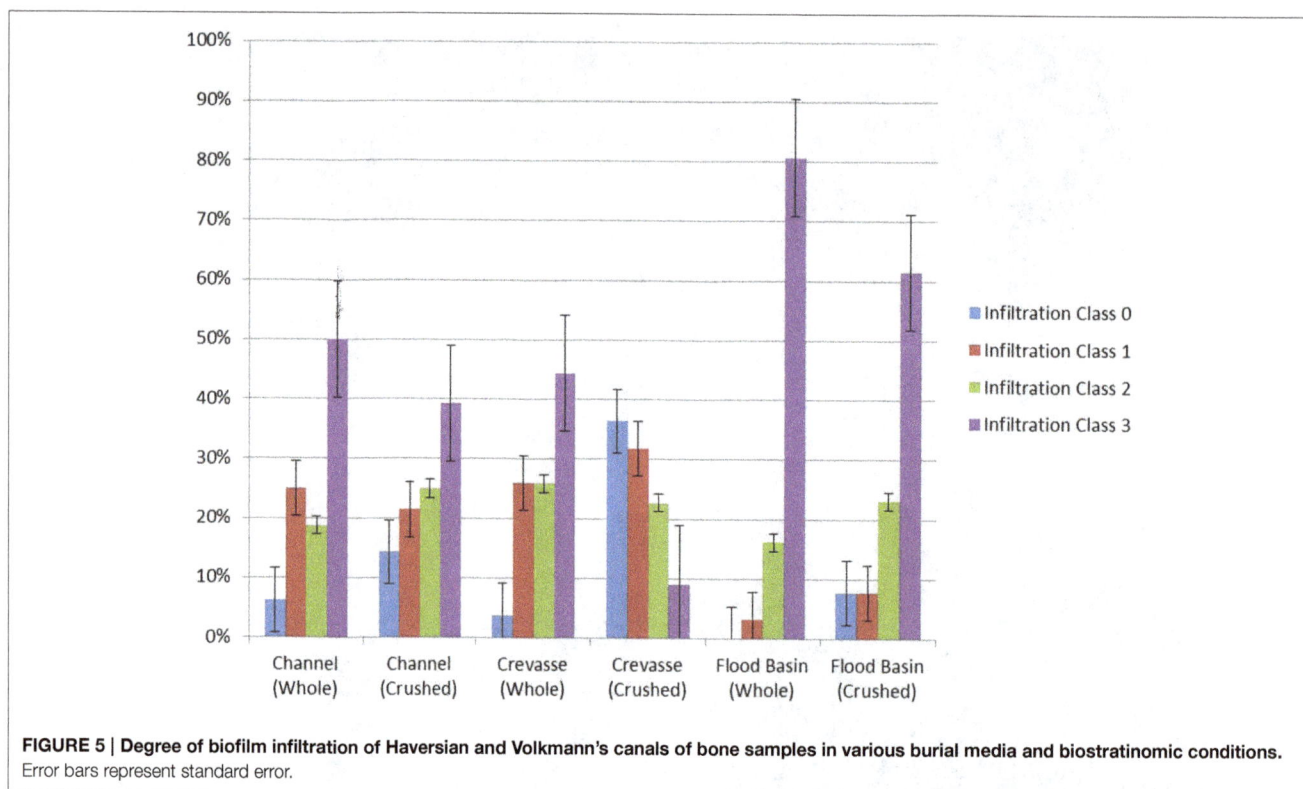

FIGURE 5 | Degree of biofilm infiltration of Haversian and Volkmann's canals of bone samples in various burial media and biostratinomic conditions. Error bars represent standard error.

in modeled flood basin deposits exhibited very high frequencies of Class 3 biofilm infiltration compared to all other samples, and very low frequencies of other infiltration Classes.

Statistical Analysis

Kolmogorov Smirnov Goodness-of-Fit test results show significant differences ($p < 0.05$) (**Tables 3A,B**) between biofilm infiltration in whole-bone vs. crushed-bone samples in modeled crevasse splay and flood basin sediments. Biofilm development in whole-bone and crushed-bone samples buried in a modeled channel sand show similar frequencies of biofilm Classes and were not found to be significantly different.

Differences between bone samples (whole and crushed) buried in modeled flood basins and channel sands also show significant differences, as do bone samples buried in modeled flood basins and crevasse splay sediments. However, differences between channel sands and crevasse splay sediments are only significant in crushed samples; whole-bone samples between channel and crevasse splay sediments were not found to be significantly different (**Table 3B**).

DISCUSSION

These results confirm previous analyses concerning the relationship between biostratinomy and biofilm growth in modern bones (Peterson et al., 2010), and further demonstrate that the grain size of burial media plays a significant role in the degree of biofilm development. However, while both biostratinomic and depositional variables significantly influence the degree of biofilm growth in bone samples, it is clear

biostratinomy plays a greater role than depositional setting. Biofilm infiltration is the highest in whole-bone samples buried in flood-basin analogs (clay-rich), and lowest in crushed samples in crevasse splay analogs (silt-rich). All samples, with the exception of crushed samples buried in crevasse splay analogs, show a similar trend of dominance Class 3 (**Table 2**), likely due to the prolonged duration of the experiment (6 months).

Peterson et al. (2010) experimentally demonstrated the potential relationship between relative degree of biofilm infilling of natural osteological vectors (e.g., Haversian and Volkmann's canals) and osteological soft tissue preservation in vertebrate fossils. The mineralization of biofilms established in osteological vectors during decomposition may promote the preservation of primary soft-tissues in extant and fossil archosaur bones by retarding further microbial metabolization of tissues in bones (Trinajstic et al., 2007; Peterson et al., 2010; Schweitzer et al., 2016). As such, bones which have experienced less biostratinomic alteration, i.e., bones which have not been crushed, are more likely to preserve soft tissues than those which have been altered (Peterson et al., 2010).

The results of this study demonstrate that well-developed biofilms infiltrate the vectors of whole bones better and more thoroughly than in crushed bones, mirroring the results of Peterson et al. (2010). Whole-bones samples buried in modeled channel sands and crevasse splay sediments exhibited relatively high frequencies of nearly complete or completely biofilm-filled Haversian and Volkmann's canals (Classes 2 and 3), and thus would be expected to yield primary soft-tissues in vertebrate fossils. Crushed bones in modeled channel and crevasse sediments display decreased frequencies of Class 3 infiltration, and are less likely to preserve soft tissues. However, whole and crushed bones in modeled flood-basin sediments both exhibited extremely high frequencies of biofilm infiltration (Class 3), demonstrating an interaction between biostratinomic condition and sediment grain size.

Samples buried in flood basin sediments possess an increased frequency of advanced biofilm development, likely due to the increased concentrations of microbial communities in mud and clay-rich sediments (Camesano and Logan, 1998). However, lithification of mudstones involves a more pronounced degree of compaction, and slickenside surfaces that develop during the early diagenetic stage can fracture and displace bioclasts, opening additional vectors for microorganisms to enter bones and metabolize soft tissues (Peterson et al., 2010). Furthermore, mud-rich flood-basin deposits are generally more likely to yield remains that have been trampled, crushed, and subaerially exposed for a considerable period of time prior to burial (Behrensmeyer, 1982; Gates, 2005), thus reducing the likelihood of preserved primary soft-tissues. The results of this experiment verify these findings. While Class 3 infiltration is most commonly exhibited by both whole and crushed bones in modeled flood basin sediments, the whole bones have a higher frequency of Class 3 (81%) compared to crushed bones (61%). Hence, while mud and clay-rich flood basin sediments are more likely to preserve soft tissues compared to deposits with larger grain sizes, biostratinomic effects still play a role in soft tissue preservation in fine-grained deposits.

Sands, while lower in concentrations of microbial communities (Camesano and Logan, 1998), provide a grain-supported framework which would reduce crushing and keep the natural vectors as the primary means for microbial integration. From these results, a prediction can be proposed that un-fractured bones in sand to silt-sized sediments are more likely to yield primary soft tissues in vertebrate fossils than crushed bones in either setting.

While previous studies on biofilm development in vertebrate fossils show a potential influence on the preservation of osteological soft-tissue in the fossil record (Trinajstic et al., 2007; Peterson et al., 2010; Schweitzer et al., 2016), it should be noted that the observations of this study do not necessarily imply that primary osteological soft tissue preservation will always occur in accordance with biofilm development. Though primary soft-tissues in vertebrate fossils are commonly associated with relatively coarse-grained facies (e.g., Schweitzer et al., 2005; Asara et al., 2007), the results of this study suggest soft tissue preservation should be more common in fine-grained sediments. The limited exposure and rapid burial of samples in this experiment may produce unique conditions for soft-tissue preservation in remains buried in finer-grained facies. However, Peterson et al. (2010) reported on the extraction of primary soft-tissues from tyrannosaurid, ceratopsian, and ornithopod dinosaurs encased in a variety of depositional settings, and only those ceratopsian bone fragments which were buried in a flood-basin deposit yielded pliable tissues, following the results of this study. The nature of the deposit studied by Peterson et al. (2010) may explain the departure from the commonly observed pattern. The taphonomy for this deposit has been thoroughly described

(Mathews et al., 2009) and the remains exhibit virtually no abrasion, crushing, or fracturing, and are attributed to three individual juvenile *Triceratops*; a unique age profile that suggests a single depositional event and limited exposure prior to burial (Mathews et al., 2009), similar to the conditions created in this study. As such, the preservation of pliable primary soft-tissues may be due to biostratinomic processes such as the rate of burial. Rapid burial of the remains could promote soft-tissue preservation rather than the usual taphonomic processes associated with floodplain deposits, such as prolonged exposure and biostratinomic reworking (Behrensmeyer, 1982; Gates, 2005).

The mechanisms required to promote preservation of soft-tissue internal to bone in vertebrate fossils are complex. Despite previously proposed alternative hypotheses for the presence of primary soft-tissues in Mesozoic vertebrate fossils, such as contaminants or recent biofilms (Kaye et al., 2008), an abundance of biomolecular evidence has confirmed their initial interpretations as primary soft-tissues (Schweitzer et al., 2005, 2009, 2016; Asara et al., 2007). Biochemical processes have been proposed to play a significant role in the preservation of soft osteological tissues; recently iron has been suggested as an alternative mechanism to explain the process by which primary soft-tissues can remain pliable over geologic time, by which iron from hemoglobin serves as a natural chelator to increase tissue immunoreactivity in vertebrate remains (Schweitzer et al., 2014; Lee et al., 2017). However, the presence of mineralized biofilms in vertebrate samples were osteological soft-tissues have also been recovered suggests a role in tissue preservation (Briggs, 2003; Peterson et al., 2010; Kremer et al., 2012; Raff et al., 2013, 2014; Schweitzer et al., 2016), where biofilm crystallization in Haversian and Volkmann's canals seals natural vectors from further microbial penetration and retards subsequent metabolization of soft-tissues. Therefore, by sealing the internal pore spaces of bones, biofilms promote the preservation of those soft tissues found within bone. Furthermore, iron has also been previously associated with soft-tissue preservation and biofilms and interpreted as pyrite framboids (Kaye et al., 2008; Peterson et al., 2010). The presence of iron associated with both preserved soft-tissues and mineralized biofilms suggests that multiple mechanisms may be involved in the preservation of primary biomolecules in fossil vertebrates (e.g., Kaye et al., 2008; Peterson et al., 2010; Schweitzer et al., 2014), including depositional chemistry, biostratinomy and sediment grain size. In order to fully explain the complex processes required to preserve primary soft-tissues in vertebrate remains over tens of millions of years, further analysis and actualistic taphonomic experimentation of biochemistry, geochemistry, and paleoenvironmental factors are needed. Furthermore, future experimentation is needed in order to understand the complex interactions between environmental factors which may lead to the preservation of those soft tissues not found within bone.

AUTHOR CONTRIBUTIONS

JP and ML conceptualized the study, designed the experiments, obtained samples, and collected data. JP, ML, SC, and JW analyzed the data, and wrote the manuscript.

ACKNOWLEDGMENTS

We thank the Department of Geology at Beloit College for access to their SEM facilities. We also thank Dr. Rongping Deng and Dr. Jim Rougvie of Beloit College for their expert guidance and instruction in the operation of instrumentation used in this study. This research was funded by the University of Wisconsin Oshkosh External Grants Expansion Program grant number 102-222051-1 and the Northern Illinois University Department of Environmental Studies. We also thank Michel Laurin for editorial assistance and three reviewers for offering constructive criticism and helpful feedback.

REFERENCES

Allison, P. A., and Briggs, D. E. G. (1991). "The taphonomy of soft-bodied animals," in *Fossilization: the Process of Taphonomy*, ed. S. K. Donovan (London: Belhaven Press), 120–140.

Asara, J. M., Schweitzer, M. H., Freimark, L. M., Phillips, M., and Cantley, L. C. (2007). Protein sequences from Mastodon and *Tyrannosaurus rex* revealed by mass spectrometry. *Science* 316, 280–285. doi: 10.1126/science.1137614

Behrensmeyer, A. K. (1982). Time resolution in fluvial vertebrate assemblages. *Paleobiology* 8, 211–227. doi: 10.1017/S0094837300006941

Behrensmeyer, A. K., Gordon, K. D., and Yanahi, G. T. (1986). Trampling as a cause of bone surface damage and pseudo-cutmarks. *Nature* 319, 768–771. doi: 10.1038/319768a0

Bertazzo, S., Maidment, S. C., Kallepitis, C., Fearn, S., Stevens, M. M., and Xie, H. (2015). Fibres and cellular structures preserved in 75-million-year-old dinosaur specimens. *Nat. Commun.* 6:7352. doi: 10.1038/ncomms8352

Briggs, D. E. G. (2003). The role of decay and mineralization in the preservation of soft-bodied fossils. *Annu. Reb. Earth Planet Sci.* 31, 275–301. doi: 10.1146/annurev.earth.31.100901.144746

Camesano, T. A., and Logan, B. E. (1998). Influence of fluid velocity and cell concentration on the transport of motile and nonmotile bacteria in porous media. *Environ. Sci. Technol.* 32, 1699–1708. doi: 10.1021/es970996m

Gates, T. A. (2005). The late Jurassic Cleveland-Lloyd Quarry as a drought-induces assemblage. *Palaios* 20, 363–375. doi: 10.2110/palo.2003.p03-22

Greenhagen, A., Lenczewski, M., and Carroll, M. (2014). Natural attenuation of pharmaceuticals and an Illicit drug in laboratory column. *Chemosphere* 115, 13–19. doi: 10.1016/j.chemosphere.2014.01.015

Kaye, T. G., Gaugler, C., and Sawlowicz, Z. (2008). Dinosaurian soft tissues interpreted as bacterial biofilms. *PLoS ONE* 3:e2802. doi: 10.1371/journal.pone.0002808

Kremer, B., Owocki, K., Królikowska, A., Wrzosek, B., and Kazmierczal, J. (2012). Mineral microbial structures in a bone of the Late Cretaceous dinosaur *Saurolophus angustirostris* from the Gobi desert, Mongolia – a Raman spectroscopy study. *Palaeogeogr. Palaeoclimatol. Palaeoecol.* 358–360, 51–61. doi: 10.1016/j.palaeo.2012.07.020

Leal-Batista, R., and Lenczewski, M. (2006). Sorption of MTBE and benzene in fine-grained materials from northern Illinois and the Chalco Basin, Mexico. *Environ. Geosci.* 13, 31–41. doi: 10.1306/eg.09280404027

Lee, Y. C., Chiang, C. C., Huang, P. Y., Chung, C. Y., Huang, T. D., Wang, C. C., et al. (2017). Evidence of preserved collagen in an Early Jurassic sauropodomorph dinosaur revealed by synchrotron FTIR microspectroscopy. *Nat. Commun.* 8:14220. doi: 10.1038/ncomms14220

Lenczewski, M., Leal-Baustista, R., and Kroll, S. (2007). Influence of Ethanol and MTBE on the biodegradation and transport of benzene in loess. *Environ. Geosci.* 14, 137–148. doi: 10.1306/eg.09150606006

Maier, R., Pepper, I., and Gerba, C. C. (2009). *Environmental Microbiology, 2nd Edn*. Amsterdam: Academic Press.

Mathews, J. C., Brusatte, S. L., Williams, S. A., and Henderson, M. D. (2009). The first Triceratops bonebed and its implications for gregarious behavior. *J. Vertebrate Paleontol.* 29, 286–290. doi: 10.1080/02724634.2009.10010382

Pawlicki, R. (1978). Morphological differentiation of the fossil dinosaur bone cells Light, transmission electron-, and scanning electron-microscopic studies. *Acta Anatomica* 100, 411–418.

Peterson, J. E., Lenczewski, M. E., and Scherer, R. P. (2010). Influence of microbial biofilms on the preservation of primary soft tissue in fossil and extant archosaurs. *PLoS ONE* 5:e13334. doi: 10.1371/journal.pone.0013334

Peterson, J. E., Scherer, R. P., and Huffman, K. M. (2011). Methods of microvertebrate sampling and their influences on taphonomic interpretations. *Palaios* 26, 81–88. doi: 10.2110/palo.2010.p10-080r

Piñeiro, G., Ramos, A., Godo, C., Scarabino, F., and Laurin, M. (2012). Unusual environmental conditions preserve a Permian mesosaur-bearing Konservat-Lagerstätte from Uruguay. *Acta Palaeontol. Pol.* 57, 299–318. doi: 10.4202/app.2010.0113

Raff, E. C., Andrews, M. E., Turner, F. R., Toh, E., Nelson, D. E., and Raff, R. A. (2013). Contingent interactions among biofilm-forming bacteria determine preservation or decay in the first steps toward fossilization in marine embryos. *Evol. Dev.* 15, 243–256. doi: 10.1111/ede.12028

Raff, R. A., Andrews, M. E., Pearson, R. L., Turner, R., Saur, S. T., Thomas, D. C., et al. (2014). Microbial ecology and biofilms in the taphonomy of soft tissues. *Palaios* 29, 560–569. doi: 10.2110/palo.2014.043

Schweitzer, M. H., Moyer, A. E., and Zheng, W. (2016). Testing the hypothesis of biofilm as a source for soft tissue and cell-like structures preserved in dinosaur bone. *PLoS ONE.* 11:e0150238. doi: 10.1371/journal.pone.0150238

Schweitzer, M. H., Wittmeyer, J. L., and Horner, J. R. (2007). Soft tissue and cellular preservation in vertebrate skeletal elements from the Cretaceous to the present. *Proc R. Soc. B Biol. Sci.* 274, 183–197. doi: 10.1098/rspb.2006.3705

Schweitzer, M. H., Wittmeyer, J. L., Horner, J. R., and Toporski, J. K. (2005). Soft-tissue vessels and cellular preservation in *Tyrannosaurus rex. Science* 307, 1952–1955. doi: 10.1126/science.1108397

Schweitzer, M. H., Zheng, W., Cleland, T. P., Goodwin, M. B., Boatman, E., Theil, E., et al. (2014). A role for iron and oxygen chemistry in preserving soft tissues, cells and molecules from deep time. *Proc. R. Soc. B* 281:20132741. doi: 10.1098/rspb.2013.2741

Schweitzer, M. H., Zheng, W., Organ, C. L., Avci, R., Suo, Z., Freimark, L. M., et al. (2009). Biomolecular characterization and protein sequences of the Campanian hadrosaur B. *canadensis. Science* 324, 626–631. doi: 10.1126/science.1165069

Trinajstic, K., Marshall, C., Long, J., and Bifield, K. (2007). Exceptional preservation of nerve and muscle tissues in Late Devonian placoderm fish and their evolutionary implications. *Biol. Lett.* 3, 197–200. doi: 10.1098/rsbl.2006.0604

Waska, K. (2014). *Hyperalkaline Aquifers of Calumet Wetlands: Environmental Interactions between Geochemistry and Microbiology*. dissertation, Northern Illinois University.

Witmer, L. M. (1995). "The extant phylogenetic bracket and the importance of reconstructing soft tissues in fossils" in *Functional Morphology in Vertebrate Paleontology*, ed J. Thomason (Cambridge: Cambridge University Press), 19–33.

Conflict of Interest Statement: The authors declare that the research was conducted in the absence of any commercial or financial relationships that could be construed as a potential conflict of interest.

The reviewer LZ and handling Editor declared their shared affiliation, and the handling Editor states that the process nevertheless met the standards of a fair and objective review.

Permissions

List of Contributors

Liping Liu
Department of Mathematics, North Carolina A&T State University, Greensboro, NC, USA

Yuh-Lang Lin
Department of Physics, North Carolina A&T State University, Greensboro, NC, USA
Department of Energy and Environmental Systems, North Carolina A&T State University, Greensboro, NC, USA

Shu-Hua Chen
Department of Land, Air and Water Resources, University of California, Davis, CA, USA

Aleah N. Sommers and Harihar Rajaram
Department of Civil, Environmental, and Architectural Engineering, University of Colorado at Boulder, Boulder, CO, USA

Eliezer P. Weber
Apex Companies, LLC, Boulder, CO, USA

Michael J. MacFerrin
Cooperative Institute for Research in Environmental Science, University of Colorado at Boulder, Boulder, CO, USA

William T. Colgan
Department of Earth and Space Science and Engineering, York University, Toronto, ON, Canada

Max Stevens
Department of Earth and Space Sciences, University of Washington, Seattle, WA, USA

Nanna B. Karlsson, Dorthe Dahl-Jensen and Lisbeth T. Nielsen
Centre for Ice and Climate, The Niels Bohr Institute, University of Copenhagen, Copenhagen, Denmark

Olaf Eisen
Alfred-Wegener-Institut Helmholtz-Zentrum für Polar-und Meeresforschung, Bremerhaven, Germany
Department of Geosciences, University of Bremen, Bremen, Germany

Johannes Freitag, Sepp Kipfstuhl and Anna Winter
Alfred-Wegener-Institut Helmholtz-Zentrum für Polar- und Meeresforschung, Bremerhaven, Germany

Cameron Lewis
Center for Remote Sensing of Ice Sheets, University of Kansas, Lawrence, KS, USA
Sandia National Laboratories, Albuquerque, NM, USA

John D. Paden
Center for Remote Sensing of Ice Sheets, University of Kansas, Lawrence, KS, USA

Frank Wilhelms
Alfred-Wegener-Institut Helmholtz-Zentrum für Polar-und Meeresforschung, Bremerhaven, Germany
Department of Crystallography, Geoscience Centre, University of Göttingen, Göttingen, Germany

Ward J. J. van Pelt and Veijo A. Pohjola
Department of Earth Sciences, Uppsala University, Uppsala, Sweden

Carleen H. Reijmer
Institute for Marine and Atmospheric Research Utrecht, Utrecht University, Utrecht, Netherlands

Sergey Marchenko
Department of Earth Sciences, Uppsala University, Uppsala, Sweden
Department of Geophysics, The University Centre in Svalbard, Longyearbyen, Norway

Ward J. J. van Pelt, Björn Claremar, Veijo Pohjola and Rickard Pettersson
Department of Earth Sciences, Uppsala University, Uppsala, Sweden

Horst Machguth
Department of Geography, University of Zurich, Zurich, Switzerland

Carleen Reijmer
Institute for Marine and Atmosphere Research, Utrecht University, Utrecht, Netherlands

Peng Zhang, Hong Ao, Lijuan Wang and Zhisheng An
State Key Laboratory of Loess and Quaternary Geology, Institute of Earth Environment, Chinese Academy of Sciences, Xi'an, China

Shan Lin
State Key Laboratory of Loess and Quaternary Geology, Institute of Earth Environment, Chinese Academy of Sciences, Xi'an, China
College of Earth Sciences, University of Chinese Academy of Sciences, Beijing, China

Xiaoyan Sun
State Key Laboratory of Loess and Quaternary Geology, Institute of Earth Environment, Chinese Academy of Sciences, Xi'an, China
College of Urban and Environment Sciences, Shanxi Normal University, Linfen, China

William J. Durkin and Matthew E. Pritchard
Department of Earth and Atmospheric Sciences, Cornell University, Ithaca, NY, USA

Timothy C. Bartholomaus
Department of Geological Sciences, University of Idaho, Moscow, ID, USA

Michael J. Willis
Department of Earth and Atmospheric Sciences, Cornell University, Ithaca, NY, USA
Geological Sciences, University of North Carolina at Chapel Hill, Chapel Hill, NC, USA
Cooperative Institute for Research in Environmental Sciences (CIRES), University of Colorado, Boulder, CO, USA

Annett Bartsch and Georg Pointner
Zentralanstalt für Meteorologie und Geodynamik, Vienna, Austria
Department of Geodesy and Geoinformation, Vienna University of Technology, Vienna, Austria

Yuri A. Dvornikov
Academic Department of Cryosophy, Earth Cryosphere Institute, Russian Academy of Sciences, Tyumen, Russia

Marina O. Leibman and Artem V. Khomutov
Academic Department of Cryosophy, Earth Cryosphere Institute, Russian Academy of Sciences, Tyumen, Russia
Academic Department of Cryosophy, Tyumen State University, Tyumen, Russia

Anna M. Trofaier
Svalbard Integrated Arctic Earth Observing System, University Centre in Svalbard, Longyearbyen, Norway

David G. Chandler and Kyotaek Hwang
Civil and Environmental Engineering, Syracuse University, Syracuse, NY, USA

Mark S. Seyfried
USDA Agricultural Research Service, Northwest Watershed Research Center, Boise, ID, USA

James P. McNamara
Department of Geosciences, Boise State University, Boise, ID, USA

Jyh-Jaan Huang, Kuo-Yen Wei, Ludvig Löwemark and Sheng-Rong Song
Department of Geosciences, National Taiwan University, Taipei, Taiwan

Chih-An Huh and Teh-Quei Lee
Institute of Earth Sciences, Academia Sinica, Taipei, Taiwan

Shu-Fen Lin
Institute of History and Philology, Academia Sinica, Taipei, Taiwan

Wen-Hsuan Liao
Research Center for Environmental Changes, Academia Sinica, Taipei, Taiwan
Earth System Science Program, Taiwan International Graduate Program, Academia Sinica, Taipei, Taiwan

Tien-Nan Yang
Exploration and Development Research Institute, CPC Corporation, Miaoli, Taiwan

Meng-Yang Lee
Department of Earth and Life Science, University of Taipei, Taipei, Taiwan

Chih-Chieh Su
Institute of Oceanography, National Taiwan University, Taipei, Taiwan

Masao Ohno, Yoshihiro Kuwahara, Asami Mizuta, Itsuro Kita and Akihiro Kano
Department of Environmental Changes, Faculty of Social and Cultural Studies, Kyushu University, Fukuoka, Japan

Tatsuya Hayashi
Mifune Dinosaur Museum, Kumamoto, Japan

Masahiko Sato
Department of Environmental Changes, Faculty of Social and Cultural Studies, Kyushu University, Fukuoka, Japan
Geological Survey of Japan, National Institute of Advanced Industrial Science and Technology, Tsukuba, Japan

Tokiyuki Sato
Faculty of International Resource Sciences, Akita University, Akita, Japan

Wasyl Drosdowsky and Matthew C. Wheeler
Research and Development, Bureau of Meteorology, Melbourne, VIC, Australia

Dirk van As, Jason E. Box and Robert S. Fausto
Department of Glaciology and Climate, Geological Survey of Denmark and Greenland, Copenhagen, Denmark

Tim Rixen
Department of Biogeochemistry, Institute of Geology, University of Hamburg, Hamburg, Germany

Antje Baum and Francisca Wit
Leibniz Center for Tropical Marine Ecology, Bremen, Germany

Joko Samiaji
Faculty of Fishery and Marine Science, University of Riau, Pekanbaru, Indonesia

Roberto Sulpizio and Silvia Massaro
Dipartimento di Scienze della Terra e Geoambientali, Università degli Studi di Bari, Bari, Italy

Antonio Costa
Istituto Nazionale di Geofisicae Vulcanologia, Bologna, Italy

Joan Martí
CSIC, Institute of Earth Sciences "Jaume Almera," Barcelona, Spain

Joseph E. Peterson
Department of Geology, University of Wisconsin Oshkosh, Oshkosh, WI, USA

Melissa E. Lenczewski
Department of Geology and Environmental Geosciences, Northern Illinois University, DeKalb, IL, USA

Steven R. Clawson
Department of Entomology, University of Wisconsin-Madison, Madison WI, USA

Jonathan P. Warnock
Department of Geoscience, Indiana University of Pennsylvania, Indiana, PA, USA

Index

A

Ablation Area, 51-53, 55, 57
Acquisition Curve, 150
Anisotropy, 15-16, 18-21, 23-26
Anthropogenic Perturbations, 144, 176
Approach Angle, 1-2, 8, 10-11, 13-14
Avian Vertebrate, 203

B

Bathymetry, 93-95, 98, 101, 114, 117
Biofilm, 203-210
Biostratinomy, 203-204, 208-209
Bubble Nucleation, 180

C

Carbon Balances, 170
Carbon Leaching, 170, 177
Carbon Loss, 170, 177-178
Circumpolar Mapping, 103, 105
Climate Change, 16, 27, 45, 58, 77-79, 88, 101, 104-105, 146, 148, 156, 169
Crystallinity, 194-195

D

Diabatic Heating, 1-3, 5-6, 8-9, 11-13

E

Elevation Profiles, 50, 52-54
Elevation Time Series, 90, 93
Environmental Perturbations, 61
Eruption Initiation, 179-180, 187
Eruptive Dynamics, 179, 190, 193-194
Evapotranspiration, 120-121, 123, 131, 134, 170, 173-178
Extensional Stress, 185, 190-191, 193-199

F

Firn Modeling, 60
Firn Permeability, 15, 17, 19, 21, 23-25
Fissure Eruptions, 195
Flow Splitting, 2, 11, 13
Freshwater Flux, 58, 173, 176

G

Glacier Advance, 90-91, 99, 101
Glacier Surface Melt, 45

Goethite, 80, 84-89
Gravimetric Water Content, 65, 72, 74, 83
Ground Fast Ice, 111, 115

H

Hematite, 80, 84-89, 203
Horizontal Vorticity Advection, 1, 5-8, 10-13
Hysteresis, 119, 124-133, 150, 152

I

Ice Lenses, 15-17, 19-20, 107
Ice Sheets, 26-28, 42, 44, 60-61, 148, 150-151, 154-155, 165-166, 169
Ice-core Drill, 27-28, 43
Ice-penetrating Radar, 27, 44
Idealized Simulations, 1-4, 14
Internal Accumulation, 58, 60, 78
Internal Glacier Stratigraphy, 27
Intraseasonal, 157-158, 163-164
Inviscid Fluid, 181

L

Landfall Location, 1-2, 9, 11, 13-14
Latent Heat, 11, 47, 52-53, 61, 64, 75-76
Lithology, 140
Lithosphere, 80, 180, 182, 193
Lithostatic Loading, 179, 194, 196, 198
Lomonosovfonna, 52, 58-60, 62-63, 67-70, 75-76, 78-79

M

Magma Ascent Dynamics, 190
Magma Chamber, 179-188, 191-200
Magma Fragmentation, 195, 199, 201
Magma Influx, 179
Magma Reservoir, 183, 190-192, 194-199
Magnetic Susceptibility, 80-82, 84-85, 87, 89, 151
Meltwater Retention, 78-79, 165-166, 168-169
Metasandstone, 137
Meteorology, 157-158, 178
Monsoon Region, 157-158, 162
Morainal Bank, 90, 101

O

Offshore Sediments, 80-82, 86
Orographic Effects, 1-2, 13

P

Paleontology, 203, 210

Peat Soils, 170-171, 174-177

Pedogenesis, 80, 87

Planetary Boundary Layer Parameterization, 5

Pleistocene, 87-88, 110, 115, 145, 148, 151, 153, 155-156

Pneumatic Testing, 15, 17-19, 25

R

Radionuclides, 136, 138, 142-146

Refreezing, 16-17, 26, 45-47, 50-55, 57-59, 61, 64-65, 70, 75-79, 165-166

River Aggradation, 137

Rock Magnetism, 80-81, 87, 89, 148

S

Secondary Vegetation, 170, 175, 177

Seismic Energy, 179-180, 183, 187

Seismic Wave, 183

Soil Hydrologic Parameters, 119, 128-129

Soil Moisture, 80, 87, 117, 119-121, 123-124, 129, 132-135

Spatial Heterogeneity, 121, 165, 167, 169

Stress Field, 179-180, 182, 185-187, 189-195

Sublimation, 47, 65, 165

Supraglacial Lakes, 165

Surface Energy Balance, 45, 47, 50, 57, 60, 63-65, 67, 73, 75-77, 131

Surface Mass Balance, 15-17, 25-26, 28, 43-44, 46

Svalbard, 45-50, 52-53, 55, 57-60, 62-64, 74, 76-79, 103

Swamp Forests, 170

Synthetic Aperture Radar, 90, 103, 105, 116-118

T

Taphonomy, 203-204, 209-210

Tc Track Deflection, 1

Thaw Lakes, 103-104, 106, 108, 115, 117-118

Tidewater Glacier, 58, 90-91, 97, 99, 101

Tropical Prediction, 157

V

Velocity Time Series, 90

Vertical Fluxes, 119

Volcano Edifice, 180, 183

Vorticity Stretching, 1-2, 5-13

W

Wetting Front, 60-61, 66, 77-78, 125, 132-133

Winter Monsoons, 137

www.ingramcontent.com/pod-product-compliance
Lightning Source LLC
Chambersburg PA
CBHW080632200326

41458CB00013B/4603